The publisher gratefully acknowledges the generous support of the Jewish Studies Endowment Fund of the University of California Press Foundation.

The Light of the World

BERKELEY SERIES IN POSTCLASSICAL ISLAMIC SCHOLARSHIP

Edited by Asad Q. Ahmed and Margaret Larkin

1. The Light of the World: *Astronomy in al-Andalus*, by Joseph Ibn Naḥmias, edited, translated and with a commentary by Robert G. Morrison

The Light of the World

Astronomy in al-Andalus

Joseph Ibn Naḥmias

Edited, Translated, and with a Commentary
by Robert G. Morrison

UNIVERSITY OF CALIFORNIA PRESS

University of California Press, one of the most distinguished university presses in the United States, enriches lives around the world by advancing scholarship in the humanities, social sciences, and natural sciences. Its activities are supported by the UC Press Foundation and by philanthropic contributions from individuals and institutions. For more information, visit www.ucpress.edu.

University of California Press
Berkeley and Los Angeles, California

University of California Press, Ltd.
London, England

Library of Congress Cataloging-in-Publication Data is available at http://www.loc.gov/.

ISBN 978-0-520-28799-0 (cloth : alk. paper)

Manufactured in the United States of America

23 22 21 20 19 18 17 16 15 14
10 9 8 7 6 5 4 3 2 1

The paper used in this publication meets the minimum requirements of ANSI/NISO Z39.48–1992 (R 2002) (*Permanence of Paper*).

CONTENTS

Acknowledgments vii
Preface xi

Introduction 1

1. Judeo-Arabic Text of *The Light of the World* 49
2. Translation of the Judeo-Arabic Text of *The Light of the World* 101
3. Hebrew Recension of *The Light of the World* 187
4. Translation of the Significant Insertions in the Hebrew Recension of *The Light of the World* 241
5. Commentary on the Judeo-Arabic Text 263
6. Commentary on the Significant Insertions in the Hebrew Recension of *The Light of the World* 359
7. The Hebrew Text of Profiat Duran's Response to *The Light of the World* 393
8. Translation of the Hebrew Text of Profiat Duran's Response to *The Light of the World* 395

Glossary of Judeo-Arabic, Hebrew, and English Technical Terms 399
Bibliography 409
Index 421

ACKNOWLEDGMENTS

My doctoral adviser at Columbia University, George Saliba, introduced me to *The Light of the World* in the spring of 1992. I owe him an immense debt of gratitude for putting the text in my hands, for his advice and encouragement over the years, and for how he has modeled the virtues of intellectual exploration and an unceasing search for fascinating material. He has been everything I could hope for in a mentor. While I was a student, Raymond Scheindlin of the Jewish Theological Seminary gave me outstanding instruction in medieval Hebrew.

Since then, I have been assisted by a number of institutions. My former home, Whitman College, provided generous sabbatical support, gave me amazing senior colleagues (Rogers Miles, Jonathan Walters, and Walter Wyman), and placed me on the same floor as members of the Mathematics Department who helped me when I was just starting to understand *The Light of the World*'s models. My current departmental colleagues (Todd Berzon, Jorunn Buckley, John Holt, and Elizabeth Pritchard) at Bowdoin College are outstanding teachers and scholars around whom it is impossible to be complacent. At Bowdoin, I have also been fortunate to be part of a cohort of associate professors (Dallas Denery, Kristen Ghodsee, Page Herrlinger, and Arielle Saiber) who have pushed and prodded one another onward through the post-tenure doldrums. Bowdoin College provided me with two summer travel grants, in 2010 and 2012, which allowed me to consult manuscripts.

The Center for Advanced Judaic Studies at the University of Pennsylvania gave me a fellowship in the spring of 2007, and the Stanford Humanities Center granted me a fellowship for my sabbatical in 2012–3. Office 209 at the Stanford Humanities Center was unforgettable. Without the support of both universities and without the camaraderie of the other fellows, this book would have taken far longer.

I published earlier versions of parts of my analysis of *The Light of the World*'s model for the sun as "The Solar Model in Joseph Ibn Nahmias' *The Light of the World*," *Arabic Sciences and Philosophy* XV (2005): 57–108, and as "Andalusian Responses to Ptolemy in Hebrew," in Jonathan Decter and Michael Rand (eds.): *Precious Treasures from Hebrew and Arabic* (Piscataway, NJ: Gorgias Press, 2007): 69–86. This book presents heavily modified versions of the conclusions found in these earlier publications, but my (hopefully) deeper comprehension would not have been possible without these earlier opportunities to publish.

I have also benefitted from the librarians at the Bodleian and Vatican libraries who provided microfilms of the manuscripts of both versions of *The Light of the World* and who generously assisted me in examining the manuscripts themselves. The assistance of the staff of the Institute of Microfilmed Hebrew Manuscripts in Jerusalem gave me prompt access to sources that informed me of the broader context of *The Light of the World*. When it came time to find a publisher, Asad Ahmed (University of California, Berkeley) proved to be an outstanding colleague, negotiating with more than one publisher before accepting this book for his own series at the University of California Press. At the University of California Press, Eric Schmidt and Maeve Cornell-Taylor have worked with me to produce a book that has exceeded all of my expectations. David Peattie at BookMatters and Mike Mollett have worked unstintingly and effectively with complex material.

Even with such an arcane subject as homocentric astronomy, the broader community of scholars has been there for me. Throughout the years, Richard Kremer (Dartmouth College) has reminded me of the importance of this book for the study of Renaissance astronomy and has exhorted me to be as clear as possible in my expositions. F. Jamil and Sally Ragep (McGill University) have provided encouragement throughout the project and guidance with respect to understanding the links between Ibn Naḥmias and the rest of the history of astronomy in Islamic societies and the importance of context even in a more technical project. Jamil's involving me in the Before Copernicus project was an additional source of motivation and an outlet for my work on *The Light of the World*. Finally, Y. Tzvi Langermann's (Bar Ilan University) advice, drawing on his astounding erudition and linguistic skills, saved me from innumerable pitfalls. His ability to offer frank, constructive criticism while never being discouraging is rare, and this book benefited greatly from his input. Despite all this help from my colleagues, the defects of this book are solely my responsibility.

The Light of the World is a text that bridges religions, cultures, regions, and the perceived gap between the sciences and the humanities. My parents (Peggy Morrison and the late Alan Morrison) and brother (Jeremy Morrison) have encouraged broad questions and have pursued them in their own way in their own lives. When I began to study *The Light of the World*, half a lifetime ago, at what was a very difficult time for my whole family, I could not have conceived of the heights of happiness that my wife Dana Gold and my daughter Aziza Gold Morrison would eventually bring me.

PREFACE

In a brief, though classic, article from 1951, entitled "The Study of Wretched Subjects," the master historian of science Otto Neugebauer argued for "the recovery and study of the texts as they are, regardless of our own tastes and prejudices."[1] Although Neugebauer's immediate concern was the eminent historian of science George Sarton's review of a study of a Mandaean treatise on astrology, Neugebauer's argument that historians will gain the most when they study early science on its own terms pertains directly to the present study of *The Light of the World*. Even to specialists, astrology's technical terms, calculations, and intertextual references are complex and demand unceasing attention. Because astrological forecasting had been criticized even in the ancient world, Sarton had described the Mandaean treatise as "'part of the superstitious flotsam of the Near East.'"[2] Similarly, *The Light of the World*'s intricacies, such as the author's tendency to revise continuously his own explanations, combined with initial assessments that *The Light of the World* exerted no real influence on the history of astronomy, have deterred a comprehensive assessment of the text. But just as Neugebauer argued with respect to Mandaean astrology, it turns out that *The Light of the World* is an important episode in Islamic intellectual history, Jewish civilization, and the history of astronomy.

The Light of the World grapples with two general ways of thinking about the

1. Otto Neugebauer: "The Study of Wretched Subjects," in *Isis* XLII (1951): p. 111; repr. in Neugebauer: *Astronomy and History: Selected Essays* (New York, Heidelberg, Berlin, and Tokyo: Springer-Verlag, 1983); p. 3.

2. Ibid.

heavens, each of which *The Light of the World*'s author, Joseph Ibn Naḥmias (fl. ca. 1400 in the Iberian Peninsula), found equally compelling. Illustrations and descriptions of the medieval cosmos, whether in Europe or in Islamic societies, frequently depicted the heavens as a series of nine nesting orbs, all of which shared the same center. It was a tidy picture, one that meshed easily with cosmological references in literature and one that provided explanations for phenomena on Earth. But one might wonder how or if everyone could have thought things were that simple. In fact, by the medieval period, scholars had been studying the heavens for centuries and recognized the complexities of celestial motions. The Babylonians were capable of predicting positions of celestial bodies with precision. In the Hellenistic age, two- and three-dimensional models of celestial motions with excellent predictive accuracy emerged. These Hellenistic and Babylonian astronomies, though different from each other, yielded a more intricate but more mathematically precise description of the cosmos than that provided by a cosmos of perfectly nesting orbs. In the cultural context of *The Light of the World*, both explanations were valuable.

The reason why there was not a single agreed-upon goal for the study of the heavens was because, in *The Light of the World*'s milieu, more than one discipline could investigate the heavens. Physics addressed the structure and material composition of the heavens. Astronomy tended to be more concerned with accurate predictions and descriptions of celestial motions. Metaphysics addressed the underlying reasons for celestial motions. The distinctions between disciplinary approaches were neither fixed nor absolute. For one, astronomy generally did not dismiss the concerns of physics and metaphysics. As well, the conclusions of physics and metaphysics were based in a general sense upon observations. Most important for *The Light of the World*, scholars of astronomy who did pay attention to physics disagreed about how to understand the physics of celestial motions. *The Light of the World* is particularly fascinating because it comes from a context in which scholars took all of the disciplines' approaches to explaining the heavens seriously. Ibn Naḥmias took on the complex question of reconciling the elegance of a cosmos modeled on nesting homocentric orbs with doing a better job of making accurate predictions. There was a tension embedded in Ibn Naḥmias's project because, on the one hand, a cosmos of nesting homocentric orbs best fit the dominant philosophic outlook, but on the other hand, accurate predictions of positions were important for any application of astronomy.

As *The Light of the World* incorporated more than one approach to explaining the heavens, it is not surprising that it is a text that challenges facile categorizations. It is a text that valued the truths of metaphysics, and thus religion, but that also aimed for mathematical precision. It is a text written for Jews that was also fully in conversation with the philosophy and science of Islamic societies. It is a text that originated in the Iberian Peninsula but which exhibited some similarities with the theoretical innovations of astronomers in Iran. And whatever the text's impact was on science in Islamic societies and in Jewish civilization, *The Light of the World*'s most lasting influence may have been exerted via its passage to Renaissance Italy.

Clearly, *The Light of the World* is important in many ways, none of which I grasped when I first encountered the text in 1992. A full understanding of the text's contents is connected to a recognition of the distinction between the two versions of the text, namely the Judeo-Arabic original and the Hebrew recension. The Hebrew recension contains significant theoretical innovations and evinces an even greater concern for the physics of celestial motions. For that reason, this book is organized so as to treat the two versions of the text separately, allowing each version to speak for itself and all the while acknowledging the strong relationship between the two versions.

The book contains eight chapters, preceded by an introduction. The introduction presents what we know about the author of *The Light of the World*, Joseph Ibn Naḥmias, and then examines the connection of *The Light of the World* to certain currents in Jewish and Islamic thought, pointing out the places where Islamic texts played a significant role in certain Jewish texts. The introduction situates *The Light of the World* in the history of texts that presumed that the cosmos was composed of nesting orbs with a common center, that is, homocentric orbs. Delving into the contents of *The Light of the World*, the introduction outlines the distinctive components of *The Light of the World*'s astronomical models, followed by an overview of the models themselves. This overview provides a more systematic outline of the contents of *The Light of the World* than is possible in the line-by-line commentary. Finally, the introduction describes the available MSS of *The Light of the World*, editorial procedures, and the history of the text's reception.

Chapter 1 is devoted to the Judeo-Arabic original of *The Light of the World*. It includes a transcription of the unique Judeo-Arabic MS with emendations and variant readings from the portions of the text where the Hebrew recension tracks the Judeo-Arabic original. Chapter 2 has a translation of the Judeo-Arabic text, followed by a technical commentary in chapter 5. The purpose of the technical commentary is to explore the full extent of *The Light of the World*'s detailed attempts to reconcile the philosophic elegance, truth, and simplicity of a cosmos of homocentric nesting orbs with the competing truth of mathematical precision. The book has two sets of figures for the Judeo-Arabic original; one set is a translation of the figures found in the MS, and the other set serves to explain points that the technical commentary makes.

Chapter 3 is devoted to the Hebrew recension of *The Light of the World*, a recension made during Ibn Naḥmias's lifetime, either by Ibn Naḥmias himself or by a scholar working in the spirit of the Judeo-Arabic original. The chapter provides a transcription of the unique Hebrew MS, with emendations and variant readings from the portions of the text where the Hebrew recension tracks the Judeo-Arabic original. Where the Hebrew recension departs from the Judeo-Arabic original, chapter 4 supplies a separate translation. The technical commentary on the Hebrew recension follows in chapter 6. Like the Judeo-Arabic original, the Hebrew recension tried to reconcile nesting, homocentric orbs with mathematical precision. But the Hebrew recension was also more concerned than the Judeo-Arabic original with specifying the physical movers for each motion. This concern led to the novel technical complexities that are explored in the separate commentary on the Judeo-Arabic recension. As was the case with the Judeo-Arabic original, the book has two sets of figures; one set is a translation of the figures found in the MS, and the other set serves to explain points that the commentary makes. As an addendum to the technical commentary on the Hebrew recension, in chapters 7 and 8 I transcribe and translate a critical response to *The Light of the World*, written by the scholar Profiat Duran (d. ca. 1415), appended to the end of the MS of the Hebrew recension.

Finally, the book concludes with a Judeo-Arabic/Hebrew/English glossary of technical terms. This glossary is intended to explain my choices in translation and also the choices that the author of the recension made in translating the Judeo-Arabic into Hebrew.

This book will show that while *The Light of the World* has been seen as an idiosyncratic curiosity, the text's interest in meshing mathematical precision with a cosmos of homocentric orbs was an agenda shared by and relevant to other scholars.

Introduction

THE IMPORTANCE OF THIS STUDY

This book contains an edition, with translation and commentary, of a text on theoretical astronomy entitled *The Light of the World* (Ar. Nūr al-ʿālam) by Joseph Ibn Naḥmias, composed in Judeo-Arabic around 1400 CE. in the Iberian Peninsula.[1] *The Light of the World* is the only text written by a Jew on theoretical astronomy in any variety of Arabic and, as such, is also evidence for a continuing relationship between Jewish and Islamic thought in the late 14th and early 15th centuries. The text's composition in Judeo-Arabic indicates the enduring cultural significance of Arabic in Iberia well after the Reconquista had made real headway.[2] *The Light of the World* is the only known Arabic text to attempt to

1. All dates, unless stated otherwise, are CE. Regarding the date of composition of *The Light of the World,* Profiat Duran's (d. 1415) response to *The Light of the World,* found in *Or ha-ʿolam,* Bodleian MS Canon Misc. 334, fols. 101a–100a (the MS is numbered backwards) referred to Ibn Naḥmias in terms that suggested he was alive when Duran wrote. Duran (fol. 100b) referred to Gersonides with a z"l (*zikrono li-b^e^ra<u>k</u>a,* indicating Gersonides was dead) but omitted the z"l for Ibn Naḥmias.

2. For more on Judeo-Arabic literature in the fourteenth century in Christian Iberia, see Nahem Ilan: "'Rᵉdip̱at ha-emet' wᵉ-'dereḵ la-rabbim' : 'iyyunim bᵉ-mishnat R. Yisra'el Yisrᵉ'eli mi-Ṭoledo" (Hebrew University, dissertation, 1999). See also Nahem Ilan: "Know What You Shall Say Back to Epiqoros" [in Hebrew], in David Doron and Joshua Blau (eds.): *Heritage and Innovation in Medieval Judeo-Arabic Culture* [in Hebrew] (Ramat Gan: Bar Ilan Press, 2000): pp. 9–26. Israel Israeli wrote a number of works in Arabic, a language in which he had been deeply educated (Ibid., pp. 51–2). He died (Ibid., p. 47) in the 1320s. For another author (Saadia Ibn Danān) who wrote in Judeo-Arabic

improve directly on al-Biṭrūjī's (Latin: Alpetragius; fl. 1185–92) *On the Principles of Astronomy*.[3] *The Light of the World*'s content is significant because the text reflected the 15th-century resurgence of the Aristotelianism in Iberia that had motivated Biṭrūjī's earlier work.[4] For all these reasons, the need for an edition, translation, and comprehensive study of *The Light of the World* had, more than two decades ago, been mentioned as a major desideratum.[5] Beyond the connection to Islamic astronomy and significance for Jewish thought, *The Light of the World* shows us how the history of astronomy in Iberia came to be connected to developments in Renaissance Italy. Astronomical theories motivated by Aristotelian philosophy, and in some cases virtually identical to those found in *The Light of the World,* proliferated in the first half of the 16th century in Renaissance Italy.[6] Recent research has explained these parallels by arguing that

---

in the 1460s, though in Muslim Granada, see Judith Olszowy-Schlanger: "The Science of Language among Medieval Jews," in Gad Freudenthal (ed.): *Science in Medieval Jewish Culture* (New York and Cambridge: Cambridge University Press, 2012): pp. 359–424, at p. 361, and pp. 409–10. Finally, see Eleazar Gutwirth: "Hispano-Jewish Attitudes to the Moors in the Fifteenth Century," in *S^eparad* XLIX (1989): pp. 237–61.

3. The source for the Hebrew and Arabic texts of Biṭrūjī's *On the Principles of Astronomy* is Bernard Goldstein's 1971 translation and commentary (Yale University Press), with the Hebrew and Arabic texts. Earlier, Francis Carmody had edited the Latin translation in *De motibus celorum; Critical Edition of the Latin Translation of Michael Scot* (Berkeley: University of California press, 1952).

4. Ruth Glasner: "The Peculiar History of Aristotelianism among Spanish Jews," in Resianne Fontaine, Ruth Glasner, Reimund Leicht, and Giuseppe Veltri (eds.): *Studies in the History of Culture and Science: A Tribute to Gad Freudenthal* (Leiden and Boston: Brill, 2011): pp. 361–81, at pp. 375–81. On p. 377, Glasner writes that the influence of the Christian environment "was subtle and not explicit." See also Mark Meyerson: *A Jewish Renaissance in Fifteenth-Century Spain* (Princeton: Princeton University Press, 2004).

5. Y. Tzvi Langermann: "The 'True Perplexity': The *Guide of the Perplexed*, Part II, Chapter 24," in Joel L. Kraemer (ed.): *Perspectives on Maimonides: Philosophical and Historical Studies* (Oxford and New York: Oxford University Press, 1991): pp. 159–74, p. 173. "A thorough study of this treatise is perhaps the chief desideratum for the history of Jewish astronomy."

6. On Girolamo Fracastoro and Giovanni Battista Amico, Italian astronomers in the early sixteenth century who pursued homocentric models, of which one was quite similar to Ibn Naḥmias's, see Mario di Bono: "Copernicus, Amico, Fracastoro, and Ṭūsī's Device: Observations on the Use and Transmission of a Model," in *Journal for the History of Astronomy* XXVI (1995): pp. 133–54; Noel Swerdlow: "Aristotelian Planetary Theory in the Renaissance: Giovanni Battista Amico's Homocentric Spheres," in *Journal for the History of Astronomy* III (1972): pp. 36–48; Michael Shank: "The 'Notes on al-Biṭrūjī' attributed to Regiomontanus: Second Thoughts," in *Journal for the History of Astronomy* XXIII (1992): pp. 15–30; and Noel Swerdlow: "Regiomontanus's Concentric-Sphere Models for the Sun and Moon," in *Journal for the History of Astronomy* XXX (1999): pp. 1–23. See also Enrico Peruzzi: *La nave di Ermete: la cosmologia di Girolamo Fracastoro* (Florence: Olschki, 1995).

the contents of *The Light of the World* were likely transmitted to Italy around 1500, and certainly by the mid-1600s.[7]

The reason why Biṭrūjī's and Ibn Naḥmias's approach to theoretical astronomy stands out in the history of science, both in Islamic societies and generally, is because both scholars aimed to bring mathematical astronomy, that is, theories that aimed to account precisely for observations, into full accord with the philosophic coherence of a cosmos of homocentric orbs (hollow spheres), in which all of the orbs were centered on the Earth. Each orb also had to revolve with its own uniform angular velocity. The interpretations of Aristotle's writings dominant in Biṭrūjī's and Ibn Naḥmias's milieus enjoined this homocentric approach to astronomy and cosmology. According to these interpretations of Aristotle's philosophy, celestial motions about multiple centers would be physically and philosophically inconceivable.

But Ptolemy (fl. 125–50), in order to have physical models correspond better with available observations without sacrificing the principle of uniform revolutions, had hypothesized in *Almagest* III.3 and III.4 orbs with different centers, some of which were embedded within the thickness of other orbs. In the first of these hypotheses, the eccentric, Ptolemy proposed that the Sun traced a path (known as the ecliptic circle or the ecliptic) against the background of the constellations of the zodiac, with the earthbound observer removed by a certain amount from the center of the circle upon which the Sun moved, if one is thinking in terms of two dimensions, or from the center of the orb that moved the Sun, if one is thinking in terms of three dimensions.[8] Once the earthbound observer was no

7. See Robert Morrison: "A Scholarly Intermediary between the Ottoman Empire and Renaissance Europe," in *Isis* CV (2014): 32–57, esp. pp. 52–3 for the connection to Ibn Naḥmias. See also Tzvi Langermann: "A Compendium of Renaissance Science: *Ta'alumot ḥokmah* of Moses Galeano," in *Aleph* VII (2007): 285–318, at pp. 291–3.

8. Gerald Toomer: *Ptolemy's "Almagest"* (London: Duckworth, 1984): pp. 153–7. In the *Almagest*, Ptolemy couched his planetary theories in two dimensions, as circles. Ptolemy described the same model in three-dimensional terms in the *Planetary Hypotheses*, a summary of the *Almagest*'s findings. For the portion of the *Planetary Hypotheses* that discusses the solar model, the topic of *Almagest* III, see Bernard R. Goldstein (trans.): *The Arabic Version of Ptolemy's Planetary Hypotheses* (Philadelphia, 1967), pp. 16–7; and J. L. Heiberg (ed.): *Claudii Ptolemaei Opera Quae Exstant Omnia* (Leipzig, 1907), vol. 2, pp. 80–1. See, now, Elizabeth A. Hamm: "Ptolemy's Planetary Theory: An English Translation of Book One, Part A of the *Planetary Hypotheses* with Introduction and Commentary" (Ph.D. Dissertation, University of Toronto, 2011). See also Régis Morelon: "La Version Arabe Du Livre Des *Hypothèses De Ptolémée*," in *Mélanges Institut Dominicain d'Études Orientales du Cairo* XXI (1993): pp. 7–85.

Ptolemy's theories for the Moon and planets cannot always be easily transferred from two dimensions to three and still function (see A. I. Sabra, "Configuring the Universe: Aporetic, Problem Solving, and Kinematic Modeling as Themes of Arabic Astronomy," in *Perspectives on Science*

longer at the center of the Sun's path, the motion would appear variable. Because observations showed that celestial motions did indeed vary, the eccentric's ability to produce (and thus account for) these variations was significant. Alternatively, Ptolemy hypothesized another orb, the epicycle, which could also account for the irregularities in the Sun's motion. In three dimensions, the epicycle was embedded in a larger, deferent orb either concentric with or eccentric to the Earth. In two dimensions, the epicycle was understood as a circle carried upon another circle, the deferent. If both the epicycle and a concentric deferent revolve with the same angular velocity about their centers, but in opposite directions, a point on the epicycle will trace the same path as the Sun revolving in an eccentric orb. Despite Ptolemy's models' correspondence with observations, the eccentric and the epicycle contradicted the philosophic doctrine of a cosmos of homocentric orbs, a truth that Ibn Naḥmias, Biṭrūjī, and others believed could not be compromised.

Despite the difficulty of Biṭrūjī's quest, as he could not come close to reconciling the predictive accuracy of mathematical astronomy with the Aristotelian philosophy of his era, he still attained a measure of renown and was one of the Arab authors cited by Copernicus in *De Revolutionibus.*[9] As an improvement upon Biṭrūjī's *On the Principles, The Light of the World* turned out to be the final attempt in an Arabic astronomy text to combine the predictive accuracy of mathematical astronomy with uniformly revolving homocentric orbs. The difficulties that Ibn Naḥmias faced in accounting for all available observations with uniformly revolving homocentric orbs is a reason why Ibn Naḥmias's work has not been a priority of historians of science. But the present detailed study of *The Light of the World* markedly alters our understanding of theoretical astronomy in the 14th, 15th, and 16th centuries in Iberia, Provence, and Italy because, in many respects, Ibn Naḥmias's models featured predictive accuracy that improved on Biṭrūjī's.

This introduction first presents what is known about Ibn Naḥmias and describes *The Light of the World* in the context of both Jewish and Islamic intellectual history as well as the history of homocentric astronomy. Then the introduction surveys the distinctive theoretical components of *The Light of the World* and how they are combined in Ibn Naḥmias's models. The introduction concludes with an account of the text's reception and a description of the available MSS and editorial principles.

---

6 (1998): 288–330, at p. 294). Fortunately, the two-dimensional solar model of the *Almagest,* when couched in the three dimensions of the *Planetary Hypotheses,* remains viable because the hypotheses of the eccentric and epicycle could be conceived of in either two or three dimensions. I use the terms *eccentric* and *epicycle* throughout, instead of *eccentric orb* and *epicyclic orb.*

9. For the reference to Biṭrūjī in the critical edition of the De Revolutionibus, see Heribert M. Nobis (ed.): *Nicolaus Copernicus Gesamtausgabe* (Hildesheim: H.A. Gerstenberg, 1974–), vol. 2 of 9: p. 17. Copernicus cited Biṭrūjī's position that Venus was above the Sun and that Mercury was below the Sun. See also Edward Rosen: "Copernicus and Al-Bitruji," in Centaurus VII (1960): pp. 152–6.

THE AUTHOR

The unique Judeo-Arabic MS of *The Light of the World*, like the unique MS of the Hebrew recension, refers to a Joseph Ibn Naḥmias as the text's author in two places. Gad Freudenthal has shown that there were two fourteenth-century scholars with that name, and he used the contents and chronology of their writings to distinguish them.[10] One wrote *The Light of the World* around 1400; three generations earlier, a different scholar with the same name wrote commentaries on Proverbs, Esther, and Jeremiah and redacted the Yom Kippur liturgy. That older Ibn Naḥmias was part of the intellectual circle of Asher ben Yeḥiel (d. 1327) in Toledo.[11] The two scholars did come from the same family in the Iberian Peninsula, perhaps from Aragon.[12] Freudenthal also cited a report from Abraham

10. Of the two passages in *The Light of the World* that mention the author's name, one was at the beginning of the text and the other was §B.4.I.1, at the beginning of the section on the planets, a section missing in the Hebrew recension. In §B.4.I.1, Naḥmias was spelled with a sin, not a samek̲. See Gad Freudenthal: "Towards a Distinction between the Two Rabbis Joseph ibn Joseph ibn Naḥmias" [in Hebrew], in *Qiryat Seip̲er* (1988–9), pp. 917–9, esp. p. 917. This article has been translated into English as "The Distinction between Two R. Joseph b. Joseph Ibn Naḥmias—the Commentator and the Astrologer," in *Science in the Medieval Hebrew and Arabic Traditions* (Aldershot, UK, and Brookfield, VT: Ashgate Variorum, 2005). The elder Joseph ben Joseph Ibn Naḥmias, an ascendant of the author of *The Light of the World,* drew on Israel Israeli's commentary on the Bible (see Ta-Shma, Israel Moses: "Israeli, Israel," in Michael Berenbaum and Fred Skolnik [eds.]: *Encyclopaedia Judaica*, 2nd ed. [Detroit: Macmillan Reference, 2007]). Mosheh Bamberger edited the published edition (Berlin, 1911) of the elder Ibn Naḥmias's commentary on Psalms. The two extant MSS of *The Light of the World* refer to the author as Joseph Ibn Naḥmias, not Joseph ben Joseph Ibn Naḥmias. It may be that the elder was known as Joseph ibn Joseph Ibn Naḥmias, whereas the younger was known as Joseph Ibn Naḥmias.

11. Ta-Shma, Israel Moses: "Naḥmias, Joseph ben Joseph," in Michael Berenbaum and Fred Skolnik (eds.): *Encyclopaedia Judaica*, 2nd ed. (Detroit: Macmillan Reference, 2007). Ilan has cautioned ("Reᵉdip̲at ha-emet," p. 50) against concluding, as Ta-Shma nevertheless did, that Israel Israeli wrote an entire commentary on the Bible.

12. I have located a Joseph Ibn Naḥmias, the recipient of a letter of condolences from Don Vidal Benveniste, in the early fifteenth century, in Aragon, in Bodleian MS 1984, fol. 124. On the connection of Profiat Duran, a critic of *The Light of the World,* to the Benveniste family and to Aragon, see Jacob S. Levinger, Irene Garbell, and Colette Sirat: "Duran, Profiat," in Michael Berenbaum and Fred Skolnik (eds.): *Encyclopaedia Judaica*,, 2nd ed. (Detroit: Macmillan Reference, 2007) pp. 56–58. See also Maud Kozodoy: "A Study of the Life and Works of Profiat Duran" (Ph.D. Dissertation, The Jewish Theological Seminary, 2006), p. 80. All of these details confirm Freudenthal's conclusion that the Ibn Naḥmias family was connected to Aragon.

Members of the Ibn Naḥmias family turn up later in the fifteenth century. There was an Abraham Ibn Naḥmias who completed a commentary on Aristotle's *Metaphysics* in Ocaña, in the province of Toledo, in the 1490s. See Eleazar Gutwirth: "History, Language, and the Sciences in Medieval Spain," in Gad Freudenthal (ed.): *Science in Medieval Jewish Culture* (New York and Cambridge: Cambridge University Press, 2012): pp. 511–28, at p. 519. Members of the Ibn Naḥmias family had established the oldest printing press in the Ottoman Empire by the end of the fifteenth century.

Zacut (d. 1515) that the younger Ibn Naḥmias, the author of *The Light of the World,* composed an astrological prediction for 1478.[13]

THE BACKGROUND OF *THE LIGHT OF THE WORLD* IN ISLAMIC AND JEWISH THOUGHT

The Light of the World, with its composition in Judeo-Arabic and background in Biṭrūjī's *On the Principles,* is evidence for the cosmopolitanism of Islamic societies and for premodern Jewish cultures' close connection with Islam. But *The Light of the World*'s connections to both Islamic and Jewish intellectual life seem to have allowed it, paradoxically, to fall through the cracks in the publications of outstanding modern scholars who are aware of the text and appreciate its contents. For example, George Saliba has written that "Biṭrūjī's account remained a curious proposition that was not pursued any further by later astronomers."[14] Given the title of Saliba's book, *Islamic Science and the Making of the European Renaissance,* he has implied that *The Light of the World* falls outside the ambit of science in Islamic societies. Taken on its own, Saliba's implication is plausible. But *The Light of the World* has also been overlooked in survey articles on science in Jewish cultures, including two in a recent, comprehensive reference work.[15] Even if the reason for the exclusion of *The Light of the World* was that the authors deemed the text too insignificant to merit inclusion, a direct explanation of how *The Light of the World* should be understood as a product of both Islamic and Jewish contexts is necessary both as a foundation for the study of the text and as

---

See A. K. Offenberg, "The First Printed Book Produced at Constantinople," *Studia Rosenthaliana* 3 (1969): pp. 96–112.

13. Freudenthal, "Towards a Distinction," p. 918.

14. George Saliba: *Islamic Science and the Making of the European Renaissance* (Cambridge and London: MIT Press, 2007): p. 121.

15. There is no mention of *The Light of the World* in Bernard R. Goldstein: "Astronomy among Jews in the Middle Ages," in Freudenthal (ed.), *Science,* pp. 136–46. Goldstein omitted *The Light of the World* from "Astronomy in the Medieval Spanish Jewish Community," in L. Nauta and A. Vanderjagt (eds.): *Between Demonstration and Imagination: Essays in the History of Science and Philosophy Presented to John D. North* (Leiden: Brill: 1999): pp. 225–41. But Goldstein had mentioned *The Light of the World* in "Scientific Traditions in Late Medieval Jewish Communities," in G. Dahan (ed.): *Les Juifs au regard de l'histoire: Mélanges en l'honneur de M. Bernhard Blumenkranz* (Paris: Picard, 1985): pp. 235–47. *The Light of the World,* which was translated into Hebrew, is also omitted from Zonta's otherwise exhaustive "Medieval Hebrew Translations," in Freudenthal (ed.) *Science,* pp. 17–73. Yet the Hebrew version of *The Light of the World* is included in Moritz Steinschneider: *Die hebräischen Übersetzungen des Mittelalters und die Juden als Dolmetscher* (Berlin, 1893; repr., Graz, 1956): p. 597 (§375). For an earlier survey of astronomy in Jewish civilization that recognized the significance of *The Light of the World,* see Y. Tzvi Langermann: "Science in the Jewish Communities of the Iberian Peninsula: An Interim Report," in *The Jews and the Sciences in the Middle Ages* (Aldershot, UK, and Brookfield, VT: Ashgate, 1999): pp. 1–54, at p. 18.

an explanation for the surprising silence in the secondary literature. The Islamic dimension of *The Light of the World* emerges with respect to the composition of *The Light of the World* in Arabic and the explicit dependence of *The Light of the World* on the astronomy and philosophy of Islamic societies. But *The Light of the World* was also connected to a tradition of Hebrew scientific texts, a longer tradition of science in Jewish cultures, and certain religious (i.e., Jewish) questions. More succinctly, *The Light of the World* was a product of Jewish culture within an Islamic society.

To begin with the Islamic context, *The Light of the World*'s goal of improving upon Biṭrūjī's *On the Principles* meant that *The Light of the World* was affected by how Biṭrūjī's desire to reject the eccentric and epicycle had placed him within an intellectual movement that A. I. Sabra has famously described as "The Andalusian Revolt against Ptolemaic Astronomy."[16] This description has been recognized by others.[17] Sabra explained that the Andalusian Revolt "was not a widely characteristic or long-lasting phenomenon of Arabic science, being definitely limited both geographically and in time. . . . Its very confinement to Andalusia under one rule, that of the Almohads,[18] gives rise to rather interesting historical questions."[19] One such question is the distinctions between the positions of the central figures in the Andalusian Revolt,[20] who were, in addition to Ibn Rushd and Biṭrūjī, Ibn

16. A. I. Sabra: "The Andalusian Revolt against Ptolemaic Astronomy: Averroes and al-Biṭrūjī," in Everett Mendelsohn (ed.): *Transformation and Tradition in the Sciences* (Cambridge and New York, 1984): pp. 133–53. Julio Samsó ("A Homocentric Solar Model by Abū Jaʿfar al-Khāzin," in *Journal for the History of Arabic Science* I [1977]: pp. 268–75) has found that the Khurasanian al-Khāzin (d. between 961 and 971) produced a homocentric model that was unrelated to the philosophy of the Andalusian Revolt and to the other homocentric models studied in this book. Khāzin's model (p. 273) produced variable speeds without varying the distance between the observer and the Sun by incorporating a point akin to the Ptolemaic equant.

17. For example, see Julio Samsó: *Las Ciencias de los Antiguos en al-Andalus* (Almeria: Fundación Ibn Ṭufayl, 2011): pp. 330–56.

18. Sarah Stroumsa discussed the Almohad influence on the Andalusian Revolt in *Maimonides in His World: Portrait of a Mediterranean Thinker* (Princeton: Princeton University Press, 2009): pp. 80–2.

19. Sabra, "The Andalusian Revolt," p. 133.

20. Astronomy in Andalusia has a much longer history that stretches back before the Andalusian Revolt. See Miquel Forcada: "Astronomy, Astrology, and the Sciences of the Ancients in Early Al-Andalus," in *Zeitschrift für Geschichte des arabisch-islamischen Wissenschaften* XVI (2005): pp. 1–74. See also Julio Samsó: "The Early Development of Astrology in al-Andalus," in *Journal for the History of Arabic Science* III (1979): pp. 228–43. Andalusian astronomers besides those associated with the Andalusian Revolt pioneered planetary theories. Examples include Ibn al-Zarqāl's solar model, which is known through Ibn al-Hāʾim's work (see Emilia Calvo: "Astronomical Theories Related to the Sun in Ibn al-Hāʾim's *al-Zīj al-Kāmil fī al-taʿālīm,*" in *Zeitschrift für Geschichte der arabisch-islamischen Wissenschaften* XII [1998]: pp. 51–111); models for trepidation; and lunar theory

Maymūn/Maimonides (d. 1204), Ibn Ṭufayl (d. 1185 or 1186), Ibn Bājja (d. 1139), and, to an extent, Jābir Ibn Aflaḥ (12th century).[21]

Although *The Light of the World* appeared two centuries after the Andalusian Revolt, many of *The Light of the World*'s important positions came from Ibn Naḥmias's deep engagement with some Arabic texts central to the Andalusian Revolt and with the interpretive debates among those texts. The most heavily referenced text from the Andalusian Revolt was Maimonides's *The Guide of the Perplexed* (Ar. *Dalālat al-ḥā'irīn*). While the *Guide* itself certainly affected Judaism more than Islam, many of the *Guide*'s foundations were in Islamic philosophy and science.[22] Maimonides explained that he had studied under a pupil of Ibn Bājja,[23] cited a number of Muslim scholars, and also argued against the *mutakallimūn* as a group.[24] The *Guide*'s goal of addressing the relationship between revelation on the one hand and philosophy and natural science on the other was shared with texts written by Muslim authors.[25] This central project of the *Guide* was Islamic (but not exclusively so) in that it engaged dimensions of Islam that are religious by any definition.[26]

Defining the epistemological limits of science was important for the *Guide* because those limits would leave room for Jewish scripture's claims about cre-

(see Roser Puig: "The Theory of the Moon in the *Al-Zīj al-kāmil fī-1-ta'ālīm* of Ibn al-Hā'im [*ca.* 1205]," in *Suhayl* 1 [2000]: pp. 71–99).

21. José Bellver Martínez: "El Lugar del *Iṣlāḥ al-Maŷisṭī* de Ŷābir b. Aflaḥ en La Llamada 'Rebelión Andalusí Contra la Astronomía Ptolemaica,'" in *al-Qanṭara* XXX (2009): pp. 83–136. Bellver argues that Jābir ibn Aflaḥ's criticisms of Ptolemy were not themselves cosmological and thus were different from those of the other Andalusian philosophers and scientists that Sabra mentioned. On Jābir ibn Aflaḥ's characterization as only a partial participant in the revolt, see Sabra, "The Andalusian Revolt," p. 135. Jacob ibn Maḵir (Lawrence Berman: "Samuel ben Judah of Marseilles," in Alexander Altmann [ed.]: *Jewish Medieval and Renaissance Studies* [Cambridge: Harvard University Press, 1967]: pp. 289–320, at p. 298) translated the *Iṣlāḥ* into Hebrew.

22. For a survey of the Muslim thinkers discussed in the *Guide*, see Shlomo Pines, "Translator's Introduction," in *The Guide of the Perplexed*, vol. 1, pp. lxxviii–cxxxii. There is also an 883 AH Arabic script MS of the *Guide* in Carullah MS 1279. See Hüseyin Atay (ed. *Dalālat al-ḥā'irīn* (Cairo: Maktabat al-thaqāfa al-dīniyya; originally Ankara, 1972): pp. xxxii–xxxiii (Turkish introduction).

23. Maimonides (trans. Pines), *Guide*, vol. 2, p. 268.

24. The *mutakallimūn* were practitioners of *kalām*, a discourse that went beyond theology to rational speculation into the nature of God. For a survey of *kalām*, see Richard Frank: "The Science of *Kalām*," in *Arabic Sciences and Philosophy* II (1992): pp. 7–37. For the classical period, see Tim Winter (ed.): *The Cambridge Companion to Classical Islamic Theology* (Cambridge: Cambridge University Press, 2008).

25. A classic example would be Ibn Rushd's *Faṣl al-maqāl* (The decisive treatise), but *kalām*, *ḥikma*, and *falsafa* (philosophy in the peripatetic tradition) all queried the boundaries and definitions of revelation and reason.

26. Cf. Marshall Hodgson's distinction between Islamic and Islamicate in *The Venture of Islam: Conscience and History in a World Civilization* (Chicago and London: University of Chicago Press, 1974): vol. 1 of 3, p. 57.

ation ex nihilo. To that end, Maimonides drew on the science of Islamic societies. He referred in Chapter 24 of Book Two (II, 24) of the *Guide* to al-Qabīṣī's (fl. second half of the 10th century) work on planetary distances in order to show that there were physical contradictions in astronomers' conclusions about the dimensions of the cosmos.[27] The computed distances of the planets were incommensurable with the physical principle of the nesting of the celestial orbs central to the Aristotelian description of the heavens. This meant, in fact, that Ptolemaic astronomy could not account for the observed sizes and distances of the planets. Another physical contradiction was that the perfectly homocentric orbs that the Aristotelian natural philosophy of the Andalusian Revolt presented as the truth of the cosmos could not easily account for the observed motions of the Sun, Moon, and planets.[28] While models of homocentric orbs were true from the perspective of philosophy, those models' inability to explain observations well meant that there was a way in which models of homocentric orbs were not true. But the only way that Ptolemy could attain the truth of predictive and retrodictive accuracy was to hypothesize orbs that were no longer centered on the Earth; from the perspective of the *Guide*, Ptolemy sacrificed philosophic truth for mathematical truth.[29] The quandary (or "true perplexity" as Maimonides put it) was either that homocentric orbs could never yield an astronomy possessed of predictive accuracy, or that such a homocentric astronomy was possible but had not yet been created.[30] Maimonides used this disjuncture between the mathematical, on the one hand, and the physical and philosophical, on the other hand, approaches to astronomy to argue that there might be limits (at least in his

27. Maimonides (trans. Pines), *Guide*, vol. 2: p. 325. This was presumably a reference to Qabīṣī's *Risāla fī al-ab'ād wa-'l-ajrām*, the unique extant MS of which is Istanbul MS Ayasofya 4832. For more on that text, see Fuat Sezgin: *Geschichte des arabischen Schrifttums* (Leiden: E. J. Brill, 1978), vol. 6: p. 209.

28. Maimonides (trans. Pines): *Guide*, pp. 322–27, especially p. 324. A scholarly debate has centered on the question of whether Maimonides thought the incommensurability of the mathematical and the physical (or philosophical) approaches was absolute. Pines, in his own introduction (p. cxi) held that such agnosticism was strategic. Langermann has noted ("Maimonides and Astronomy: Some Further Reflections," in *The Jews and the Sciences*, p. 2) that while Pines changed his position in a later essay, the original position has much to recommend it. Langermann pointed out (pp. 6–7) that scholars of Maimonides's era, believing that true, real knowledge of the heavens was possible, opened up another path to salvation and to a measure of theosophy.

29. Maimonides (trans. Pines), *Guide*, vol. 2, p. 323. See also p. 326 for Maimonides's acknowledgment that the calculations of the mathematical astronomers were always true even if the epicycle and eccentric were not: "The truth of this is attested by the correctness of the calculations—always made on the basis of these principles—concerning the eclipses and the exact determination of their times . . ."

30. For a recent interpretation of Maimonides's position, with references to earlier scholarship, see Y. Tzvi Langermann: "My Truest Perplexities," in *Aleph* VIII (2008): pp. 301–17. See also Langermann, "The 'True Perplexity,'" p. 173.

time) to what humans could investigate with confidence, limits that did not affect information transmitted by prophets. Again, these epistemological concerns of *Guide* II, 24 were important to Islam as well as Judaism. Ghazālī's *Incoherence of the Philosophers* addressed which of astronomy's conclusions depended on Aristotelian philosophy and which did not.[31] *The Light of the World*'s goal of reconciling mathematical accuracy and philosophic consistency meant that it aimed to alleviate certain criticisms of astronomy and resolve Maimonides's perplexity. Ibn Naḥmias's response to the epistemological challenge of *Guide* II, 24 brought *The Light of the World* into conversation with earlier scientific and philosophic texts from Islamic societies.[32]

In fact, the very act of the composition of *The Light of the World* was an outgrowth of a debate among the Arabic and Islamic texts that made up the Andalusian Revolt. Maimonides had disagreed with Ibn Rushd in that Ibn Rushd had aimed to recover a lost astronomy of Aristotle.[33] Like Maimonides, Ibn Naḥmias aspired instead to a *new* homocentric astronomy with better predictive accuracy.[34] Never in *The Light of the World* did Ibn Naḥmias conclude that the shortcomings of his models, shortcomings he acknowledged, were evidence of astronomy's epistemological limits. Nor did he think that the project of homocentric astronomy was inherently flawed when it came to predictive accuracy.[35] To improve homocentric astronomy's predictive accuracy, Ibn

31. Ghazālī (trans. Michael Marmura): *The Incoherence of the Philosophers* (Provo, UT, and London: Brigham Young University Press, 1997): pp. 6–7, 18, 24–5, 27–9, 48–9, and 75. A classic analysis of later Islamic critiques of astronomy's metaphysical foundations is A. I. Sabra: "Science and Philosophy in Medieval Islamic Theology," in *Zeitschrift für Geschichte der arabisch-islamischen Wissenschaften* XIII (1994): pp. 1–42.

32. This response begins in *The Light of the World*, §0.8.

33. Langermann, "The 'True Perplexity,'" p. 170, n. 39. Langermann was referring to Ibn Rushd's references, in his *Great Commentary on the Metaphysics*, to Aristotle's descriptions of Eudoxus's astronomy. See also Sabra, "The Andalusian Revolt," p. 142: Averroes "postulates a perfect astronomy that antedated Aristotle, that Aristotle only hinted at in his writings, and that perished as a result of the triumph of Ptolemaic astronomy. . . . It is the duty of the astronomy who is interested in truth as well as computation to try to discover anew that ancient and forgotten astronomy."

34. Joel Kraemer: "Maimonides and the Spanish Aristotelian School," in Mark D. Meyerson and Edward D. English (eds.): *Christians, Muslims and Jews in Medieval and Early Modern Spain* (Notre Dame, IN: University of Notre Dame Press, 1999): pp. 40–68, at p. 51. The contrast between Maimonides and Ibn Rushd would be mostly with respect to Ibn Rushd's commentary on Aristotle's *Metaphysics*, not with respect to all of Ibn Rushd's writings.

35. In fact, in *The Light of the World*, Ibn Naḥmias questioned observations relevant to explaining planetary distances that might conceivably be marshaled as evidence against models of homocentric orbs. These were the well-known and widespread, since antiquity, observations of variations in the observed diameter of the Moon (§B.2.IV.1). In many cases, though not all, Ibn Naḥmias's criticisms of these observations went against the conclusions of other astronomers (but cf. Maimonides [trans. Pines], *Guide*, pp. 322–7; here, Maimonides was criticizing Ptolemy's shortcomings

Naḥmias's astronomy incorporated a distinctive feature, absent from Biṭrūjī's work, which had its foundations in the writings of Muslim scholars associated with the Andalusian Revolt. Ibn Naḥmias proposed small circles, revolving on their own, at the equator of an orb, which he recognized as taking the place of the epicycle (*al-tadwīr*). Ibn Rushd had taken positions on the permissibility of the eccentric and epicycle other than his famous rejection of them in the long commentary on the *Metaphysics*.[36] In the earlier *The Compendium of the Almagest* (Ar. *Mukhtaṣar al-Majisṭī*; survives only as the Hebrew *Qiṣṣur al-Magisṭi*), Ibn Rushd had accepted the existence of eccentrics and epicycles.[37] And as it turns out, not all of Maimonides's writings on astronomy had rejected epicycles along with eccentrics.[38]

Besides the connection to the Andalusian Revolt, a second way in which *The Light of the World* was part of the enterprise of science in Islamic societies was through *The Light of the World*'s attention to a physical inconsistency of the *Almagest*, in which Ptolemy transgressed his own principle of the celestial orbs' uniform motions in revolution. Ptolemy had found that the revolutions of some orbs were uniform about a point other than their center, a point that came to be known as the equant. Even if one accepted the epicycle and eccentric, it was inconceivable (and contrary to Aristotelian philosophy) that orbs might revolve uniformly in place about a point other than their center. Because Ibn Naḥmias's characterization of small circles on the surface of an orb as the replacement for the epicycle might indicate some familiarity with Ibn Rushd's *Compendium*, it is worth noting that in the *Compendium*, Ibn Rushd evinced an awareness of Ibn al-

---

in accounting for planetary distances). Thus, to Ibn Naḥmias, if even the mathematical astronomy of Ptolemy was flawed when it came to planetary distances, the shortcomings of homocentric astronomy in explaining planetary distances might not be so severe.

36. For this famous rejection, see Ibn Rushd (Maurice Bouyges ed.): *Tafsīr Mā ba'd al-ṭabī'a* (Beirut: Dar el-Machreq, 1948–67), 3 vols.: vol. 3, pp. 1647–8. See also Ibn Rushd (intro. and trans. Charles Genequand): *Ibn Rushd's Metaphysics: A Translation with Introduction of Ibn Rushd's Commentary on Aristotle's Metaphysics, Book Lam* (Leiden: E. J. Brill, 1984): p. 172. In his different commentaries on the *De Caelo*, Ibn Rushd gave different explanations for why the Earth did not move. See Henri Hugonnard-Roche: "Remarques sur l'évolution doctrinale d'Averroès dans les commentaires au De Caelo; Le problème du mouvement de la terre," in *Mélanges de la Casa de Velázquez* XIII (1977): pp. 103–17. See, too, Miquel Forcada: "La ciencia en Averroes," in *Averroes y los averroísmos: actas del III Congreso de Filosofía Medieval* (Zaragoza: Sociedad de Filosofía Medieval, 1999): pp. 49–102, at pp. 73–7.

37. Juliane Lay: "*L'Abrégé de l'Almageste;* Un Inédit d'Averroès en Version Hébraïque," in *Arabic Sciences and Philosophy* VI (1996): pp. 23–61, at pp. 47–8 (for the *Shukūk*) and, for example, pp. 46–8 for a discussion of eccentrics and epicycles. For more on Ibn Rushd's *Mukhtaṣar al-Majisṭī*, see Fuat Sezgin, *Geschichte des arabischen Schrifttums* (Leiden: E. J. Brill, 1978): vol. 6: p. 93. For more on the admissibility of epicycles in *The Light of the World*, see the comment on §0.1

38. Langermann, "The 'True Perplexity,'" p. 173.

Haytham's (d. ca. 1040) enumeration and analysis of these internal inconsistencies of Ptolemaic astronomy in his own *al-Shukūk ʿalā Baṭlamyūs* (Doubts about Ptolemy).[39] Ibn Bājja had referred to the *Shukūk* in a letter to Abū Jaʿfar Yūsuf ibn Ḥasday.[40] Still, Ibn al-Haytham's critique of the equant itself resurfaced neither in *On the Principles* nor in *The Light of the World*.[41] Instead, Ibn Naḥmias pointed to the Moon's prosneusis point, another inconsistency of Ptolemy's *Almagest* related to the nonuniformity of an orb's motion that Ibn al-Haytham also addressed.[42] Ptolemy had postulated that the starting point of the lunar epicycle's revolution would have to oscillate, meaning that the measure of the revolution of that orb would not be uniform with respect to a fixed point. This was another way in which Ptolemy achieved predictive accuracy only by making statements about the cosmos that contradicted his own ground rule of uniformly revolving orbs. Ibn Naḥmias wrote that Ptolemy's explanation was neither possible nor feasible (§B.2.V.7; for an explanation of this notation, see p. 39). Ibn Naḥmias also wrote in that passage that Ptolemy's second lunar model featured a "repugnancy" not that of the epicycle or eccentric. Presumably, Ibn Naḥmias meant that the lunar epicycle's motion, in the second lunar model, was not uniform about its own center. George Saliba has found evidence of an eleventh-century scholar in Andalusia who hoped to expound a number of criticisms of the Ptolemaic lunar model.[43] Clearly, there was at least one critic of the Ptolemaic models in Andalusia before the time of the Andalusian Revolt.

The Light of the World was translated into Hebrew in Ibn Naḥmias's own lifetime, indicating that astronomy in Jewish cultures was linked to the astronomy of Islamic societies. Such a connection had existed for several centuries. Jews living in Islamic societies, for whom Arabic was their native language, had participated in those societies' scientific enterprise for a long time.[44] The first texts in Arabic

39. See Lay, "L'Abrégé," p. 25 and pp. 47–8 for Ibn Rushd's knowledge of Ibn al-Haytham's *al-Shukūk* and p. 31 for Ibn Naḥmias's knowledge of the *Qiṣṣur*. Ibn Naḥmias's remarks on the lunar model (B.2.V.7) indicated familiarity with the prosneusis problem.

40. Shlomo Pines: "Ibn al-Haytham's critique of Ptolemy," in *Actes du dixième congrès international des sciences* I, no. 10 (Ithaca, 1962; Paris, 1964): pp. 547–50. Ibn Bājja's Arabic text is now published in Jamāl ʿAlawī (ed.): *Rasāʾil falsafiyya li-Abī Bakr Ibn Bājja* (Beirut: Dār al-thaqāfa; Casablanca: Dār al-nashr al-Maghribiyya, 1983), pp. 77–81, esp. p. 78), cited in Sabra, "Configuring the Universe," p. 305.

41. Sabra, "The Andalusian Revolt," p. 134.

42. *The Light of the World*, 79r (§B.2.V.7). The prosneusis point was described in that paragraph as the third anomaly. Biṭrūjī's description of his lunar model did not attempt to solve, or mention, the prosneusis point.

43. George Saliba: "Critiques of Ptolemaic Astronomy in Islamic Spain," in *al-Qanṭara* XX (1999): pp. 3–20, esp. p. 13.

44. For more, see Robert Morrison, (Norman A. Stillman, ed.), "Science (Medieval)," *Encyclopedia of Jews in the Islamic World*, Brill Online, 2013.

on the astral sciences by Jews were due to Māshā'allāh (d. ca. 815) and Sahl ibn Bishr al-Isrā'īlī (d. ca. 845).[45] In the discipline of medicine, Ḥasday ibn Shapruṭ (fl. tenth century) helped translate Dioscorides's writings into Arabic in Andalusia.[46] Goldstein has explained that Jews were able to participate in an enterprise instigated by the dominant religion of Islam because science could transcend a particular religion, even if science was relevant to religion.[47] Or Judaism and Islam could have overlapping concerns. In any case, Arabic scientific texts were part of Jewish cultures. Foundational works of astronomy, such as Abraham Ibn Ezra's (d. 1164 or 1167) translation of Ibn al-Muthannā's commentary on al-Khwārizmī's tables, appeared in Hebrew translation only in the 12th century.[48] These translations led to astronomy texts being composed in Hebrew, at first in the Iberian Peninsula and then in Provence. The proliferation of original scientific texts in Hebrew has led scholars to recognize a tradition separate from Islamic astronomy.[49] In addition to Abraham Bar Ḥiyya's (d. ca. 1136) *Ṣurat ha-areṣ* (The form of

45. For Māshā'allāh, see Moritz Steinschneider: *Mathematik bei den Juden* (Hildesheim: G. Olms, 1964): pp. 35–6; and Steinschneider, *Die arabische Literatur der Juden* (Frankfurt: J. Kauffmann, 1902), pp. 15–23. See also David Pingree, "Māshā'allāh," in *Complete Dictionary of Scientific Biography*, (Detroit: Charles Scribner's Sons, 2008). pp. 159-62.. For Sahl ibn Bishr al-Isrā'īlī, see *Geschichte des arabischen Schrifttums*, vol. 6, p. 138, and vol. 7, pp. 125–8; and Bernard Goldstein: "Astronomy and the Jewish Community in Early Islam," *Aleph* I (2001): pp. 17–57 at p. 24.

46. Ashtor, Eliyahu, "Ḥisdai (Ḥasdai) Ibn Shaprut," in Michael Berenbaum and Fred Skolnik (eds.), *Encyclopaedia Judaica*, 2nd ed. (Detroit: Macmillan Reference, 2007.

47. Goldstein, "Astronomy and the Jewish Community," p. 24. And even a scholar, Maimonides, who argued vociferously against the Karaites, held (pp. 35–6) that the authority of the rules for calendar calculation rested on the fact that they were transparent to any intelligent person, whatever his or her religion.

48. On the cultural role of the translations from Arabic in science in Jewish civilization, see Goldstein, "Astronomy among Jews," p. 136. For a recent survey in English of translations of scientific texts into Hebrew, including but not limited to texts originally in Arabic, see Mauro Zonta: "Medieval Hebrew Translations of Philosophical and Scientific Texts: *A Chronological Table*," in Freudenthal (ed.), *Science*, pp. 17–73. An earlier classic work on translations into Hebrew is Moritz Steinschneider: *Die Hebräischen Übersetzungen des Mittelalters* (Berlin, 1893).

On the translation of Ptolemy's *Almagest* into Hebrew in the 1230s by Jacob Anatoli, see Mauro Zonta: "La Tradizione Ebraica dell'*Almagesto* di Tolomeo," in *Henoch* XV (1993): pp. 325–50, at pp. 329–31. Anatoli drew on both the Ḥajjāj and Isḥāq-Thābit Arabic translations of the *Almagest* (Zonta, "La Tradizione Ebraica," pp. 331–43). About a century later, in 1331, Samuel of Marseilles completed a paraphrase of Anatoli's translation (ibid., pp. 347–8); see also Lawrence Berman: "Samuel ben Judah of Marseilles," in Alexander Altmann (ed.): *Jewish Medieval and Renaissance Studies* (Cambridge: Harvard University Press, 1967): pp. 289–320 at 297; unfortunately, the MS is interrupted at the end of Book Three. The *Planetary Hypotheses* was translated into Hebrew by Qalonymos ben Qalonymos around 1317 (Zonta, "Medieval Hebrew Translations," p. 53).

49. See Bernard Goldstein: "The Medieval Hebrew Tradition in Astronomy," in *Journal of the American Oriental Society* LXXXV (1965): pp. 145–8, at p. 145. "I have become increasingly aware of a Hebrew tradition in astronomy separate from, though dependent on, Islamic astronomy."

the Earth), which followed the Ptolemaic system, other important original compositions on astronomy in Hebrew in Iberia were the tables of Jacob ben David Bonjorn (early 14th century) and the more detailed theoretical work *Yesod ʿOlam* (Foundation of the cosmos) composed by Isaac Israeli in 1310.[50] *Yesod ʿOlam* summarized and then rejected Biṭrūjī's theories, as did Levi ben Gerson (Gersonides; d. 1344) in the section on astronomy in his lengthy philosophical work *Milḥamot ha-shem* (The Wars of the Lord), which was composed in nearby Provence.[51] Levi ben Gerson has an uncontested reputation as the most gifted astronomer in premodern Jewish cultures, an innovator both in theoretical astronomy and in instrument construction, and he was influential in matters of observation and their intersection with theory.[52]

The Light of the World is connected to this tradition of science in Jewish cultures because the original's composition in Judeo-Arabic limited its readership to those familiar with the Hebrew alphabet, and the Hebrew translation produced while Ibn Naḥmias was alive signaled that there was a significant Jewish readership unfamiliar with Arabic. Although *The Light of the World* did not cite a single text in Hebrew besides the Bible, there are a number of important thematic parallels between *The Light of the World* and other texts in Hebrew. First, Ibn Naḥmias's position on astronomy's epistemological challenges makes sense in light of Levi ben Gerson's position in the debate over Maimonides's perplexity.[53] True, Ibn Naḥmias's definition of a demonstrably true astronomy foreswore orbs whose centers were not those of the Earth; Gersonides had rejected epicycles, though he relied on eccentrics. Gersonides believed that he had produced a demonstrably true astronomy in that there was no other known model that could explain the observations.[54] While Ibn Naḥmias did not refer to Gersonides's work, Gersonides's position that humans could increase their certain knowledge

50. For Jacob ben David Bonjorn, see Langermann, "Science in the Jewish Communities," p. 7. Isaac Israeli was the brother of R. Israel Israeli, the author of the Judeo-Arabic commentary on *Aḇot*. See Ilan, "Redipat ha-emet," p. 49.

51. See also Biṭrūjī (trans. and comm. Goldstein), *On the Principles*, pp. 40–3 and p. 51, where Goldstein explained that Isaac Israeli held that Biṭrūjī's concerns belonged to metaphysics and not astronomy.

52. Bernard Goldstein: "Astronomy among Jews in the Middle Ages," in Freudenthal (ed.), *Science*, pp. 136–46, at p. 141. Book five, part one of *The Wars of the Lord* (which was devoted to astronomy) existed in MS as an independent treatise of over 200 folios. On the intersection between theory and observation in Gersonides's astronomy, see Bernard Goldstein: "Theory and Observation in Medieval Astronomy," in *Isis* LXIII (1972): pp. 39–47.

53. José Luis Mancha: "Demonstrative Astronomy: Notes on Levi ben Geršom's Answer to *Guide* II.24," in Fontaine et al., *Studies in the History*, pp. 323–46.

54. Mancha, "Demonstrative Astronomy," p. 324. On humans' ability to increase their certain knowledge of the heavens, see pp. 328–9, where Mancha quoted Levi ben Gerson: "Indeed, it *has been* possible for us to attain the starting points of demonstrations that lead to conclusions about

of the heavens accorded with Ibn Naḥmias's view of the possibility of devising a homocentric astronomy endowed with predictive accuracy.

A second dimension of *The Light of the World,* namely its being an example of the reception of the *Guide,* fell squarely within the realm of Jewish thought. The *Guide* was translated into Hebrew at the end of Maimonides's lifetime, and debates about the existence of the epicycle and the motions of the celestial orbs entered early Hebrew commentaries on *Guide* II, 24.[55] While Ibn Naḥmias did not name any of these commentators, the content of these commentaries and their parallels with *The Light of the World* show how such debates were part of Jewish intellectual life. Certainly at least one commentator, Joseph Kaspi (d. ca. 1340), had little to say about *Guide* II, 24;[56] another, Mosheh Narboni (d. after 1362), understood the challenge of reconciling predictive accuracy with a cosmos of homocentric orbs to be beyond the grasp of reason (Heb. *bi-mah she-lo yussag ba-heqqeish*).[57] Yet we find in other early commentaries on the *Guide* an openness to a less stringent interpretation of the rejection of epicycles and eccentrics. For example, Shem Ṭoḇ ibn Falaquera (d. 1295) began by writing that there was a lost ancient astronomy that rejected epicycles and eccentrics.[58] It was unclear from his comment whether there were scholars of his time actively engaged in trying to recover it. Ibn Falaquera suggested, though, that what Maimonides might have meant by his definition of the heavens as orbs moving about a stationary center was that only the uppermost orb responsible for the main motion (and not the other orbs) needed to move about the center of the world. Ibn Falaquera noted that Maimonides had written in *Guide* I, 70 that "a sphere is counted as one though there be several heavens contained in it."[59] Therefore, if one interpreted *Guide* II, 24 to stipulate only that all of the planets be carried about the Earth through the daily motion, much of the difficulty of reconciling mathematical astronomy with homocentric orbs would disappear.

---

the inclination of the Sun's orb or the solar and lunar distances, despite the heavens are 'too far away and high.'"

55. The *Guide* was translated into Hebrew by Samuel ibn Tibbon, who incorporated Maimonides's advice, in 1204. Samuel Ibn Tibbon's translation engaged key debates (see Gad Freudenthal: "Maimonides on the Knowability of the Heavens and Their Mover [*Guide* II, 24]," in *Aleph* VIII (2008): pp. 152–4), and Ibn Tibbon consulted Maimonides. Jews in France were immediately interested.

56. See Joseph Kaspi (Werbluner ed.): *Commentaria hebraica in R. Mosis Maimonidis* (Frankfurt, 1848: repr. in *Sh^{e}losha qadmonei m^{e}p̱ar^{e}shei* ha-Moreh [Jerusalem, 1968]): p. 107.

57. See Moses Narboni (Goldenthal ed.): *Der Commentar des Rabbi Moses Narbonensis zu dem Werke "More Nebuchim" des Maimonides* (Vienna, 1862; repr. in *Sh^{e}losha qadmonei m^{e}p̱ar^{e}shei* ha-Moreh [Jerusalem, 1968]): pp. 36a–b.

58. Shem Ṭoḇ ben Joseph ibn Falaquera (Yair Shiffman ed. and comm.): *Moreh ha-Moreh* (Jerusalem: World Union of Jewish Studies, 2001): pp. 284–5.

59. Maimonides (trans. Pines), *Guide*, p. 172.

In another commentary on the *Guide,* Ibn Naḥmias's contemporary Profiat Duran (d. ca. 1415) addressed Maimonides's statement that "that motion [of epicycles and eccentrics] is likewise not a motion taking place around an immobile thing."[60] Duran commented that there were many cases in which one body moved about another body in motion, such as the case of a human walking while on a moving ship.[61] But to take one example, the ship, in turn, moved on the water, and the water could be said to be at rest. Thus, Duran's interpretation of Maimonides's doctrine of homocentricity left space for the epicycle. Duran's direct response to *The Light of the World* indicates that he accepted the eccentric as well, but he did not defend the existence of the eccentric in his comments on *Guide* II, 24.

Ibn Naḥmias's implication that the motion of an upper orb did not necessarily affect the motion of an orb beneath it, a topic with which Biṭrūjī had been, on his part, quite concerned, also had a foundation in the Hebrew commentaries on *Guide* II, 24. Responding to the presumption that if an upper orb moved, then the lower orb must also be moved, the only exception being when the upper orb's motion is on an axis passing between the two centers, Maimonides wrote, "neither of the two spheres, the containing and the contained, is set in motion by the movement of the other nor does it move around the other's center or poles, but that each of them has its own particular motion. Hence necessity obliges the belief that between every two spheres there are bodies other than those of the spheres. Now if this be so, how many obscure points remain?"[62] On the question of whether the motions of the orbs that moved an upper planet might affect the motions of the orbs that moved a lower planet, both Duran and Shem Ṭob̲ ibn Shem Ṭob̲ (d. ca. 1441) accepted Maimonides's explanation that, in most cases, there would have to be nonspherical bodies intervening between the orbs so that the motion of the upper was not transmitted to the lower.[63] *The Light of the World*'s presumption that celestial motions were not transmitted all the way down contrasted with Biṭrūjī's interest in explaining the motions of the orbs as part of the organic motions of the cosmos, with progressively lower orbs lagging ever more behind the uppermost. Even if Ibn Naḥmias did not cite the Hebrew

60. Maimonides (trans. Pines), *Guide,* p. 323.

61. Maimonides: *Seip̱er u-p̱eirush Moreh nᵉb̲uk̲im/la-Rab̲ Mosheh bar Maimon, zal ; niṣṣab̲ peirusho . . . peirush Shem Ṭob̲ u-p̱eirush Ep̱odi* (Żolkiew: Gedrucht be Leib Matfes & Berl Lorie, 1860), vol. 2 of 3: 52d/1–2. The rest of the paragraph depends on this citation.

62. Maimonides (trans. Pines), *Guide,* p. 324.

63. Maimonides, *Seip̱er Moreh nᵉb̲uk̲im* (with commentaries): 52d/2. Shem Ṭob̲ wrote, "He wants to say that when the upper orb moves, with its motion not being on the diameter passing through the two centers, it is not necessary that that which is beneath it moves. But the matter does not exist like this, with bodies different from the body of the orbs [intervening between orbs]. He wants to say that this is through a body that does not retain its form."

commentaries on the *Guide* penned by his contemporaries, their existence shows that other scholars in the thirteenth, fourteenth, and fifteenth centuries took similar positions on the question of whether and how celestial motions were transmitted all the way down through the orbs, and on the implications of Maimonides's interpretation of homocentricity.

There are less technical ways in which *The Light of the World* drew on currents in Jewish thought. The earliest Hebrew responses to Biṭrūjī's *On the Principles* ascribed a religious value to any new development in astronomy, irrespective of whether the author of those responses understood those developments to be correct.[64] James Robinson has found that *On the Principles* attracted serious attention from authors who nevertheless dismissed Biṭrūjī's conclusions, perhaps a testament to how compelling Biṭrūjī's goal was.[65] Specifically, Samuel Ibn Tibbon (d. 1232) discussed Biṭrūjī's *On the Principles* in three of his works, beginning in 1213.[66] Then, Moses Ibn Tibbon (fl. 1244–83) produced the Hebrew translation of *On the Principles* in 1259.[67] Biṭrūjī's theories were most relevant to commentaries on Ezekiel and Ecclesiastes. Though Maimonides had linked the vision of the chariot in Ezekiel 1 to Aristotelian metaphysics, Ibn Tibbon identified "the work of the chariot" with astronomy.[68] Ecclesiastes 1:5 implied that everything under the Sun is eventually destroyed, so for Ibn Tibbon, astronomy (and astrology) explained the processes by which that destruction occurred.[69] These early responses to Biṭrūjī provide a context for the four references to the Hebrew Bible in *The Light of the World*.[70] Such connections between science and scripture had an earlier history, as the first Jewish scientific text, the late eighth-century CE Hebrew composition *Barayta di-Sh^{e}mu'el,* which focused on astronomy and astrology, mirrored the Mishnah's structure; authorship was ascribed, fictitiously, to a Talmudic sage.[71] Subsequently, other identifiably Jewish

64. James T. Robinson: "The First References in Hebrew to al-Biṭrūjī's *On the Principles of Astronomy*," in *Aleph* III (2003): pp. 145–63, at p. 148.

65. Robinson, "The First References," p. 146. Robinson noted that the authors who rejected Biṭrūjī referred to him with seemingly complimentary cognomens. Robinson also translated passages (ibid., p. 154, p. 160) in which authors writing in Hebrew approved of Biṭrūjī's theories.

66. Robinson: "The First References," pp. 145–6.

67. Goldstein, "Medieval Hebrew Tradition," p. 146.

68. Robinson, "The First References," p. 148.

69. Robinson, "The First References," p. 150. See also James T. Robinson: *Samuel Ibn Tibbon's Commentary on Ecclesiastes: The Book of the Soul of Man* (Tübingen: Mohr Siebeck, 2007): pp. 249–50. Ibn Tibbon described Biṭrūjī as "possessed by the spirit of God" and saw the absence of opposite motions as a distinguishing feature of Biṭrūjī's models.

70. Of the four biblical passages quoted in *The Light of the World,* three had appeared in the Guide. The only exception is the excerpt from Ecclesiastes 1:11.

71. Y. Tzvi Langermann: "On the Beginnings of Hebrew Scientific Literature and on Studying History through Maqbilot (Parallels)," in *Aleph* II (2002): pp. 169–89, at pp. 170–5. The first half of

concerns, such as Rabbanite-Karaite debates, could bring about the composition of scientific texts. Saadia Gaon's (d. 942) treatise on calendar calculation argued that scientific calculations of the Moon's positions, the Rabbanite preference, would always be confirmed by (and thus agree with) observations, the basis of the Karaites' calendrical calculations.[72] In the Andalusian context, examples of astronomy being brought to bear on Jewish concerns, other than the *Guide*, include the references to astronomy in Ibn Gabirol's (d. 1058) *Keter malkut* and in Abraham Ibn Ezra's (d. 1164 or 1167) scriptural commentary and thought.[73]

In sum, *The Light of the World* was a text written for Jewish readers but was also an outgrowth of an earlier episode in the science of Islamic societies, the Andalusian Revolt.[74] As the thinkers associated with the Andalusian Revolt asserted that mathematical astronomy alone did not fully portray reality[75] and attempted to place both philosophy and mathematics in the proper relationship to reality,[76] they exerted the primary influence on *The Light of the World*. Because one of Ibn Naḥmias's cited sources, Ibn Rushd, stated his desire to recover an earlier homocentric astronomy, and because the true inspiration for Biṭrūjī's homocentric astronomy has been a source of modern scholarly debate, the intro-

---

the text, on astronomy, discussed questions related to timekeeping such as eclipse limits and oblique ascensions. The second half covered the positions of the planets, including information about how to make astrological predictions. The *Barayta di-Sh^{e}mu'el* also had no clear connection to science in Islamic societies. For Langermann's understanding of the place of science in Jewish civilization, see "Science in the Jewish Communities," p. 3, where he framed science "as a phenomenon of Jewish culture(s)."

72. Y. Tzvi Langermann: "Sa'adya and the Sciences," in Y. Tzvi Langermann (ed.): *The Jews and the Sciences in the Middle Ages* (Aldershot, UK, and Brookfield, VT: Ashgate, 1999): pp. 1–21 (second article), at p. 5. See also Saadia Gaon, *The Book of Beliefs and Opinions* (New Haven: Yale University Press, 1989): pp. 16–26. Sa'adya, in addition, composed an Arabic treatise on inheritance calculations, *Kitāb al-Mawārīth*. See also Goldstein, "Astronomy and the Jewish Community," pp. 32–4. Judah Ha-Levi's (d. 1135) *Kuzari* (trans. Slonimsky, pp. 241–2) contained similar pro-Rabbanite positions on astronomy.

73. On Ibn Gabirol, see Goldstein, "Astronomy among Jews," p. 137. On Abraham Ibn Ezra, see Y. Tzvi Langermann: "Some Astrological Themes in the Thought of Abraham Ibn Ezra," in Isadore Twersky and Jay M. Harris (eds.): *Rabbi Abraham Ibn Ezra: Studies in the Writings of a Twelfth-Century Jewish Polymath* (Cambridge, MA, and London: Harvard University Press, 1993): pp. 28–85.

74. Robert Morrison: "The Solar Model of Joseph ibn Joseph ibn Naḥmias," in *Arabic Sciences and Philosophy* XV (2005): pp. 57–108, at pp. 58–9 and 108. Sabra himself pointed out ("The Andalusian Revolt," p. 133) that the Andalusian Revolt, despite its impact, was limited in its chronological scope.

75. Gerhard Endress: "Mathematics and Philosophy in Medieval Islam," in Jan P. Hogendijk and A. I. Sabra (eds.): *The Enterprise of Science in Islam: New Perspectives* (Cambridge and London: MIT Press, 2003): pp. 121–76, at p. 148. "Starting with Ibn Bājja, they would attack Ibn al-Haytham for trespassing into foreign territory." In this context, Ibn al-Haytham's astronomy was perceived to have focused too much on mathematical astronomy.

76. Endress, "Mathematics and Philosophy," 122.

duction turns to the history of homocentric models and *The Light of the World*'s place in it.

THE HISTORY OF HOMOCENTRIC MODELS

The history of homocentric models begins with Eudoxus of Knidos (d. ca. 338 BCE), whose proposed models based on homocentric orbs were, in addition, the first three-dimensional models of planetary motions.[77] Calippus of Cyzicus (d. ca. 300 BCE) and Aristotle both tried to improve upon Eudoxus's system.[78] Unfortunately, despite the importance of Eudoxus's models for the history of astronomy, the models themselves do not survive. All that modern scholars possess are reports from Aristotle and Simplicius and modern reconstructions based on those reports. One's choice of reconstruction has implications for the perceived similarities (or differences) between Eudoxus's models and those of Biṭrūjī and Ibn Naḥmias. Any perceived similarities and differences would affect one's conclusion about whether Biṭrūjī's and Ibn Naḥmias's main concern was recovering lost models or pioneering new ones. In Schiaparelli's classic reconstruction, Eudoxus's models for the upper planets (Mars, Jupiter, and Saturn) involved four orbs.[79] One represented the heavens' daily motions, and another represented the planets' motion in longitude, through the plane of the ecliptic. The axis of one of the final two orbs was inclined to the axis of the other one; they revolved in opposite directions at equal speeds. Ragep has described this combination of concentric orbs as a Eudoxan couple.[80] With the appropriate

77. See Thomas Heath: *Aristarchos of Samos: The Ancient Copernicus* (Oxford: Clarendon Press, 1913; repr., Mineola, NY: Dover Publications, 1981, 2004): pp. 190–248. See also Menso Folkerts: "Eudoxus," in Hubert Cancik and Helmuth Schneider (eds.): *Brill's New Pauly*, Brill Online, 2014. http://referenceworks.brillonline.com/entries/brill-s-new-pauly/eudoxus-e404350.

78. Otto Neugebauer, *History of Ancient Mathematical Astronomy* (New York and Berlin: Springer-Verlag, 1975): pp. 683–5. The meaningful details of Calippus's models have been lost. Aristotle's accounts of Calippus's models do not provide technical details.

79. Giovanni Schiaparelli: *Scritti sulla storia della astronomia antica* (Bologna: N. Zanichelli, 1925–7): vol. 2 of 3: pp. 3–112 (the essay appeared originally in 1877). See also Neugebauer, *History of Ancient Mathematical Astronomy*, pp. 677–83, which follows Schiaparelli. But see, now, Ido Yavetz: "On the Homocentric Spheres of Eudoxus," in *Archive for History of Exact Sciences* LII (1998): pp. 221–78; and Yavetz: "A New Role for the Hippopede of Eudoxus," in *Archive for History of Exact Sciences* LVI (2001): pp. 69–93. See also Yavetz, "On Simplicius' Testimony Regarding Eudoxan Lunar Theory," in *Science in Context* XVI (2003): pp. 319–29. One of Yavetz's observations was that scholars have prematurely foreclosed other possible reconstructions by focusing on hippopedes and by assuming that the planets' retrograde motion was due to the third and fourth orbs.

80. F. J. Ragep (ed., trans., and comm.): *Naṣīr al-Dīn al-Ṭūsī's "Memoir on Astronomy (al-Tadhkira fī 'ilm al-hay'a)"* (New York and Berlin: Springer-Verlag, 1993): pp. 451–2. Ragep's term reflects the regnant scholarly consensus before Yavetz wrote. I use the term Eudoxan couple as a shorthand for these concentric orbs.

speeds and inclination of one orb to the other, the result would be a curve that could account approximately, in Schiaparelli's reconstruction, for the planets' retrograde motion. In Schiaparelli's reconstruction, this curve turned out to be in the shape of a hippopede, a figure eight.

Recently, Ido Yavetz has pointed out that other reconstructions, with better predictive accuracy, are possible, but he has cautioned that his reconstructions are not necessarily a more accurate picture of what Eudoxus had actually meant.[81] Rather, Yavetz has argued simply that these newer reconstructions are just as plausible as Schiaparelli's. Yavetz pointed out that Aristotle never mentioned the word *hippopede* in his account of Eudoxus's astronomy, meaning that the range and, hence, potential predictive accuracy of reconstructions are greater.[82] Some of Yavetz's reconstructions do more to mirror the asymmetries of observed retrograde arcs than the hippopede had.[83] The increased number of potential interpretations means, paradoxically, that we are now less sure of what Eudoxus actually meant. There is the associated issue of whether medieval astronomers could have had the technical proficiency necessary for understanding the interpretations.[84]

Yavetz's argument that Eudoxan principles could have been understood in varied and broad ways nuances the earlier scholarly debate over whether Biṭrūjī's models developed out of a quest to reconstruct Eudoxus's earlier lost homocentric models or out of an attempt to bring astronomy into conformity with Aristotelian philosophy by modifying models for trepidation, that is, oscillations in the rate of the equinoxes' motion, to account for the planets' motions. The debate began when E. S. Kennedy argued that Biṭrūjī's *On the Principles* was an attempt to reconstruct Eudoxus's lost astronomy.[85] Goldstein responded that Biṭrūjī's models were based on Ibn al-Zarqāl's (d. 1100) model for trepidation, in which the equinoxes move back and forth with respect to the ecliptic, not

81. Yavetz, "On the Homocentric Spheres," p. 221.

82. Yavetz, "On the Homocentric Spheres," pp. 237–9.

83. Yavetz, "On the Homocentric Spheres," for example, p. 266.

84. Langermann ("The 'True Perplexity,'" p. 170, n. 39) had asked this question. But there is no explicit textual evidence that gives us reason to doubt the ability of medieval astronomers to reconstruct Eudoxus's models. That is, medieval astronomers could bring remarkable mathematical sophistication to theories such as trepidation, the observational evidence for which was contested. See, for example, Jerzy Dobrzycki: "The Medieval Theory of Precession," in *Studia Copernicana* XLIII (2010): pp. 15–60.

85. For the argument that Biṭrūjī was trying to recover Eudoxus's models, see E. S. Kennedy, review of Carmody, *al-Biṭrūjī's "De motibus celorum,"* in *Speculum* XXIX (1954): pp. 246–51; and E. S. Kennedy: "Alpetragius's Astronomy," in *Journal for the History of Astronomy* IV (1973): pp. 134–6. The resolution of the debate about Biṭrūjī's motivation would affect our understanding of *The Light of the World*, as Ibn Naḥmias was motivated by a desire to improve on Biṭrūjī.

by a quest to reconstruct a lost Eudoxan model.[86] But José Mancha has recently found that Biṭrūjī's model for the fixed stars included a Eudoxan couple, two concentric orbs moving about the same axis in opposite directions with the same speed, causing the equinoxes to trace a hippopede, that is, moving back and forth with respect to the ecliptic.[87] Mancha cautioned, however, that the presence of a Eudoxan couple in Biṭrūjī's *On the Principles* did not mean that Biṭrūjī was motivated by a quest to recover Eudoxus's lost models: "We know that some medieval scholars (e.g., Ibn al-Haytham, Ibrāhīm b. Sīnān, Levi ben Gerson) understood the working of Eudoxan couples; however, how they conceived Eudoxus's and Aristotle's astronomical theories is another matter, for Simplicius's commentary on *De Caelo* was apparently never translated into Arabic."[88] In addition, Mancha has been able to locate the principles for a Eudoxan couple *only* in Biṭrūjī's model for the fixed stars, suggesting that there is still much scope to consider the role of the trepidation models of Ibn al-Zarqāl as a foundation of Biṭrūjī's models. Yavetz's contributions serve to caution scholars from concluding too hastily that the presence of a hippopede indicated a desire to recover Eudoxus's models, as Eudoxus's models may not have produced hippopedes.

Still, Mancha's research, by showing that there are hippopedes in Biṭrūjī's astronomy, has raised the question of the proximate source(s) for Biṭrūjī's knowledge of Eudoxan couples if not Eudoxus's models themselves. Ibn al-Haytham had incorporated concentric epicycles, that is, a Eudoxan couple, into his *Maqāla fī ḥarakat al-iltifāf* (Treatise on the motion of iltifāf) as a physical mover for one of the planets' motions in latitude known as *iltifāf*.[89] That treatise of Ibn al-Haytham represented an initial attempt to remedy a glaring lacuna in the *Almagest*,

86. Biṭrūjī (trans. and comm. Goldstein), *On the Principles,* pp. 9–10. See also Bernard R. Goldstein: "On the Theory of Trepidation according to Thābit ibn Qurra and al-Zarqāllu and Its Implications for Homocentric Planetary Theory," in *Centaurus* X (1964): 232–47, at p. 233. Goldstein favored the attribution to Ibn al-Zarqāl. Goldstein was responding to E. S. Kennedy's argument that Biṭrūjī's motivation was to recover the homocentric models of Eudoxus. See, now, Forcada, "La ciencia," pp. 86–8.

87. J. L. Mancha: "Al-Biṭrūjī's Theory of the Motions of the Fixed Stars," in *Archive for History of Exact Sciences* LVIII (2004): pp. 143–82, at p. 144. Mancha relied primarily on the Latin version of Biṭrūjī's *On the Principles,* but he did not observe differences between the Latin, Arabic, and Hebrew versions that would be substantial enough to discredit his conclusions for the Hebrew or Arabic versions. Richard Lorch's review of Goldstein's study of *On the Principles* (*Archives internationales d'histoire des sciences* XXIV [1974]: pp. 173–5, at p. 174) pointed out some places where the Latin version's account of the model for the motion of the fixed stars made more sense than the Arabic text. See also J. L. Mancha: "Right Ascensions and Hippopedes: Homocentric Models in Levi ben Gerson's Astronomy I. First Anomaly," in *Centaurus* XL (2003): 264–83.

88. Mancha, "Al-Biṭrūjī's Theory," p. 144, n. 5.

89. A. I. Sabra: "Ibn al-Haytham's Treatise: Solutions of Difficulties Concerning the Movement of *Iltifāf*," in *Journal for the History of Arabic Science* III (1979): pp. 388–422.

as Ptolemy had not proposed physical movers for any of the planets' motions in latitude. Ragep has noted that "the Eudoxan *technique* was clearly understood and utilized at least as early as Ibrāhīm b. Sinān (d. AH 335/CE 946) in his *Kitāb fī ḥarakāt al-shams* [Book on the motion of the Sun]."[90] Thus, Biṭrūjī could have learned of concentric epicycles (i.e., a Eudoxan couple) through earlier Islamic sources without ever having been motivated to recover Eudoxus's lost models. The possibility of a connection between Biṭrūjī and Ibrāhīm b. Sinān cautions us against limiting homocentric astronomy in Islamic societies to Andalusia.[91]

The history of the Eudoxan couple is relevant for *The Light of the World* because the Hebrew recension of *The Light of the World* includes a Eudoxan couple (§B.1.II.26/X.10–13). It would be hasty to presume that the recension extracted the Eudoxan couple from Biṭrūjī's model for the fixed stars since Ibn Naḥmias's own model for the fixed stars did not have a Eudoxan couple; we need to consider other explanations. Aspects of Ibn al-Haytham's treatise on *iltifāf* appeared in the Latin West during the 14th century, and given the history of contacts between Christian and Jewish scholars in Iberia, it is possible that Ibn Naḥmias learned of the Eudoxan couple in Ibn al-Haytham's model from Christian scholars.[92] Or, in light of how *The Light of the World* contained an innovative hypothesis involving two circles similar to the Ṭūsī couple of Naṣīr al-Dīn Ṭūsī (d. 1274), perhaps

90. Ragep, *Tadhkira*, pp. 450–3. In his comment, Ragep raised the possibility of whether Biṭrūjī (and hence Ibn Naḥmias) might have been influenced by Eudoxus and concluded (p. 452) that the idea of having combinations of homocentric orbs was known as early as Ibrāhīm ibn Sinān's (d. 946) *Kitāb fī ḥarakāt al-shams*. Ibn al-Haytham used a Eudoxan couple in order to provide a mover for the small circles that were part of Ptolemy's model for planetary latitudes. For an edition and translation of the text in which Naṣīr al-Dīn Ṭūsī responded to Ibn al-Haytham's solution, see F. Jamil Ragep: "Ibn al-Haytham and Eudoxus: The Revival of Homocentric Modeling in Islam," in Charles Burnett, Jan P. Hogendijk, Kim Plofker and Michio Yano (eds.): *Studies in Honour of David Pingree* (Leiden: E. J. Brill, 2004): pp. 786–809. See also F. Jamil Ragep: "Al-Battānī, Cosmology, and the Early History of Trepidation in Islam," in Josep Casulleras and Julio Samsó (eds.): *From Baghdad to Barcelona: Studies in the Islamic Exact Sciences in Honor of Prof. Juan Vernet* (Barcelona, 1996); vol. 1 of 2: pp. 291–2 for Ibrāhīm ibn Sinān's work on the question of trepidation.

91. There is some evidence for knowledge of homocentric astronomy in Egypt. One of the two surviving Arabic MSS (Escurial Arab. 963) of Biṭrūjī's *On the Principles* was copied by an Egyptian Christian, Abū Shākir, b. Abū Faraj Ibn al-'Assāl in 1281. See Biṭrūjī (trans. and comm. Goldstein), *On the Principles*, vol. 1, p. 46. See also David A. King and Julio Samsó, with Bernard R. Goldstein: "Astronomical Handbooks and Tables from the Islamic World (750–1900): An Interim Report," in *Suhayl* II (2001): pp. 9–105, at p. 73 for the suggestion that Maimonides introduced Biṭrūjī's *On the Principles* in Egypt.

92. José Mancha: "Ibn al-Haytham's Homocentric Epicycles in Latin Astronomical Texts of the XIVth and XVth centuries," in *Centaurus* XXXIII (1990): pp. 70–89. For the text of the treatise itself, see Sabra: "Ibn al-Haytham's Treatise." On the history of contacts between Jewish and Christian scholars in Andalusia, see, now, José Chabás, "Interactions between Jewish and Christian Astronomers in the Iberian Peninsula," in Freudenthal (ed.), *Science*, pp. 147–54.

other contacts between the author of the recension and the Islamic East led him directly to Ibn al-Haytham's Eudoxan couples, without the intermediation of Christian scholars.[93] Finally, we know that Gersonides, whose career intervened between those of Biṭrūjī and Ibn Naḥmias, examined at length, and then rejected, homocentric models that incorporated Eudoxan couples.[94] The numerous possible sources for the Eudoxan couples in the Hebrew recension of *The Light of the World* mean that there is a difference between detecting Eudoxan elements in homocentric models and concluding that those homocentric models were motivated by a desire to rediscover Eudoxus's models.[95] There was also much more to *The Light of the World*'s models than the Eudoxan couple.

DISTINCTIVE INNOVATIONS OF IBN NAḤMIAS'S MODELS

The Light of the World's models featured five innovations. Ibn Naḥmias referred to each of these innovations at one time or another as hypotheses (Ar. *uṣūl,* sing. *aṣl;* Heb. *'iqqar, yᵉsod*).[96] One could also translate the Hebrew and Arabic as "principle" or "model." By categorizing these innovations as *uṣūl,* Ibn Naḥmias communicated that they could apply to the configuration of homocentric orbs (or model) for more than one planet. This introduction will provide, in a way that

93. See Y. Tzvi Langermann: "Science in the Jewish Communities of the Byzantine Cultural Orbit: New Perspectives," in Freudenthal (ed.): *Science,* pp. 438–53, at p. 444. There Langermann referred to a Hebrew treatise on logic, probably by a Jew named Elisha the Greek, that mentioned Naṣīr al-Dīn Ṭūsī. Langermann has also discovered that the philosopher and polemicist (and apostate later known as Alfonso de Valladolid) Avner de Burgos (ca. 1270–1340) proved a theorem identical to a planar Ṭūsī couple; on Avner de Burgos, see Y. Tzvi Langermann: "Medieval Hebrew Texts on the Quadrature of the Lune," in *Historia Mathematica* XXIII (1996): pp. 31–53, at pp. 33–5. On other connections between the Islamic East and Christian Iberia, see M. Comes, "The Possible Scientific Exchange between the Courts of Hulaghu of Maragha and Alphonse 10th of Castille," in Nasrallah Pourjavady and Živa Vesel (eds.): *Sciences, techniques et instruments dans le monde iranien* (Tehran: Presses Univ. d'Iran, 2004): pp. 29–50. Recently, Taro Mimura has written on the knowledge of Ibn Rushd's work in the Islamic East. See Taro Mimura: "Quṭb al-Dīn Shīrāzī's Medical Work, *al-Tuḥfa al-Sa'diyya* [Commentary on volume 1 of Ibn Sīnā's *al-Qānūn fī al-Ṭibb* and its Sources]," in *Tārīkh-e ᶜElm* X (2013): no. 2, pp. 1–13, at pp. 10–13.

94. J. L. Mancha: "Right Ascensions and Hippopedes: Homocentric Models in Levi ben Gerson's *Astronomy;* I. First Anomaly," in *Centaurus* XLV (2003): pp. 264–83. This article is reprinted in *Studies in Medieval Astronomy and Optics* (Ashgate, 2006).

95. *The Light of the World,* §0.9. Ibn Naḥmias wrote that Biṭrūjī's teacher Ibn Ṭufayl had come across (Ar. *'athara 'alā*/Heb. *pagash*) a homocentric astronomy, suggesting that Ibn Naḥmias's view was that homocentric models had a history that stretched back before Biṭrūjī but maybe not as far back as Eudoxus. Or this could mean that Ibn Ṭufayl (but not Ibn Naḥmias) was engaged in the project of discovering a lost homocentric astronomy.

96. An argument for translating *aṣl* as "hypothesis" is found in the commentary on §0.1.

would be difficult within the line-by-line commentary, a systematic overview of these theoretical innovations in the order in which they appear in *The Light of the World.*

The first innovation that Ibn Naḥmias introduced was that the motions of the orbs could appear to be in opposite directions. Biṭrūjī's astronomy held that the heavenly orbs should appear to move in a single direction, due to their essential unity and simplicity.[97] In order to explain how all the celestial motions could vary yet be moved in a single direction by a single mover, Biṭrūjī described the motions in terms of lag (*taqṣīr*).[98] Each of the orbs that was below the uppermost orb of the fixed stars would move in the same direction and around the uppermost orb's poles, although the orbs beneath the uppermost would lag (*qaṣṣar, yuqaṣṣir*) behind the uppermost because of the dissipation of the force of the uppermost owing to the distance between the uppermost orb and any orb beneath it. The orbs would move owing to their desire to imitate the uppermost orb and not due to the action of one orb upon another.[99] There turn out to be, however, a number of places where Biṭrūjī seems to have compromised his commitment to the principle that all of the celestial motions should appear to be in the same direction.[100]

Ibn Naḥmias described each point of view, both the lag of celestial motions that appeared to be in the same direction and apparently opposite celestial motions, at times as an opinion (*ra'y*) and at other times as a hypothesis (*aṣl*). Ibn Naḥmias acknowledged the hypothetical plausibility of Biṭrūjī's position proscribing opposite motion: "He [Biṭrūjī] began by mentioning the manner (*kayfiyya*) of the motions and how all that which is in the heavens moves in one

97. Biṭrūjī (trans. and comm. Goldstein), *On the Principles,* vol. 1, p. 5

98. Biṭrūjī (trans. and comm. Goldstein), *On the Principles,* vol. 1, p. 63. Here Biṭrūjī explained the principle of desire: "What is close to the mover is necessarily more powerful and swifter moving than what is far from it. This agrees with the natural order and is the basic principle on which this treatise will be built, namely that what is far away has less power and speed than what is close." The commentary on §0.12 discusses the history of the concept of lag in more detail.

Yet even Biṭrūjī proposed motions, such as in his lunar model, that could be interpreted to be in opposite directions (ibid., p. 150). As well, Biṭrūjī's model for the motion of the fixed stars incorporated both the lag of the pole of the ecliptic and the motion of the ecliptic about its own poles through a motion of completion. See ibid., p. 80. The lag and the motion of completion did not always appear to be in the same direction.

99. For Ibn Naḥmias's understanding of *tashawwuq* (desire), see *The Light of the World,* §0.11–0.15. For Ibn Naḥmias, *tashawwuq* did not mean that all of the motions were in the same direction. On the earlier history of *tashawwuq* and *taqṣīr,* see Julio Samsó, "Biṭrūjī and the *Hay'a* tradition," reprinted in Samsó: *Islamic Astronomy and Medieval Spain* (Aldershot, UK, 1994), article XII, pp. 1–13, at pp. 9–13. Samsó explained that the concept of lag derived from the forces, possibly Neoplatonic in origin, of impetus and *tashawwuq.* Orbs moving in one direction could transmit these forces to lower spheres moving in the opposite direction, perhaps so that these lower orbs' desire for perfection could compensate for the lag of the upper orbs.

100. Lorch, "Review of Biṭrūjī (trans. and comm. Goldstein), *On the Principles,*" p. 174.

direction and that all that follows necessarily from these motions is in agreement with that which appears to the senses. This extent of what he has said is correct; I mean that it is possible that it be as he has said."[101] Even though Ibn Naḥmias did not stipulate that celestial motions be in the same direction, the motive power of an orb's desire for the uppermost orb, which underpinned the concept of lag, remained necessary for Ibn Naḥmias's models.[102] For example, Ibn Naḥmias constructed his model for the Sun so that as many as possible of the orbs' various motions were of the same measure, suggesting that all of the orbs that made up the model of a given celestial body shared the same mover.

Ibn Naḥmias argued, however, that the constraints of having all celestial motions appear to be in a single direction were an unnecessary imposition on astronomy, given the difficulty of determining when motions were and were not in opposite directions.[103] *The Light of the World* pointed out that since celestial motions occurring at a 90° angle to each other could not be described as either opposite or unidirectional, there was no need to preserve the appearance of heavenly motions in a single direction.[104] For Ibn Naḥmias, allowing apparently opposite motions was the more worthy (*aḥaqq;* §B.1.I.4–5) position. According to the second opinion, in which the motions had to appear to be in the same direction, the solar and lunar models for mean motion depended on one orb (or great circle) lagging in one direction by a certain amount about one pole (the pole of the universe's daily motion) and another orb (or great circle) moving about a different pole in the direction of the daily motion by about half that amount. But if one admitted celestial motions in apparently opposite directions, the first opinion, then the models for mean motion became much simpler and more precise. A single orb could revolve about its own pole, with the solar or lunar mean motion, in a direction apparently opposite to the daily motion of the universe. Still, Ibn Naḥmias attempted, whenever possible, to explain things both ways in order to accommodate readers who, in his view mistakenly, believed that heavenly motions had always to appear to be in the same direction (§0.11).

A second innovation characteristic of the models of *The Light of the World* was that Ibn Naḥmias introduced, at the equator of an orb, a small circle on the

101. *The Light of the World*, §0.9.

102. *The Light of the World*, §0.13.

103. *The Light of the World*, §0.5.

104. *The Light of the World*, §0.6. Ibn Naḥmias's position (which diverged from Biṭrūjī's) that opposite revolutions in the heavens were possible had support, to which Ibn Naḥmias did not refer, in other writings by Maimonides. See Langermann: "The 'True Perplexity,'" p. 173. Maimonides stated in chapter three, paragraph two of *The Book of Knowledge*, the first book of the *Mishneh Torah*, that some orbs may revolve in one direction and some in another. See Maimonides (ed. and trans. Moses Hyamson): *The Book of Knowledge* (Jerusalem, 1965): p. 37a. The pagination of the printed edition matches that of the Bodleian codex.

surface of the orb called the circle of the path of the center. He explained that this small circle, which he sometimes described as a hypothesis (e.g., §B.1.II.3), took the place of the Ptolemaic epicycle in the solar and lunar models.[105] The pole of the circle of the path of the center moved on the inclined circle carrying the circle of the path of the center, a great circle inclined to the ecliptic or Moon's path. This inclined circle would move with the solar or lunar mean motion. Some of Biṭrūjī's models had included a small circle, called "the circle of the path," the pole of which revolved on another small circle.[106] Though the function of Biṭrūjī's circle of the path was to cause the motion of the planet in a circle near the ecliptic, Biṭrūjī had located the circle of the path near the orb's pole, not at the orb's equator. Goldstein interpreted Biṭrūjī's model as having transferred the Ptolemaic epicycle to the pole from the vicinity of the orb's equator. In *The Light of the World,* the circle of the path of the center was close to the orb's equator. In *The Light of the World,* there was no corresponding circle near the pole of the orb that moved the circle of the path of the center, as a pole moving in a small circle does not cause a perfectly circular motion 90° away at the orb's equator. Rather, the connection that Ibn Naḥmias made between the circle of the path of the center and the epicycle, as well as the location of the circle of the path of the center at the orb's equator, suggests Ibn Naḥmias's interest lay in transferring the *Almagest's* models, and their predictive accuracy, to homocentric orbs. The difference was that in locating the circle of the path of the center at the orb's equator, Ibn Naḥmias intended that the Sun or Moon's motion on the small circle would not only (a) account for the motion in anomaly but also (b) keep the Sun or Moon in the ecliptic or at the appropriate inclination to it.[107]

Given the Hebrew recension's interest in eliminating the circle of the path of the center and the Hebrew recension's greater interest in proposing physical movers, it is worth noting that the circle of the path of the center was the one hypothesis found in *The Light of the World* that neither version of the text could explain through the motions of poles 90° from the circle of the path of the center.

A third innovation found in Ibn Naḥmias's models, intended to create an oscillation along a great circle arc, was what the Hebrew recension named the "hypothesis of the two circles."[108] Ibn Naḥmias introduced these two circles in the

105. See also *The Light of the World*, §B.1.II.1, where Ibn Naḥmias described the circle of the path of the center as the hypothesis for the solar anomaly.

106. Goldstein had translated the term for this circle (*dā'irat al-mamarr*) as "epicycle." See Biṭrūjī (trans. and comm. Goldstein), *On The Principles*, vol. 1, p. 111.

107. *The Light of the World*, §B.1.II.7. There Ibn Naḥmias wrote that "accordingly it is necessary that the center of the Sun have a motion about the poles of the circle of its path," for without that motion, the Sun would not be seen in the ecliptic.

108. See *Or ha-'olam*, MS Bodleian Canon Misc. 334, 113b (§B.1.II.26/X.32). The Judeo-Arabic original referred to the innovation simply as "the two circles."

Judeo-Arabic original in order to eliminate an outstanding displacement of the Sun from the ecliptic (§B.1.II.17–25). He proposed two small circles of the same size traced on the surface of the orb, with the center of one being carried upon the diameter of the other. The Sun would be carried in the second circle, and the center of the second circle was a pole that traced the first circle. The double circles were carried by the circle of the path of the center. In the version of the double-circle hypothesis that Ibn Naḥmias preferred, as the first circle revolves a certain amount, the second circle revolves simultaneously twice that amount in the opposite direction. The effect of the combined motions is that the Sun, while tracing the circumference of the second circle, is also oscillating almost on a great circle arc.[109] Those additional oscillations could eliminate the displacement of the Sun from the ecliptic that was a by-product of the circle of the path of the center. This hypothesis of the two circles was physically similar to the rudimentary Ṭūsī couple found in Naṣīr al-Dīn Ṭūsī's (d. 1274) 1247 *Taḥrīr al-Majisṭī* (The recension of the *Almagest*).[110] The commentary on the Judeo-Arabic version demonstrates that the double-circle hypothesis was mathematically equivalent to the spherical Ṭūsī couple from Ṭūsī's later *Tadhkira*. Mathematical equivalence means that the models predict the same positions, whereas physical equivalence would mean that the configuration of orbs was also identical.[111] One of the most interesting parts of *The Light of the World* is where it proposed combining the hypothesis of the two circles, perhaps with increased diameters, with a single uniformly revolving orb in order to account for all of the variations observed with the Sun.[112]

109. *The Light of the World*, §B.1.II.20. At first, Ibn Naḥmias had proposed that the motion of the center of the first circle about the center of the second would be the same amount as the motion of the center of the second about the center of the first. Subsequently, he proposed doubling the measure of the second motion.

110. On the rudimentary version of the Ṭūsī couple that most closely resembled Ibn Naḥmias's double-circle device, see George Saliba: "The Role of the *Almagest* Commentaries in Medieval Arabic Astronomy: A Preliminary Survey of Ṭūsī's Redaction of Ptolemy's *Almagest*," in *Archives Internationales d'Histoire des Sciences* XXXVII (1987): pp. 3–20, at p. 19. The difference between Ibn Naḥmias's hypothesis of the two circles and the rudimentary version of the Ṭūsī couple was that Ṭūsī was, in *The Recension of the "Almagest,"* describing motion in the plane, whereas Ibn Naḥmias was describing motions on the curved surface of an orb. For the later, fully-formed versions of the Ṭūsī couple, see F. J. Ragep (ed., trans., and comm.): *Naṣīr al-Dīn al-Ṭūsī's "Memoir on Astronomy" (al-Tadhkira fī 'ilm al-hay'a)* (New York and Berlin: Springer-Verlag, 1993). On the spherical Ṭūsī couple, see George Saliba and E. S. Kennedy: "The Spherical Case of the Ṭūsī Couple," in *Arabic Sciences and Philosophy* I (1991): pp. 285–91.

111. The difference between the hypothesis of the two circles in *The Light of the World* and the fully-formed Ṭūsī couple in Ṭūsī's *Tadhkira* is that the Ṭūsī couple had one circle (or orb) that was *twice* the size of the other.

112. For Ibn Naḥmias's brief mention in the Judeo-Arabic version of the possibility of using the double-circle hypothesis without the circle of the path of the center, see *The Light of the World*, §B.1.II.26. On 113b of the Hebrew recension (*Or ha-'olam*, Bodleian MS Canon Misc. 334), one finds

Accordingly, the circle of the path of the center, the second hypothesis, would no longer be necessary to keep the Sun in the ecliptic. Hence, *The Light of the World* has proposed a solar model virtually identical to that proposed later by Giovanni Battista Amico (d. 1538) in Padua.

A fourth innovation found in Ibn Naḥmias's models was the hypothesis of the slant (Heb. *'iqqar ha-'iqqum*).[113] A slant is the change of the alignment of the equator of a small circle, such as the circle of the path of the center, with respect to another great circle, such as the equator of the orb. The equator of the small circle was itself a great circle arc. The concept had emerged in the Judeo-Arabic original (§B.2.I.7–9) in models that incorporated the hypothesis of lag as a corollary of the motion of the circle of the path of the center. For example, in the solar model, the pole of the orb that carries the pole of the circle of the path of the center itself moves in a small circle about the pole of the cosmos. These two poles begin on a given great circle arc. As the pole of the orb that carries the circle of the path of the center moves, it is no longer on that great circle arc, and the equator of the circle of the path of the center is slanted with respect to the inclined circle carrying the circle of the path of the center.[114] After the pole of the orb that carries the circle of the path of the center has revolved 90°, the slant of the equator of the circle of the path of the center to the equator upon which it began will be at a maximum, equal to the angular distance between the two poles. One can add the inclination from the slant to the uniform angular motion of the circle of the path of the center. This additional motion imparted by the slant to the circle of the path of the center could be helpful because, in the case of both the solar and lunar models, the motion of the luminary on the circle of the path of the center alone was insufficient to place the luminary at the position in which it was observed.[115]

Only the Hebrew recension identified and analyzed the slant as an independent hypothesis (§B.1.II.26/X.14).[116] And only the Hebrew recension theorized and analyzed discrete physical movers to account for the slant. Because the Hebrew

---

the same suggestion (§B.1.II.26/X.32). In the homocentric astronomy of Giovanni Battista Amico in sixteenth-century Padua, the double-circle device is used, in the same way, to account for all of the variations in the Sun's motion. On Amico, see Noel Swerdlow: "Aristotelian Planetary Theory in the Renaissance: Giovanni Battista Amico's Homocentric Spheres," in *Journal for the History of Astronomy* III (1972): pp. 36–48.

113. Another possible translation for *'iqqum* would be "inclination." I have chosen *slant* to avoid confusion with the prosneusis point, that is, the inclination of the Moon's epicycle, which Ibn Naḥmias does discuss. In several places in the Hebrew recension, *'iqqum* is spelled with a yod.

114. See, for example, *Or ha-'olam*, MS Bodleian Canon Misc. 334, §B.1.II.26/X.5.

115. An analysis of the effect of the hypothesis of the slant on predictive accuracy follows in the commentary on *The Light of the World*, §B.1.II.10 and on §B.1.II.26/X.14 (a passage found only in the Hebrew recension).

116. See also *The Light of the World*, §B.1.II.9.

recension explored the possibility of accounting for the solar anomaly with only the hypothesis of the slant and without the circle of the path of the center, it is not clear that either Ibn Naḥmias or the author of the Hebrew recension fully understood the potential of the hypothesis of the slant.[117]

Only the Hebrew recension of *The Light of the World* contained a fifth innovation, which was introduced to produce an oscillation along a great circle arc (§B.1.II.26/X.10–13), something which the double-circle hypothesis also did. But this fifth hypothesis was different from the double-circle hypothesis. Two concentric orbs, with the pole of one inclined to the pole of the other, would revolve at the same speed in opposite directions. This was a Eudoxan couple, and the result was that a point at the equator of one of the orbs traced a hippopede. The width of the hippopede, and hence the deviation of the point from an oscillation on a great circle arc, depended on the degree of the inclination of one orb to the other. Most likely, the Hebrew recension introduced this final hypothesis as a way to produce, at the equator of an orb through the revolution of orbs about their poles, an oscillation similar to that caused by the double-circle hypothesis. The only physical movers that the Hebrew recension could propose for the double-circle hypothesis that resembled the Ṭūsī couple were the motions of the poles themselves. Thus, though the double-circle hypothesis that resembled the Ṭūsī couple was *neither* physically *nor* mathematically equivalent to the Eudoxan couple, the Hebrew recension may not have distinguished between the two.

AN OVERVIEW OF IBN NAḤMIAS'S MODELS

This section of the introduction will provide an overview of the models that *The Light of the World* proposed to explain the motions of the Sun, Moon, and fixed stars and thus a framework for the more specific line-by-line commentary on *The Light of the World.* Although both versions of *The Light of the World* mentioned models for the motions of the five planets, and although the Judeo-Arabic version began to discuss these models, not enough of *The Light of the World* has survived to make possible even a general description of Ibn Naḥmias's planetary models. Within this overview is an outline of how the five aforementioned innovations functioned within Ibn Naḥmias's models. In each case, this overview sets the stage by describing the observed celestial motions that Ibn Naḥmias had to explain, followed by the solutions that Biṭrūjī had proposed, as Biṭrūjī's solutions were the starting point for Ibn Naḥmias's astronomy. This overview refers to figures found in the technical commentary for the Judeo-Arabic original of *The Light of the World.*

117. *Or ha-ʿolam*, MS Bodleian Canon Misc. 334, §B.1.II.26/X.41. "[O]ur thesis is that the greatest slant is equal to the degrees of the greatest anomaly."

1. *The Solar Model*

Ibn Naḥmias aspired to the mathematical precision of the *Almagest*.[118] In §B.1.II.27, he wrote, "It also results from what we have said about the Sun that the tables which Ptolemy used for the motions and the anomalies are themselves the tables needed for calculations with these hypotheses." The *Almagest* had found that the Sun revolved once around the Earth approximately every 365 1/4 days.[119] Its path, known as the ecliptic, traced against the background of the zodiac, was inclined to the Earth's equator by about 24°.[120] Observations had led Ptolemy to posit that the heavens were spherical.[121] He calculated the Sun's average daily motion about the Earth to be about 59/60 of a degree (0;59°).[122] If the seasons, the Sun's path from solstice to equinox or from equinox to solstice, were the same length, Ptolemy would have been able to theorize that the Sun moved uniformly through the ecliptic. Such a theory could not, however, account for how the observed motion of the Sun through the ecliptic is slower than the average daily motion at the summer and winter solstices and faster than the average daily motion at the autumnal and vernal equinoxes.

In Biṭrūjī's solar model (see figure 5.9), the radius of the circle of the pole of the Sun was equal to the inclination of the ecliptic to the celestial equator. The pole of the Sun lagged, due to the dissipation of motion from the uppermost sphere, behind the cosmos's daily motion by an amount equal to twice the mean daily motion of the Sun. That is, as the pole of the Sun moved 180° from *A* to *B*, the Sun moved 90° from *C* to *D* on the ecliptic.[123] In order to account for the unequal length of the seasons, Biṭrūjī removed the pole of the circle of the path of the Sun's pole from the pole of the equator by an amount and direction equal to the solar eccentricity in the Ptolemaic system.[124] Biṭrūjī's solar model placed the Sun in the ecliptic at the solstices and equinoxes, endowing his solar model with greater predictive and retrodictive accuracy than his planetary models.[125] Still, there was much room for improvement, as at points other than the solstices and equinoxes, the predicted position of the Sun could be outside

118. Parts of this overview of the solar model are excerpted from Robert Morrison: "The Solar Theory of Joseph Ibn Nahmias," in *Arabic Sciences and Philosophy* XV (2005): 57–108.

119. The *Almagest* III.1 (see Gerald Toomer, *Ptolemy's "Almagest"* [London, 1984]: p. 140) gives 365; 14, 48 days for the length of a year.

120. *Almagest* (Toomer: *Ptolemy's "Almagest,"* p. 63) gives 23; 51, 20° for the obliquity of the ecliptic.

121. *Almagest* I.3 (Toomer: *Ptolemy's "Almagest,"* pp. 38–40). See also Aristotle's *Metaphysics*, Book 12, Chapter 8.

122. *Almagest* III.1 (Toomer: *Ptolemy's "Almagest,"* p. 140).

123. Biṭrūjī (trans. and comm. Goldstein), *On the Principles*, vol. 1, p. 133.

124. Biṭrūjī (trans. and comm. Goldstein), *On the Principles*, vol. 1, p. 13 and pp. 136–7.

125. Biṭrūjī (trans. and comm. Goldstein), *On the Principles*, vol. 1, pp. 12–3.

of the ecliptic by more than 1°. Ibn Naḥmias considered that deviation to be unacceptably large.

Ibn Naḥmias was able to improve upon the predictive accuracy of Biṭrūjī's model because Ibn Naḥmias had reconsidered the admissibility of apparently opposite motions in the heavens.[126] My summary descriptions of the solar and lunar models present the motions sequentially, but the motions actually occur simultaneously. Ibn Naḥmias began by following the second opinion in which the Sun's motions had to appear to be in the same direction as the direction of the daily motion of the universe. He accounted for the Sun's mean motion (see figure 5.1) by having the pole of the Sun's orb be inclined to the pole of the universe (§B.1.I.2). Then the Sun's orb, with its poles, lagged behind the daily motion by twice the mean solar motion, about the pole of the universe, while the Sun's orb revolved about its own poles in the same direction as the universe's daily motion by, supposedly, the measure of the mean solar motion (§B.1.I.3). This second motion was intended to return the mean Sun to the ecliptic. The commentary will show that for the mean Sun to be in the ecliptic, the second motion cannot be precisely the mean solar motion. Ibn Naḥmias added that if one proceeded according to the first opinion (or hypothesis), in which there was no barrier to apparently opposite motions, then one could hypothesize simply that the Sun's orb revolved about its own poles, with the mean solar motion, in a direction apparently opposite to the universe's daily motion. There would be no need for two motions about two different poles.

Ibn Naḥmias accounted for the variations in the Sun's motion in longitude, the solar anomaly, with the circle of the path of the center of the Sun (B.1.II.3), akin to an epicycle and an application of the second innovation. The model for the solar anomaly (see figure 5.2) presumed that the pole of the circle of the path of the center moved on a great circle, called the circle carrying the circle of the path of the center, through the proposal for the solar mean motion. The circle carrying the circle of the path of the center is inclined to the ecliptic by the radius of the circle of the path of the center, the maximum solar anomaly. The motion of the circle of the path of the center, 0;59°/day, was supposed to account for the variations in the Sun's longitude and keep the Sun in the ecliptic. Ibn Naḥmias did not propose a mover for the circle of the path of the center; nor did he specify whether the circle of the path of the center was a pole of another orb or merely a small circle traced on the surface of an orb. The second opinion, in which the celestial motions had to appear to be in the same direction, entailed a slant of the diameter of the circle of the path of the center of the Sun. The first opinion, which Ibn Naḥmias preferred, did not impart this slant.

126. *The Light of the World*, §0.9.

Because the motion of the circle of the path of the center did not keep the Sun precisely in the ecliptic, Ibn Naḥmias incorporated the hypothesis of the double circles (§B.1.II.17; see figure 5.7), the third innovation, in order to resolve the remaining displacement of the Sun from the ecliptic. In the Judeo-Arabic original, Ibn Naḥmias added the near oscillation of a point through the double circles to the motion of the Sun on the circle of the path of the center. Thus, the Sun would be carried both by the circle of the path of the center and by the double circles. The Hebrew recension considered using the hypothesis of the slant and the hypothesis resembling the Eudoxan couple, the fourth and fifth innovations, as alternative ways to eliminate the remaining displacement of the Sun from the ecliptic. Finally, the Hebrew recension proposed combining a uniformly revolving orb, to account for the Sun's mean motion, with either the third, fourth, or fifth hypotheses (the double circles, slant, or Eudoxan couple), to account for the variations in the Sun's motion. In these cases, the circle of the path of the center would no longer be necessary.

2. *The Lunar Model*

The Moon's motions are more complicated than the Sun's. Again, Ibn Naḥmias stipulated Ptolemy's observations (§B.2.I.1), meaning that a successful lunar model would be one that could account for those observations. According to Ptolemy (*Almagest*, IV.2), the Moon revolves around the Earth and catches up to the Sun once every 29;31,50 days (a synodic lunar month). The Moon, therefore, in a month of mean motion in longitude, covers about 389;30°, since the Moon had to complete a circle and catch up to the moving Sun. The Moon's path is inclined by 5° to the ecliptic, which explains why there is not a solar eclipse at the beginning of every lunar month and why there is not a lunar eclipse every month around the time of the full Moon. Ptolemy, in *Almagest* book IV, computed three key parameters for the Moon (§B.2.I.4). The Moon had a mean daily motion in anomaly (on its epicycle) of 13;3,54°, a mean daily motion on its circle of latitude of 13;13,46°, and a mean daily motion in longitude (with respect to the ecliptic) of 13;10,35°.[127] Just as Ptolemy did, Ibn Naḥmias used the parameter for mean motion in longitude with respect to the ecliptic of 13;10,35° to compute the lunar mean longitude, in place of the parameter for the Moon's mean motion on the circle of its latitude (13;13,46°), because the relatively slight 5° inclination of the Moon's circle of latitude to the ecliptic rendered the effect of any distinction between the circles negligible.

According to Ptolemy, the first lunar anomaly was the variation from the mean motion that could be explained by the Moon's motion on its epicycle (again

127. These parameters are taken from *The Light of the World* (§B.2.I.4) and are rounded from the parameters given in *The Almagest* (IV.4).

13;3,54° per day).[128] Ptolemy detected a second lunar anomaly when he found that the correction entailed by the motion on the epicycle could not on its own account for the Moon's longitude at quadratures (i.e., at 45°, 135°, etc., of elongation from the Sun).[129] An additional source of complexity was that along with these variations in longitude came variations in the observed lunar diameter, though Ptolemy's model for the second lunar anomaly predicted greater variations in the Moon's diameter than those actually observed. The third Ptolemaic anomaly was that the Moon's motion on its epicycle had to be reckoned from a moving point in order to account for the second anomaly.[130] That moving point was known as the prosneusis point.

Biṭrūjī had proposed a model (see figure 5.12 and figure 5.13) in which the pole of the ecliptic lagged with the Moon's mean motion in longitude, and the pole of the center of the Moon lagged with the motion in anomaly about another circle, the center of which was carried on the circle of the path of the pole of the ecliptic. Biṭrūjī's model attempted to explain only the first anomaly in longitude; the second and third anomalies as well as the predicted variations in the Moon's size went unaddressed. The commentary will show, in addition, that Biṭrūjī's model did not necessarily keep the Moon in its observed latitude on a great circle inclined 5° to the ecliptic. That said, an interesting feature of Biṭrūjī's lunar model was that Biṭrūjī's model has proposed motions in seemingly opposite directions.[131]

Following the second opinion, which proscribed apparently opposite motions, to account for the Moon's mean motion, Ibn Naḥmias proposed that the pole of the circle of the Moon's latitude lagged behind the motion of the universe by twice the Moon's final lag (probably in longitude) (§B.2.I.5–10; see figure 5.14). That is, the pole of the Moon moved from *M* to *R*, which was twice the Moon's mean daily motion less 3′, the daily motion of the nodes (K to R). At the same time, the circle of the Moon's latitude revolved about its own pole in the direction of the motion of the universe by a measure (arc *SZ*) equal to the Moon's mean motion in longitude, bringing the Moon back to its observed latitude. This model for the Moon's mean motion resembled Ibn Naḥmias's model for the Sun's mean motion and suffered from the same drawback, that the motion of the circle of the Moon's latitude about its own poles, in order to return the Moon to its observed latitude, had to vary and was most often not precisely equal to the Moon's daily

128. Olaf Pedersen (ed. Alexander Jones): *A Survey of the Almagest* (New York, Heidelberg, Berlin: Springer-Verlag, 2011): pp. 167–84.

129. Pedersen, *Survey of the Almagest*, pp. 184–92.

130. Pedersen, *Survey of the Almagest*, pp. 192–5.

131. Biṭrūjī (trans. and comm. Goldstein), *On the Principles*, vol. 1, p. 148. Biṭrūjī wrote that the second "motion of the Moon takes place in the direction of the signs, though the first [motion] takes place in the direction of the universal motion and on the inclined circle." The universal motion would be in the opposite direction of the signs.

mean motion. Ibn Naḥmias also proposed a model for the lunar mean motion founded on the first opinion (or hypothesis) in which the circle of the Moon's latitude and the ecliptic each revolved about their own poles, sometimes in a direction apparently opposite to the motion of the universe with the Moon's mean motion and the motion of the nodes, respectively.

For the first anomaly, Ibn Naḥmias proposed a model with the circle of the path of the center (see figure 5.15). Its pole was carried on a deferent circle inclined to the circle of the Moon's motion in latitude (§B.2.II.2–7). Ibn Naḥmias began by following the second opinion (i.e. motions through lag). After pole *B* lags to point *H*, the pole of the circle of the Moon's latitude, that is, the pole of the mean Moon, lagged from *M* to *R* to account for the motion of the nodes. That motion would carry the pole of the Moon's deferent circle to point *X*. Then, pole *X* moves in the direction of the motion of the universe to point *V* by 10′, which was the surplus of the Moon's mean motion on the circle in latitude over the motion in anomaly. Those motions move the pole of the circle of the path of the center from *Q* to *Z* to *P*. Subsequently, the pole of the circle of the path of the center moves from *P* to *Q*, in the direction of the motion of the universe, by the measure of the daily motion in anomaly. Finally, the Moon revolves on the circle of the path of the center by the measure of the daily motion in anomaly. As was the case with the model for the solar anomaly, Ibn Naḥmias went on to hypothesize double circles, the motions of which would be added to the Moon's motion on the circle of the path of the center in order to eliminate the Moon's remaining displacement from the circle of its latitude with respect to its observed position (§B.2.II.5). Ibn Naḥmias did not detail the operation of the double-circle hypothesis in the lunar models; instead he referred the reader to the solar model.

Like Ptolemy's first lunar model, Ibn Naḥmias's homocentric model for the first lunar anomaly turned out to be more successful at accounting for the lunar positions around syzygies, when the Moon was 0° or 180° from the Sun, that is, at the beginning or middle of a lunar month, than for the Moon's positions near quadratures, when the Moon was 90° from the Sun. In the fifth chapter on the Moon's motion, Ibn Naḥmias took on the second and third lunar anomalies (or variations in the Moon's motion). Ibn Naḥmias modified his lunar model to account for the second lunar anomaly, which was at its maximum at quadratures, by adding an additional small circle (see figure 5.17) to carry the center of the Moon (§B.2.V.4). The inclination of the deferent circle of the circle of the path of the center (*EHTS*) has increased by the diameter of the additional small circle *EWDV*. In the figure, the Moon is at point *D* at quadratures, and because small circle *EWDV* revolved at twice the rate of *EHTS*, the effect of small circle *EWDV* disappears at syzygies, that is, at points *A* and *G*. Indeed, Ibn Naḥmias's model for the second lunar anomaly accounted most successfully for the minimum and maximum values of the second anomaly, which occurred when the Moon was at syzygies and quadratures respectively. But the increased lunar equation could not affect the lunar longitudes in the same way it did in the *Almagest*. Ptolemy's second lunar model (i.e., his model for the second anomaly) also predicted variations in the Moon's diameter as, at times, it cranked the center of the lunar epicycle closer to Earth. Ibn Naḥmias's second lunar model did not entail any variations either in the Earth–Moon distance or the lunar diameter, because the second small circle was on the surface of an orb that had the Earth at its center, implying a constant Earth–Moon distance.

There were systematic observations of variations in the Moon's diameter with which Ibn Naḥmias had to contend. There are two reasons why, in the fourth chapter on the Moon, Ibn Naḥmias argued that these widely observed changes in the Moon's diameter did not mean that the distance of the Moon to the Earth actually varied.[132] First, conceding that the Earth–Moon distance could vary would be the downfall of homocentric models, because if the Earth was at the center of all of the orbs, then there could be no real variation in the Earth–Moon distance. Second, Ptolemy's lunar model necessitated changes in the Earth–Moon distance that were *not* observed (as it attempted to account for changes in the Earth–Moon distance that were observed). Thus, from Ibn Naḥmias's perspective, if Ptolemy predicted variations in the Moon's diameter that could not be observed, then Ptolemy's models did not perfectly account for observations either.[133] So with respect to variations in the Moon's size, Ibn Naḥmias could conceivably hold that his lunar model's predictive accuracy was no worse than Ptolemy's.[134]

The third and final Ptolemaic lunar anomaly (or variation) that Ibn Naḥmias addressed was that of the prosneusis point. Ibn Naḥmias noted, correctly, that the measure of the third anomaly or variation was correlated with the measure of the second.[135] This oscillation of the starting point of the Moon's mean motion

132. *The Light of the World*, §B.2.IV.3.

133. Still, Ibn Naḥmias did not specify whether his comments about how putative variations in the Moon's observed size were directed at how the Ptolemaic lunar model predicted unobserved variations in the lunar diameter or at how homocentric models could not tolerate observed variations in the lunar diameter beyond those due to parallax.

134. These shortcomings in the *Almagest*'s ability to explain observations of the lunar size may explain how "paradoxical though it may seem, Regiomontanus was very interested in the homocentric tradition, in spite of the fact that he was an exceptionally competent mathematical astronomer." See Shank, "The 'Notes on al- Biṭrūjī,'" p. 15.

135. See Pedersen, *Survey of the Almagest*, p. 192. The prosneusis point could be understood as a third anomaly, but Ibn Naḥmias explained that the prosneusis point could be dealt with simply by thinking differently about the second lunar model (as the second and third lunar anomalies were closely related). Pedersen had explained (p. 192) that "perhaps one would think that in order to account for this discrepancy at the octants, Ptolemy had to introduce a third 'anomaly' of the Moon. . . . But this would have been unnecessary and misleading."

on the epicycle was, to Ibn Naḥmias, a complication,[136] but of a different sort than the epicycle and eccentric. As the effect of the third anomaly was the increase or decrease of the Moon's motion (or lag) over what the first two anomalies would otherwise require, Ibn Naḥmias said that the third anomaly could be accounted for without any additional small circle, orb, or motion (§B.2.V.7–9). Instead, he alleged that the variations in motion that constituted the Ptolemaic third anomaly followed from his (Ibn Naḥmias's) model for the second anomaly. The commentary evaluates this claim.

3. The Model for Fixed Stars

Ibn Naḥmias's lunar model depended on skepticism of observations of variations of the lunar size and distance. Likewise, his model for the motions of the fixed stars also reflected skepticism of existing observations and thus lends insight into why he might have thought that a homocentric astronomy could attain a level of predictive accuracy on a par with Ptolemy's. The motions for which models of the fixed stars accounted were themselves a subject of some dispute. Astronomers of Ibn Naḥmias's era agreed that the equinoxes moved slowly in the opposite direction of the zodiacal signs (i.e., from Aries to Pisces to Aquarius) producing a phenomenon known as the precession of the equinoxes. Because of precession, the spring equinox occurs each year when the Sun is at a slightly different point against the background of the zodiac. Some significant debates in the history of astronomy were over the rate of precession itself, whether the rate of precession could vary, and what the nature and causes of any of those variations were.[137] The theory that the rate of precession might be changing arose because astronomers of different eras had found different measurements for the rate of precession.[138] If the rate of precession did not vary, meaning that the observed variations were simply due to observational imprecisions (or increasing precision), then precession could easily be accounted for with a single orb. This single orb could carry the fixed stars with a motion equal to the rate of precession. Models to explain variations in the rate of precession, however, posited additional orbs. Astronomers such as Ibn al-Zarqāl (d. 1100) proposed models for a phenomenon called trepidation (Ar. *al-iqbāl wa-'l-idbār;* literally, accession and recession), which was when the variations in the rate of precession were sufficient to cause the equinoxes to move sometimes in the opposite direc-

136. *The Light of the World*, §B.2.V.7. He wrote that this oscillation was one of the repugnancies (*shināʿāt*) of the Ptolemaic lunar model.

137. On the origins of trepidation, see Ragep: "Al-Battānī, Cosmology," vol. 1, pp. 267–98.

138. Modern astronomy has also detected variations in the rate of precession.

tion (i.e., eastward).[139] It was also possible that the motion of trepidation could be added to the rate of precession.[140]

Biṭrūjī, like most astronomers of his era, presumed that all of the fixed stars shared the same motion, whatever that motion was. He proposed, for the fixed stars, a single-orb model that engendered a variable rate of precession (but not trepidation).[141] The pole of the ecliptic was inclined to the pole of the equinoctial at an angle equal o the obliquity of the ecliptic. Then, the pole of the ecliptic lagged at an equal velocity on a small circle about the pole of the equinoctial (figure 5.20). Because the ecliptic was inclined to the equinoctial, the motion of the equinox with respect to the ecliptic could not be uniform. As a result, the motion of the equinoxes sped up and slowed down, though always in the same direction.[142]

Models for trepidation after Biṭrūjī endured but were debated. Recent scholarship on Gersonides has concluded that his position on trepidation was nuanced in that he felt that he did not know enough to come to a conclusion.[143] Gersonides's reticence is notable given his skill with observations and with instruments, as well as his significant theoretical insights. In this context, it is not surprising that Ibn Naḥmias did not accept the existence of trepidation. Ibn Naḥmias also did not mention a parameter for precession; the astronomers of his time measured it to be 1° every 66 or 70 years.[144] Apart from the question of whether trepidation existed, there was also a measure of skepticism about whether observations of

139. See Julio Samsó: "Trepidation in al-Andalus in the 11th Century," in Samsó: *Islamic Astronomy,* article VIII.

140. F. Jamil Ragep: "Al-Battānī, Cosmology," vol. 1, pp. 267–98, at p. 272. See also Mercè Comes: "The Accession and Recession Theory in al-Andalus and the North of Africa," in Casulleras and Samsó: *From Baghdad,* pp. 349–64. For Ibn al-Hā'im's (fl. early 13th century) trepidation model, which survives in an Arabic text, see Mercè Comes: "Ibn al-Hā'im's Trepidation Model," in *Suhayl* II (2000): pp. 291–408.

141. My summary of Biṭrūjī's model for the fixed stars follows Biṭrūjī (trans. and comm. Goldstein): *On the Principles,* vol. 1: pp. 15–8.

142. Mancha's recent analysis has argued that Biṭrūjī's model incorporated a Eudoxan couple and has found that, actually, the equinoxes moved up and down more than they moved to the east and west. Mancha also concluded ("Al-Biṭrūjī's Theory," p. 157) that Biṭrūjī had no clear understanding of his own model. For more information, see the commentary on §B.3.I.1.

143. See, now, J. L. Mancha: "The Theory of Access and Recess in Levi ben Gerson's Astronomy and Its Sources," in *Aleph* XII (2012): pp. 37–64, at p. 58: "Gersonides' silence about trepidation and the historical material transmitted in the *Kitāb fī 1-hay'a* is striking." Gersonides seems not to have known (pp. 38–9) enough about trepidation to reject it. Bernard Goldstein ("Levi ben Gerson's Analysis of Precession," in *Journal for the History of Astronomy* VI [1975]: pp. 31–41, at p. 31) had found that Gersonides had rejected trepidation because there were no reliable observations of a change in the obliquity of the ecliptic, a corollary of the known models for trepidation.

144. Ptolemy had measured it at 1° every 100 years. See Ragep, *Tadhkira,* p. 396, for a survey of some of the parameters for precession. See José Chabás and Bernard R. Goldstein: *The Alfonsine*

the fixed stars' motions were correct, and, for some astronomers, skepticism about the existence of any single orb for the motion of the fixed stars given their number and distance.[145]

Ibn Naḥmias's model for the motions of the fixed stars took a different approach to predictive accuracy than his models for the Sun and the Moon. While Ibn Naḥmias had accepted and attempted to account for Ptolemy's observations of the solar and lunar longitudes, Ibn Naḥmias went to great lengths to preserve the possibility of heretofore unobserved variations in the motions of the fixed stars. Ibn Naḥmias went as far as to argue that what previous astronomers had perceived to be trepidation was actually the result of some stars moving at a different rate than others.[146] Thus, each star (or pair of stars according to Ibn Naḥmias) would have to be in a separate complex of orbs (figure 5.21). In a sense, this proposal had a precedent in how some of Ibn Naḥmias's most adept predecessors had pointed out that it was conceivable that there be multiple orbs carrying sets of stars (or even individual stars) at the same rate or even at different rates.[147] But the difference was that these scholars had, in the end, founded their models for the fixed stars' motions on the presumption that the fixed stars *did* move with a single motion because there were no observations to suggest the contrary. Ibn Naḥmias's model melded a skepticism of observations that were, after all, complex, into the foundation of his model for the fixed stars' motions.

Ibn Naḥmias's willingness to go to such lengths to question the reliability of observations in his model for the fixed stars furnishes a reason why he might have pursued a homocentric astronomy. The complex model for the fixed stars suggests that he was not ignoring observations of the Moon's varying distances simply because he could not account for them in a homocentric system (though he could not). Rather, Ibn Naḥmias's inclination not to rely on the available observations to conclude safely that there was a single orb moving the fixed stars suggests that he did not want to make observations the foundation for a statement about the physical structure of the heavens. Instead, observations were just the criterion for his model for the motion in longitude. If we recall that Ibn

Tables of Toledo (Dordrecht, Boston, and London: Kluwer Academic Publishers, 2003): p. 257. The value of 1° in 72;30 years was ascribed to King Alfonso.

145. Rāzī expressed skepticism about whether the fixed stars were truly in one orb in *al-Tafsīr al-kabīr* (Beirut: Dār iḥyā' al-turāth, n.d.), vol. 4: pp. 180–1. Ibn Sīnā ([ed. Madkūr]: *al-Shifā,' al-Samā' wa-'l-'ālam* [Cairo: General Egyptian Book Organization, 1969], ch. 6: p. 46) acknowledged the possibility of the fixed stars being in more than one orb. Ṭūsī ([Ragep ed.], *Tadhkira*, p. 389) acknowledged the same thing. Maimonides, in *Guide* II.11 ([trans. Pines], vol. 2, p. 274) said multiple orbs for the fixed stars were certainly possible and claimed agnosticism. Maimonides ([trans. Pines], *Guide*, vol. 2, p. 326) was skeptical of whether humans could know the distances of the planets.

146. *The Light of the World*, §B.3.II.1–2.

147. See note 144 above.

Naḥmias's motivation for writing *The Light of the World* was to resolve the apparent incommensurability between the physical and mathematical approaches to astronomy, then the evidence of observations, in the case of planetary distances and sizes, may have remained an insufficiently stable basis of demonstration to Ibn Naḥmias.[148] With planetary longitudes, the physical (or philosophical) and mathematical approaches could coincide, or, at the least, were on the way to coinciding from Ibn Naḥmias's perspective, so there was no need to confront the question of whether epicycles and eccentrics, and their attendant physical implications, were essential for accurate predictions of planetary longitudes.

ANNOTATED TABLE OF CONTENTS OF *THE LIGHT OF THE WORLD*

The Light of the World is divided into an introduction (which I have designated 0) and two sections (which I have designated A and B). I divide the introduction into paragraphs. Section B is further divided into treatises. Each treatise of section B is divided into chapters, as is section A as a whole. I have divided each chapter into paragraphs. For example, B.1.III.4 is the fourth paragraph in the third chapter of the first treatise of section B; A.II.5 is the fifth paragraph in the second chapter of section A. The positions of the lengthy insertions found in the Hebrew recension are indicated with an /X. For example, B.3.II.4/X indicates the insertion in the fourth paragraph of the second chapter of the third treatise of the second section. In B.2.V.12 there are two insertions; the second is marked with an /Xb. Because §B.1.II.26/X is so long, I have numbered the paragraphs (§B.1.II.26/X.1, §B.1.II.26/X.2, etc.).

Introduction (begins with §0.1)

First Part

Chapter One (begins with §A.I.1): Premises necessary for astronomy

Chapter Two (begins with §A.II.1): Geometric preliminaries

Chapter Three (begins with §A.III.1): Instruments

Chapter Four (begins with §A.IV.1): Partial declinations and rising times

Chapter Five (begins with §A.V.1): Measure of the solar year

Second Part

Treatise One

Chapter One (begins with §B.1.I.1): Model for the solar mean motion

Chapter Two (begins with §B.1.II.1): Model for the solar anomaly

148. Endress, "Mathematics and Philosophy," pp. 153–4.

§B.1.II.13 Error analysis

§B.1.II.15 The double-circle hypothesis makes the model for the solar anomaly more precise

§B.1.II.28 Critique of Biṭrūjī's solar model

Chapter Three (begins with §B.1.III.1): Calculating the greatest solar anomaly

Treatise Two

Chapter One (begins with §B.2.I.1): Lunar motions and model for mean motion

Chapter Two (begins with §B.2.II.1): Model for the first lunar anomaly

Chapter Three (begins with §B.2.III.1): Calculating the greatest lunar anomaly

Chapter Four (begins with §B.2.IV.1): Visible variations in the lunar diameter do not necessitate variations in the Earth–Moon distance

Chapter Five (begins with §B.2.V.1): On the second and third lunar anomalies

§B.2.V.4 The model for the second lunar anomaly

§B.2.V.7 On the third lunar anomaly

Treatise Three

Chapter One (begins with §B.3.I.1): Biṭrūjī's model for the motion of the fixed stars

Chapter Two (begins with §B.3.II.1): Ibn Naḥmias's model for the motion of the fixed stars

Treatise Four

Chapter One (begins with §B.4.I.1): Motions of the five planets

Chapter Two (begins with §B.4.II.1): Hypothesis for the first anomaly in the upper planets

Chapter Three (begins with §B.4.III.1): Things known via rising times (these pages were most likely not originally part of Treatise Four)

THE RECEPTION OF *THE LIGHT OF THE WORLD*

The only known direct response to *The Light of the World* was composed by Profiat Duran (d. ca. 1415).[149] Chapter 7 contains a transcription of the Oxford MS of Duran's response, and chapter 8 contains a translation. Turns of phrase in this

149. See also Kozodoy, "A Study," pp. 206–65, for Profiat Duran's writings on astronomy and arithmology. Kozodoy briefly discussed Duran's response to *The Light of the World* on pp. 207–8.

brief text indicate that Ibn Naḥmias was alive and respected when Profiat Duran wrote, so Duran's response helps us determine the period of Ibn Naḥmias's life and work. Duran criticized Ibn Naḥmias for ignoring observed variations in the sizes of the Moon and planets, particularly the variations in the size of the Moon that could not be attributed simply to parallax or to distortions resulting from atmospheric phenomena. To Profiat Duran, these variations in size must have resulted from variations in the planets' distances from the Earth, meaning that a cosmos of homocentric orbs centered on the Earth was not possible.[150] Moreover, Profiat Duran's commentary on the *Guide* shows us that he saw no philosophical advantage in adhering to an astronomy that forbade epicycles and eccentrics. Duran pointed out, in his response to *The Light of the World,* that Gersonides had shown, by observing the Sun's diameter, that the solar orb was eccentric. Ibn Naḥmias, in Profiat Duran's judgment, was pursuing a hopeless course in light of this additional observational evidence.[151]

Despite Duran's negative assessment of Ibn Naḥmias's acumen in astronomy, an assessment that has influenced modern scholarship on Ibn Naḥmias, Duran's response had another dimension, a dimension that has attracted much less attention.[152] Near the end of his response, Duran wrote that Ibn Naḥmias was aware that he (Ibn Naḥmias) was engaged in a particular enterprise, that of homocentric astronomy, and that within the boundaries of homocentric astronomy, Ibn Naḥmias may well have succeeded to a greater extent than his predecessors.[153] Thus, Duran's response might be understood not only as a preference for other (nonhomocentric) astronomies but also as a recognition that homocentric astronomy was a distinct enterprise, whether or not he considered it worthwhile.

I have written about the subsequent history of *The Light of the World,* focusing on the probable path of its passage to the Veneto during the Renaissance, in three articles and will summarize the results here.[154] The key link in the passage of

150. *Or ha-'olam*, Bodleian MS Canon Misc. 334, 100b.

151. Bernard R. Goldstein: *The Astronomy of Levi ben Gerson (1288–1344)* (New York, Berlin, Heidelberg, Tokyo: Springer-Verlag, 1985): p. 9. Gersonides's argument (see Mancha, "Demonstrative Astronomy," pp. 35–6) had not been that the eccentric was demonstrable but that it was necessary and true since it was the only known way of producing the available observations.

152. An example of Duran's apparent influence on modern scholarship is Freudenthal, "Towards a Distinction," pp. 917–9.

153. *Or ha-'olam*, Bodleian MS Canon Misc. 334, 100a. "And just as others besides him endeavored in this, perhaps he completed more of an astronomy (*t*ᵉ*ḵunah yoteir*)." On this measured approval, see Kozodoy, "A Study," p. 208, note 12, where she cites Steinschneider's assessment that Duran approved of Ibn Naḥmias's efforts, but not the resulting level of predictive accuracy.

154. Robert Morrison: "An Astronomical Treatise by Mūsā Jālīnūs alias Moses Galeano," in *Aleph: Historical Studies in Science and Judaism* X/2 (2011): 315–53; Morrison: "The Position of the Jews as Scientific Intermediaries in the European Renaissance," in F. Jamil Ragep and Rivka Feldhay (eds.): *Before Copernicus* (forthcoming, submitted to McGill Queens University Press); Morrison,

the text to the Veneto was the 15th- and 16th-century rabbi and physician Moses Galeano (a.k.a. Mūsā Jālīnūs).[155] He became familiar, probably in Istanbul, with the specific contents of *The Light of the World,* as Galeano composed a text in Arabic on homocentric astronomy that survives in a unique MS in the Topkapi Library.[156] That text reproduced, verbatim, large portions of Ibn Naḥmias's description of his solar model and is evidence for how the contents of *The Light of the World* must have already passed from the Iberian Peninsula to the Ottoman Empire by 1500, most likely via the Sephardic diaspora. Then, between 1497 and 1502, Galeano was in the Veneto. This was the likely time frame of the passage of the contents of *The Light of the World* to the Veneto.[157] The circumstantial evidence that Galeano or a contemporary effected the passage of *The Light of the World* to scholars at the University of Padua is the presence of double circles similar to those hypothesized in *The Light of the World* in the homocentric astronomy of Giovanni Battista Amico (d. 1538).[158] Moreover, Amico incorporated the double circles into his solar model in a way that Ibn Naḥmias had suggested. A reason why Amico would have been interested in *The Light of the World* in the first place was because the Hebrew recension of *The Light of the World* contained hypotheses, for example the Eudoxan couple, that could resolve shortcomings of the homocentric astronomy of Regiomontanus (d. 1476), who had worked at Padua.[159] Therefore, the earlier history of homocentric astronomy at Padua created a context for the interest in *The Light of the World.*

"A Scholarly Intermediary between the Ottoman Empire and Renaissance Europe," *Isis* CV (2014): pp. 32–57.

155. Other important recent articles devoted to Galeano are Y. Tzvi Langermann: "A Compendium of Renaissance Science: *Ta'alumot Ḥokmah* by Moshe Galeano," in *Aleph: Historical Studies in Science and Judaism* VII (2007): pp. 283–318; and Langermann: "Medicine, Mechanics and Magic from Moses ben Judah Galeano's *Ta'alumot Ḥokmah,*" in *Aleph: Historical Studies in Science and Judaism* IX (2009): pp. 353–77.

156. For the text, translation, and analysis, see Morrison, "An Astronomical Treatise." I have not found a reference to Ibn Naḥmias by name in any of Galeano's writings. In *Puzzles of Wisdom* (see Langermann, "A Compendium," p. 291), Galeano named R. Joseph ibn Ya'ish as the author of *The Light of the World.*

157. Langermann, "A Compendium," p. 295.

158. On Amico, see Mario di Bono: "Copernicus, Amico, Fracastoro, and Ṭūsī's Device: Observations on the Use and Transmission of a Model," in *Journal for the History of Astronomy* XXVI (1995): pp. 133–54. On the connection with Galeano, see Morrison, "A Scholarly Intermediary," pp. 50–4,

159. S ee Michael Shank: "Regiomontanus as a Physical Astronomer: Samplings from *The Defence of Theon Against George of Trebizond,*" in *Journal for the History of Astronomy* XXXVIII (2007): pp. 325–49. See also Michael Shank: "The 'Notes on al-Biṭrūjī' attributed to Regiomontanus: Second Thoughts," in *Journal for the History of Astronomy* XXIII (1992): pp. 15–30. On the shortcomings of Regiomontanus's homocentric astronomy, see Swerdlow: "Regiomontanus's C oncentric-

There is, too, direct evidence that the entire text of *The Light of the World* eventually arrived at the University of Padua by the mid-1600s, as there is a report in Bartolocci's *Bibliotheca magna rabbinica de scriptoribus* of the text, probably its Hebrew recension, being seen at the University of Padua.[160] There is little doubt that Galeano had contact with Christian scholars in Italy because he was part of a network of scholars who sold books to representatives of the famous Fugger family.[161] He also knew Latin well enough to translate the canons of the *Almanach Perpetuum* into Arabic in 1505, shortly after his return from Italy.[162]

Galeano's role in the exchange of *The Light of the World* is important also because it increases the likelihood that he passed other important information to scholars in the Veneto. A comment in another text by Galeano, *Taʿalumot Ḥokma* (Puzzles of Wisdom), described key features of the models of Ibn al-Shāṭir (d. 1375), along with those of Ptolemy, Gersonides, Biṭrūjī, and Ibn Naḥmias.[163] Recent and not-so-recent research has argued that the models of Ibn al-Shāṭir and other astronomers of the Islamic world were a source for Copernicus's *De Revolutionibus*.[164] Copernicus himself studied medicine at the University of

---

Sphere Models." Swerdlow argued (p. 4) that Regiomontanus's homocentric models, particularly the feature of the reciprocation mechanism that solved a problem with Biṭrūjī's models, were due to Regiomontanus himself.

More generally, Biṭrūjī's astronomy generally met with a more favorable reception in the Latin West than in Islamic societies, and Avi-Yonah has noted a resurgence of interest in Biṭrūjī in the sixteenth century in the West. See Reuven S. Avi-Yonah: "Ptolemy vs. al-Biṭrūjī: A Study in Scientific Decision-Making in the Middle Ages," in *Archives internationales d'histoire des sciences* XXXV (1985): pp. 124–47, at pp. 144–7.

160. Giulio Bartolocci: *Bibliotheca magna rabbinica de scriptoribus* (Rome: Sacrae Congregationis de Propaganda Fide, 1675–94), vol. 4: p. 501. The report came via a certain Petrus Rivier, who was associated with the Collegium Neophytorum (where Bartolocci had also been a professor) and was a convert from Judaism (ibid., vol. 4: p. 228). Petrus Rivier was likely a contemporary of Bartolocci (ibid., vol. 4: p. 229). The Collegium Neophytorum was founded in 1543. On its founding, see Robert Wilkinson: *Orientalism, Aramaic, and Kabbalah in the Catholic Reformation* (Leiden, Boston: Brill, 2007): p. 42.

161. See Morrison, "A Scholarly Intermediary," pp. 54–6.

162. Julio Samsó: "Abraham Zacut and José Vizinho's *Almanach Perpetuum* in Arabic (16th–19th C.)," in *Centaurus* XLVI (2004): pp. 82–97, at p. 83.

163. Langermann, "A Compendium," p. 291.

164. These studies include (but are not limited to) Victor Roberts: "The Solar and Lunar Theory of Ibn al-Shāṭir: A Pre-Copernican Copernican Model," in Isis XL (1957): pp. 428–32; E. S. Kennedy and Victor Roberts: "The Planetary Theory of Ibn al-Shāṭir," in Isis L (1959): pp. 227–35; and Otto Neugebauer: The Exact Sciences in Antiquity (Providence, RI: Brown University Press, 1957; repr. Dover Publications, 1969): pp. 202–5. These results and some subsequent findings are presented in Otto Neugebauer and Noel Swerdlow: *Mathematical Astronomy in Copernicus' "De Revolutionibus"* (New York and Berlin: Springer-Verlag, 1984), vol. 1: pp. 41–64. The most important recent studies, studies which include additional direct historical evidence for exchange, are George Saliba: *Islamic*

Padua between 1501 and 1503, a period that could easily have overlapped with Galeano's visit to the Veneto. Thus, further research on the passage of *The Light of the World* to Renaissance Italy may enhance our understanding of the passage of the contents of Ibn al-Shāṭir's work.[165]

In sum, the reception of *The Light of the World* was dichotomous in that it, on the one hand, was critiqued by Profiat Duran for its lack of agreement with observations, but, on the other hand, it contained ideas that seem to have been taken seriously in Renaissance Italy. This book will investigate how this dichotomous reception could be the case. Until now, as concerns about predictive accuracy, including questions of observed variations in planetary distances, have led to dismissals of *The Light of the World* and other works of homocentric astronomy by some scholars and some of the historical actors, explanations for the attractiveness of these works have been limited to an acknowledgment of their philosophical elegance. Indeed, *The Light of the World* itself conceded that the *Almagest* was sufficient for calculating positions: "It also results from what we have said about the Sun that the tables which Ptolemy used for the motions and the anomalies are themselves the tables needed for calculations with these hypotheses."[166] But even though Ibn Naḥmias's explicit focus was not mathematical astronomy alone, this book will argue that Ibn Naḥmias's homocentric astronomy succeeded in improving upon Biṭrūjī's predictive accuracy. Homocentric astronomy attracted a host of talented scholars in late medieval and Renaissance Europe; in order to determine why homocentric astronomy exerted such an attraction upon a scholar well versed in mathematical astronomy such as Regiomontanus, the commentary includes, in certain cases, quantitative analysis. Such analysis will show where the predictive accuracy of *The Light of the World* did improve on that of *On the Principles.*

Science and the Making of the European Renaissance (Cambridge and London: MIT Press, 2007): pp. 193–232; F. Jamil Ragep: "Copernicus and His Islamic Predecessors: Some Historical Remarks," in History of Science XLV (2007): pp. 65–81; and Ragep "'Ali Qushji and Regiomontanus: Eccentric Transformations and Copernican Revolutions," in *Journal for the History of Astronomy* XXXVI, no. 4 (2005): pp. 359–371.

165. Cf. Saliba, "Critiques," p. 17. As both Ibn Naḥmias and Ibn al-Shāṭir excluded eccentrics, Saliba's observation that Ibn al-Shāṭir saw the epicycle as a philosophic nonproblem is fascinating. "He first stated categorically that 'the existence of small spheres, like the epicyclic spheres, which do not encompass the center of the world, is not impossible (*ghayr mumtani'*) except in the ninth sphere. The proof for this is that since each sphere has a planet and the eighth has several spherical stars, each of which is larger than some of the epicycles of the planets, and since the star (or planet, *kawkab*) is different from the rest of the sphere, then the existence of epicyclic spheres and the like would not be impossible.'" To Ibn al-Shāṭir, the presence of some sort of composition (*tarkīb*^{un} *mā*) in the heavens did not compromise the heavens' overall simplicity.

166. *The Light of the World*, §B.1.II.27.

DESCRIPTIONS OF THE MSS AND EDITORIAL PROCEDURES[167]

There are two versions of *The Light of the World,* one in Judeo-Arabic and one in Hebrew. The earlier Judeo-Arabic original is Vatican MS Ebr. 392, folios 51r–88r, which, according to Steinschneider, contained the text in Arabic in Hebrew characters.[168] Langermann's survey of Arabic writings transcribed into Hebrew characters stated that *Nūr al-ʿālam* (The Light of the World) is in Judeo-Arabic.[169] This book accepts Langermann's characterization of *Nūr al-ʿālam* as a Judeo-Arabic text because divergences in the text from classical Arabic match Blau's descriptions of Judeo-Arabic.[170] There is hypercorrection with the dual, for example, אן האתין אלחרכתין אלתאן for إنّ هاتين الحركتين التين on folio 51v and confusion with the relative pronoun הדא אלקדר אלתי תרי for هذا القدر الذي يرى on folio 52r. In addition, there are numerous cases of where the jussive voice is missing, such as ולתסמי for ولتسمّ on folio 63v and passim and לם יכון for لم يكن on folio 64r and passim. There are also examples of orthographic hypercorrection, such as תכ֗לוא for تخلو on folio 68v, and orthographic variations, such as רג֗ול for رجل on folio 68v. Indefinite active participles of verbs with final weak radicals frequently end with a long vowel, such as מסאוי for مساوٍ on folio 63v. Finally, with a noun-adjective combination, the definite article on the noun is sometimes eliminated. For instance, the text reads *taqṣīr al-akhīr* (*al-taqṣīr al-akhīr*) for "the final lag."[171]

The first 50 folios of the Vatican MS contain a recension of the *Almagest* in 12 books in a variety of Arabic in Hebrew characters in the same hand as *The Light of the World.* The Vatican MS of *The Light of the World* begins:

> אן אהל אלעלם אלריאצ֗י אדא תכלמוא פי אלאצול אלתי עליהא יג֗רי אלאמר פי אלחרכאת אלסמאויה באצטלאח, מנהם בנוא אלאמר עלי אצלין אתנין קד כ֗אלפוהא בהמא ג֗מיעא֗ בעץ֗ אהל אלעלם אלטביעי. . . .

> When the practitioners of mathematics spoke in conventional, technical language about the hypotheses according to which the heavenly motions proceed, some of them founded the matter upon two hypotheses with regard to which some of the practitioners of physics completely disagreed. . .

167. Here I follow Morrison, "The Solar Theory," pp. 60–3.

168. Steinschneider, *Der Mathematik bei den Juden,* p. 114.

169. Y. Tzvi Langermann, "Arabic Writings in Hebrew Manuscripts: A Preliminary Re-Listing," in *Arabic Sciences and Philosophy,* 6 (1996): pp. 137–60, p. 139.

170. Joshua Blau, *A Grammar of Mediaeval Judaeo-Arabic* [in Hebrew] (Jerusalem, 1961), p. 87 (*verba mediae wāw / yā* in the jussive mood), p. 91 (*verba tertiae infirmae* in the jussive mood), and active participles (p. 93), p. 104 (hyper-correction with dual), and p. 233 (confusion with the relative pronoun). On p. 91, too, there is a description of orthographic irregularities with *verba tertiae infirmae,* although there is no example that perfectly matches אנהא ליסת תכ֗לוא

171. Cf. Blau, *A Grammar,* p. 161.

The MS lacks a colophon and ends abruptly in the preliminaries of the treatise on the three planets:

> פנעמל ג̇דולא̇ ונבתדי מן מעאדל אלנהאר אלי אלדאירה אלתי אטול נהארהא י̇ז̇ סאעה בזיאדה נצף סאעה. פי אלצף אלאול אלברוג̇ אלי̇ב̇ ופי אלב̇ עשראתהא ופי אלג̇ מטאלעהא ופי אלד̇ ג̇מעתהא.

> So we make a table and we begin from the equinoctial to the circle whose longest day is 17 hours by increments of a half hour. In the first column are the twelve zodiacal signs and in the second are their ten [degree interval]s, in the third are the rising times, and in the fourth their accumulated time-degrees.

The Hebrew version of the text, entitled *Or ha-'olam,* exists in Bodleian Canon. Misc. MS 334 on folios 127a to 101a (the manuscript is numbered backward).[172] Steinschneider wrote that the Hebrew was a translation of the original Arabic in Hebrew characters (or Judeo-Arabic).[173] My earlier study of the Hebrew version revealed extensive expansions on the models found in the Judeo-Arabic version, which is why I refer to the Hebrew version as a recension and not as a translation.[174] Chapters 4 and 6 study those expansions, particularly the several additional pages on the solar model, and include translations and technical analysis. It is possible that the Hebrew reflects a lost Judeo-Arabic version, or that the producer of the Hebrew MS expanded on the original Judeo-Arabic. Freudenthal suggested that the Hebrew MS could be the work of Ibn Naḥmias himself.[175] Although the precise relationship between the Judeo-Arabic and Hebrew versions is unclear, I sometimes use the Hebrew MS, in the cases where it follows the Judeo-Arabic, to improve my English translation. Though the Hebrew recension lacks any discussion of the planets, a topic included in the Judeo-Arabic MS, the presence of improved models in the Hebrew MS supports the conclusion that the extant Hebrew text reflects developments of Ibn Naḥmias's models beyond those found in the Judeo-Arabic version.

The incipits of both the Hebrew and Judeo-Arabic versions are preceded by a stanza of poetry, which is studied in the commentary.

The Hebrew MS begins:

> אומר כי בעלי החכמה הלמודית כשדברו בעקרי תנועות השמימיות הסכימו לבנות הענין על שני עקרים כבר חלקו עליהם בשניהם בכלל קצת בעלי החכמה הטבעית

172. See Robert Morrison, "The Solar Theory," p. 62. See also Steinschneider, *Der Mathematik bei den Juden*, p. 114. Duran's response runs on to folio 100a.

173. Steinschneider, *Der Mathematik bei den Juden*, p. 114

174. Robert Morrison: "Andalusian Responses to Ptolemy in Hebrew," in Jonathan Decter and Michael Rand (eds.): *Precious Treasures from Hebrew and Arabic* (Piscataway, NJ: Gorgias Press, 2007): pp. 69–86.

175. Freudenthal, "Towards a Distinction," p. 917.

> I say that the practitioners of mathematics, when they spoke of the hypotheses of the celestial motions, agreed to establish the matter on two principles about which some of the practitioners of physics, upon hearing them, disagreed.

The Hebrew MS concludes, lacking a detailed colophon:

> ואיפשר כי זה החילוף הטעה האחרונים עד ששמו תנועת האיחור והקדימה. ומה שאמרנו יספיק בהשלמת העיקר האמיתי לזאת ההעתקה. והשבח הגדול לאל אשר הגיענו לזאת התכלית

> It is possible that this anomaly caused the recent astronomers to err to the extent that they propounded the motion in accession and recession. What we have said is sufficient to complete the correct hypotheses for this movement. Great praise to God Who has brought us to this end.

MS SIGLA, PARAGRAPH NUMBERING SCHEME, AND EDITORIAL PRINCIPLES

Oxford: א.
Vatican: ו.

This book provides separate transcriptions of the Judeo-Arabic and Hebrew MSS.

My transcriptions of the Judeo-Arabic are not corrected according to the grammar of Modern Standard Arabic. Orthographic emendations are so noted in the apparatus. A dot over a letter transforms a Hebrew consonant into an Arabic consonant absent from the Hebrew alphabet. A dot over an aleph indicates the accusative ending (although the MS never indicates the accusative on *ayḍ*^{an} and *idh*^{an}, so neither has my transcription with those two words). The Judeo-Arabic MS did not indicate the letters *dhāl* and *thā'* with dots, so I have not either. The Judeo-Arabic MS did not indicate doubling of consonants (*shadda*) or the *tā' marbūṭa*, so I have not either. When the Judeo-Arabic consistently deviates from standard orthography, only the first such deviation is noted in the apparatus. Because vowels are indicated inconsistently with consonants in the Hebrew MS, I have inserted some vowel points into my transcription of the Hebrew in order to resolve ambiguities. The critical apparatus takes the Judeo-Arabic version to be the base MS and notes variants only when one version adds or omits words or when the Hebrew departs from a plausible translation of the Judeo-Arabic. I note omissions with a minus sign (–) and repetition with a number two (2). Illegible portions are indicated by a question mark (?). I have occasionally chopped up long sentences but have more frequently tried to reproduce the original syntax. Insertions in the Hebrew recension that are so extensive as to be translated in full and commented upon are not reproduced in full in the critical apparatus; rather, there is a notation in the text, involving an *X*, that there is a long insertion.

The figures for chapters 2 and 4 represent the figures found in the Judeo-Arabic and Hebrew MSS of *The Light of the World*. The MSS figures, and consequently the figures for chapters 2 and 4, do not convey a three-dimensional perspective. The figures for chapters 5 and 6 illustrate the commentary and often do portray three-dimensional orbs or their surfaces. I have translated the letters from the Hebrew alphabet on each figure according to the following key:

א = A, ב = B, ג = G, ד = D, ה = E, ו = W, ז = Z, ח = H, ט = T, י = Y, כ = K, ל = L, מ = M, נ = N, ן = U, ס = S, ע = O, פ = P, צ = C, ק = Q, ר = R, ש = X, ת = V

CONCLUSIONS

Because I composed the commentary in a line-by-line format, I list below the most important conclusions that this book reaches:

1. *The Light of the World* does improve, in many places, on the predictive accuracy of Biṭrūjī's *On the Principles of Astronomy.*
2. One component of Ibn Naḥmias's models is a hypothesis of double circles that resembles the rudimentary Ṭūsī couple from the 1247 *Taḥrīr al-Majisṭī* and which is mathematically equivalent to the spherical Ṭūsī couple presented in the *Tadhkira.*
3. Although Ibn Naḥmias's stated intention was to continue Biṭrūjī's project and not to recover a lost astronomy from classical antiquity, a hypothesis introduced in the Hebrew recension of *The Light of the World* is identical to a central component of Schiaparelli's reconstruction of Eudoxus's models.
4. Given the overwhelming evidence that models from *The Light of the World* were known by astronomers at the University of Padua in the early 1500s, the role of Moses Galeano/Mūsā Jālīnūs in the exchange of *The Light of the World,* and Galeano/Jālīnūs's knowledge of Ibn al-Shāṭir's models, homocentric astronomy is an important space for contacts between scholars of astronomy in the Veneto during the Renaissance and scholars in the eastern Mediterranean.

1

Judeo-Arabic Text of *The Light of the World*

51r

כתאב נור אלעאלם [ל̇][1]ר̇ יוסף ן̇ נחמיס

ספר אשר הוציא לאור נעלם   יצב גבולות רום לכל חילם
משם דבר חפץ מצוא בקש   [יהיה לך תמיד לאור עולם][2]
[הורני יי דרכך ונחני באורח מישור][3]   [מלפניך משפטי יצא עיניך תחזינה מישרים][4]

0.1 [][5] אן אהל אלעלם אלריאצ̇י אדא תכלמוא פי אלאצול [אלתי עליהא יג̇רי אלאמר פי][6] אלחרכאת
אלסמאויה באצטלאח، [מנהם][7] בנוא אלאמר עלי אצלין אתנין קד כ̇אלפוהא בהמא ג̇מיעא̇ בעץ̇ אהל
אלעלם אלטביעי מן גיר אן יקולוא פי סבב מא יט̇הר מן אלחרכאת ואלאכ̇תלאפאת פי אלסמא קול
[מקנע][8]، פאן קצדהם אנמא כאן אן יבינוא אסתחאלה האדין אלאצלין פי חק אלאג̇ראם אלסמאויה
בדלאיל טביעיה פקט. אחד האדין אלאצלין וצ̇עהם פלך כ̇ארג̇ אלמרכז ופלך אלתדויר. ואלתאני
וצ̇עהם פי אלסמא חרכאת מתצ̇אדה עלי זעמהם. ויסתבעדוא כון חרכאת מתצ̇אדה פי אלסמא
מתלמא יסתבעדוא אלפלך אלכ̇ארג̇ אלמרכז ופלך אלתדויר. ואנמא ערץ̇ להם דלך לאנהם ירון אן

1. א: בתכונה חברו.
2. Cf. Isaiah 60:20.
3. Cf. Psalms 27:11.
4. Cf. Psalms 17:2.
5. א: אומר.
6. א: -.
7. א: -.
8. א: מספיק.

חרכה מסתדירה תצֹאד חרכה מסתדירה, כמא אן חרכה מסתקימה תצֹאד חרכה מסתקימה, עלי כֹלאף מא בינה ארסטו פי כתאב אלסמא ואלעאלם.

0.2 וגרצֹנא פי הדא [אלמוצֹע][9] אן נבין אן מא קאלוה פי אמתנאע אלפלך אלכֹארג׳ אלמרכז ופלך אלתדויר חק בחית ליס יחתאג׳ אלי ברהאן אכֹר [][10]. תם נעטי אצול יקיניה יטֹהר מנהא מא כאן יטֹהר פי [כל מא][11] מן קבל [][12] האדין אלפלכין עלי אתם מא ימכן. ואמא מא קאלוא פי אמתנאע חרכה מתצֹאדה פי אלסמא פבאטל לאנה קד יתבין [בעד קליל][13] אן ליס פי אלחרכה אלמסתדירה תצֹאד בכלאם ברהאני, אעני אן חרכה מסתדירה ליסת תצֹאד חרכה מסתדירה, פאן ארסטו אנמא בין דלך בדלאיל פקט. פנקול אמא אלפלך אלכֹארג׳ אלמרכז פטֹאהר מן אמרה אלאסתחאלה ואלבעד ען אלאתפאק ואלתכֹלי באלנטֹאם, פאנה ליס האהנא שי אשנע מן וצֹע פלך כֹארג׳ אלמרכז ען מרכז אלעאלם דאירא חול מרכזה ומרכזה דאירא חול מרכז אכֹר.

51v

0.3 וילזם ען דלך אן יכון בין אלפלכין אמא כֹלא וקד תבין אמתנאע וגוֹדה ואנה מתאל אמתנאע וגוֹד אלעדם אלמחץֹ, ואמא מלא מן גֹסם גריב יבדל אמאכנה באלחרכה. ואן כאנא האדין אלפלכין מתקאטעתין, ילזם [][14] מא קלנאה איצֹא אדא תחרך אלמרכז אלואחד חול אלאכֹר אן תתשתת אלפלכין [גֹמיעאֹ][15] ותתשדי חתי תכון טביעה הדה אלאפלאך מתל טביעה אלמא או אלהוא תכֹלי מכאנאֹ ותמלי מכאנא, וסאיר מא ילזם הדה אלאוצֹאע מן אלמחאלאת ואלאחואל אלמכֹאלפה ללחק אלתי דכרהא רבי משה עלֹשֹ פי כתאב דלאלה אלחאירין.

0.4 ואשנע מן דלך וצֹע פלך תדויר דאיראֹ חול מרכזה ומרכזה איצֹא דאיראֹ פי נפס ערץֹ פלך אכֹר, פאן הדה אלמחאלאת תלזם ענה [פכיף][16]? וקד אצֹטר עליהם אלאמר פי כתיר מן אלמואצֹע אלי וצֹע אלאצלין [גֹמיעאֹ][17], ואן חרכה אלכוכב חול מרכז פלכה ואלכוכב מחמול פי הדה אלחרכה עלי פלך אכֹר זאיל אלמרכז ענה. פהדא אלקול כאף פי אבטאל [כון][18] האדין אלפלכין פי אלסמא. ואמא מא קאלוה פי אבטאל כון חרכתין מתצֹאדתין פי אלסמא לאן חרכאתהא טביעיה ולמחרך [טביעי][19] ואחד בסיט, פאנה מן נפס קולהם הדא יתבין אחד אמרין. אמא אלא יכון פי אלחרכה אלמסתדירה תצֹאד אעני אלא תצֹאד חרכה מסתדירה לחרכה מסתדירה, ואמא אן תצדראן באלצֹרורה [חרכתין][20] מתצֹאדתין מן הדא אלמחרך אלואחד אלבסיט. אן סלמנא אן פיהא תצֹאד מן קבל אן אלחרכה טביעיה ולמחרך טביעי ואחד בסיט דלך יתבין עלי הדא [אלוגֹה][21].

9. א: -.
10. א: על זה.
11. א: בשמים.
12. א: הנחת.
13. א: -.
14. א: עם.
15. א: -.
16. א: כל שכן.
17. א: -.
18. א: -.
19. א: -.
20. ו: חרכה. א: שתי תנועות.
21. א: המופת.

0.5 אן מן כ̇אצה אלכרה אנהא ימכן אן תתחרך חול כל קטבין יתכ̇ילאן פי סטחהא. פלננזל מתלא̇ אן קטבי פלך אלזהרה מוצ̇ועין עלי דאירה נטאק אלברוג̇، אחדהמא פי נקטה ראס אלסרטאן ואלתאני פי נקטה ראס אלג̇די. ולננזל אן פלך אלברוג̇ מנתקל חול קטביה מן אלמשרק אלי אלמגרב، וינקל מעה פלך אלזהרה וקטביה. ולננזל איצ̇א אן פלך אלזהרה מנתקל נחו תלך אלג̇הה חול קטביה. פאנה ליס פרק בין הדא אלוצ̇ע ובין וצ̇ע פלך אלברוג̇ מע קטביה מנתקל מע אלחרכה אליומיה

52r

ומנתקלה איצ̇א חול קטביה. [פאדא כאן דלך כדלך פט̇אהר אנה][22] מא דאם אלקטב אלואחד אעני אלשמאלי מן פלך אלזהרה מנתקלא̇ מן ראס אלסרטאן אלי ראס אלחמל באנתקאל פלך אלברוג̇، ואלקטב אלג̇נובי אלרבע אלמקאבל، פאן אלחרכתין אעני חרכה [מעדל][23] אלנהאר וחרכה פלך אלזהרה תכונא אלי ג̇הה ואחדה.

0.6 ופי וקת חלול אלקטבין עלי נקטתי אלאעתדאל، תכונא אלחרכתין מתקאטעתין עלי זאויה קאימה. ומא דאם אלקטב אלשמאלי מנתקלא̇ מן ראס אלחמל אלי ראס אלג̇די، ואלתאני אלרבע אלמקאבל، פקד תכון אלחרכתין מתצ̇אדתין עלי ראיהם. פיא לית שערי כיף תכון אלחרכאת אלמתשאבהה אלטביעיה אלג̇אריה עלי נצ̇אם ואחד פי נצף אלזמאן מתשאבהה אלי ג̇הה ואחדה ופי נצפה מתצ̇אדה? בל אלחרכה אלואחדה בעינהא תצ̇אד נפסהא פי אזמנה מכ̇תלפה מע כונהא מתשאבהה אבדא̇? הדא פי ג̇איה אלשנאעה، והדא אלברהאן [מאכ̇וד][24] מן ג̇הה לואחק אלחרכה. פקד ט̇הר מן דלך אן חרכה מסתדירה ליסת תצ̇אד חרכה מסתדירה. ולו סלמנא דלך، עלי אנה מסתחיל، פקד תבין מן דלך אן האתין אלחרכתין אלתאן המא מתצ̇אדתין עלי מדהבהם המא מן קבל צ̇רורה חפט̇ הדא אלמחרך אלטביעי אלואחד אלבסיט נט̇אמה אבדא̇. פאנה לו כאן מן שאן הדא אלמחרך אלא תצדר ענה חרכאת מתצ̇אדה، לוג̇ב אדא וצלת אלקטבין עלי נקטתי אלאעתדאל אן תקף אלחרכה ודלך מסתחיל.

0.7 פאד קד תבין דלך פאלאמתנאעאת אללאזמה להם אנמא הי מן קבל וצ̇עהם אלפלך אלכ̇ארג̇ אלמרכז ופלך אלתדויר פקט، ופי הדא מן [אלאמתנאעאת][25] הדא אלקדר אלתי תרי. ואלעג̇ב מן [עט̇ם][26] מא אסתנבטה מן תלך אלאצול כ̇אצה רייס [הדא][27] אלפן בטלמיוס ומא אפאד למן בעדה מן הדא אלעלם אלג̇ליל חתי יסהל למן שא וג̇וד אזמאן עודאת אלכואכב ומערפה מואצ̇עהא פי אפלאכהא פי אלטול ופי אלערץ̇، ומערפה מקאדיר עט̇מהא ונסבה בעצ̇הא אלי בעץ̇ ונסבתהא אלי אלארץ̇، ובאוקאת כסוף מא ינכסף מנהא ומקדאר אלמנכסף מן ג̇רמהא פי כל זמאן، מע כון אלאצול אלמבני עליהא פי הדא אלבעד ען אלתצויר פי חקהא، עלי אנה ליס יג̇ב אן יט̇ן [][28] במתל בטלמיוס

52v

אנה יצ̇ע הדה אלאצול עלי אנה ירי אן אלאמר פי נפסה כדלך، לכון מא יט̇הר ללחס תאבעא̇ למא יכ̇רג̇ ען תלך אלאצול.

22. א: אכ̇ ידוע.
23. ו: מעאדל.
24. ו: מאכ̇ד. א: נלקח.
25. ו: אלאמתנאע. א: המניעות.
26. ו: עצ̇ים. א: מרוב.
27. א: -.
28. א: כאיש.

0.8 ואנא עאג̇ב בכת̇יר ממא קאל אלפילסוף אלעט̇ים רבי משה ע̇אס̇ פי פצל כ̇ד̇ מן אלג̇ז אלת̇אני מן כתאב דלאלה אלחאירין ענדמא עדד אלמחאלאת אללאזמה להאדין אלאצלין. קאל [פכיף יוג̇ד ללכואכב הדה אלחרכאת אלמכ̇תלפה והל תם וג̇ה ימכן [מעה][29] אן תכון אלחרכה אלדוריה מסתויה כאמלה וירי פיהא מא ירי אלא באחד אלאצלין או במג̇מועהמא][30]، חתי אנה קאל אן אללה תעלי מכן אלאנסאן מן מערפה מא דון אלסמא פקט. ואמא צור אלסמא וחרכאתהא פהו וחדה תעלי אלעאלם. [ולדלך][31] קאל דוד המלך ע̇אס̇ השמים שמים [לש̇][32] וכון . ולנרג̇ע אלי חית כנא.

0.9 ולמכאן ט̇הור הדה אלאשיא ללחס עלי מא כונהא [צארת הדה אלאצול מתסלמה מן אלג̇מיע][33]. ולמכאן דלך צאר באב אלפחץ ען אלאצול אלחקיקיה מסתגלקא̇ זמאנא̇ כתירא̇. וכאן דלך עאיקא̇ ען תתמים הדה אלחכמה אלג̇לילה אלי אן ג̇א אלבטרוג̇י ונבה עלי הדה אלמחאלאת אללאזמה לתלך אלאצול. וזעם אנה אבתדע [היאה][34] להדה אלחרכאת ואלאכ̇תלאפאת דון אן יצ̇ע פלך כ̇ארג̇ אלמרכז ולא פלך תדויר. ודכר אנה אנתבה אלי דלך מן וקת סמע לאבי בכר בן אלטפיל אנה עתר עלי [היאה][35] ואצול גיר דינך אלמוצ̇ועין، אלא אנה תבע פי דלך פרקה אלטביעיון פי אסתבעאד אלחרכאת אלמתצ̇אדה. ואבתדא בדכר כיפיה אלחרכאת וכיף יכון כל מא פי אלסמא מתחרכא̇ אלי ג̇הה ואחדה ויכון מא ילזם מנהא מטאבק ללט̇אהר ללחס. ואלי הדה אלקדר מן קולה קד אצאב פיה، אעני אנה ממכן אן יכון כמא קאלה.

0.10 ואמא אדא אתי באלאסבאב אלמוג̇בה ללאכ̇תלאפאת פאנה ליס יטאבק שי ממא ילזם ען [][36] תלך אלאצול אלמוצ̇ועה אלי אלאן אעני למא יט̇הר ללחס، אלא אן כאן קצדה ליעלמנא כיף יוג̇ד אכ̇תלאף מא פי הדה אלחרכאת גיר אלאכ̇תלאף אלט̇אהר ללחס. פאני אחסב אן הדא כאן מדהבה ואלא למדא אחתאג̇ אלי תג̇דיד רצד כמא קאל. וקד יתבין ללנאט̇ר פי כתאבה מן מואצ̇ע אֻכ̇ר כתירה אן קצדה כאן דלך [בעינה][37] והדא קצד כ̇ארג̇

53r

ען אלואג̇ב، פאנה קד אדכ̇לנא פי חירה ומחאלאת אעט̇ם ואשנע מן אלתי כנא פיהא. ומע מא קלנא מן נקצאן אצולה פי דלך וכתרה זואלהא ען אלחק כמא יתבין [בעד][38] אדא וצלנא אלי הדה אלמואצ̇ע בעון אללה، פאנה קד יסתחק שכר ליס באלקליל לכונה אול מן נֻבה עלי דלך ווצ̇ע פיה כתאב.

29. בהנחתו.

30. Both passages follow Maimonides (ed. S. Munk): *Dalālat al-ḥā'irīn* (Jerusalem: Y. Yunovitz, 1929): p. 227 and Maimonides (ed. Hüseyin Atay): *Dalālat al-ḥā'irīn* (Ankara: Ankara University Press, 1972; repr., Cairo: Maktabat al-thaqāfa al-dīniyya, n.d.): p. 348, except that they read *dawriyya*, not *al-dawriyya*.

31. א: ועל דומה לזה.

32. א: להש̇.

33. א: הודו הכל באלו העקרים.

34. ו: היה. א: תכונה.

35. ו: האיה. א: תכונה.

36. א: מעקריו לדבר מה שיתחייב.

37. א: -.

38. א: -.

0.11 פאני מן וקת ראית כתאבה לם אזל זמאנא̇ כתירא̇ אתפכר פי ד̇לך אלי אן וצלת אלי חקיקה אלאצול אלמוג̇בה ללאכ̇תלאפאת פי הד̇ה אלחרכאת כמא יתבין כל ד̇לך [פימא בעד אן שא אללה][39] תעלי. ואעלם אן גרצ̇נא פי הד̇א אלכתאב אנמא הו וצ̇ע אצול ללאכ̇תלאפאת בד̇ל אלאצלין אלמד̇כורין, אעני אלכ̇ארג̇ אלמרכז ופלך אלתדויר. ואמא אמר אלחרכאת אלמתצ̇אדה פבאלואג̇ב כאן אן נתרכהא כמא הי עלי ראי בטלמיוס למא בינה מן עדם אלתצ̇אד פיהא. ולאכן מן אג̇ל אנתשאר הד̇א אלראי ענד ג̇מהור אלנאס, אעני אן פי אלחרכה אלמסתדירה תצ̇אד, ראינא אן נצ̇ע אלאצול להד̇ה אלחרכאת ולו עלי אן תכון נחו אלג̇הה אלתי הי ענדהם ואחדה, אלא חית יצ̇טר אלאמר אלי וצ̇ע חרכה מתל אלתי וצ̇ענאהא פי אלמתאל אלד̇י תקדם, אעני אלחרכאת אלד̇י יתוהם אנהמא ואחדה פי זמאן ומתצ̇אדה פי זמאן אכ̇ר מע כונהמא אבדא̇ מתשאבהה ג̇אריה עלי נט̇אם ואחד.

0.12 פנקול אן אלשי אלד̇י אצ̇טרהם אלי וצ̇ע אלחרכתין אלתאן המא ענדהם מתצ̇אדתין אנמא הו אנתקאל אלשמס ואלקמר וסאיר אלכואכב נחו אלשמאל ואלג̇נוב. ולולא ד̇לך ליס יכון ענדהם שי יוג̇ב אן יכון אנתקאלהא נחו אלמשרק חרכה אלי כלף אכתר מן אן יכון תקצירא̇, וקד צרח בטלמיוס בד̇לך פי אלמקאלה אלאולי מן כתאבה. פמאד̇א אלשי אלד̇י ימנע [מן אן נקול אן][40] כל ואחד מן אלאפלאך אלתי דון אלאעלי יתחרך נחו חרכה אלאעלי תאבעא̇ לחרכתה ועלי קטביהא, ויכון מקצרא̇ בהד̇ה אלחרכה [ען חרכה][41] אלאעלי לאכ̇תסאר אלקוה אלוארדה עליה מן אלאעלי לאג̇ל בעדה ען אלמחרך אלאול? וליס פי טביעתה אן יקבל אכתר מן ד̇לך אד̇ ליס הו מחמול פי הד̇ה אלחרכה באלאעלי עלי אנהמא שי ואחד ומתחרכאן חרכה מתסאויה.

53v

0.13 ויתחרך [איצ̇א][42] עלי קטביה אלמכ̇צוצה בה נחו תלך אלג̇הה כאן קצדה אן יתמם מא קצר ען אלאעלי באלחרכה אלאולי אד̇ קצדה וגאיתה אנמא הו קצד אלאעלי ואלתשבה בה אד̇ הי גאיתה. ובתלך אלחרכה [ינתקל][43] נחו אלשמאל ואלג̇נוב בקדר בעד קטביה ען קטבי אלאעלי. ומע הד̇ה אלחרכה איצ̇א ינקץ̇ לכל פלך מנהא קדר מחדוד ען אלוצול אלי מוצ̇ע אלאעלי וען אלתשבה בה, וד̇לך בקדר בעד כל ואחד מנהא ען אלמחרך אלאול אלמפיד לה אלחרכה.

0.14 ולד̇לך כלמא יכון וצ̇עה אקרב אלי אלאעלי יכון תקצירה אלאכ̇יר אקל. ובהד̇ה אלקול ירתפע אלכ̇לאף אלד̇י בין אהל הד̇א אלעלם פי תרתיב אלאפלאך לאנה ט̇אהר אן תרתיבהא סבב תקצירהא. פאד̇א̇ כל פלך מן אללואתי דון אלאעלי יקצראן קטביה אלמכ̇צוצה בה פי כל דורה קדר מחדוד פי דאירתי ממרהמא וד̇לך בקדר בעד כל ואחד מנהמא ען אלמחרך אלאולי. ואמא אלכואכב אלתי עלי הד̇ה אלאפלאך פאן תקצירהמא אקל מן ד̇לך במקדאר חרכה כל ואחד [][44] עלי קטביה חתי אן אלפלך אלמכוכב יסתפי בחרכתה [אלכ̇אצה][45] לה מא קצר פלכה ענד קום, ואמא אן יבקי לה תקציר קליל עלי מד̇הב אכ̇רין. ואלד̇י תחתה יקצר אכתר מן ד̇לך חתי ינתהי אלי פלך אלקמר אלאספל אלד̇י תקצירה אכתר מן תקציר כל ואחד ממא פוקה לכתרה בעדה ען אלאעלי.

39. א: בה̇.
40. א: מלהיות.
41. א: -.
42. א: -.
43. ו: ינתקאל. א: יעתק.
44. א: מהם.
45. ו: אלכ̇אציה.

0.15 פאד̇א אלפלך אלמחרך אלחרכה אליומיה והו אלמדרך באלעקל אד ליס לה [][46] ידרך באלחס מן קבלהא לאנה בסיט תאם אלבסאטה ט̇אהר מן אמרה אנה אעלי מן אלג̇מיע אד חרכתה ואחדה בסיטה שאמלה ללג̇מיע. ואמא סאיר אלאפלאך אלתי תחתה פליסת חרכאתהא בסיטה חתי אן אלפלך אלמכוכב לה חרכתין כמא לסאיר אלאפלאך אלתי תחתה. ואיצ̇א פאן אעט̇ם גרצ̇נא אן נבין אן ליס האהנא שיא יוג̇ב אן יכון פי אלסמא פלך אלא ומקערה ממאס למחדב אלדי דונה ממאסה צחיחה، מע כונה כרי תאם אלאסתדארה מן כ̇ארג̇ ומן דאכ̇ל ומחדבתה ממאס למקער אלדי פוקה، אן כאן [פוקה][47] פלך אכ̇ר כאנת עדד אלאפלאך כמא כאנת. פאן

54r

דלך לא יצ̇ר. והדא הו אלדי יוג̇ב עלי אלנאצ̇ר פי כתאבנא הדא אן יטלבה מן הדא אלכתאב פאנה [יג̇דה][48] פיה משרוחה בחית ליס פיה שך، והדא הו אלאמר אלדי קצרת אלעקול ען אדראכה [אלי הדה אלגאיה][49]. ונקול אנה ואן כאן ואג̇בא בחסב הדא אלנט̇ר אן נבדא באלקול פי אלפלך אלאעלי וננתהי באלנט̇ר אלי אלאספל، פאנא ג̇עלנא אלאבתדא מן חרכה אלשמס [][50] לנט̇אם אלאצול אלמסלוכה אלי אלאן לכי יתיסר ללנאט̇ר [אלתמיז בין הדה אלאצול ובין תלך][51] ויסהל לה אלתבין בין [אלטרק][52] אלמסלוכה פי סהולה אחתמאלהא אלעקלי ופי אלקרב ען אלחקיקה.

0.16 פאן [כלאמנא][53] פי הדא אלכתאב אנמא הו מע מן יתקדם לה אלנט̇ר פי אלכתב אלמוצ̇ועה פי הדא אלעלם אלי אלאן ואן כאן ליס יחתאג̇ פי קראתה אלי הדה אלכתב. ואני ארי אן מן תקדם לה אלנט̇ר פי אלארא אלמוצ̇ועה אצול ואסבאב להדא אלעלם ען בטלמיוס וגירה ואפני זמאנה פיהא. אדא נט̇ר פי כתאבנא הדא יערץ̇ לה מא יערץ̇ למן אטאל אללבת פי אלט̇לם מן אנה לא [יסתטיע][54] אלכרוג̇ דפעה אלי אלצ̇ו. ואמא אדא תכ̇רג̇ פיה פאנה יסתלד בה וידרך תפאצ̇ל אלצ̇יא עלי אלט̇לם וען [מתל][55] הדא [ ][56] אלנבי ע̇א̇ס̇ העם ההולכים בחשך ראו אור גדול، יושבי בארץ צלמות אור נגה עליהם.

0.17 ויג̇ב אן תעלם אנה אדא קלנא אן מרכז כוכב כדא מתחרך עלי דאירה כדא קטבהא עלי דאירה כדא، פאן כל דאירה מנהא מרסומה פי סטח פלך כאץ מן אלאפלאך אלתי עלי אלצפה אלמדכורה. וכדלך אדא סמינא לחרכה אלכוכב אלכ̇אצה לה תקצירא̇ פאן דלך [מנא][57] לכי נתבת אלאצול ולו עלי מדהב אלדין ירון אן פי אלחרכה אלמסתדירה תצ̇אד، ואנת קד עלמת ראינא פי דלך. ואלואג̇ב עלי ראינא אן תסמי מסירא̇ ולא תקצירא̇ ואנת תאכד מא שית מן אלמדהבין פאן אצולנא פי אלאכ̇תלאפאת קד תואפק כלי אלמדהבין.

46. ו: ? .א: מקדים.
47. א: שם.
48. ו: יוג̇דה
49. א: עד עתה.
50. א: להמשך.
51. א: השמש היו שני מיני העקרים.
52. ו: אלטרוק.
53. א: כונתינו.
54. ו: יסתטע.
55. א: כיוצא.
56. היוצא.
57. א: -.

0.18 וקסמנא הדא אלכתאב אלי ג̇מלתין. אלג̇מלה אלאולי אג̇מל פיהא ג̇מיע אלמקדמאת אלהנדסיה וגירה אלואג̇ב תקדימהא עלי אלאכבאר בכיפיה אלחרכאת וכלהא משתרכה בין בטלמיוס ובינגא אד ליס ישובהא כיפיה

54v

חרכה [אלבתה].[58] נאתי בהא באוג̇ז אלטריק ואקרבה אלי אלפהם ופי הדה אלג̇מלה כמסה פצול. אלפצל אלאול פי אלמקדמאת אלמסלמה אלואג̇ב תקדימהא להדא אלעלם ומא ינתג̇ מנהא. אלפצל אלתאני פי ג̇מיע אלמקדמאת [אלהנדסיה][59] אלתי יג̇ב אן תתקדם קבל אלאכבאר בחרכאת אלכואכב. אלפצל אלתאלת פי אלאלאת אלמחתאג̇ה ללארצאד אלתי בהא יתם הדא אלעלם. אלפצל אלראבע פי מערפה אלמיול אלג̇זייה ואלמטאלע פי אלכרה אלמנתצבה ואלמאילה ואשיא אכר ג̇זיה תלזם מנהא. אלפצל אלכאמס פי מקדאר זמאן סנה אלשמס. אלג̇מלה אלתאניה אדכר פיהא כיפיה אלחרכאת ואלאצול אלתי עליהא ינבגי אן יעמל פי אלאכ̇תלאפאת אלמחסוסה ללאג̇ראם אלסמאויה פי חרכאתהא ומואצ̇עהא, ואלי דלך כאן אול אלקצד פי הדא אלכתאב. והנא חין אבתדי בדכר מא ועדת אליה.

A.I.1 אלפצל אלאול מן אלג̇מלה אלאולי אלמקדמאת אלמסתלזמה אלואג̇ב תקדימהא להדא אלעלם. הי כ̇ד̇ מקדמה. א̇ אן אלשמס ואלקמר וסאיר אלנג̇ום תתחרך אבדא̇ מן אלמשרק [אלי אלמגרב][60] עלי דואיר מתואזיה, תשרק תם תרתפע עלינא תם תהבט ותכפי ענה זמאנא̇ תם תרג̇ע אלי מטאלעהא מן ראס ודלך עלי תשאבה דאימא̇. ב̇ אן אלקטב אלשמאלי מרתפע עלי אלאפק ואלג̇נובי מנכפץ̇ ענה פי כל ואחד מן אלמיול פי אלנצף אלשמאלי מן אלארץ̇, ודלך [][61] מיל כל אפק ען כ̇ט אלאסתוא נחו אלשמאל אלא חית אלכרה מנתצבה. ואן אלכואכב אלתי בעדהא ען אלקטב אלשמאלי אקל מן אלארתפאע ליסת תגרב אבדא̇ בל תדור חול אלקטב עלי דואיר מתואזיה עלי קדר בעד אלכואכב אלדאירה עליהא מן אלקטב. ג̇ אן אלכואכב אלקריבה מן אלאבדיה אלט̇הור תמכת פי אלגיבובה אקל מן אלבעידה מנהא. ד̇ אן עט̇ם אלכואכב ונורהא הי אבדא̇ עלי חאל ואחד אלא אנהא תעט̇ם ענד אלגרוב ודלך לאג̇ל בכ̇אר אלרטובה [][62] אלדאכ̇ל חיניד בין אלבצר ובינהא. ה̇ אנהא תנקטע בבסיט

55r

אלארץ̇ ענד אלגרוב. ו̇ אן אלאבעאד אלתי מן אלארץ̇ אלי אלמואצ̇ע אלעלויה מתסאויה [][63]. ז̇ צורה אלג̇סם ינבגי אן תכון תאבעא̇ לחרכתה. ח̇ אן חרכה אלסמא אסלס אלחרכאת ואסהלהא. ט̇ אן אלג̇סם אלאסלס חרכה הו אלכרי. י̇ אן אלאשכאל אלמכ̇תלפה אלתי אחאטתהא מתסאויה מא כאן מנהא אעט̇ם זואיא פהו אעט̇ם מסאחה. יא̇ אן ג̇רם אלסמא יג̇ב אן יכון אעט̇ם ממא סואהא. יב̇ אן ג̇רם אלפלך אשד שבהא̇ בעצ̇ה לבעץ̇ מן ג̇מיע אלאג̇סאם. יג̇ אן אלאג̇ראם אלסמאויה מתל אלכואכב תט̇הר מסתדירה מן נואחי מכ̇תלפה פי וקת ואחד ענד כל מן יראהא [][64]. יד̇ אן

58. א:-.
59. א: הלמודיות.
60. א: -.
61. א: בשיעור.
62. א: הארץ.
63. א: וזאת תתבאר בהקדמה הרביעית וכאן הן הקדמה אחת.
64. א: וטבע המקיף אותם דומה לטבעם.

אלקיאסאת באלאלאת [][65] תתפק אלא בהדא אלוג̇ה [][66]. [טו̇][67] אן טלוע ג̇מיע אלכואכב וגרובהא
עלי אלמשרקיין יתקדם טלועהא וגרובהא עלי [אלמגרביין][68]. יו̇ אן מן כתב [][69] מן [אלמשרקיין][70]
וקת כסוף קמרי אלדי הו פי וקת ואחד ללג̇מיע [][71] אכתר תאכרא̇ ען נצף אלנהאר ממן כתבה מן
[אלמגרביין][72] [][73] ודלך בקדר אכ̇תלאף אלבעד בינהמא. יז̇ כל מן סאר פי אלבחאר נחו אלג̇באל או
מוצ̇ע משרף כלמא קרב אליה יראה יתזיד [][74] וכאנה נאבת מן אלמא. יח̇ אן פי כל מוצ̇ע מן אלארץ̇
יסתוי אלנהאר ואלליל מרתין פי אלסנה והדא אלאסתוא יכון ענד אלמג̇אז אלאוסט בין אלאנקלאבין
אעני בין אלנהאר אלאקצר ואלאטול. יט̇ אן נצף אלנהאר הו אבדא̇ ענד כון אלשמס פי וסט אלסמא
אעני פי אלדאירה אלסמתיה אלקאימה עלי מעאדל אלנהאר. כ̇ אן כל ואחד מן אלאפאק אעני אלסטח
אלכ̇ארג̇ מן אלבצר יקטע אלכרה בנצפין ודלך לאנא נרי דאימא̇ סתה ברוג̇ ט̇אהרה תם תטלע אלסתה
אלג̇איבה ותגרב תלך. כא̇ אן ט̇ל אלמקיאס אלשרקי מע ט̇ל אלמקיאס אלגרבי יקעאן ג̇מיעא̇ עלי כ̇ט
ואחד מסתקים פי אסתוא אלנהאר [][75]. כב̇ אן אלכסופאת אלקמריה פי כל נואחי אלסמא אנמא תכון
ענד אלמקאבלה עלי אלקטר. כג̇ אן חכם מקיאס אלט̇ל פי אי נאחיה פרץ̇ מן סטח אלארץ̇ כחכמה לו
פרץ̇ פי מרכז אלארץ̇ וכדלך מראכז אלאלאת [אלמסתעמלה][76] פי הדא אלעלם. כד̇ אן אלאנסאן אדא

55v

תחרך בחרכה אסרע מן סאיר אלחרכאת אלמוגו̇דה פואג̇ב לה אן לא ירי שי מתחרך אלי ג̇הה חרכתה,
לכנה ירי כל שי כאנה מתחרך אלי כ̇לאף חרכתה.

A.I.2 פאדא כאנת תלך אלאשיא מפרוצ̇ה, פבין ממא קלנאה אן אלסמא כריה וחרכתה איצ̇א כריה ודלך
יתבין באלמקדמה אלא̇ ובאלב̇ ובאלג̇ ובאלד̇ ובאלה̇ ובאלו̇ ובאלז̇ ובאלח̇ ובאלתאסעה ובאלי̇ ובאליא̇
ובאליב̇ ובאליג̇ ובאליד̇. פאן אלקיאס אנמא יתם אמא במקדמה ואחדה מן הדה עלי אן אלתאניה
מעלומה ואמא במקדמתין אג̇מע כאלח̇ ואלט̇ ואלי̇ מע אלי̇[ז̇][77]. ומן בעץ̇ הדה אלמקדמאת יתבין
אלמטלוב אן אלחרכה כריה מן קבל אעראץ̇ לאחקה להא ומן בעצ̇הא יתבין דלך מן קבל אן אלצורה
כריה מן קבל כיפיה אלחרכה. ומן בעצ̇הא יתבין אלמטלוב בקיאס מסתקים ומן בעצ̇הא בקיאס
כ̇לף וכל דלך ט̇אהר פיהא. ויתבין אן אלארץ̇ בג̇מיע אג̇זאיהא כריה ענד אלחס באלקיאס אלי אלכל
באלמקדמה אלב̇ ובאלט̇ו ובאליו̇ ובאל [כב̇][78]. [][79] פאן בבעץ̇ הדה אלמקדמאת יתבין אנהא ליסת
בזאילה אלי פוק או אספל. ובבעצ̇הא אנהא ליסת בזאילה נחו אלגרב או אלשרק ופי בעצ̇הא אנהא
ליסת בזאילה נחו ואחד מן אלקטבין. ובדלך יתבטל אלראי אלמרכב מן הדה, ויתבין אן אלארץ̇ ליסת

65. א: לא.
66. א: רצוני לומר בשום התמונה כדורית.
67. א: -.
68. ו: אלמגרבין. א: המערביים
69. א: מי שכתב.
70. ו: אלמשרקין. א: המזרחיים.
71. א: ימצא.
72. ו: אלמגרבין.
73. א: יו̇.
74. א: בגובהו.
75. א: והלילה.
76. ו: אלמסתמעמלה. א: אשר ישתמשו.
77. א: א̇.
78. א: יז̇.
79. א: ויתבאר כי הארץ באמצע השמים בהקדמה הד̇ ובה̇ ובכ̇ ובכא̇ ובכב̇.

להא חרכה אנתקאל באלמקדמאת אלתי תבינת מנהא אן אלארץ̇ פי אלוסט. ויתבין אנהא ליסת במתחרכה חול מרכזהא ואלסמא [][80] כמא קד ט̇ן קום באלמקדמה אלכ̇ד̇.

A.I.3 ויתבין אן אצנאף אלחרכאת אלאֻול פי אלסמא אתנתין לאנא נג̇ד ג̇מיע מא פי אלסמא יטלע ויתוסט אלסמא ויגרב פי כל יום עלי דואיר מתואזיה אלתי אעט̇מהא [][81] מעאדל אלנהאר אד אלאפק יקסמהא בנצפין והדה אלחרכה אלאולי אלמדירה ללכל באסתוא. ונג̇ד אלשמס ואלקמר ואלכ̇מסה אלמתחירה תתחרך נחו אלמשרק, אעני כלאף אלאולי, חרכאת מכ̇תלפה או תקציר עלי אחד אלמדהבין כמא דכרנא, ובאנתקאלהא נחו אלמשרק או תקצירהא [תנתקל][82] איצ̇א נחו אלשמאל ואלג̇נוב אנתקאל מתשאבה. [][83] פחכמנא עלי אן הדה אלחרכה אלתאניה הי עלי קטבי דאירה מאילה

56r

עו מעאדל אלנהאר ותסמי נטאק אלברוג̇ וחרכה אלשמס תרסמהא ועליהא. ועו ג̇אנבתיהא [ממר][84] אלקמר ואלכואכב אלכמסה אלמתחיירה נחו אלשמאל ואלג̇נוב. ונתוהם דאירה תמר בקטבי אלכל ובקטביהא, והי תקטעהא עלי זואיא קאימה עלי נקטתין, אלשמאליה עו מעאדל אלנהאר תסמי אלאנקלאב אלציפי ואלתאני שתוי. ונקטה תקאטע אלמאילה מע מעאדל אלנהאר אלתי אלממר בהא אלי אלשמאל תסמי אסתוא רביעי ואלתאניה [][85] כ̇ריפי, פקטבי אלמאילה לאזמאן מואצ̇עהא מן אלמארה באלאקטאב מע אנהמא מתחרכאן באלחרכה אלאולי. ודאירה נצף אלנהאר הי דאירה קאימה עלי אלאפק אבדא̇ מארה בקטבי אלכל.

A.I.4 ואד פרגנא מן דלך פלנשרע פי אלבראהין אלג̇זיה ונבני אלאמר עלימא קד אצטלח עליה מן אלג̇מיע והו תג̇זיה מחיט אלדאירה עלי ש̇ס̇ ג̇ז ותג̇זיה אלקטר אלי ק̇כ̇ ג̇ז לאג̇ל סהולה תלך אלאעדאד ענד אלחסאב, וכדלך תג̇זיה אלג̇ז אלי סתין דקיקה ואלדקיקה אלי ס̇ תאניה וכדלך אלי גיר נהאיה אלי חית לא יצ̇ר תרך אלתג̇זיה.

A.II.1 אלפצל אלתאני פי ג̇מיע אלמקדמאת אלהנדסיה אלתי יג̇ב אן תתקדם קבל אלאכ̇באר בחרכאת אלכואכב. ולנקדם אולא̇ אלקול פי מערפה אקדאר אותאר אג̇זא אלדאירה. ולנקדם לדלך אדא ארדנא אן נקסם עדד מעלום עלי נסבה דאת וסט וטרפין פנאכ̇ד מרבעה ומרבע נצפה ונג̇מעהם. ונאכ̇ד ג̇דר אלמג̇תמע וננקץ ענה נצף אלעדד ואלבאקי הו אלקסם אלאכבר. ולנצ̇ע אלמתאל פי אלכ̇טוט. וליכון אלעדד אלמעלום ב̇ג̇ וליכן א̇ג̇ מתל נצף ב̇ג̇ ויחיט מע כ̇ט ג̇ב̇ בקאימה. ונצל א̇ב̇ וננקץ מן ב̇א̇ נצף ב̇ג̇ וליכן ב̇ד̇. ונפצל מן ב̇ג̇ מתל ד̇א̇ וליכן ג̇ה̇ פאקול אן ג̇ה̇ הו אלקסם אלאכבר. ברהאנה לאן צ̇רב ב̇ג̇ פי נפסה בל פי ב̇ה̇ ופי ה̇ג̇ וג̇א̇ פי נפסה מתל ב̇א̇ פי נפסה בל ב̇ד̇ פי נפסה וד̇א̇ פי נפסה וצ̇עף ב̇ד̇ פי ד̇א̇. נסקט ב̇ד̇ פי נפסה. אעני ג̇א̇ פי נפסה. פיבקי ב̇ג̇ פי ב̇ה̇ ופי ה̇ג̇ מתל ד̇א̇ אעני ה̇ג̇ פי נפסה וצ̇עף ב̇ד̇ פי ד̇א̇ אעני ב̇ג̇ פי ה̇ג̇. יבקי ב̇ג̇ פי ב̇ה̇ מתל ה̇ג̇ פי נפסה וד̇מ̇א̇. והדא בעינה קד ברהנה אקלידס פי אלתאניה מן כתאבה בברהאן קותה קוה הדא אלברהאן.

80. א: שוקטים.
81. א: גלגל.
82. ו: תנתקאל. א: יעתקו.
83. א: ולפי זה.
84. ו: ממאר. א: מעבר.
85. א: השואת.

56v

A.II.2 פאד קדמנא דלך פאדא̇ אדא עלמנא קדר צ̇לע אלמסדס אלואקע פי אלדאירה אלדי הו ותר קוס ס̇ ג̇ז̇، והו מעלום לאנה נצף קטר אלדי בקיאסה נריד אן נעלם אקדאר אלאותאר، פקד נעלם [][86] קדר צ̇לע אלמעשר אלדי פי הדה אלדאירה، והו ותר קוס ל̇ו̇ ג̇ז̇ אד הו אלג̇ז̇ אלאכבר מנה כמא תבין פי אלט̇ו̇ מן אקלידס. ולנקדם איצ̇א אדא כאן פי דאירה שכל דו ארבעה אצ̇לאע כג̇ד̇ב̇א̇ פאן צ̇רב ב̇ד̇ פי ג̇א̇ מתל צ̇רב ד̇א̇ ג̇ב̇ וב̇א̇ פי ג̇ד̇ מג̇מועין. פנג̇על זאויה א̇ב̇ה̇ מתל ד̇א̇ג̇ אן כאנת א̇ב̇ד̇ אעט̇ם מנהא، פא̇ב̇ד̇ מתל ה̇ב̇ג̇ וב̇ד̇א̇ מתל [ב̇ג̇ה̇][87] פמתלתא א̇ב̇ד̇ ב̇ג̇ה̇ מתשאבהאן. פנסבה ב̇ג̇ אלי [ג̇ה̇][88] כב̇ד̇ אלי ד̇א̇ פצ̇רב ב̇ג̇ פי א̇ד̇ מתל צ̇רב ב̇ד̇ פי ג̇ה̇. וכדלך מתלתא א̇ה̇ב̇ ב̇ג̇ד̇ מתשאבהאן פצ̇רב ב̇א̇ פי ד̇ג̇ מתל צ̇רב ב̇ד̇ פי ה̇א̇. פצ̇רב ב̇ד̇ פי א̇ג̇ כלה מתל צ̇רב א̇ב̇ פי ד̇ג̇ וא̇ד̇ פי ג̇ב̇ מג̇מועין. פאדא כאנת א̇ב̇ א̇ג̇ מעטיאן פב̇ג̇ אלדי בינהמא מעטי לאן ב̇ד̇ ג̇ד̇ מעטיאן לתמאם נצפי אלדאירתין. וא̇ד̇ אלקטר מעטי פג̇ב̇ מעטי. פאדא̇ קד נעלם ותר כ̇ד̇ ג̇ז̇ מן אלתפאצ̇ל בין צ̇לע אלמסדס וצ̇לע אלמעשר.

A.II.3 ואקול איצ̇א אנה אן עלמנא ותר קוס א̇ג̇ פקד נעלם ותר נצפהא וליכן א̇ד̇ לאן א̇ה̇ נצפה מעטי וא̇ז̇ מעטי פה̇ז̇ מעטי. יבקי ה̇ד̇ מעטי וא̇ד̇ איצ̇א. פאדא̇ קד נעלם ותר י̇ב̇ ג̇ז̇ וותר סתה וותר תלאתה וותר ג̇ז̇ ונצף וותר נצף ורבע [בתנציף][89]. ויכ̇רג̇ בהדא אלטריק אן ותר ג̇ז̇ ונצף יכון ג̇ז̇ ול̇ד̇ דקיקה וט̇ו̇ תאניה וותר נצף ורבע יכון [מ̇ז̇][90] דקי̇ וח̇ תואני. ועלמנא בדלך כל אלאותאר [אלמתפאצ̇לה][91] בג̇ז̇ ונצף באן נבתדי מן אלתפאצ̇ל בין ג̇ז̇ ונצף ובין נצף אלדאירה.

A.II.4 ונחתאל [אלאן][92] פי וג̇וד ותר ג̇ז̇ ואחד מן קבל [][93] ג̇ז̇ ונצף [ומן קבל ג̇ז̇][94] ורבע אד ליס ימכננא אן [נברהן][95] דלך עלי טריק אלכ̇טוט. ויכון עלי הדה אלצפה. אקול אנא אדא אכ̇רג̇נא פי דאירה כ̇טין גיר מתסאויין אן נסבה אלאטול אלי אלאקצר אצגר מן נסבה אלקוס אלדי עלי אלאטול אלי אלקוס אלדי עלי אלאקצר، ודלך קד ברהנה [][96] אקלידס. פלנצ̇ע

57r

א̇ג̇ ותר ג̇ז̇ וא̇ד̇ ותר ג̇ז̇ ונצף ורבע. פכ̇ט א̇ג̇ אקל מן מתל ותלת כ̇ט א̇ד̇ [][97] אד קוס א̇ג̇ מתל ותלת א̇ד̇. פכ̇ט א̇ג̇ אקל מן ג̇ז̇ ודקיקתין ונ̇ תאניה ומ̇ תאלתה. ואיצ̇א ננזל א̇ד̇ ותר ג̇ז̇ וא̇ג̇ ותר ג̇ז̇ ונצף פכ̇ט ג̇א̇ אקל מן מתל ונצף כ̇ט א̇ד̇، פכ̇ט א̇ד̇ אכתר מן ג̇ז̇ ודקיקתין ונ̇ תאניה אעני תלתי א̇ג̇، פאלאקל ואלאכתר יקסמאן [][98] אלתאלתה. פותר ג̇ז̇ מן קוס הו ג̇ז̇ ודקיקתאן ונ̇ תאני̇ וכ̇ תאלתה. פאדא̇ קד עלמנא אלאותאר כלהא אלמתפאצ̇לה בנצף ג̇ז̇ לאן קד נעלם ותר נצף אלדאירה אעני אלקטר. ונעלם ותר נצף

86. א: ג̇כ̇.
87. א، ו: ג̇ד̇ה̇.
88. ו: ג̇ז̇. א: ג̇ו̇. Euclid (ed., trans. and intro. Heath), *The Elements*, vol. 2, p. 225, reads GE
89. א: -.
90. א: מ̇ח̇.
91. ו: אלמתפצ̇לה. א: העודפים.
92. א: אחר זה.
93. א: ידיעת.
94. א: וחצי.
95. ו: נברהאן. א: להביא מופת.
96. א: בספר.
97. א: יחד.
98. א: המ̇.

גז، פקד נעלם ותר קע̇ט̇ גז ונצף מן אלתפאצ̇ל אלדי בינהמא، וכדלך נעלם סאירהא בתנקץ נצף גז. וקד
נעלם איצ̇א ג̇יב כל קוס לאן אלג̇יב הו נצף ותר צ̇עף אלקוס ודמ̇א.

A.II.5 ונקדם איצ̇א אדא וקעת פי דאירה כדאירה אב̇ תלאת נקט כאב̇ג̇ עלי אן תכון כל ואחד מן קוסא ג̇ב̇
ב̇א̇ אצגר מן נצף דאירה، ונצל אה̇ג̇ ודה̇ב̇ מן אלמרכז פאקול אן נסבה ג̇יב קוס אב̇ אעני אס̇ אלעמוד
[A.II.5/א [עלי][99] דב̇ אלי ג̇יב קוס בג̇، אעני ג̇ח̇ אלעמוד، איצ̇א כנסבה אה̇ אלי הג̇ לאן אלמתלתין מתשאבהאן.

A.II.6 ולנקדם איצ̇א מא יחתאג אלי תקדימה ללבראהין עלי אלמעאני אלכריה. כל דאירתאן עט̇ימתאן
פי בסיט כרה כאד̇ב̇ג̇ ואל̇ב̇ תפצל מן אחדהמא קוסין כאד̇ אג̇ מן אחדי נקטתי אלתקאטע، וכל ואחד
מנהמא אצגר מן נצף דאירה. ואכ̇רג̇נא מן נקטתי דג̇ עמודין עלי סטח אלדאירה אלאכ̇רי. פאקול
אן נסבה [][100] ג̇יב קוס אג̇ כנסבה אלעמוד אלכ̇ארג̇ מן נקטה ד אלי אלאכ̇ר. ברהאנה ליכן אלפצל
אלמשתרך ללדאירתין אעני קטריהמא כ̇ט אב̇ ונכ̇רג̇ עליה עמודי דז̇ גה̇ והמא ג̇יוב אלקוסין. פאן כאנא
איצ̇א עמודין עלי סטח דאירה אל̇ב̇ פקד בינא מא ארדנא. ואן לם [יכונא כדלך][101] פנכ̇רג̇ מן נקטתי דג̇
עמודי דט̇ גח̇ עלי סטח דאירה אל̇ב̇. ונצל הח̇ טז̇ פלאן כ̇טי דז̇ גה̇ מתואזיאן، וכדלך דט̇ גח̇، פזאויתא
זדט̇ הגח̇

57v
מתסאויתאן וח̇ וט̇ קאימתאן. פאלמתלתאן מתשאבהאן פנסבה דז̇ אלי גה̇ אעני אלג̇יוב כנסבה דט̇ אלי
גח̇ אעני אלאעמדה ודמ̇א. וכדלך יתבין לו פצלא אלקוסין מן גהה ואחדה מן נקטה אלתקאטע.

A.II.7 פאד קד תקדם דלך [פלינקטעא][102] פימא בין קוסי אב̇ אג̇ מן דואיר עט̇אם והמא פי בסיט כרה קוסא
גד̇ בה̇ מן דואיר עט̇אם איצ̇א עלי נקטה ז̇. אקול אן נסבה ג̇יב קוס גה̇ אלי ג̇יב קוס הא̇ מולפה מן נסבה
ג̇יב קוס גז̇ אלי ג̇יב קוס זד̇ ומן נסבה קוס בד̇ עלי ג̇יב קוס בא̇. פלנכ̇רג̇ מן נקט גאד̇ אעמדה עלי סטח
דאירה בזה̇ והי גל̇ אח̇ דט̇. פלאן כל תלאתה כ̇טוט، פאן נסבה אלאול אלי אלתאלת מולפה מן נסבה
אלאול אלי אלתאני ומן נסבה אלתאני אלי אלתאלת، פנג̇על עמוד דט̇ וסט פי אלנסבה בין גל̇ ואח̇.
פנסבה עמוד גל̇ אלי עמוד אח̇ [אעני][103] נסבה ג̇יב קוס בה̇ אלי ג̇יב קוס הא̇ מולפה מן נסבה גל̇ אלי
דט̇ [אעני][104] נסבה ג̇יב קוס גז̇ אלי ג̇יב קוס [גד̇][105] ומן נסבה דט̇ אלי אח̇ אעני ג̇יב קוס בד̇ אלי ג̇יב
קוס בא̇. וכדלך יתבין איצ̇א אן נסבה ג̇יב קוס גז̇ אלי ג̇יב קוס זד̇ מולפה מן נסבה ג̇יב קוס גה̇ אלי ג̇יב
קוס הא̇ ומן נסבה ג̇יב קוס אב̇ אלי ג̇יב קוס בד̇ אדא ג̇עלנא עמוד אח̇ וסט גל̇ ודט̇. וכדלך יתבין עלי
אלתרכיב אן נסבה ג̇יב קוס גא̇ אלי ג̇יב קוס אה̇ מולפה מן נסבה ג̇יב קוס גד̇ אלי ג̇יב קוס דז̇ ומן נסבה
ג̇יב קוס זב̇ אלי ג̇יב קוס בה̇ באן נכ̇רג̇ אלאעמדה מן נקט אה̇ז̇ עלי סטח דאירה בד̇א̇. וכדלך יתבין
[A.II.7/א אלתרכיב פי כל ואחד מן אלאצ̇לאע. והדא אלברהאן אנמא הו עלי אן יכון כל צ̇לע מע קרינה פי קוס
ואחדה. ויתבין סאיר [אלנסב][106] אלתי תאתלף מן הדה אלג̇יוב אלסתה באלברהאן אלדי אתי בה אבן
מעאד אלג̇יאני והו הכדא:

99. א: וגם.
100. א: בקע קשת אד̇ אצל.
101. א: -.
102. ו: פלינקאטעא. א: יתחתכו.
103. א: -.
104. א: -.
105. א: זד̇.
106. ו: אלנסאב. א: הערכים.

A.II.8 אקול אן צ̇רב אלג̇יב אלאול פי אלראבע ומא יג̇תמע פי אלסאדס אעני אלמג̇סם אלדי יחיט בה הדה אלג̇יוב אלתלאתה מתל צ̇רב אלתאני פי אלתאלת ומא יג̇תמע פי אלכ̇אמס. פנצ̇ע כ̇ט ג̇ב̇ מתל אלג̇יב אלראבע ובח̇ מתל אלסאדס, יחיט מע ג̇ב̇ בקאימה. ונכ̇רג̇ מן נקטה ב̇ כ̇ט בא̇ קאים עלי סטח אלכ̇טין, ויכון מתל אלג̇יב אלאול, ונתמם סטחי

58r

ג̇ח̇ ג̇א̇ [][107] אלמתואזיה [אלאצ̇לאע][108] יחיטאן במג̇סם ד̇ח̇. ונעמל מג̇סם [מ̇ה̇][109] איצ̇א עלי אן יכון כ̇ט מ̇ל̇ מתל אלג̇יב אלתאלת, ול̇ס̇ מתל אלכ̇אמס ולכ̇ מתל אלתאני. פלאן נסבה סטח מ̇ס̇ אלי סטח ג̇ח̇ מולפה מן למ̇ אלי ג̇ב̇ ומן [לה̇][110] אלי בח̇ פקאעדתי מ̇ס̇ ג̇ח̇ [מכאפיתין][111] ללארתפאעין פאלמג̇סם מסאוי ללמג̇סם.

A.II.9 ואד קדמנא [][112] פנקול אן נסבה אי צ̇לע אֻכ̇ד מן אלמג̇סם אלואחד אלי אי [צ̇לע][113] אֻכ̇ד מן אלאכר מולפה מן אלארבע אלבאקיה, מתל אן נסבה בא̇ והו אלג̇יב אלאול אלי למ̇ והו אלג̇יב אלתאלת אעני ג̇יב קוס ג̇ז̇ פי אלצורה אלמתקדמה מולפה מן נסבה כ̇ט לכ̇, והו אלתאני, אלי [ג̇יב][114] בג̇, והו אלראבע, ומן נסבה [לה̇][115] והו אלכ̇אמס אלי בח̇ והו אלסאדס. ודלך לאן נסבה סטח כ̇ס̇ אלי סטח ג̇ח̇, בל נסבה כ̇ט בא̇ אלי כ̇ט למ̇, [ללתכאפו][116], מולפה מן הדה אלארבע נסב אלמדכורה. וכדלך יתבין סאירהא. ואמא אדא אֻכ̇ד מע קרינה מן דלך אלמג̇סם פליסת תאתלף מן אלבאקיה ודמ̇א̇.

A.II.10 ונקול נחן אן כל ארבע אקדאר פאן אלנסבה אלמולפה מן אלאול אלי אלתאני ומן אלתאלת אלי אלראבע הי בעינהא אלנסבה אלתי תאתלף מן אלאול אלי אלראבע ומן אלתאלת אלי אלתאני. מתאלה ליכון סטחין מתואזיה אלאצ̇לאע, והמא סטחא [ד̇ג̇][117] אז̇, זאויה ב̇ משתרכה להמא. פנסבה סטח אז̇ אלי סטח ד̇ג̇ מולפה מן נסבה כ̇ט אב̇ אלאול אלי בג̇ אלתאני ומן נסבה זב̇ אלתאלת אלי בד̇ אלראבע. אקול אנהא מולפה איצ̇א מן נסבה אב̇ אלאול אלי בד̇ אלראבע ומן נסבה זב̇ אלתאלת אלי בג̇ אלתאני. ברהאנה לנכ̇רג̇ בז̇ עלי אסתקאמה חתי יכון מתל בא̇ וליכון בה̇ ונעמל בח̇ מתל בז̇ ונתמם סטח חה̇. פתכון נסבה סטח הח̇ אעני אז̇ אלי סטח ד̇ג̇ מולפה מן נסבה צ̇לע הב̇ אעני אב̇ אלי דב̇ ומן נסבה חב̇ אעני בז̇ אלי בג̇ ודמ̇א̇.

A.II.11 פיט̇הר מן דלך אן כל נסבה תאתלף מן ארבעה מן תלך אלאקדאר פאנהא תתאלף מנהא

58v

עלי וג̇הין. פנעמל לדלך ג̇דולא̇ צגירא̇ ונג̇על פי אלצף אלאול עדד אלנסב אלמתאלפה אלתי ימכן אתלאפהא פי אלסתה אקדאר. ופי אלצף אלתאני ואלתאלת אלנסב אלתי ימכן אן תאתלף מן אלבאקיה ואלתי לא ימכן. ופי אלארבע צפוף אלבאקיה נג̇על באזא כל נסבה מולפה אלנסב אלתי

107. א: אח̇.
108. ו: אלאצ̇לע. א: הצלעות
109. א: מ̇ס̇.
110. א: ל̇ס̇.
111. א: מתהפכות בערך.
112. א: זה.
113. ו: וצ̇ע. א: צלע.
114. א: -.
115. א: ל̇ס̇.
116. א: להפוך הערך.
117. א: כ̇ג̇.

מנהא איתלאפלאת פי אלתי פיהא איתלאף. ואלתי לא ימכן פיהא דלך נתרכהא וליס נכתב פיהא [חרפא̇][118] אצלא̇. וכל נסבה מולפה מן הדה קד ימכן אן תנעכס חתי יכון עדד אלנסב אלמולפה צ̇עף אלמכתובה פי אלג̇דול. והכדא רסם אלג̇דול.

A.II.12 פאדא כאן ואחד מן הדה אלסתה מג̇הולא̇ פאנא נסתכ̇רג̇ה אד קד עלמנא אן צ̇רב אלאול פי אלראבע ומא יג̇תמע פי אלסאדס מתל צ̇רב אלתאני פי אלתאלת תם פי אלכ̇אמס. [פאדא כאן מתלא̇ אלמג̇הול אלסאדס צ̇רב אלתאני פי אלתאלת תם פי אלראבע][119], ונקסם אלמג̇תמע עלי צ̇רב אלאול פי אלראבע, יכ̇רג̇ אלסאדס ועלי הדא אלמתאל ינבגי אן נעמל פי אלבאקיה.

A.III.1 אלפצל אלתאלת פי אלאלאת אלמחתאג̇ה ללארצאד אלתי בהא יתם הדא אלעלם. פמן אג̇ל תבין מקדאר מיל דאירה אלברוג̇ ען מעאדל אלנהאר אעני מיל כל ואחד מן אלאנקלאבין ענהא, והדא הו מסאו למא בין אלקטבין, נעמל חלקתין מן נחאס

59r

מרבעתי אלבסיט ונדכ̇ל אלצגרי פי ג̇וף אלכברי בחית תדור פי באטנהא נחו אלשמאל ואלג̇נוב. ונקסם אלכברי בש̇ס̇ ג̇ז̇ ונג̇על עלי קטר אלצגרי שטבתין [מתואג̇התין][120] פי וסטיהמא [לסאנין][121] ילקיאן וג̇ה אלכ̇ארג̇ה. ונקים אלחלקתין עלי עמוד קאים עלי אלאפק ועלי כ̇ט̇ נצף אלנהאר לכי תכונא פי סטח דאירה נצף אלנהאר. פאדא אדרנא אלחלקה אלדאכ̇לה חתי תסתט̇ל אלשטבה אלספלי באלעליא ענד כון אלשמס פי נצף אלנהאר, דלנא טרף אלמקיאס עלי מקדאר בעד מרכז אלשמס מן נקטה סמת אלראס פי דאירה נצף אלנהאר. פבהדה אלאלה עלמנא מקדאר אלקוס אלתי בין סמת אלראס [ובין][122] כל ואחד מן אלאנקלאבין. פאדא̇ מא בין אלאנקלאבין מעלום. וקד עמלוא לדלך בעינה אלה תאניה באן רסמוא פי לבנה מן חג̇ר או כ̇שב רבע דאירה יסתכ̇רג̇וא̇ בהא מא יכ̇רג̇ מן תלך אלאלה אלמדכורה [לא גיר][123]. ולדלך לם נדכרהא אד קצדנא אלאיג̇אז במא לא יודי אלי נקצאן פי הדא אלעלם.

A.III.2 אלאלה אלתאניה והי אלתי תסמי דאת [אלחלק][124]. נתכד קרנין מרבעי אלסטוח מתסאויתין ונרכב אחדהמא באלאכ̇רי עלי זואיא קאימה עלי אן אחדהא דאירה אלברוג̇. ואותדנא עלי קטביהא ותדין ורכבנא עליהמא חלקה מן כ̇ארג̇ ימאסהא, ותכון סלסה אלמדאר פי אלטול, ואכרי מן דאכ̇ל תדור איצ̇א̇ פי אלטול. וקסמנא הדה אלדאכ̇לה [ופלך][125] אלברוג̇ אלי אג̇זאיהא. ורכבנא פי אלדאכ̇לה חלקה לטיפה תדור נחו אלקטבין פי סטחהא לאג̇ל רצד אלערץ̇ ופיהא תקבאן יתקטראן נאשזאן. ורכבנא פי קטבי דאירה מעאדל אלנהאר חלקה כבירה מן כ̇ארג̇ חלקה אלקיאס לכי אדא אצבנאהא מואזיה [][126] לנצף אלנהאר אלחקיקי עלי ארתפאע קטב אלמוצ̇ע אלדי נחן פיה, כאן מדאר [אלחלקה אלדאכ̇לה][127] חול הדה אלאקטאב תאבע לחרכה אלכל אלאולי.

118. ו: חרף̇.
119. א: -.
120. א: קטנות.
121. ו: אלסאנין. א: שתי יתדות.
122. א: -.
123. א: לא פחות ולא יתר.
124. ו: אלחלאק.
125. א: ועגולת.
126. א: לעגולת.
127. א: הטבעות הפנימיות.

A.III.3 פמתי ארדנא נרצד בהדה אלאלה ג̇עלנא אלחלקה אלכ̇ארג̇ה אלתי עלי קטב אלברוג̇ עלי ג̇ז אלשמס ואדרנאהא חתי [יציר][128] תקאטע אלחלקתין בחדא אלשמס אעני חין תסתט̇ל באנפסהא. ואן ג̇עלנא בדל אלשמס כוכב תאבת נעלם ג̇זה

59v

ג̇עלנא איצ̇א אלתקאטע בחדאיה בוצ̇ע אלבצר פי אחדי ג̇אנבי [אלחלקה][129] חתי ירי אלכוכב כאנה לאצק בבסיטאתהא. תם נדיר אלחלקה אלדאכ̇לה נחו אלקמר או אי כוכב נריד [קיאסה][130] חתי נראה באלתקבין אלמתקאטרין. פבדלך נעלם מוצ̇ע אלכוכב פי אלטול באג̇זא דאירה אלברוג̇ ובאלערץ̇ באג̇זא אלדאכ̇לה.

A.III.4 אלאלה אלתאלתה והי אלתי תסמי דאת אלמסאטר [][131]. נתכ̇ד מסטרתין מרבעתין טולהמא ליס באקל מן ארבעה אדראע ובתכן מא לא יתעוג̇אן، תם נתקבהמא ענד אלטרפין פי וסטי כ̇טין נכ̇טטהמא פי וסטי ערצ̇יהמא. ונרתבטהמא במחור ידכ̇ל פי אלתקבין בחית ימכן אן תדור אחדהמא עלי אלמחור תכון אלתאניה מרכוזה פי קאעדה، וננצבהא [מרכוזה כדלך][132] קאימה עלי סטח אלאפק מואזיה לכ̇ט נצף אלנהאר [][133] עלי אן [נצור אנצ̇מאם][134] אחדהמא באלאכ̇ר מואג̇הא̇ ללג̇נוב. ונרכב פי אלמסטרה אלגיר מרכוזה שטבתין פי וסטיהמא תקבין עלי וסט אלכ̇ט אלמוסט ללעצ̇אדה، ואלתקב אלדי ילי עין אלנאט̇ר אצגר מן אלתאני. ונג̇על פי וסטי אלכ̇טין אלמוסטין נקטתין בעדהמא מן מרכז אלמחור באלסוא ואכתר מא ימכן. ונקסם אלכ̇ט אלמנפרז פי דאת אלקאעדה בס̇ ג̇ז. ונרכב פי אלנקטה אלתי פי אלמסטרה אלמרכוזה מסטרה לטיפה סלסה אלמדאר טולהא בחית תבלג טרף אלאכ̇רי ענד תפארקהא ללמרכוזה. פמתי ארדנא אן נרצד בעד אלקמר אלמרי מן נקטה סמת אלראס אדרנא אלמסטרה דאת אל[][135] פי סטח דאירה נצף אלנהאר، אדא כאן אלקמר פי דאירה נצף אלנהאר، אלי אן ירי אלקמר כלה. [ונתעלם][136] פי אלמסטרה אללטיפה אלבעד [אלדי חצל][137] בין טרפי כ̇טי אלמסאטר אלמתסאויתין ונצ̇עה עלי אלכ̇ט אלמקסום בסתין. ונאכ̇ד אלקוס אלדי עליה והו בעד מרכז אלקמר אלמרי מן נקטה אלסמת [פי אלפלך][138] אלמכ̇טוט עלי קטבי אלאפק והדא כלה קד ימכן אדראכה בדאת [אלחלק][139].

A.III.5 אלאלה אלראבעה והי אלתי תסמי דאת אלשטבה אלסיארה. נעמל מסטרה טולהא ארבעה אדראע ונכ̇ט פי וסטהא כ̇ט ונג̇על פי טרפהא ותד צגיר. תם נרכב פי הדה אלמסטרה שטבה בחית ימכנהא אן תתחרך עלי טול אל

128. ו: תציר.
129. א: הטבעות.
130. א: לדעת מקומו.
131. א: בערבי דאת אלמסאטר.
132. א: -.
133. א: נעוץ בתושבת.
134. א: שיהיה דבוק.
135. ו: ?. א: השתי ידות.
136. א: ונרשום.
137. א: הנפרש.
138. א: בעגולה.
139. ו: אלחלאק.

60r

מסטרה מע כונהא קאימה אבדא̇ עלי סטח אלמסטרה. ונג̇על פי וסט אלשטבה תקב מסתדיר יכון בעד חרפה אלאספל ען סטח אלמסטרה מתל בעד ראס אלותד ענהא. תם נקסם אלכ̇ט אלדי פי וסט אלמסטרה באלאג̇זא אלתי שינא בחית תכון קטר דאירה אלתקב מעלום בתלך אלאג̇זא. פאן בתלך אלאלה נעלם מסאחה קטר אלקמר או אי כוכב שינא، אדא כאן בעדה [מעלומא̇][140]، באן נג̇על נקטה אלבצר ענד ראס אלותד ונחרך אלשטבה עלי טול אלמסטרה חתי ירי אלכוכב כלה، לאן נסבה אלכ̇ט אלדי בין אלבצר ואלתקב אלי נצף קטר אלתקב כנסבה בעד אלכוכב אלי נצף קטרה. ונעלם איצ̇א אלכוכב אלדי ג̇רמה אעט̇ם מן כוכב אכ̇ר ענד אלבצר או מסאוי לה.

A.IV.1 אלפצל אלראבע פי מערפה אלמיול אלג̇זיה ואלמטאלע פי אלכרה אלמנתצבה ואלמאילה ואשיא אכ̇ר ג̇זיה תלזם מנהא. אנא קד עלמנא אלקוס אלדי בין אלאנקלאבין באלאלה אלתי ג̇עלת לדלך [וגד][141] הדה אלקוס מז̇ ג̇ז מג̇ דקי̇ מן אלדאירה אלמארה באלקטבין עלי ראי בטלמיוס. וכדלך נעלם בהדה אלאלה מיל אלמסאכן אעני בעד סמת אלראס ען [][142] מעאדל אלנהאר בל ארתפאע אלקטב.

A.IV.2 פלתכון אלדאירה אלמארה באלקטבין אב̇ג̇ד̇ [][143] ומעאדל אלנהאר אה̇ג̇ ונצף דאירה אלברוג̇ אלדי מן אלג̇די אלי אלסרטאן בה̇ד̇ פי אלצורה אלאולי ואלנצף אלתאני בה̇ד̇ פי אלתאניה. וקטב [][144] מעאדל אלנהאר אלג̇נובי ז̇ ונג̇על קוס ה̇ח̇ מן דאירה אלברוג̇ מן ג̇הה אלאסתוא מפרוץ̇. ונכ̇רג̇ קוס ז̇ח̇ט̇ פי אלצורתין עלי הדא אלברהאן הו מעמול עלי כלתי אלצורתין. פלאן נסבה ג̇יב קוס ז̇א̇ אלרבע אלי ג̇יב קוס א̇ב̇، והו גאיה אלמיל، מולפה מן נסבה ג̇יב קוס ז̇ט̇ אלי ג̇יב קוס ט̇ח̇ אלמג̇הול ומן נסבה ג̇יב קוס ח̇ה̇ אלמפרוץ̇ אלי ג̇יב קוס ה̇ב̇. פיכון אלמג̇הול אלראבע ויכון מעטי פימא תקדם פי אלפצל אלתאני.

[A.IV.2/ ואדא כאן אלג̇יב מעטי פאלקוס מעטי.

A.IV.3 נריד אן נעלם קדר מא יטלע מן [][145] מעאדל אלנהאר מע אג̇זא מפרוצ̇ה מן דאירה אלברוג̇ פי אלכרה אלמנתצבה ופי אלמאילה אעני קדר קוס ה̇ט̇ אלטאלע פי אלאפק אלמנתצב וליכן ז̇ח̇ט̇ מע קוס ה̇ח̇. ודלך יכון עלי הדא אלוג̇ה לאן נסבה ג̇יב קוס ז̇ב̇ אלי ג̇יב קוס ב̇א̇

60v

אלמעלומין מולפה מן נסבה ג̇יב קוס ז̇ח̇ אלי ג̇יב קוס ח̇ט̇ אלמעלומין ומן נסבה ג̇יב קוס ט̇ה̇ אלמג̇הול אלי ג̇יב קוס ה̇א̇. פאלמג̇הול אלכ̇אמס ונסתכ̇רג̇ה במא תקדם. ואיצ̇א נכ̇רג̇ קוס כ̇ח̇מ̇ אעני אלאפק פי אלכרה אלמאילה ונקצד אן נסתכ̇רג̇ קוס ה̇מ̇ אלטאלע פי הדא אלאפק מע קוס ה̇ח̇ מן אלברוג̇. ודלך באן נסתכ̇רג̇ קדר קוס ט̇מ̇ אעני פצ̇ל אלמטאלע בין אלכרה אלמנתצבה ואלמאילה. ויכון עלי הדא אלוג̇ה، פלאן נסבה ג̇יב קוס ז̇כ̇ אעני אנכ̇פאץ̇ אלקטב אלי ג̇יב קוס כ̇א̇ תמאמה אלמעלומין מולפה מן נסבה ג̇יב קוס ז̇ח̇ אלי ג̇יב קוס ח̇ט̇ אלמעלומין، ומן נסבה ג̇יב קוס ט̇מ̇ אלמג̇הול אלי ג̇יב קוס מ̇א̇ אלרבע. פקוס ט̇מ̇ מעטי וננקצה מן מטאלע אלקוס אלמפרוץ̇ פי אלכרה אלמנתצבה פי אלנצף אלדי פי אלצורה אלאולי، או נזידה עליה פי אלנצף אלדי פי אלתאניה، חצלת מטאלע תלך אלקוס פי אלכרה אלמאילה. וכדלך נעמל הדה אלנסבה פי אלג̇אנב אלאכ̇ר מן אלאסתוא באן נכ̇רג̇ קוס ל̇ח̇ט̇ מן אלקטב

140. ו: מעולמא̇.
141. ו: ווג̇ד.
142. א: גלגל.
143. א: גלגל.
144. א: גלגל.
145. א: גלגל.

אלשמאלי, וקוס כ̇ח̇מ̇ קטעה מן אלאפק אלמאיל, ונסתכ̇רג̇ כדלך מטאלע כל ג̇ז מן אג̇זא אלדאירה [פי][146] אלמיל וד̇מ̇א.

A.IV.4 ובהדא אלטריק ימכן רסם ג̇דאול ללמטאלע [פי כל][147] מיל מפרוץ̇. ויתבין מן הדא אנה אדא כאן ג̇ז אלשמס פי יום מא מעלום כאן מקדאר דלך אליום מעלום באן נאכ̇ד מטאלע אלסתה ברוג̇ אלטאלעה פי דלך אליום או אללילה פי הדא אלאקלים ונאכ̇ד מנהא ג̇ז מן טו̇, יחצל עדה אלסאעאת אלאסתואיה. ואן אכ̇דנא ג̇ז מן יב̇ חצל אג̇זא אלסאעה אלזמאניה. ואדא עלמנא כם סאעה זמאניה [מצ̇ת̇][148] מן אלנהאר או אלליל, צ̇רבנאהא באג̇זאיהא ונזידהא עלי ג̇מיע אלמטאלע אלתי מן ראס אלחמל אלי ג̇ז אלשמס. וטלבנא אלג̇ז מן אלברוג̇ אלדי ג̇מיע מטאלעה פי אלאקלים עלי תואלי אלברוג̇ מן ראס אלחמל [הו הדה

61r

אלג̇מלה][149] והו אלג̇ז אלטאלע.

A.IV.5 ואדא צ̇רבנא אלסאעאת אלזמאניה מן נצף אלנהאר אלמאצ̇י אלי הנא פי אג̇זאיהא כל בנט̇ירה ואלקינא מא אג̇תמע מן מטאלע ג̇ז אלשמס עלי תואלי אלברוג̇ פי אלכרה אלמנתצבה, פינתהי אלי אלג̇ז מן אלברוג̇ אלדי פי וסט אלסמא פי הדא אלוקת. ואדא אכ̇דנא אלמג̇מועה ללג̇ז אלטאלע מן ראס אלחמל פי אלאקלים ואטרחנא מנה תסעין [ואן כאן אקל][150] פבעד אן נזיד עליה אג̇זא דאירה [][151] פחית ינתהי מן אלברוג̇ [פי][152] אלמנתצבה פהו אלג̇ז אלדי פי וסט אלסמא. ובאלעכס נאכ̇ד ג̇מאעאת [][153] וסט אלסמא [פי][154] אלמנתצבה ונזיד עליהא תסעין, פחית ינתהי מן אלברוג̇ פי אלאקלים פהו אלג̇ז אלטאלע פי הדא אלוקת. ומן יריד אסתכ̇ראג̇ שי מן דלך יחתאג̇ אן תכון ענדה ג̇דאול אלמטאלע אלתי רסמהא בטלמיוס והי אלמסתכ̇רג̇ה בהדה [אלטרוק][155] אלמדכורה פי הדא אלפצל וחיניד יכון דלך כלה סהל אלמאכ̇ד.

A.V.1 אלפצל אלכ̇אמס פי מקדאר סנה אלשמס. מן אלואג̇ב אן יכון קיאס עודה אלשמס פי דאירה אלברוג̇ אדא אבתדא מן נקטה גיר מנתקלה חתי יעוד אליהא, ואחרי אלנקט בדלך נקט אלאסתואין ואלאנקלאבין לאן פיהא יט̇הר אלאנפצאל ואלתגאיר בין אוקאת אלסנה אכתר מן גירהא. וליס יג̇ב אן יכון קיאס אלעודה באלקיאס אלי כוכב תאבת לאג̇ל נקלתה אלבטיה ואיצ̇א קד ימכן אן יגלט אלאנסאן פיקיס בכוכב מתחיר. ואברכס וכתיר מן אלקדמא אתפקוא במקדאר זמאן אלסנה באן רצדוא וקת אסתוא רביעי וכ̇ריפי בחלקה נחאס כאנת מנצובה פי סטח [][156] מעאדל אלנהאר. ויעלם דלך מנהא באן תצ̇י פי וקת חלול אלשמס פי נקטה אלאסתוא מן גאנביהא. וקאלוא אן הדא אלזמאן

146. א: בכל.
147. א: בלא.
148. ו: מצ̇תא. א: שעברו.
149. א: שוה לנקבץ.
150. א: -.
151. א: שלמה אם לא היו שם תשעים.
152. א: בכדור.
153. א: לחלק.
154. א: בכדור.
155. ו: אלטרוק. א: הדרכים
156. א: גלגל.

הו ש̇ס̇ה̇ ורבע אלא ג̇ז מן ש̇ מן יום באן רצדוא אסתוא רביעי [][157] ורצדוא מרה אכ̇רי בעד תלאת מאיה סנה מצריה וכל סנה מנהא ש̇ס̇ה̇ יום. ווג̇דוה בעד ארבעה וסבעין יום [מן תמאם][158] אלש̇ סנה מכאן אלכ̇מסה וסבעין יום אלתי תצוב ללרבע אלזאיד פי אלש̇ סנה. וקסמוא הדא אלזמאן עלי אג̇זא אלדאירה אעני ש̇ס̇ ג̇ז כ̇רג̇ת אלחרכה אלוסטי פי אליום אעני תקציר אלוסט עלי חסב אחד [אלאתנין][159]. ועמלוא ג̇דול ללחרכה אלוסטי פי

61v

אלאיאם ואלסאעאת ואלשהור ואלסנין באן ג̇עלוא פי אלצף אלאול עדד אלזמאן מן אי נוע כאן ובאזאיהא פי סאיר אלצפוף עדד אלחרכה אלתי תצובה.

A.V.2 קאל ן̇ סינא מא הדא מענאה. ואדא כאן וקת אלאסתוא באלליל פנחן נעלם פי אי וקת כאן מן אלליל באן ננט̇ר מקדאר [אלתפאות][160] בין ארתפאעי נצפי אלנהארין [אלמתקדם][161] ואלמתאכ̇ר ללאסתוא. פנסבה הדא [אלתפאות][162] אלי [אלתפאות][163] בין ארתפאע נצף אלנהאר אלמתאכ̇ר ובין נצף נהאר אלאסתוא כנסבה אלזמאן אלדי בין נצפי אלנהארין אלי אלזמאן אלדי ביו וקת אלאסתוא ונצף אלנהאר אלמתאכ̇ר.

A.V.3 כמלת אלג̇מלה אלאולי ואלחמד [ללה][164] אלואחד

[לא רב גירה ולא כ̇אלק סואה][165]

B.O.1 אלג̇מלה אלתאניה פי כיפיה אלחרכאת ואלאצול אלואג̇ב וצ̇עהא ללאכ̇תלאפאת אלמחסוסה ללאג̇ראם אלסמאויה וקסמת הדה אלג̇מלה [אלי][166] מקאלאת. אלמקאלה אלאולי פי חרכאת אלשמס ופי אכ̇תלאפה [ופיה תלאתה][167] פצול. אלפצל אלאול פי אלאצל אלדי יוצ̇ע ללחרכה אלמסתויה. אלפצל אלתאני פי אלאצל אלדי יוצ̇ע ללאכ̇תלאף. אלפצל אלתאלת פי תבין [קדר מקדאר][168] אכתר אלאכ̇תלאף אלדי ירי פי אלשמס ווג̇ה חסאב אלאכ̇תלאפאת אלג̇זיה ותבין אלנקטה מן אלברוג̇ אלתי פיהא אקל אלתקציר או אלסיר.

62r

B.O.2 אלמקאלה אלתאניה פי תבין אלאצול אלדי עליהא ינבגי אן [יעמל][169] פי חרכאת אלקמר ואלאכ̇תלאפאת אלתי ירי פיה, ופיהא כ̇מסה פצול. אלפצל אלאול פי אלאצול אלמעטאה לחרכאת

157. א: אחת.
158. א: יותר.
159. ו: אלנתין. א: הדעות.
160. א: מרחק.
161. א: -.
162. א: המרחק.
163. א: המרחק.
164. ו: לללה. א: האל.
165. א: אין זולתו.
166. א: לארבעה.
167. ו: ופי תאלתה. א: ובו שלשה.
168. א: שעור.
169. ו: יעלמל. א: לעשות

אלקמר אלדוריה. אלפצל אלתאני פי תבין אכׄתלאף אלאול אלחקיקי [אלטׄאהר][170] פי אלקמר. אלפצל אלתאלת פי מקדאר אכתר הדא אלאכׄתלאף. אלפצל אלראבע [פי][171] אן אכׄתלאף מקאדיר אקטאר אלקמר ענד אלבצר ליס ינבגי ענה ולא בד אכׄתלאף אבעאד אלקמר ען אלארץׄ. אלפצל אלכׄאמס פי אלאכׄתלאפין אלבאקיין ללקמר אעני אלאכׄתלאף אלמנסוב ענד בטלמיוס אלי אלשמס، והו אלדי יעמל עליה בכׄארגׄ אלמרכז ואלאכׄתלאף אלמנסוב ענדה אלי אנחראף קטר אלתדויר.

B.0.3 אלמקאלה אלתאלתה פי אנתקאל אלכואכב אלתאבתה ופיהא פצלאן. אלפצל אלאול פי ראי
[B.0.3/x אלבטרוגׄי פי הדה אלנקלה ואבטאלה. אלפצל אלתאני פי וגׄה נקלה [][172] אלכואכב.

B.1.1.1 אלפצל אלאול מן אלמקאלה אלאולי פי אלאצל אלדי יוצׄע ללחרכה אלמסתויה. פנקול אן אלשמס וגׄד מקצראׄ ען אלפלך אלאעלי، אעני ען אלחרכה אליומיה، במקדאר מא יתם דורה פי תקצירה פי כל סנה מרה ואחדה. ומע אן חרכתה עלי דואיר מואזיה פי אלחס לדאירה [][173] מעאדל אלנהאר אלתי עלי קטביהא אלחרכה אלאולי פאן בתקצירה יוגׄד מנתקלאׄ נחו אלשמאל ואלגׄנוב. ומקדאר תבאעדה ענהא מחדוד והו מתל מיל דאירה נטאק אלברוגׄ ענהא.

B.1.1.2 ודלך יתוהם עלימא אצף. והו אנא ננזל דאירה [][174] מעאדל אלנהאר גׄהׄד קטבהא אׄ ודאירה מיל אלשמס גׄטׄד קטבהא בׄ. ואלדאירה אלמארה בנקטתי אלאנקלאבין ובקטבי [][175] מעאדל אלנהאר טׄהׄבׄאׄ ואלמארה באלאסתואין ובקטבי [][176] מעאדל אלנהאר גׄאׄדׄ. ונפרץׄ פי יום ואחד תחרך קטב בׄ עלי דאירה ממרה והי דאירה בׄקׄשׄתׄ שאפעאׄ

62v

לחרכה אלכל מן אלמשרק אלי אלמגרב קוס בׄשׄכׄ، ויקצר קוס כׄבׄ. פאדא אנזלנא אלשמס מתלאׄ עלי נקטה טׄ، אעני נקטה אלמנקלב، צאר קד תחרך אלשמס [מתלאׄ][177] מע גאיה מילהא נחו אלמגרב בחרכה אלקטבין ונקץ ען דורה תאמה קוס מן אלדאירה אלמואזיה [][178] למעאדל אלנהאר אלמארה בנקטה אלמנקלב שביה בקוס כׄבׄ، וליכון קוס טׄחׄ.

B.1.1.3 פיכון חיניד וצׄע דאירה אלשמס אדא כאן קטבהא עלי כׄ כוצׄע דאירה לׄזׄחׄ. פתציר נהאיה אלמיל עלי נקטה חׄ מן דאירה טׄחׄ פיכון קד קצר ען תמאם דורה תאמה קוס חׄטׄ מן אלמואזיה. וננזל איצׄאׄ אן פי הדא אלזמאן אלדי יתחרך אלקטב קוס בׄשׄכׄ ונהאיה אלמיל [אלדי][179] כאן פיהא אלשמס תמאם קוס חׄטׄ לדאירה תאמה، תתחרך דאירה אלשמס חול קטביהא. ותנתקל אלשמס מעהא איצׄאׄ נחו תלך אלגׄהה קוס חׄזׄ טלבאׄ ללכמאל ואלתשבה באלאעלי، ואשתאק מנה ליסתפי בעץׄ מא קצרת קטביהא ען אלאעלי מן אגׄל אנכסאר אלקוה אלנאזלה עליה. ובתלך אלחרכה יערץׄ לה אן יכון מרה שמאליאׄ [ען][180] מעאדל אלנהאר ומרה גׄנוביאׄ. ולדלך תסמי הדה אלחרכה חרכה אלערץׄ. ובקי תקציר

170. א: -.
171. א: יבאר.
172. א: אלו.
173. א: גלגל.
174. א: גלגל.
175. א: גלגל.
176. א: גלגל.
177. א: -.
178. א: גלגל.
179. ו: אל. א: אשר.
180. א: לעגולת.

אלשמס בעד האתין אלחרכתין מג̇מועתין ינקץ ען אלוצול אלי אלמוצ̇ע אלדי מנה אבתדא אלמקדאר אלדי אדרכה בטלמיוס והו נ̇ט̇ דקי̇ באלתקריב. ותכון חרכתה פי אלערץ̇ פי דלך אליום בקדר מא ינקץ ערץ̇ נקטה ז̇ ען ערץ̇ נקטה ח̇.

B.1.I.4 ולאן חרכה אלטול פי אלשמס אעני תקציר אלטול אלאכ̇יר מסאוי לחרכה אלערץ̇, פיג̇ב לדלך אן תכון קוס ח̇ז̇, אעני חרכה אלפלך עלי קטביה, מסאויא̇ לתקציר אלשמס פי אלאכיר אעני נ̇ט̇ דקי̇. פתקציר אלקטב אדא̇, אעני קוס כ̇ב̇, מתל חרכה אלערץ̇ ותקציר אלטול אלמתסאויתין אג̇מע. פג̇מיע מא ירי פי אלשמס מן אנה ינקץ כל יום ען דורה תאמה אלקדר אלמדכור, ואנה ירי מתחרך דאימא̇ עלי

63r

דואיר מואזיה [][181] למעאדל אלנהאר, ויתחרך איצ̇א אלי אלשמאל ואלי אלג̇נוב אלקדר אלמדרך. פבהדא פקט יתם מן גיר אחתיאג̇ פיה אלי וצ̇ע חרכתין מתצ̇אדתין, ודלך אנה סוא יסתכמל אלאנקלאב אעני נקטה ט̇ דורה ואחדה פי יום ואחד ותכון חרכה אלערץ̇ מן נקטה ט̇ אלי נקטה ז̇ אלי כ̇לאף או תכון אלחרכתין אלי ג̇הה ואחדה פאן באלאצלין ג̇מיעא̇ יתם ג̇מיע מא ירי פיה מן אלחרכאת עלי חאלה ואחדה מע כון אלשמס אבדא̇ עלי דאירה אלברוג̇. וליס יכון בינהמא אכ̇תלאף אלא פי אמר אלמטאלע, פאנה עלי הדה אלאצול עלי כ̇לאף מא פי תלך לאנה ילזם ען הדא אלמדהב אן תכון חרכה אלשמס, ענד כונה פי אלברוג̇ אלתי ענד אלאנקלאב, אלי אלנקצאן מן קבל אלמטאלע. ואדא כאנת ענד אלאסתוא באלעכס חתי אן הדא אלאכ̇תלאף יסתכמל פי אלרבע אעני אלזיאדה באלנקצאן.

B.1.I.5 וליס להם דליל עלי אן קולהם פי דלך אחק מן הדא אלקול מן הדה אלג̇הה. פהדא הו אלאצל אלדי ינבגי אן יוצ̇ע ללחרכה אלמסתויה ענד אלואצ̇עין פי אלחרכה אלמסתדירה תצ̇אד. ואנת תעלם ממא סלף מן קולנא אן אלאצל אלאול אחק למא בינא מן עדם אלתצ̇אד פי אלחרכה אלמסתדירה.

B.1.II.1 אלפצל אלתאני פי אלאצל אלדי יוצ̇ע ללאכ̇תלאף. ואקול אן אלאכ̇תלאף אלדי ירי פי אלשמס אעני אנה ירי יתחרך פי אקסאם אלברוג̇ מרה אסרע ומרה אבטא ומרה באלוסט בין האדין. והדא אלאכ̇תלאף אלמרי̇ פיה יחפט̇ אבדא̇ נט̇אמהא עלימא וג̇דוה, אעני אלחרכה אלבטיה אנמא תוג̇ד לה פי אג̇זא [ואחדה][182] באעיאנהא דאימא̇ וכדלך אסרע אלחרכה ואוסטהא. פינבגי אן נצ̇ע להדא אלאכ̇תלאף אצול חקיקיה תואפק מדהבנא.

B.1.II.2 פנקול אן מרכז אלשמס לו כאן יתחרך חרכתהא אלכ̇אציה לה עלי דאירה אלברוג̇ או עלי דאירתה אלמסאויה אלמיל למיל דאירה אלברוג̇, פליס יג̇ב אן ירי האהנא אכ̇תלאף פי אלטול אצלא̇, אד אלשמס כאן יג̇ב אן יתחרך פי אלאזמאן אלמתסאויה קסי מתסאויה מן דאירה מילה לאנה מחמול עליהא פי אלחרכה. אעני אנה יחד זואיא מסאויה ענד קטבי אלברוג̇ פי אלאזמאן אלמתסאויה. לאכן אלשמס ליס יקטע אלקסי אלמתסאויה מן דאירה מילה פי אזמנה מתסאויה, פינתג̇ מקאבל אלמקדם והו אן מרכז אלשמס ליס הו מחמול פי

63v

חרכתה אלכ̇אציה לה עלי אלדאירה אלמאילה עלי אנה ליס יוג̇ד כ̇ארג̇א̇ ענהא.

181. א: לעגולת.
182. א: ידועים.

B.1.II.3 פיט̇הר מן דלך אן מרכז אלשמס מתחרך עלי דאירה צגירה קטב תלך אלדאירה הו מחמול אבדא̇ עלי דאירה עט̇ימה מאילה עלי דאירה מיל אלשמס בקדר נצף קטר תלך אלדאירה אלצגירה. ולתסמי תלך אלדאירה אלצגירה דאירה ממר אלמרכז ותקום תלך אלדאירה מקאם פלך אלתדויר אלדי וצ̇עה בטלמיוס ועמל עליה, וליכן עלי הדה אלצפה.

B.1.II.4 לתכון דאירה אלברוג̇ ג̇טד̇ קטבהא ב̇ וקטב מעאדל אלנהאר א̇. ואלדאירה אלמאילה אלחאמלה לקטב דאירה אלממר ג̇הד̇ קטבהא ח̇. ודאירה ממר הדא אלקטב חכ̇ קטבהא קטב מעאדל אלנהאר. ולנזל דאירה ממר אלמרכז למ̇נט̇ קטבהא ה̇ ותכון ממאסה לדאירה אלברוג̇ חתי יכון אלקוס אלכ̇ארג̇ מן קטבהא אלי מחיטהא מסאוי לגאיה מיל דאירה ג̇הד̇ עלי דאירה ג̇טד̇. ולנצ̇ע מתלא̇ אן נקטה ט̇ הי נקטה אלאנקלאב, ולנצ̇ע איצ̇א אנה אדא כאן קטב דאירה ממר אלמרכז עלי נקטה ה̇ עלי מא פי אלצורה, יכון מרכז אלשמס עלי נקטה ט̇ אלממאסה לדאירה אלברוג̇.

B.1.II.5 פאדא אנתקל קטב ח̇ פי יום ואחד מתלא̇ תאבעא̇ לחרכה אלאעלי וקצר ען תמאם דורה תאמה אלקדר אלמדכור וליכן מתל קוס חכ̇ וצאר קטב ח̇ עלי נקטה כ̇. צארת איצ̇א נקטה ה̇ אעני קטב דאירה ממר אלמרכז עלי נקטה ס̇ מקצרא̇ איצ̇א קוס שביה בקוס חכ̇. ומתל הדא אלקדר יקצר אלשמס איצ̇א אד הו מחמול עלי דאירה ממרה. ובהדא אלזמאן יתחרך קטב דאירה ממר אלמרכז עלי נקטה כ̇ וינקל אלשמס מעהא מן נקטה ס̇ אלי נקטה ז̇ מתלא̇, ויכון קדר קוס סז̇ נצף אלתקציר אלאול. פיבקי תקציר אלקטב בל תקציר אלשמס אלאכ̇יר נצף אלתקציר אלאול אעני נט̇ דקי̇ באלתקריב.

B.1.II.6 הדא לו כאן ·

64r

מרכז אלשמס עלי נקטה ט̇ מן דאירה ממרה אבדא̇, ולם יכון למרכז אלשמס חרכה עלי קטב ה̇. לאכן אן אנזלנא דלך כאן יג̇ב אן לא יחפט̇ אלשמס ערץ̇ דאירה אלברוג̇ אבדא̇. ודלך אנה אדא כאן [קטב][183] דאירה ממר אלמרכז עלי נקטה ז̇ כאן סיחדת ללשמס ערץ̇ ען דאירה אלברוג̇ לאן ערץ̇ נקטה ז̇ ען אלנקטה אלמחאדיה להא מן דאירה אלברוג̇ אצגר מן קוס הט̇. וכלמא אנתקל נחו אלעקדה כאן אנקץ̇ פיעט̇ם ערץ̇ אלשמס ען דאירה אלברוג̇.

B.1.II.7 ולאן אלאמר פי אלשמס בכ̇לאף דלך, ודלך אנה ליס יוג̇ד כ̇ארגא̇ ען ערץ̇ דאירה אלברוג̇ אבדא̇, פיג̇ב עלי הדא אן יכון למרכז אלשמס חרכה עלי קטבי דאירה ממרה. פאדא נחן וצ̇ענא הדה אלחרכה מן לדן נקטה ט̇ נחו נקטה נ̇, פאן הדה אלחרכה ימכן וצ̇עהא אלי אי ג̇הה שינא לאנהא מן אלחרכאת אלתי יתוהם פיהא אנהא מרה אלי אלג̇הה אלאולי מרה באלעכס. ווצ̇ענא קדר תלך אלחרכה מתל תקציר אלאכ̇יר, אעני נט̇ דקי̇ פי אליום. כאן בדלך אלשמס חאפט̇א̇ לערץ̇ דאירה אלברוג̇ לאנהא אדא אנתקל אלשמס מן נקטה ט̇ נחו נקטה נ̇ קוס שביה בקוס סז̇, כאנת נסבה נקצאן ערץ̇ נקטה ז̇ ען ערץ̇ נקטה ס̇ אלי ערץ̇ נקטה ס̇ כנסבה נקצאן ערץ̇ אלנקטה אלדי ינתהי אליהא מרכז אלשמס פי הדא אליום, מבתדיא̇ מן נקטה ט̇ נחו נקטה נ̇ ען ערץ̇ נקטה ה̇ אלי ערץ̇ נקטה ה̇. פאדא כאנת נקטה ה̇ עלי נקטה ז̇ כאן אלשמס איצ̇א חאפט̇א̇ לערץ̇ דאירה אלברוג̇ לאג̇ל אנתקאלה עלי דאירה ממרה.

B.1.II.8 פיחדת ללשמס לאג̇ל הדא אלאנתקאל זיאדה פי תקצירה אלאכ̇יר פאנה לו כאן עלי נקטה ט̇ דאימא̇ כאן תקצירה באסתוא עלי מא הו עליה תקציר אלקטב. ומא דאם אלשמס מנתקלא̇ עלי קוס לט̇נ̇ יתזיד אלתקציר אלאכ̇יר. פתרי לאג̇ל דלך [חרכתה][184] אעני תקצירה אלי תואלי אלברוג̇ אסרע, ומא

183. ו: קטה. א: קטב.
184. א: -.

דאם ינתקל עלי קוס למ̇נ̇ צאר תקצירה אלאכ̇יר ינתקץ [][185], פירי לדלך תקצירה אבטי. ויכון קדר אלזיאדה ואלנקצאן במקדאר קוס לנ̇ מן אלדאירה אלעט̇ימה. ולך אן תצ̇ע הדא אלאכ̇תלאף באן תכון חרכה קטב דאירה ממר אלמרכז אלי כ̇לאף, מתלא̇ מן נקטה ה̇ נחו נקטה ז̇, ותטרח חיניד דאירה ס̇ז̇ מן אלצורה. ולתסמי נקטה ט̇ מן דאירה אלממר [נקטה][186] אעט̇ם אלתקציר או אלסיר, ונקטה מ̇ אקל אלתקציר או אלסיר. והי אלנקט אלקסימה לאנצאף נ̇ט̇ל̇ נ̇מ̇ל̇ בנצפין.

B.1.II.9 וקד כאן ימכן אן יתצור הדא

64v

אלאכ̇תלאף פי אלנצף אלואחד עלי גיר הדא אלוג̇ה, אעני ואן לם נצ̇ע ללשמס חרכה עלי קטבי דאירה ממרה פי אלאצל אלתי וצ̇ענא פיה אלחרכה אלי ג̇הה ואחדה. ודלך לאן נקטתי תקאטע דאירה ממר אלמרכז מע אלדאירה אלמאילה, אעני נקטתי לנ̇, יט̇הר פי באדי אלראי אנהא תכונא אבדא̇ עלי דאירה ס̇ז̇ט̇ לאן אנתקאל קטבי דאירה אלממר אנמא הו עלי תלך אלדאירה. פאדא כאן קטב ה̇ עלי נקטה ז̇, ונקטתי לנ̇ עלי נקטתי חכ̇ מן דאירה ס̇ז̇ פי אלצורה אלתי בעד הדה, כאן תקציר אלשמס אכתר מן תקציר אלקטב ואן כאן אלשמס לם יפארק אבדא̇ נקטה ט̇, לאן נקטה ט̇ תכון חיניד מאילה ען אלקטב אלדי הו עלי נקטה ז̇ נחו נקטה ד̇. ויזדאד דאימא̇ הדא אלמיל חתי יכון אלקטב פי נקטה אלתקאטע. ויתם חיניד אכתר אלמיל ולדלך יט̇הר פי תקצירה אלאכ̇יר זיאדה. ואדא אנתקל פי אלרבע אלתאלי לה צאר יתראג̇ע הדא אלמיל [][187] וינתקץ אלי אן יכמל אלרבע, פתרג̇ע נקטה ט̇ אלי מוצ̇עהא וצאר מוצ̇ע נקטה ט̇ ונקטה ה̇ מן אלברוג̇ ואחד.

B.1.II.10 ולדלך ירי פי הדא אלרבע נקצאן פי אלתקציר וכדלך פי אלרבע אלתאלי לה. ופי אלרבע אלאכ̇ר יט̇הר איצ̇א זיאדה פי אלתקציר מתל מא כאן פי רבע ה̇ז̇ד̇ פכאן יתצור אלאכ̇תלאף עלי מא הו עליה באלוצ̇ע אלאכ̇ר. אלא אנה לו כאן אלאכ̇תלאף יג̇רי עלי הדא אלנט̇אם כאן אלשמס ליס יחפט̇ אבדא̇ ערץ̇ דאירה אלברוג̇ ואנמא יחפט̇ הדא אלערץ̇ פי נצף ג̇ט̇ד̇ פקט. ואמא פי אלנצף אלתאני פאנה קד יזול ען הדא אלערץ̇ במקדאר קוס מ̇ה̇ט̇ ודלך יערץ̇ לוצ̇ענא אלשמס גיר מתחרך עלי קטבי דאירה ממרה. פלדלך יג̇ב אן נצ̇ע אן אלאכ̇תלאף אנמא הו מן קבל חרכה אלשמס חול קטבי דאירה ממרה, לאן בדלך פקט יוג̇ד אלשמס עלי כל ג̇ז מן דאירה ממרה, וצאר בדלך חאפט̇א̇ ללערץ̇ אלמדכור. ואנמא דכרת לך דלך כלה עלי טריק אלריאצ̇ה וליסדד דהן אלנאט̇ר פי הדה אלאשיא לכי יתבת פכרה פיהא, ולאן נקטתי לנ̇ לאזמתין לדאירה ס̇ז̇ עלי הדא אלאצל ובחסב דלך כאן אלאכ̇תלאף אכ̇תלאפין.

B.1.II.11 אחדהמא אלדי יכון בחסב אנתקאל אלפלך אלדי פיה דאירה אלממר חול קטביה. ואלתאני הדא אלמדכור פנחן

65r

נבין כיף יג̇רי אלאמר פי דלך [פימא בעד בעון אללה תעלי][188]. ואן וצ̇ענא אצל הדא אלאכ̇תלאף אלאצל אלאול, אעני בחסב אצול בטלמיוס פי אלחרכאת אלמתצ̇אדה באן תכון חרכה קטב דאירה אלממר אלי כ̇לף מן נקטה [ה̇][189] נחו נקטה ז̇, פבחסב דלך תכונא [][190] נקטתי תקאטע לנ̇ חאפט̇תין לדאירה ג̇ה̇ד̇ אד ליס בנא חיניד חאג̇ה אלי וצ̇ע דאירה ס̇ז̇.

185. א: והולך.
186. א: -.
187. א: לאחוריה.
188. א: ב̇ה̇.
189. א: -.
190. א: ג̇כ̇.

B.1.II.12 ונקול איצ̇א אן נחן וצ̇ענא אן נקטתי ל̇נ̇ ליסת תפארקאן דאירה ד̇ז̇ה̇ אבדא̇ ולא בשי יסיר אצלא̇ כמא הו ואג̇ב בחסב הדא אלאצל، פאנה כאן ילזם אן לא יחפט̇ אלשמס ערץ̇ דאירה מילה אבדא̇ מע כונה מתחרכא̇ עלי דאירה ממרה עלי אלג̇הה אלמדכורה. ואלכ̇לאף אלדי יקע מן קבל דלך עט̇י̇ם לאנה אדא אג̇תמע אלזואל ען אלערץ̇ ענד נקטה אלאסתוא כאן יקע מן קבלה פי רצד וקת אלאסתוא، והו אלדאל עלי אלאכ̇תלאף זלל ליס באלקליל.

B.1.II.13 ואמא כיף יט̇הר דלך פיכון עלי הדא אלוג̇ה. אנא נעד אלצורה אלסאלפה וננזל קטב דאירה ממר אלמרכז עלי נקטה ז̇. ויוצ̇ע אן מרכז אלשמס תחרך עלי דאירה ממרה מן לדן נקטה ט̇ נחו נקטה נ̇ קוס שביה בקוס ה̇ז̇. פנקול אנה ליס ינתהי אלשמס אלי נקטה ג̇ ואן קוס ט̇ג̇ גיר שביה בקוס ה̇ז̇. ברהאן דלך אן נסבה נקצאן ערץ̇ נקטה ז̇ מן אלדאירה אלמאילה ען ערץ̇ נקטה ה̇ אלי ערץ̇ נקטה ה̇ כנסבה נקצאן ערץ̇ אלנקטה אלדי אליהא ינתהי אלשמס מן דאירה ממרה מבתדיא̇ מן נקטה ט̇ ען קוס ז̇ט̇، אלי קוס ז̇ט̇، וקוס ז̇ט̇ מסאוי לערץ̇ נקטה ה̇. פנקצאן ערץ̇ נקטה ז̇ ען ערץ̇ נקטה ה̇ הו קוס ט̇ב̇ מן אלדאירה אלעט̇ימה. פיג̇ב לדלך אן יכון נקצאן ערץ̇ אלנקטה אלדי אליהא ינתהי אלשמס מן דאירה ממרה מן קוס ז̇ט̇ הו קוס ט̇ב̇ איצ̇א.

B.1.II.14 ולאכן נקצאן ערץ̇ נקטה ג̇ מן דאירה מיל אלשמס ען ערץ̇ נקטה ה̇ אכתר מן קוס ט̇ב̇ לאן נקצאן ערץ̇ נקטה ג̇ אכתר מן נקצאן ערץ̇ נקטה ז̇. פלו אנתהי אלשמס פי דלך אלזמאן פי דאירה ממרה אלי נקטה ג̇، אעני

65v

נקטה תקאטעה מע דאירה [][191] אלשמס، כאן קד נקץ ען אלערץ̇ אכתר מן קוס ט̇ב̇. הדא כ̇לף. פאדא̇ אלקוס אלשביה בקוס ה̇ז̇ אלדי אליה ינתהי אלשמס הו אצגר מן קוס ט̇ג̇، פאלשמס אדא̇ ליס ינתהי אלי דאירה מילה، אעני דאירה ד̇ב̇א̇. ובאלג̇מלה קד ינתהי אלשמס אלי ערץ̇ מוצ̇עה אלוסט מן אלברוג̇، אעני נקטה ב̇، לא אלי ערץ̇ מוצ̇עה אלחקיקי.

B.1.II.15 פאד קד תבין דלך פיג̇ב עלינא אן נבין כיף יחפט̇ מרכז אלשמס מחיטי האתין אלדאירתין، אעני מחיט דאירה אלברוג̇ ומחיט דאירה אלממר אבדא̇ עלי חסב כל ואחד מן אלאצלין، אעני בחסב וצ̇ע אלחרכאת מתצ̇אדה עלי ראי בטלמיוס ובחסב וצ̇עהא אלי ג̇הה ואחדה. ובדלך יתבין מא וג̇דוא אלקדמא ועמל עליה בטלמיוס פי הדא אלאכ̇תלאף [מן][192] אלזמאן אלדי מן אעט̇ם אלחרכה אלי אוסטהא אצגר מן אלזמאן אלדי מן אוסטהא אלי אצגרהא. ונבין הדא אלמעני אולא̇ עלי ראי בטלמיוס، אעני בוצ̇ענא אלחרכאת אלמתצ̇אדה، תם נתלו ביאן דלך עלי חסב אלראי אלתאני.

B.1.II.16 ולנזל אלצורה עלי חאלהא וקד תבין אנה אן כאן מרכז אלשמס מחמולא̇ עלי דאירה ממרה، אעני דאירה ל̇ט̇נ̇מ̇، ותחרך קטב ז̇ קוס ז̇ה̇ ואנתקל אלשמס מן נקטה ט̇ נחו נקטה נ̇ קוס שביה בה، אנה ליס ינתהי מרכז אלשמס פי הדה אלנקלה פי הדא אלרבע אלי מחיט דאירה אלברוג̇، אעני תקאטע דאירה אלממר מע דאירה אלברוג̇ והו נקטה ש̇ פי הדה אלצורה. [ואנה לו אנתקל כדלך][193] קד יג̇תמע ענד תמאם אלרבע זואל ען אלערץ̇ כתיר. פלאן נקטתי אלתקאטע בחסב הדא אלאצל חאפט̇תאן אבדא̇ מוצ̇עיהמא מן דאירה ג̇ה̇ד̇ וליסת יתפרקאן ענהא، פיט̇הר מן דלך אן מרכז אלשמס ליס הו מחמול עלי

191. א: נטיית.
192. א: כי.
193. א: ואם היה כן הנה.

דאירה ממרה, אעני דאירה לט֗נמ֗ נפסה עלי אנה ליס יוג֗ד כ֗ארג֗א ענהא. ואמא כיף יכון אלאמר פי דלך פהו עלי הדא אלוג֗ה.

B.1.II.17 אנא נתוהם אדא כאנת נקטה ז֗ עלי נקטה ה֗ ונקטה ט֗ מן דאירה לט֗נמ֗ עלי נקטה ט֗ מן דאירה אלברוג֗, אן קטב א֗ מן דאירה חצ֗ מנטבק עלי נקטה ט֗, וקוס אח֗ אלדי מן דאירה עט֗ימה יכון חיניד עלי קוס טה֗ אלדי מן דאירה עט֗ימה איצ֗א. ונתוהם עלי קטב ח֗ איצ֗א דאירה סר֗כ֗ מסאויה להא ויכון [][194] קוס כס֗ אלדי מן דאירה עט֗ימה איצ֗א מנטבק עלי קוס טה֗. ויכון מקדאר אלקוס מן אלדאירה אלעט֗ימה אלכ֗ארג֗ מן

66r

קטב כל דאירה מן האתין אלי מחיטהא מסאוי לנצף קוס אעט֗ם אכ֗תלאף יוג֗בה אלקוס מן דאירה ממר אלמרכז אלשביה בקוס אעט֗ם אכ֗תלאף אלשמס אלדי מקדארה ג֗זאן, אעני נצף קוס דאירה אלממר אלשביה בקוס אעט֗ם אלאכ֗תלאף. וננזל אן מרכז אלשמס מחמול פי חרכתה עלי דאירה סר֗כ֗ ואן קטב דאירה סר֗כ֗ מחמול פי חרכתה איצ֗א עלי דאירה חו֗צ֗. ואדא כאן קטב א֗ עלי נקטה ט֗ מן דאירה לט֗נמ֗ ועלי נקטה ט֗ מן דאירה אלברוג֗, כאן חיניד קטב דאירה סר֗כ֗ אעני ב֗ עלי נקטה ח֗. וכאן חיניד מרכז אלשמס עלי נקטה כ֗ מן דאירה סר֗כ֗ אלתי כאנת חיניד מנטבקה עלי קטב א֗ אלדי עלי נקטה ט֗ מן דאירה אלברוג֗. פאדא תחרך קטב ז֗ קוס זה֗ אנתקל קטב א֗ עלי דאירה לט֗נמ֗ חול קטב ז֗ מן נקטה ט֗ אלי נקטה א֗, קוס שביה בקוס זה֗. ואנתקל איצ֗א קטב דאירה סר֗כ֗, אעני ב֗, חול קטב א֗ מן נקטה ח֗ אלי נקטה ב֗ קוס שביה בקוס זה֗ איצ֗א.

B.1.II.18 ואפתרקת חיניד נקטה כ֗ אלתי פיהא מרכז אלשמס מן נקטה א֗ נחו נקטה נ֗, וחדת בדלך ללשמס ערץ֗ מן דאירה לט֗נמ֗, פאן קוס כ֗בס֗ מואזי אבדא לקוס ח֗אצ֗ בהדה אלנקלה. ואנתקל מרכז אלשמס איצ֗א חול קטב ב֗ מן נקטה כ֗ אלי נקטה ש֗ מן דאירה סר֗כ֗ קוס שביה בקוס הז֗ איצ֗א. ואנתהי בדלך מרכז אלשמס אלי מחיט דאירה לט֗נמ֗ ואלי מחיט דאירה אלברוג֗ איצ֗א, חתי אנה אדא תחרך קטב ז֗ אג֗זא אלרבע אלא אג֗זא אעט֗ם אלאכ֗תלאף כאן חיניד קוס ט֗א֗ נאקצ֗א ען רבע קוס שביה בקוס אעט֗ם אלאכ֗תלאף. וכאן קטב ב֗ קריבא֗ מן נקטה ו֗ מן דאירה חו֗צ֗, ומרכז אלשמס קריבא֗ מן נקטה [כ֗][195], אעני קריבא֗ מן רבע מבתדיא֗ מן לדן נקטה כ֗.

B.1.II.19 ואן ג֗עלנא כל ואחדה מן האתין אלנקלתין גיר שביהה בנקלה קטב ז֗ ולאכן אכתר מנהא בקדר

66v

אעט֗ם אלאכ֗תלאף פי אלרבע כלה, כאן חיניד מרכז אלשמס עלי נקטה [כ֗][196] אעני [פי רבע נקטה][197] כ֗. וצאר בדלך מרכז אלשמס עלי רבע תאם מן דאירה לט֗נמ֗ מבתדיא֗ מן לדן נקטה ט֗, אעני עלי נקטה נ֗, ונקטה נ֗ תכון חיניד עלי נקטה תקאטע אלדאירה אלחאמלה מע דאירה אלברוג֗, אלא אנה ליס ילזם דלך פי סאיר אלארבאע. ולדלך וצ֗ענא אלחרכאת מתשאבהה לאן אלזואל אלחאדת מן קבל דלך גיר מחסוס, פמרכז אלשמס בהאתין אלנקלתין אדא קד יחפט֗ דאירה אלברוג֗ אבדא֗, ויחפט֗ מחיט דאירה

194. א: אז.
195. א: ה֗.
196. א: ה֗.
197. א: על רביע שלם מנקודת.

ממרה איצ̇א. וכדלך בעינה ילזם פי כל רבע מן [אלארבאע][198] אלבאקיה, אעני אנה יוג̇ד בהאתין אלנקלתין חאפט̇א למחיטי האתין אלדאירתין אבדא̇ וכדלך כאן [קצדנא][199] לתבינה.

B.1.II.2 ודלך בעינה יט̇הר ולו אנזלנא קוס ס̇ב̇כ̇ גיר מואזי בהדה אלחרכה אבדא̇ לקוס ח̇א̇צ̇, באן תכון נקטה כ̇ מן דאירה ס̇ר̇כ̇ מנטבקה אבדא̇ עלי קטב א̇, כמא הו אלואג̇ב בחסב קולנא אן קטב ב̇ מנתקל חול קטב א̇, פאן נקטתי תקאטע דאירה ס̇ר̇כ̇ מע דאירה ח̇ו̇צ̇ לאזמא̇ אבדא̇ מוצ̇עיהמא מן דאירה ח̇ו̇צ̇, ודאירה ח̇ו̇צ̇ אנתקלת חול קטביהא, ואנתקלת מעהא דאירה ס̇ר̇כ̇ וקטבהא אלקדר אלמדכור. ומרכז אלשמס אנתקל פי הדא אלזמאן מן נקטה כ̇ נחו נקטה ר̇ צ̇עף אלקדר אלמדכור לאן אדא אנתקל קטב ב̇ עלי דאירה ח̇ו̇צ̇ רבע דאירה נחו נקטה ו̇, צארת נקטה ס̇ עלי מחיט דאירה לט̇נמ̇. פיחתאג̇ אן תכון חרכה מרכז אלשמס חול קטב ב̇ פי הדא אלזמאן צ̇עף אלקדר אלמדכור, ובדלך יוג̇ד חאפט̇א̇ אבדא̇ למחיטי
[B.1.II.2 אלדאירתין כמא תבין קבל והדא הו אלאגו̇ד.

B.1.II.2 ולנזל תאניא̇ עלי חסב אלראי אלתאני אן נקטתי אלתקאטע חאפט̇תין אבדא̇ מוצ̇עיהמא מן דאירה ס̇ז̇ והי אלדאירה אלדי עלי קטביהא ידור אבדא̇ קטב דאירה ממר אלמרכז, ודלך פקט הו אלואג̇ב בחסב הדא אלאצל. ונקול אן

67r

ואדא כאנת נקטה ס̇, אעני אלנהאיה אלג̇נוביה, עלי נקטה ה̇, כאנת נקטתי תקאטע ל̇נ̇ עלי דאירה ה̇ד̇ איצ̇א, וכאנת נקטה ט̇ מן דאירה אלממר אלתי הי נקטה אעט̇ם אלתקציר עלי נקטה ט̇ מן דאירה אלברוג̇. פאדא קצרת נקטה ס̇ ען נקטה ה̇ נצף אלדאירה אלמואזיה אלתי תמר בהא ובנקטה ה̇, ואנתקל [קטב][200] דאירה אלממר רבע אלדאירה אלחאמלה צאר עלי נקטה ד̇, והי נקטה תקאטע דאירה ה̇ד̇ מע דאירה אלברוג̇.

B.1.II.2 וננזל ג̇ז̇ מתל ז̇כ̇ פנקטה אעט̇ם אלתקציר אלתי כאן מן אלואג̇ב אן תכון עלי נקטה ז̇ לו כאנת נקטתי אלתקאטע חאפט̇תין מוצ̇עיהמא מן דאירה ה̇ד̇, צארת עלי נקטה ט̇ והי אלקאטעה לקוס ל̇ט̇נ̇ בנצפין, קדר קוס ז̇ט̇ מתל קוס ג̇ל̇, [והו][201] מתל צ̇עף מיל נקטה ה̇ ען דאירה [][202] מעאדל אלנהאר אלדי הו מעלום. פקד יחתאג̇ אן יתחרך מרכז אלשמס אלדי כאן אולא̇ עלי נקטה ט̇, אעני נקטה אעט̇ם אלתקציר, חול קטב דאירה ממרה אעני ד̇ פי הדא אלזמאן מן נקטה ט̇ מן דאירה אלממר אלי נקטה א̇ מן דאירה אלברוג̇ פקט. וחרכה אלמסתויה פי הדא אלזמאן חול הדא אלקטב אנמא הי רבע דאירה אלממר. פלדלך יחתאג̇ אן ננזל אן מרכז אלשמס אנתקל באלפלכין אלמוצופין [אל][203] פי הדא אלזמאן נחו נקטה ז̇ קדר מא ינקץ ט̇א̇ ען רבע.

B.1.II.2 וליכון עלי הדה אלצפה אן מרכז אלשמס מנתקלא̇ אבדא̇ עלי דאירה ט̇ו̇ש̇ קטבהא ב̇, וקטב ב̇ מנתקל אבדא̇ עלי דאירה ח̇ב̇ו̇ אלתי קטבהא ט̇ אלדי הו מחמול עלי דאירה אלממר. וקדר אלקוס מן דאירה אלממר אלכ̇ארג̇ה מן קטב כל ואחד מן האתין אלדאירתין אלי מחיטהא מסאוי לנצף פצ̇ל קוס צ̇עף אלמיל אלמדכור עלי קוס אעט̇ם אלאכ̇תלאף, אעני פצ̇ל [ט̇ה̇ עלי כ̇א̇][204]. [לאכן][205] כ̇א̇ שביה בקוס

198. א: -.
199. ו: קדצנא.
200. א: קטר.
201. ו: הוה. א: והו.
202. א: גלגל.
203. א: -.
204. א: ט̇ז̇ על ט̇ו̇.
205. א: לפי ש.

אעטׄם אלאכׄתלאף, פיכון אדא קוס שׁזׄ מתל כׄאׄ. פאדא כאנת נקטתי סׄ ודׄ אלדי הו קטב דאירה אלממר עלי נקטה הׄ כאנת נקטה טׄ, אעני נקטה אעטׄם אלתקציר, עלי

67v

נקטה טׄ מן דאירה אלברוגׄ. ופיהא כאן מרכז אלשמס וכאן קטב בׄ עלי נקטה חׄ מן דאירה חׄבׄוׄ.

B.1.II.24 ופי אלזמאן אלתי קצרת נקטה סׄ נצף אלדאירה אלמואזיה ואנתקל קטב דאירה אלממר רבע דאירה, אנתקל קטב בׄ חול קטב טׄ רבע דאירה איצׄאׄ, אעני מן נקטה חׄ אלי נקטה בׄ. ואנתקל מרכז אלשמס אלדי עלי נקטה טׄ חול קטב בׄ נצף דאירה, אעני מן נקטה טׄ אלי נקטה וׄ ואלי נקטה שׁ וצאר מרכז אלשמס עלי נקטה שׁ. אנתקל איצׄא בהדא אלזמאן חול קטב דאירה אלממר רבע דאירה פצאר עלי נקטה אׄ לאן קוס שׁטׄאׄ רבע.

B.1.II.25 ופי אלרבע אלתאלי לה [][206] רגׄע מרכז אלשמס בהדה אלחרכאת אלי נקטה טׄ, אעני מן דאירה טׄוׄשׁ לא אלי נקטה אעטׄם אלתקציר. ורגׄעת נקטה טׄ אלי נקטה זׄ, אעני אן נקטה טׄ קסמת חיניד אלנצף אלדי יחוזה דאירה הׄדׄ בנצפין לאן נקטתי אלתקאטע רגׄעת איצׄא אלי דאירה הׄדׄ. ואנתקל אלשמס איצׄא חול קטב דאירה אלממר רבע תאני פצאר עלי נקטה מׄ, ונקטה מׄ תכון חיניד עלי אלנקטה מן דאירה אלברוגׄ אלמקאבלה לנקטה טׄ. ויכון אלשמס חאפטׄאׄ אבדאׄ למחיט דאירה אלברוגׄ ודׄמׄאׄ.

B.1.II.26 ובדלך [בעינה][207] יטׄהר אן אלזמאן אלדי מן אעטׄם אלתקציר אלי אוסטה אצגר מן אלזמאן אלדי מן אוסטה אלי אצגרה. פאנא למכאן דלך אחתגׄנא אלי וצׄע אלפלך אלדי פיה דאירה אלממר מע האדין אלפלכין אלדאן פיהמא האתין אלדאירתין פי כל ואחד מן אלאצלין. ולולא דלך קד כנא נכתפי מן אגׄל אלאכׄתלאף בוצׄע האתין אלדאירתין פקט עלי דאירה אלברוגׄ מן גיר וצׄע דאירה אלממר ואלחאמלה להא. פקד ראית חקיקה כל ואחד מן האדין אלאצלין ואנה ליס [יבקי][208] פיהמא אלא אכׄתיאר [אלמאכׄד][209] פקט, עלי אנה ארי אן אלאצל אלאול אחרי אן יעמל עליה מן אגׄל סהולה

[B.1.II.26/X] אלעמל פיה [ודׄמׄאׄ][210].

B.1.II.27 פקד אתינא בגׄמיע אלאצול אלחקיקיה לחרכה אלשמס ואלאכׄתלאפאת אלתי תטׄהר פיה, מע לזומנא ארא בטלמיוס [פי כל דלך][211]. וליס ישוב שי מן דלך עדם אלאמכאן אצלאׄ לאן הדה אלאצול מבניה עלי אן אלחרכאת כלהא עלי אכר תאמה אלאסתדארה. וליס תם חרכה אלא והי מתסאויה מתשאבהה פי אזמאן מתסאויה ועלי אסתדארה חול אקטאב מחדודה, או תאבעה לחרכה מסתדירה שאפעה מעהא חול

68r

אקטאבהא איצׄא מן גיר אחתיאגׄ פי שי מן דלך אלי כׄארגׄ אלמרכז ולא פלך תדויר. וטׄאהר איצׄא ממא קלנא פי אלשמס אן אלגׄדאול אלתי [עמל עליהא][212] בטלמיוס ללחרכאת ואלאכׄתלאפאת הי באעיאנהא אלגׄדאול אלמחתאגׄה ללחסאב בתלך אלאצול, אלא אן אלדי יסמי ענדה חרכה אלי כׄלף יתסמי ענדנא פי אלאצל אלתאני תקצירא. וקד יתבין דלך פימא יסתאנף [][213].

206. א: ונמשך לו.
207. א: -.
208. ו: יבק. א: ישאר.
209. א: -.
210. א: -.
211. א: -.
212. א: חקק אותם.
213. א: בׄהׄ.

B.1.II.2 ואמא אלבטרוג̇י פאן ראיה פי הדא אלאכ̇תלאף ראי בעיד ען אלחק, פאנה ירי אן מרכז אלשמס יתחרך דאימא̇ עלי דאירה מילה עלי אנה מחמול עליהא פי אלחרכה. ואן סבב אלאכ̇תלאף אלמרי פיה אנהא הו לאג̇ל אנתקאל קטבי דאירה מילה עלי דאירה ממר קטבי דאירה אלברוג̇. וזעם אנה אדא וצ̇ע קטבי דאירה מיל אלשמס יקצראן קטאע דאירה ממר קטבי אלברוג̇ אלמתסאויה פי אזמנה מכ̇תלפה, אן כדלך יערץ̇ פי אלשמס פי דאירה אלברוג̇ אנה יקצר פי אזמנה מכ̇תלפה קסי מתסאויה. והדא אנמא יכון עלי מא קאלה לו כנא לסנא נקצד באלאכ̇תלאף אלא אן יכון אלשמס יקצר אג̇זא אלברוג̇ פי אזמנה מכ̇תלפה פי אלטול ואן כאן זאילא ענהא פי אלערץ̇ אד כדלך ילזם מן וצ̇עה לאנה ואן קצר אלשמס פי אלטול אלרבע אלדי מן אלאנקלאב אלשתוי אלי אלאסתוא אלרביעי מתלא̇ פי אקל מן רבע איאם אלסנה, פאנה ליס יקטע מן דאירה מילה רבע תאם פי הדא אלזמאן חתי ינתהי אלי עקדה הדה אלדאירה מע [][214] מעאדל אלנהאר כמא אנתהי פי אלברוג̇, אד ליס האהנא שי יוג̇ב אן יקטע ארבאע הדה אלדאירה פי אזמנה גיר מתסאויה.

B.1.II.2 פאדא ליס יכון אלשמס אבדא̇ עלי דאירה אלברוג̇ ואדא לם יכון פי וקת אלאסתוא עלי דאירה אלברוג̇, אעני פי וקת כון אלשמס פי ראס אלחמל כמא ילזם ען וצ̇עה, פכיף יסלם אלארצאד אלתי וצ̇עהא בטלמיוס לאוקאת אלאסתואין אלתי רצדהא בחלקה אלנחאס אלמנצובה פי סטח דאירה מעאדל אלנהאר, [][215] באנהא אצ̇את אלחלקה מן אלג̇אנבין פי הדא אלוקת ועמל עלי תלך אלארצאד איצ̇א. פאנה ליס ימכן עלי וצ̇עהא אן יוג̇ד בהדה אלאלה אלאזמאן אלדי בין כון אלשמס פי נקט אלאסתואין ואלאנקלאבין מכ̇תלפא̇ אצלא̇,

68v

לאנה ליסת תצ̇י אלחלקה [מן ג̇אנביהא][216] אלא פי וקת כון [][217] אלשמס פי נקט תקאטע דאירה מיל אלשמס מע מעאדל אלנהאר, וליס ינתהי אלי הדה אלנקט אלא פי אזמנה מתסאויה.

B.1.II.3 ואיצ̇א פאנה יצ̇ע לאג̇ל תקציר אלקטב דאירה ממרה פי אזמנה מכ̇תלפה אנה ינתקל עלי דאירה זאילה אלקטב. ובחסב דלך ליס יוג̇ד אלשמס דאימא̇ עלי דאירה אלברוג̇, והו יקר אן אלשמס ליס יכ̇רג̇ ען דאירה אלברוג̇. פאלעג̇ב כל אלעג̇ב כיף אגפל דלך כלה. ואנא אחסב אן גרצ̇ה אנמא הו אן יעטי אצול לאכ̇תלאף מא יבתדעה [][218] מן נפסה ליס בלאזם לארא אלקדמא פיה. ומע דלך פאנה יבני [בראהינה][219] עלי ארצאד אלקדמא פיה, פוקע מן דלך פי אצ̇טראב בין אלזלה והו לא ישער. וקד אסתכ̇רג̇נא להדה אלאצול אלדי וצ̇עהא הדא [אלרג̇ל][220] וג̇וה ימכן אן יט̇הר מנהא אלמטלוב, אלא אנהא ליסת תכ̇לוא עו [אמתנאעאת][221] אשד ואשנע עו אלאצול אלמסלוכה אלי אלאן. ולדלך אסקטנא כל דלך מן הדא אלכתאב ונרג̇ע אלי חית כנא ונבין מקדאר הדא אלאכ̇תלאף.

B.1.III. אלפצל אלתאלת פי תבין קדר אכתר אלאכ̇תלאף אלדי ירי פי אלשמס ווג̇ה חסאב אלאכ̇תלאפאת אלג̇זיה ותבין אלנקטה מן אלברוג̇ אלתי פיהא אקל אלתקציר או אלסיר. ודלך יתהיא באן נבין אולא̇ נסבה נצף קטר פלך אלשמס אלי נצף קטר דאירה ממר אלמרכז. וכדלך יוג̇ב אן נבין אלנקט מן

214. א: גלגל.
215. א: וזה ידוע.
216. ו: מן ג̇אנבהא. א: משני צדדיה.
217. א: מרכז.
218. א: וברא אותו.
219. ו: בראהנה.
220. ו: אלרג̇ול. א: האיש.
221. ו: אמתנעאת. א: מניעיות.

אלברוג̇ אלדי פיהמא אעט̇ם אלמסיר ואקלה, אעני [][222] אלנקטה מן אלברוג̇ אלדי יכון מרכז אלשמס פיהא אדא כאן עלי נקטה אקל אלמסיר. ודלך יתבין באלארצאד אלדי בהא בין בטלמיוס קדר אלכ̇ט אלדי בין אלמרכזין ומוצ̇ע אלאוג̇ מן אלברוג̇. פאנה בין דלך באן וג̇ד אלזמאן אלדי מן וקת חלול אלשמס פי נקטה אלאסתוא אלרביעי אלי [נקטה][223] חלולהא פי נקטה אלאנקלאב אלציפי צ̇ד̇ יומא̇ ונצף. [ומן][224] אלציפי אלי [][225] אלכ̇ריפי צ̇ב̇ יומא̇ ונצף ואלשמס יקטע עלי אסתוא פי צ̇ד̇ יומא̇ ונצף צ̇ג̇ ג̇ז ט̇ דקיקה ויקטע פי צ̇ב̇ יום ונצף צ̇א̇ ג̇זא וי̇א̇ דקיקה.

B.1.III.2 פלתכן דאירה ממר אלמרכז א̇ב̇ג̇ ואלפצל אלמשתרך להא ולסטח אלדאירה אלחאמלה להא

69r

כ̇ט נ̇ל̇כ̇. ונצ̇ע אלשמס פי וקת אלאסתוא אלרביעי עלי נקטה א̇ ותחרך אלי וקת אלאנקלאב אלציפי קוס א̇ב̇ מבלגה צ̇ג̇ ט̇ [דקי̇][226] יותר מן אלברוג̇ ג̇ ט̇ ללנקצאן. ותחרך מן אלאנקלאב אלציפי אלי אלאסתוא אלכ̇ריפי קוס ב̇ג̇ מבלגה צ̇א̇ י̇א̇ יותר מן אלברוג̇ א̇ י̇א̇ ללנקצאן איצ̇א̇. ונכ̇רג̇ מן נקטה ב̇ עמוד עלי אלפצל אלמשתרך ינתהי אלי מחיט דאירה אלממר פי אלג̇הה אלאכ̇רי והו עמוד ב̇ל̇ה̇. ונכ̇רג̇ מן נקט א̇ג̇ עלי כ̇ט ב̇ה̇ עמודי א̇מ̇ ג̇ז̇ ונצל א̇ה̇ ג̇ה̇ א̇ג̇, ונכ̇רג̇ מן נקטה ג̇ כ̇ט ג̇ט̇ [][227] עלי כ̇ט א̇ה̇, אדא אכ̇רג̇ עלי אסתקאמה, לאן זאויה א̇ה̇ג̇ מנפרג̇ה. ונכ̇רג̇ מן נקטה ד̇ אעני מרכז אלברוג̇ אלי נקט א̇ג̇ כ̇טי
[B.1.III.2/x] ד̇ג̇ ד̇א̇ ואלי נקטתי מ̇ז̇ כ̇טי ד̇מ̇ ד̇ז̇.

B.1.III.3 [פלאן מוצ̇ע נקטה ב̇ מן אלברוג̇ הו נקטה מ̇][228] פתכון זאויה א̇ד̇מ̇, אעני אלתי תנקץ ען אלתקציר בחסב אלראי אלתאני או אלתי תנקץ ען אלמסיר בחסב אלראי אלאול, בחסב קוס ב̇א̇ מעטאה.
פנבסה כ̇ט ד̇א̇, אעני נצף קטר דאירה אלברוג̇ אלמתוהמה פי פלך אלשמס אלי אלעמוד אלכ̇ארג̇ מן נקטה א̇ עלי כ̇ט ד̇מ̇ מעטאה. והדא אלעמוד אמא אדא כאן כ̇ט ב̇ה̇ קטרא̇ לדאירה אלממר יכון הו בעינה כ̇ט א̇מ̇. ואמא אדא כאן גיר אלקטר פליס הו כ̇ט א̇מ̇. ונחן נעמל פי אסתכ̇ראג̇ אלנסבה עלי אנה כ̇ט א̇מ̇ ולא [נבאלי פי אלעאג̇ל][229] באלפצ̇ל אלדי יכון בינהמא אד אלפרק בינהמא קליל. תם נעד בתבין הדא אלפצ̇ל ונצלח בה אלנסבה.

B.1.III.4 פתכון אדא̇ נסבה ד̇א̇ אלי א̇מ̇ מעטאה, ולאן קוס ב̇א̇ בל זאויה ב̇ה̇א̇ מעטאה פנסבה א̇ה̇ אלקטר אלי א̇מ̇ מעטאה פנסבה ד̇א̇ אלי א̇ה̇ מעטאה. ואיצ̇א̇ פלאן מוצ̇ע נקטה ב̇ מן אלברוג̇ הי נקטה ז̇ פזאויה ג̇ד̇ז̇ מעטאה. פנסבה ד̇ג̇ אעני ד̇א̇ אלי אלעמוד אעני ג̇ז̇ עלי אלשרט אלמדכור מעטאה. ולאן קוס ב̇ג̇ בל זאויה ב̇ה̇ג̇ מעטאה פנסבה ג̇ה̇ אלקטר אלי ג̇ז̇ מעטאה. פאדא̇ נסבה ד̇א̇ אלי ג̇ה̇ מעטאה. [ולאן זאויה א̇ה̇ג̇ בל ג̇ה̇ט̇ מעטאה][230] פנסבה ג̇ה̇ אלקטר אלי ג̇ט̇ בל אלי ה̇ט̇ מעטאה, פתכון נסבה ג̇ה̇ אלי ט̇א̇ בל אלי

222. א: הנקודה מהמזלות אשר יהיה מרכז השמש בה כשיהיה על נקודת המהלך הגדול.
223. א: עת.
224. א: ומהיפוך.
225. א: השואת.
226. א: -.
227. א: עמוד.
228. א: -.
229. א: נחוש.
230. ו: 2.

69v

גׄאׄ אלקוי עלי גׄטׄ טׄאׄ מעטאה. פנסבה דׄאׄ אלי [אׄגׄ][231] מעטאה וקד כאנת נסבה גׄאׄ אלי קטר דאירה אלממר מעטאה לאן קוס אׄבׄגׄ מעטאה. נסבה דׄאׄ אעני נצף קטר פלך אלשמס אלי נצף קטר דאירה אלממר מעטאה. ויכֿרג בדלך אלנסבה אלתי אסתכֿרגהא בטלמיוס והו אן נצף קטר דאירה אלממר גׄזׄ מן כׄדׄ באלתקריב מן נצף קטר דאירה אלשמס. ודלך בעד אלאצלאח אלדי נדכרה ען קריב.

B.1.III. ואמא כיף נעלם אלגֿז מן אלברוג אלדי יכון אלשמס פיה אדא כאן פי נקטה אקל אלתקציר, פדלך יכון באן נעלם בעד נקטה בׄ מן נקטה אקל אלתקציר, והדא מעלום לאן נסבה גׄהׄ אלי גׄאׄ מעטאה, פקוס גׄהׄ מעטי בל קוס בׄגׄהׄ פנצפה אעני בׄנׄ מעטי. פבעד נקטה בׄ מן נקטה אקל אלתקציר אעני נקטה וׄ מעטי. והדא הו בעד נקטה אלאנקלאב אלציפי מן אלנקטה אלתי תכון אלשמס פיהא אדא כאן פי נקטה אקל אלתקציר אלי תואלי אלברוג [][232], ותכון הדה אלנקטה עלי הׄ אגֿזא באלתקריב מן אלתומין כמא כֿרג לבטלמיוס. ותכון נקטה אעטׄם אלתקציר פי אלגֿז אלמקאבל מן אלברוג.

B.1.III. פאד קד בינא דלך פנרגע ונבין מקדאר אלפצֿל אלדי בין כֿטי אׄמׄ גׄזׄ ובין אלעמודין אלכֿארגין מן נקטתי אׄגׄ עלי כֿטי דׄמׄ דׄזׄ לכי נצלח כדלך אלנסבה. ונקדם לדלך אלמקדמה אלתי בינאהא פי אלגֿמלה אלאולי והי הדה. אדא וקעת פי דאירה תלאת נקט בחית יכון כל ואחדה מן אלקוסין אלמנפצלה בהדה אלנקט אצגר מן נצף דאירה ווצל כֿט בין אלנקטה אלאולי ואלתאלתה, אעני ותר אלקוס אלדי הו מגֿמוע אלקוסין, תם וצל כֿט בין מרכז אלדאירה ובין אלנקטה אלוסטי מן אלתלאתה יקטע אלכֿט אלמדכור בקסמין, פאן נסבה גיב אחדי אלקוסין אלמנפצלה אלי גיב אלאכֿרי כנסבה [אלקוס][233] מן אלותר אלמדכור אלדי ילי הדה אלקוס אלי קסמה אלתאני.

B.1.III. פאד קדמנא דלך פלנתוהם אן כֿט אׄמׄ פצל אלמשתרך לדאירה אלממר ולדאירה אלברוג, פלנכֿרגה אלי נקטה חׄ פיציר כֿט אׄמׄחׄ ותראׄ ללקוס מן דאירה אלברוג אלדי יותרה קוס אׄבׄחׄ מן דאירה אלממר. ואדא אכֿרג כֿט דׄמׄ עלי אסתקאמה צאר יקטע הדא אלקוס מן דאירה אלברוג, פקד צאר פי הדא אלקוס גׄ [נקט][234] אלאתנין המא נקטתי אׄחׄ ואלוסטי אלנקטה אלתי תקטע עליהא כֿט דׄמׄ להדא אלקוס. ואלקוסין אלמנפצלה בהדה אלנקט מעטאה לאן אלקוס אלדי מן נקטה אׄ אלי אל

70r

נקטה אלוסטי הו במקדאר זאויה [אׄדׄמׄ][235], ואלקוס אלתאני הו במקדאר אלזאויה אלתי תזיד או תנקץ בסבב קוס בׄחׄ. ותכון מעטאה לאן קוס בׄוׄ מעטי, ובחסב אלנסבה אלקוס מן דאירה אלברוג אלדי יותרה בׄוׄ מעטי וכדלך אלקוס אלדי יותרה קוס וׄחׄ. פיבקי אלקוס אלדי יותרה בׄחׄ מעטי. פבחסב אלמקדמה נסבה גיב אלקוס אלדי מן נקטה אׄ אלי אלנקטה אלוסטי, והו אלעמוד אלכֿארג מן נקטה אׄ עלי כֿט דׄמׄ והו מעטי, אלי גיב אלקוס אלתאני והו אלעמוד אלכֿארג מן נקטה חׄ עלי הדא אלכֿט, והו מעטי איצׄאׄ, כנסבה כֿט אׄמׄ אלי כֿט מׄחׄ. ואד כל כֿט אׄחׄ מעטי פכֿט אׄמׄ מעטי. ובמתל דלך יתבין מקדאר כֿט גׄזׄ ודׄמׄאׄ.

231. א: דׄגׄ.
232. א: אחר תקונו בחלוף אשר יחייב אותו.
233. א: החלק.
234. ו: נקאט. א: נקודות
235. א: אׄכׄמׄ.

B.1.III.8 פאדא נחן כררנא אסתכ̇ראג̇ אלנסבה בהדה אלאקדאר וכ̇רג̇ת אלנסבה אלדי בין [נצפי][236] אלקטרין, ואסתכ̇רג̇נא מקדאר אלפצ̇ול מרה תאניה. וכררנא אסתכ̇ראג̇ אלנסבה מרה תאניה בעד אן נזיד או ננקץ עלי אלקסי מקדאר אלאכ̇תלאף אלתי תוג̇בהא תלך אלקסי מן אג̇ל אנתקאל אלשמס חול
[B.1.III.8/x] קטבי אלדאירתין אלמדכורתין. כ̇רג̇ת אלנסבה אלצחיחה. וט̇אהר אנה ימכן וצ̇ע [אצל לתקייד][237] אלחרכאת אלוסטי באוקאת אלארצאד אלתי דכרהא ללאסתואין או אלאנקלאבין ונסתכ̇רג̇ פי כל וקת שינא אלמואצ̇ע אלוסטי ללשמס פי דאירה אלברוג̇ פי דאירה אלממר.

B.1.III.9 וכימא יתבין וג̇ה חסאב אלאכ̇תלאפאת אלג̇זיה פלתכן דאירה ממר אלמרכז אה̇ג̇מ̇ מרכזהא ז̇ ואלפצל אלמשתרך להא ולסטח אלדאירה אלחאמלה אדא כאנת ממאסה לדאירה אלברוג̇ כ̇ט ב̇ז̇מ̇. וליכן כ̇ט א̇ז̇ג̇ קאימא̇ עליה, פנקטתי א̇ג̇ אחדהמא נקטה אעט̇ם אלתקציר ואלתאניה נקטה אקל אלתקציר. ולתכן נקטה אקל אלתקציר נקטה ג̇ ולננזל נקטה אלמג̇אז אלאוסט ב̇ ונצל ב̇ז̇ ודב̇ ודז̇. ולאן נסבה ד̇ב̇ אלי ב̇ז̇ מעטאה וזאויה ז̇ קאימה, פזאויה ב̇ד̇ז̇ והי זאויה אעט̇ם אלאכ̇תלאף מעטאה. ולנצ̇ע מרכז אלשמס עלי נקטה ח̇ עלי אן יכון בעדה [מן נקטה][238] ג̇, אעני נקטה אעט̇ם אלתקציר, מעטאה. ונכ̇רג̇ מן נקטה ח̇ עמוד ח̇ט̇ עלי כ̇ט א̇ג̇ ונצל ד̇ח̇ ד̇ט̇. פלאן קוס ח̇ג̇ בל זאויה ח̇ז̇ג̇ אעני זאויה אלחרכה אל

70v

מסתויה מעטאה פנסבה ז̇ח̇ אלי ח̇ט̇ מעטאה. [פאדא̇ נסבה ד̇ח̇ אעני נצף אלקטר אלי ח̇ט̇ מעטאה][239] וזאויה ח̇ט̇ד̇ קאימה. פזאויה ח̇ד̇ט̇ אעני זאויה אלאכ̇תלאף אלתי תוג̇בהא קוס ח̇ג̇ מעטאה, וכדלך יתבין פי סאיר אלאוצ̇אע. ובמתל דלך יתבין אנהא לו כאנת זאויה אלאכ̇תלאף מעטאה פאן זאויה אלחרכה אלמסתויה מעטאה. ותכון מן קבל דלך זאויה אלחרכה אלחקיקיה, אעני אלמריה, מעטאה באן נזיד זאויה אלאכ̇תלאף עלי אלחרכה אלוסטי אדא כאן אלשמס עלי קוס ג̇ה̇א̇ וננקצהא ענהא אדא כאן עלי קוס א̇ב̇ג̇.

B.1.III.10 ולאן אדא כאן אלשמס עלי נקטה אעט̇ם אלתקציר אעני נקטה ג̇, ואנתקל אלי נקטה ח̇ בקדר אלחרכה אלוסטי, אנתקל איצ̇א מרכז אלשמס זאידא̇ אלי תלך אלג̇הה ללמעני אלמדכור באנתקאלה חול קטבי אלדאירתין אלמדכורתין. ומקדאר הדה אלנקלה הו במקדאר קוס אלאכ̇תלאף. פלדלך נזיד עלי קוס ח̇ג̇ [][240] מתל קוס אלאכ̇תלאף אלדי יוג̇בה ונחסב תאניא̇ קוס אלאכ̇תלאף אלדי יוג̇בה הדא אלקוס אלמג̇מוע והו קוס אלאכ̇תלאף אלחקיקי. ולא נזאל נזיד הדא אלקוס עלי אלחרכה אלוסטי ליסתכ̇רג̇ בה אלאכ̇תלאף מא דאם אלשמס עלי אלקוס אלזאיד פי אלתקציר, אעני אלקוס אלדי בין אלמג̇אזין אלוסטין מן נאחיה אעט̇ם אלסיר. ולא נזאל ננקצה פי באקי אלדאירה ונחסב אלאכ̇תלאף אלדי יוג̇בה אלבאקי והו אלאכ̇תלאף אלחקיקי. פקד אתינא עלי ג̇מיע מא ינבגי אן יבני עליה מן אמר חרכה אלשמס ואכ̇תלאפה וכלה מטאבק למא יכ̇רג̇ ען אצול בטלמיוס אלא אלשד אלדי הו כ̇אצה תלך אלאצול אלדי ליס ידכ̇ל צ̇רר מן קבלה מן גיר אן יחתאג̇ פי שי מנה אלי וצ̇ע פלך כ̇ארג̇ אלמרכז ולא
[B.1.III.10/x] פלך תדויר. ואללה תעאלי אלעאלם. כמלת אלמקאלה אלאולי [][241] מן כתאב נור אלעאלם ואלחמד ללה תעאלי.

236. א: -.
237. א: עקר שורש מונח.
238. א: -.
239. א: -.
240. א: קשת.
241. א: מן הכלל השני.

71r

B.2.0.1 אלמקאלה אלתאניה פי תבין אלאצול אלדי עליהא ינבגי אן יעמל פי חרכאת אלקמר ואלאכ̇תלאפאת אלתי ת̇רי פיה ופיהא כ̇מסה פצול.

B.2.1.1 [אלפצל][242] אלאול פי אלאצול אלמעטאה לחרכאת אלקמר אלדוריה. אמא נקלה אלקמר בחסב אלראי אלתאני פהי [שביהה][243] בנקלהִ אלשמס אלא אן תקצירה אכתר מן תקציר אלשמס מן אג̇ל אנכסאר אלקוה לבעדה ען אלמחרך [אלאול][244]. ותט̇הר פיה איצ̇א כתרה אלאכ̇תלאפאת ויג̇ב אן תכון אצולנא אלמעטאה פי דלך מבניה עלי ארצאד בטלמיוס כמא עמלנא פי אלשמס. פנקול אן בטלמיוס וג̇ד אלקמר קד יבעד נחו אלשמאל ונחו אלג̇נוב ען דאירה אלברוג̇ באג̇זא מתסאויה. פאדא̇ אלקמר ואג̇ב אן ינתקל עלי דאירה מאילה עלי דאירה אלברוג̇ בקדר תבאעד אלקמר פי אלערץ̇ ען דאירה אלברוג̇ אלדי וג̇ד נחו מן ה̇ אג̇זא. ווג̇ד איצ̇א אן עודה אלערץ̇ אסרע מן עודה אלטול. ודלך אנה וג̇ד בארצאדה אנה יסתכמל אלעודה פי אלערץ̇ פי זמאן אקל ממא יסתכמל בה אלעודה פי אלטול. ודלך ליס ימכן בחסב אצולהם אלא באן תנתקל אלעקדה [][245] עלי אג̇זא פלך אלברוג̇ נחו אלמגרב، אעני אלי כ̇לאף תואלי אלברוג̇. ואסתדל איצ̇א עלי אן אלעקדה מנתקלה אנה וג̇ד אכתר אלערץ̇ נחו אלשמאל מרה ונחו אלג̇נוב מרה פי ג̇ז [ואחד][246] מן אלברוג̇ מע כון אלקמר פי כליהמא שמאלי ען מעאדל אלנהאר או ג̇נובי. ווג̇ד מרה לא ערץ̇ לה פי דלך אלג̇ז בעינה.

B.2.1.2 ואסתדל עלי דלך איצ̇א לאנה וג̇ד קד ינכ̇סף אלשמס ואלקמר פי כל ואחד מן אג̇זא אלברוג̇ חאפט̇א להדא אלתרתיב، פהאהנא אדא̇ חרכתין פי אלערץ̇. אחדהמא חרכה אלקמר עלי קטבי [אפלאכה][247] אלמכ̇צוצה בה אעני עלי קטבי דאירה מיל אלקמר. פאן בהדה אלחרכה ירי אלקמר מרה ג̇נובי ען דאירה אלברוג̇ ומרה שמאלי، ליס בג̇ז ואחד מן אלברוג̇ חתי אנה אן כאן אכתר ערץ̇ אלג̇נוב פי ג̇ז ואחד יכון אכתר ערץ̇ אלשמאל באלג̇ז אלמקאטר לה ומן אלנאחיה אלאכ̇רי מן מעאדל אלנהאר. והדא לו כאנת נקטתי אלעקדתין סאכנה חאפט̇ה לאג̇זא ואחדה מן דאירה אלברוג̇. פאן בתלך אלחרכה יציר ללקמר ערץ̇ באלקיאס אלי דאירה אלברוג̇ ובאלקיאס אלי מעאדל אלנהאר ג̇מיעא̇.

B.2.1.3 ואלחרכה אלתאניה אנתקאל דאירה אלערץ̇ מע נקטתי אלעקדתין עלי אג̇זא אלברוג̇ אלי

71v

כ̇לאף ג̇הה תואליהא חאפט̇ה לערצ̇הא מן דאירה אלברוג̇. פאן בתלך אלחרכה מכ̇תלטה באלאולי יציר ללקמר ערץ̇. ויציר אלקמר מרה פי גאיה בעדה פי אלג̇נוב ומרה פי גאיה בעדה פי אלשמאל [פי][248] ואחד מן אלברוג̇ מע כון אלקמר [פי אלנקטתין][249] פי נאחיה ואחדה מן מעאדל אלנהאר. ודלך לאן אנתקאל אלעקדה מע חפט̇ גאיה מילהא ען דאירה אלברוג̇ ליס הו סוי אנתקאל קטבי פלך אלקמר، אעני קטבי דאירה ערץ̇ אלקמר עלי דאירה ממר קטבהא קטב פלך אלברוג̇، ואלקוס אלתי מן קטבהא אלי מחיטהא מסאוי לגאיה מיל דאירה אלערץ̇ ען דאירה אלברוג̇. פאדא אתפק אן יכון

242. א: -.
243. ו: שביה. א: דומה.
244. א: -.
245. א: הנקרא ראש אלתלי.
246. א: אשה.
247. ו: פלכה. א: גלגליו.
248. א: בחלק.
249. א: בשני העתים.

קטב פלך אלקמר פי גאיה בעדה ען קטב אלכל וכאן אלקמר פי אלנהאיה אלג̇נוביה, מתלא̇ מן דאירה מילה, כאן אלקמר חיניד פי גאיה אלמיל אלדי ימכן אן יכון לה ען דאירה אלברוג̇ וען מעאדל אלנהאר. פיכון חיניד מיל אלקמר ען מעאדל אלנהאר אלמילין ג̇מיעא̇ נחו אלג̇נוב, ואן כאן פי אלנהאיה אלשמאליה פנחו אלשמאל עלי אלג̇ז אלמקאטר לה.

B.2.1.4 ואמא אן אתפק אן יכון קטב פלך אלקמר פי גאיה קרבה מן קטב אלכל פי דאירה ממרה וכאן אלקמר פי אלנהאיה אלשמאליה, כאן אלקמר חיניד פי גאיה מילה נחו אלשמאל ען דאירה אלברוג̇ פי דלך אלג̇ז בעינה אלדי מאל ענה נחו אלג̇נוב אולא̇. וכאן חיניד ערץ̇ אלקמר ען מעאדל אלנהאר אלי ג̇הה אלג̇נוב פצ̇ל אעט̇ם אלמילין עלי אלאצגר. ואן כאן פי אלנהאיה אלג̇נוביה פבאלעכס. ולדלך סמינא הדה אלחרכה חרכה אלערץ̇ אלדי באלקיאס אלי אלברוג̇ וסמינא אלאולי חרכה אלערץ̇ אלכליה. ובטלמיוס חקק הדה אלחרכאת בארצאדה אלמתואליה בכסופאת אלקמר אד ליס ידאכ̇להא אכ̇תלאף מנט̇ר. ווג̇ד חרכה אלטול אלוסטי פי אליום יג̇ י̇ לה̇ באלתקריב וחרכה אלערץ̇ יג̇ יג̇ מו̇ באלתקריב, אעני [אלערץ̇][250] אלכלי. ווג̇ד חרכה אלאכ̇תלאף אעני חרכה אלקמר פי פלך אלתדויר אלדי וצ̇עה יג̇ ג̇ נד̇ באלתקריב פי אליום. פיכון זיאדה חרכה אלערץ̇ עלי חרכה אלטול ג̇ דקי̇ פי אליום. ובהדא אלקדר יכון אנתקאל אלעקדה [][251] אלי כ̇לאף פאן אלחרכה אלתי תכון עלי דאירה אלערץ̇ הי בעינהא תפעל חרכה אלטול סוי הדה אלתלאת דקי̇ אלתי תנתקץ ען אלטול למכאן אנתקאל אלעקדה אלי כ̇לאף. ודלך לאן מיל הדה אלדאירה ען דאירה אלברוג̇ קליל בחית ליסת תוג̇ב זיאדה ולא נקצאן פי חרכה אלטול.

72r

B.2.1.5 ולנצ̇ע אולא̇ אלאצל ללחרכאת אלוסטי פי אלטול ואלערץ̇. ולתכן דאירה אלברוג̇ אלמתוהמה פי אלפלך אלמכוכב דאירה ג̇ט̇ד̇ קטבהא ב̇ אלדאיר עלי דאירה לב̇נ̇ אלתי קטבהא קטב [][252] מעאדל אלנהאר והו א̇. ולתכן אלדאירה אלמאילה אלתי ללקמר אלמתוהמה פי הדא אלפלך איצ̇א דאירה ג̇ה̇ד̇ קטבהא מ̇ והדא אלקטב הו דאיר אבדא̇ עלי דאירה ממר למ̇נ̇ קטבהא קטב דאירה אלברוג̇ אעני ב̇. ואלדאירה אלמארה בנקטתי אלאנקלאבין ובאלאקטאב דאירה ה̇ט̇ב̇א̇. פאן בחסב אלראי אלאול ליס יחתאג̇ אלי שי סוא אן נתוהם האתין אלדאירתין בהדה אלצפה פי פלך אלקמר עלי אן יכון מקסום כמא קלנא. ואמא בחסב אלראי אלתאני פלנצ̇ע פי פלך אלקמר דאירתין [אכ̇ריין][253] מקאם דאירה אלברוג̇ ואלדאירה אלמאילה, כל ואחדה מנהמא פי סטח נט̇ירתהמא מן [אללואתי][254] פי אלפלך אלאעלי, וקטב דאירה אלערץ̇ דאירא̇ [][255] עלי דאירה ממר קטבהא קטב אלדאירה אלממתלה בדאירה אלברוג̇ שביהה באלמתוהמה פי אלפלך אלאעלי.

B.2.1.6 ונצ̇ע מתלא̇ אן אלקמר פי [נהאיתה][256] אלג̇נוביה ולתכן נקטה ה̇ ומוצ̇עה מן אלברוג̇ נקטה אלאנקלאב אלשתוי. ותכן חיניד דאירתי אלברוג̇ ואלערץ̇ אלמתוהמה פי פלך אלקמר פי סטחי נט̇אירהא אלמתוהמה פי אלפלך אלאעלי. וכדלך אקטאבהא אלנט̇איר עלי מחור ואחד וערץ̇ אלקמר ען מעאדל

250. א: הקירוב.
251. א: ר̇ל̇ ראש התלי וזנבו.
252. א: גלגל.
253. ו: אכ̇רין. א: אחרות.
254. א: המדומות.
255. א: גם כן.
256. ו: נהאתיה. א: בתכליתו.

אלנהאר אלמילין אג̇מע. ונצ̇ע אן פי יום ואחד מתלא̇ אנתקל פלך אלקמר בכליתה תאבע ללחרכה אליומיה וקצר כל ואחד מן קטבי דאירה אלברוג̇ ודאירה אלערץ̇ אלמתוהמה פיה ען קטב אלאעלי קוס מן דאירה ממר קטב דאירה אלברוג̇, מקדארה צ̇עף אלתקציר אלאכ̇יר.

72v

וצאר קטב דאירה אלברוג̇ עלי נקטה ח̇ מן דאירה ממר אלקטב וקטב דאירה אלערץ̇ עלי נקטה כ̇. ומתאל דלך קצר נקטה אלאנקלאב אלשתוי אלמתוהמה פי פלך אלקמר ונקטה אלנהאיה אלג̇נוביה מן דאירה אלערץ̇ קוסין מן אלדאירתין אלמתואזיתין למעאדל אלנהאר מקדאר כל ואחדה מנהא צ̇עף אלתקציר אלאכ̇יר אעני [][257] מתל קוס ב̇ח̇. וצארת מתלא̇ דאירה אלערץ̇ אלתי כאנת אולא̇ מתוהמה פי סטח דאירה ג̇ה̇ד̇ הי דאירה ס̇ז̇ ואלנהאיה אלג̇נוביה אלתי כאנת עלי נקטה ה̇ צארת עלי נקטה ס̇.

B.2.1.7 ינבגי אן נבין כיף יקטע מרכז אלקמר אלרבע מן דאירה ערצ̇ה אלדי מן נקטה ס̇ אלי תקאטע דאירה אלערץ̇ מע דאירה אלברוג̇, אעני אלעקדה, באלזמאן אלדי יקטע רבע ה̇ד̇ אכ̇דא מן נקטה ה̇ אלי כ̇לאף חרכה אלכל עלי ראי בטלמיוס ועלי ראינא בחסב אלראי אלאול. ומע דלך ליס יקטע רבע תאם מן דאירה אלברוג̇ אלמתוהמה פי פלך אלקמר פי הדא אלזמאן כמא אנה ליס יקטע רבע ט̇ד̇ מן דאירה אלברוג̇ פי הדא אלזמאן עלי אלראי אלאול. ואנמא לזם דלך עלי אלוצ̇ע אלאול לאן נצ̇ע פיה אן [עקדה][258] ד̇ מנתקלה אלי ג̇הה חרכה אלכל אעני נחו נקטה ט̇, פאלקמר יקטע רבע ה̇ד̇ פי אלזמאן אלדי יקטע פיה אקל מן רבע ט̇ד̇.

B.2.1.8 ודלך יכון בחסב הדא אלאצל אלתאני עלי הדה אלצפה. אן מרכז אלקמר יתחרך עלי קטב דאירה אלערץ̇, אעני קטב ב̇, פי יום ואחד קוס ס̇ז̇ מבלגה אלאג̇זא אלדי יקול בטלמיוס אנהא חרכה אלערץ̇ והי י̇ג̇ מ̇ו̇ באלתקריב. והדה אלחרכה ליסת בהדא אלמקדאר פי דאירה אלברוג̇ אלמתוהמה פי פלך אלקמר בל אקל מנהא בג̇ דקי̇. ודלך יכון אדא וצ̇ענא אן נקטה ס̇, אעני אלנהאיה אלג̇נוביה מן דאירה אלערץ̇ אלמתוהמה פי פלך אלקמר, קצרת פי יום ואחד ען אלוצול אלי מוצ̇עהא אכתר ממא קצר נקטה אלאנקלאב אלשתוי אלמתוהמה פי הדא אלפלך ג̇ דקי̇, פאנחרפת [אלמחאדאה][259] אלתי כאנת בין נקטה ס̇ ובין נקטה אלאנקלאב. וצאר נקטה קטב ס̇ עלי ג̇ דקי̇ מן ראס אלג̇די מע חפט̇ גאיה מילהא. אעני אן מיל נקטה ס̇ פי אל

73r

ערץ̇ ען הדה אלנקטה אלתי הי עליהא מסאויא̇ ללמיל אלתי כאן להא ען ראס אלג̇די אולא̇.

B.2.1.9 פאדא קצר קטב דאירה אלברוג̇ אלמתוהם פי פלך אלקמר אעני ב̇ קוס ח̇ב̇ כאן ואג̇בא̇ אן לא יצל קטב דאירה אלערץ̇ אעני מ̇ אלי נקטה כ̇. אעני באן תכון אלאקטאב אלתלאתה עלי קוס ואחד מן דאירה עט̇ימה מתל קוס א̇ח̇כ̇ס̇, לאנה לו כאן כדלך כאן תקציר אלנהאיה אלג̇נוביה מתל תקציר אלאנקלאב. פיג̇ב לדלך אן לא ינתקל קטב מ̇ חאפט̇א̇ לוצ̇עה מנה, בל יכון כאן קטב דאירה אלברוג̇ יתקדמה ויכ̇לפה וראה עלי מחיט דאירה למ̇נ̇ מן לדן נקטה מ̇ נחו נקטה נ̇. פאדא וצל קטב ב̇ אלי נקטה ח̇ ותוהמנא דאירה למ̇נ̇ עלי קטב ח̇, ונקטה מ̇ עלי קטב כ̇, ליס יצל קטב דאירה אלערץ̇ אלדי כאן עלי נקטה מ̇ אלי נקטה כ̇ [בל][260] אלי נקטה ר̇ מתלא̇. ולא יזאל ינתקל הדא אלקטב עלי הדה אלדאירה

257. א: קשת.
258. א: קצוה.
259. ו: אלמחדאה. א: הנכחות
260. א: כי אם.

נחו תלך אלג̇הה תאבעא̇ לחרכה אקטאב דאירה אלברוג̇ אלמתוהמה פיה، חתי יציר אבדא̇ תקציר אלנהאיה אכתר מן תקציר אלאנקלאב בהדא אלמקדאר פי כל יום.

B.2.I.10 ויתחרך מרכז אלקמר קוס ס̇ז̇ מן דאירה אלערץ̇ וירי קד סאר מן דאירה אלברוג̇ קוס אקל מנהא באלקדר אלמדכור، וינתהי אלקמר אלי ערץ̇ אלג̇ז מן אלברוג̇ אלדי ינתהי אליה בחסב אצול בטלמיוס. וינתהי אלי אלעקדה איצ̇א̇ באלזמאן אלדי ינתהי אליהא ואלי ערץ̇ דלך אלג̇ז בעינה אלדי עליה אלעקדה בחסב אצולה. ויבקי בעד דלך תקציר אלטול אלאכ̇יר אלקדר אלמדכור לצ̇עף̇ אלקוה אלוארדה עליה، ויכון אנתקאל [אלעקדתין][261] מע חפט̇ גאיה אלמיל אבדא̇. ודלך לאן אנתקאל קטבי דאירה אלערץ̇ אנמא הו עלי דאירה קטבהא קטב דאירה אלברוג̇ והי דאירה למ̇נ̇. וליס הדא אלאנתקאל [][262] אלי כ̇לף̇ לאן הדא אלאנתקאל אנמא הו תאבע לחרכה אקטאב דאירה אלברוג̇ אד קצדהא אן יכון תקצירהא אקל מן תקציר דאירה אלערץ̇، אעני תקציר אלפלך אלדי הי פיה، פאן אלחרכה אנהא הי ללפלך לא ללדואיר אלמג̇רדה. וינבגי אן תכון דאכרא̇ למא קלנאה פי אלמקאלה אלאולי מן אנקסאם אלפלך אלי פלכין או תלאת או אכתר למא

73v

יחתאג̇ אליה ענד וצ̇ענא ללפלך אלואחד חרכתין או תלאתה עלי אקטאב מכ̇תלפה.

B.2.II.1 אלפצל אלתאני פי תבין אלאכ̇תלאף̇ אלאול אלחקיקי אלט̇אהר פי אלקמר. פאד קד אתינא באלאצול אלתי בהא תט̇הר אלחרכאת אלוסטי פי אלטול ואלערץ̇ עלי מא הי עליהא، [פלנבין אלאן][263] אלוגה [אלאול][264] אלדי עליה יג̇רי אלאמר פי אלאכ̇תלאף̇ אלאול והו וחדה אלמחתאג̇ עלמה פי אלכ̇סופאת. ודלך יכון בנט̇יר מא עמלנא פי אלשמס עלי הדה אלצפה.

B.2.II.2 לתכון אלדאירה אלמאילה אלתי ללקמר אלמתוהמה פי אלפלך אלאעלי אלמכוכב، אעני דאירה אלערץ̇ דאירה ג̇ה̇ד̇ קטבהא מ̇ דאירא̇ עלי דאירה קטבהא קטב פלך אלברוג̇. ונצ̇ע פי הדא אלפלך דאירה מאילה עליהא בקדר נצף̇ קטר דאירה אלממר אלתי סנצ̇עהא והי דאירה ג̇ט̇ד̇ קטבהא דאירא̇ איצ̇א̇ עלי דאירה ממר קטבהא קטב דאירה אלערץ̇. ויכון קטב דאירה אלברוג̇ ב̇ דאירא̇ עלי דאירה ב̇ח̇ אלתי קטבהא קטב מעאדל אלנהאר וליכון א̇. ולנתוהם פי פלך אלקמר דאירתין מקאם דאירה אלערץ̇ ואלדאירה אלמאילה עליהא כל ואחדה מנהא פי סטח נט̇ירתהא [][265] מן אלמתוהמה פי אלפלך אלמכוכב [ואקטאב][266] אלנט̇איר עלי מחור ואחד.

B.2.II.3 ונצ̇ע מתלא̇ אן נקטה ה̇ אלנהאיה אלג̇נוביה מן דאירה אלערץ̇

74r

וכדלך נקטה ט̇ מן אלדאירה אלחאמלה ואן מוצ̇עהמא מן דאירה אלברוג̇ נקטה אלאנקלאב אלשתוי. ונצ̇ע אן מרכז אלקמר דאירא̇ עלי דאירה [צגירה][267] קטבהא מחמולא̇ עלי אלדאירה אלמתוהמה פי פלך אלקמר פי סטח דאירה ג̇ט̇ד̇ ודאירא̇ עליהא. ולתכון הדה אלדאירה דאירה כ̇ל̇נ̇ע̇ וקטבהא ק̇ עלי

261. א: ראש התלי וזנבו.
262. א: תנועה.
263. א: נשוב לבאר.
264. א: -.
265. א: ח̇ בירח.
266. א: -.
267. א: -.

אנא ננזל קטב ק̇ עלי נקטה ט̇, ואלקוס אלדי מן קטבהא אלי מחיטהא מסאוי ללקוס אלשביה בקוס ה̇ט̇, אעני גאיה מיל אלדאירה אלחאמלה עלי דאירה אלערץ̇. פאדא וצ̇ענא קטב דאירה ממר אלמרכז כאנה עלי נקטה ט̇, ומרכז אלקמר פי דאירה ממרה פי אלנהאיה אלתי תלי דאירה אלברוג̇ אעני עלי נקטה כ̇, כאן מרכז אלקמר פי אלנהאיה אלג̇נוביה מן דאירה אלערץ̇. וכאן מוצ̇עה מן אלברוג̇ נקטה אלאנקלאב אלשתוי.

B.2.II.4 ולנבין אולא̇ הדא אלאמר עלי אלראי אלתאני. ונצ̇ע מתלא̇ אן פי יום ואחד קצר קטב דאירה אלברוג̇ אלמתוהמה פי פלך אלקמר אלדי כאן עלי נקטה ב̇ ען אלוצול אלי מוצ̇עה קוס ח̇ב̇ מן דאירה ממר אלקטב מקדארה צ̇עף̇ תקציר אלטול אלאכ̇יר. ואנקל מעה [דאירה ממר][268] קטבי דאירה אלערץ̇ ואלדאירה אלחאמלה אעני דאירה מ̇ר̇ ודאירה ש̇ת̇. ובהדא אלזמאן אנתקל קטב דאירה אלערץ̇ אלי כ̇לף̇ קוס מ̇ר̇ מקדארה אלג̇ דקי̇. פצאר קטב דאירה אלערץ̇ עלי נקטה ר̇ וקטב אלדאירה אלחאמלה עלי נקטה ש̇. פתכון אדא̇ דאירה אלערץ̇ אלתי כאנת אולא̇ פי סטח דאירה ג̇ה̇ד̇ הי דאירה צ̇ס̇ נהאיתהא אלג̇נוביה אלתי כאנת אולא̇ עלי נקטה ה̇ צארת עלי נקטה ס̇. ואלנהאיה אלג̇נוביה מן אלדאירה אלחאמלה אלתי כאנת עלי נקטה ט̇ צארת עלי נקטה ז̇ וקטב דאירה ממר אלמרכז אלדי כאן עלי נקטה ט̇ צאר עלי נקטה ז̇ איצ̇א̇. פקד קצרת אדא̇ נהאיתי אלג̇נוב אלתי לדאירה אלערץ̇ ולדאירה אלחאמלה אלתי עלי נקטתי ס̇ז̇, וקטב דאירה ממר אלמרכז, ען אלוצול אלי מוצ̇עהמא אכתר ממא קצר נקטה אלאנקלאב אלשתוי אלמתוהמה פי פלך אלקמר בתלאת דקי̇ מע חפט̇הא לערצ̇יהמא מן דאירה אלברוג̇. ולנצ̇ע איצ̇א̇ אן פי דלך אליום אנתקל קטב דאירה ממר אלמרכז חול קטב אלדאירה אלחאמלה לה תאבעא̇ לחרכה אלכל מן נקטה ז̇ אלי נקטה ק̇ מקדאר קוס ז̇ק̇ מתל חרכה אלערץ̇ פי אליום ואן כאנת נקטתי ל̇ע̇, אעני נקטתי אלתקאטע, חאפט̇תין אבדא̇ למחיט דאירה צ̇ז̇.

B.2.II.5 פאן מן אג̇ל אלפלכין אל

74v

חאמלין ללדאירתין אלמדכורתין פי אלשמס אלדי מתלהמא יחתאג̇ אן תוצ̇ע פי אלקמר כמא קלנא אנפא̇ יכון [כאן][269] נקטתי אלתקאטע חאפט̇תין למחיט דאירה ג̇ט̇ד̇ באלקיאס אלי מרכז אלקמר. פיבקי̇ אדא̇ תקציר אלקמר אלמקדאר אלדי קלנא אנהא תקציר אלטול אלאכ̇יר. ונצ̇ע איצ̇א̇ פי הדא אליום אנתקל מרכז אלקמר [אלדי][270] כאן עלי כ̇ עלי קטב דאירה אלממר אלי ג̇הה חרכה אלכל איצ̇א̇ קוס כ̇ו̇ מקדארה אלאג̇זא אלתי קלנא אנהא ללאכ̇תלאף̇. לאכן אן נחן אנזלנא אלאצל עלי הדה אלצפה כאן ילזם אן לא יצל מרכז אלקמר אלי מחיט דאירה אלערץ̇ אלמתוהמה פי אלפלך אלמכוכב. ודלך לאן אג̇זא אלערץ̇ אכתר מן אג̇זא אלאכ̇תלאף̇ פאן קוס ז̇ק̇ יג̇ יג̇ מ̇ו̇ וקוס כ̇ו̇ יג̇ ג̇ נ̇ד̇.

B.2.II.6 ואלאמר פי אלקמר ליס כדלך פאנה קד יט̇הר מן קבל אלכ̇סופאת ומקדארהא אן אלקמר ליס יכ̇רג̇ אבדא̇ ען ערץ̇ דאירה אלערץ̇. פיט̇הר מן דלך אן חרכה קטב דאירה ממר אלמרכז עלי קטב אלדאירה אלחאמלה לה מסאויה לחרכה אלאכ̇תלאף̇, ואן קטב אלדאירה אלחאמלה ונהאיתהא אלג̇נוביה ינתקל אלי תלך אלג̇הה איצ̇א̇ חאפט̇א̇ לערצ̇הא מן דאירה אלערץ̇ אבדא̇ עלי קטב דאירה אלערץ̇ במקדאר פצ̇ל חרכה אלערץ̇ עלי חרכה אלאכ̇תלאף̇ והו י̇ דקי̇ באלתקריב. ויתוהם בהדה אלצפה, אנא נתוהם קטב דאירה ממר אלמרכז [][271] ואלג̇נוביה מן אלדאירה אלחאמלה עלי נקטה ז̇, וחיניד

268. א: שתי עגולות מעבר.
269. א: -.
270. ו: אלתי.
271. א: והתכלית.

כאן קטב אלדאירה אלחאמלה עלי נקטה שׁ מן דאירה ממרה. ופי הדא אליום אנתקל קטב אלדאירה אלחאמלה חול קטב דאירה אלערץׄ אעני רׄ אלי גׄהה חרכה אלכל מן נקטה שׁ אלי נקטה [תׄ][272] מקדאר הדא אלקוס מתל פצׄל חרכה אלערץׄ עלי חרכה אלאכׄתלאף והו עשרה דקיקה. ואנתקלת אלנהאיה אלגׄנוביה אלתי כאנת עלי נקטה זׄ וצארת עלי נקטה פׄ במתל הדא אלקדר.

B.2.II.7 ובהדא אלקדר זאד אלקמר פי חרכה אלטול. ופי הדא אליום אנתקל קטב דאירה ממר אלמרכז עלי קטב אלדאירה אלחאמלה אלי תלך אלגׄהה מתל חרכה אלאכׄתלאף וליכון קוס פקׄ. ובמתל הדה אלאגׄזא אנתקל מרכז אלקמר אלי גׄהה חרכה אלכל איצׄא מן נקטה כׄ מן דאירה ממרה אלי נקטה וׄ אעני במקדאר אגׄזא אלאכׄתלאף. ויציר נקצאן ערץׄ נקטה וׄ ען ערץׄ נקטה כׄ מתל נקצאן

75r

ערץׄ נקטה קׄ מן אלדאירה אלחאמלה ען ערץׄ נקטה פׄ. ויצל בדלך מרכז אלקמר אלי ערץׄ אלגׄז מן דאירה אלערץׄ אלמתוהמה פי אלפלך אלמכוכב אלנטׄיר ללגׄז אלדי פיה מן דאירה אלערץׄ אלמתוהמה פי פלך אלקמר. ומע דלך יחתאגׄ אן נצׄיף אלי הדא אלמוצׄע אלמעני אלדי דכרנאה פי אלשמס מן אלדאירתין פי נטׄיר הדא אלמעני.

B.2.II.8 ונבין הדא אלאמר עלי אלראי אלאול. [ונקול אנא][273] נתוהם בחסב אלראי אלאול אן אלנהאיה אלגׄנוביה מן אלדאירה אלחאמלה אעני נקטה טׄ תנתקל אלי כׄלאף חרכה אלכל עלי אגׄזא דאירה אלערץׄ חאפטׄאׄ לערצׄהא מנהא אבדאׄ במקדאר פצׄל חרכה אלערץׄ עלי חרכה אלאכׄתלאף. וליכון מתלאׄ הדה אלנקלה פי יום ואחד מן טׄ אלי פׄ פי אלצורה אלתאניה באנתקאל קטבהא מן שׁ אלי תׄ, וינתקל קטב דאירה אלממר אלי תלך אלגׄהה עלי אלדאירה אלחאמלה במקדאר חרכה אלאכׄתלאף. ותנתקל [אלעקדתין][274] אלי כׄלאף איצׄא אלמקדאר אלמדכור מן מׄ אלי רׄ פי הדה אלצורה אלמדכורה. וינתקל אלקמר פי דאירה ממרה ועלי אלדאירתין אלמדכורתין במקדאר חרכה אלאכׄתלאף.

B.2.II.9 פאן אלדי ילזם מן הדא אלוצׄע הו מתל אלדי ילזם ען וצׄענא אלתאני, לאנא סוא וצׄענא אלנהאיה אלגׄנוביה אלתי כאנת עלי נקטה זׄ מע דאירתהא ואנתקלת אלי גׄהה חרכה אלכל אלמקדאר אלמדכור חאפטׄאׄ לערצׄהא מן דאירה אלערץׄ באנתקאל קטבהא עלי דאירתי ממרהמא חול קטבי דאירה אלערץׄ, ואנתקל קטב דאירה ממר אלמרכז עלי קטבי אלדאירה אלחאמלה במקדאר חרכה אלאכׄתלאף

75v

אלי תלך אלגׄהה, ותנתקל [אלעקדתין] [275] באנתקאל אלקטבין עלי דאירה ממר מׄרׄ אלתי קטבהא קטב דאירה אלברוגׄ אלי כׄלאף אלמקדאר אלמדכור, פאן קטב דאירה ממר אלמרכז ינתהי אלי נקטה [ואחדה][276] באלאצלין גׄמיעאׄ. וליס יכון בינהמא פרק פי אלערץׄ אצלאׄ. והדא הו אלמוגׄב [לחדות][277] אלכסוף ואלמוגׄב איצׄא מקאדיר אלכׄסופאת בחסב אלבעד מן אלעקדה.

272. א: כׄ.
273. א: -.
274. א: ראש התלי וזנבו.
275. א: ראש התלי וזנבו.
276. א: -.
277. א: -.

B.2.II.1 ואמא אלפרק אלדי יכון בינהמא פהו מן קבל אלזיאדה או אלנקצאן פי אלטול אלדי מן קבל אלמטאלע כמא בינא פי אלמקאלה אלאולי. ואלצ׳רר אלדי ימכן אן יקע מן קבל דלך פאנה יקע אכתר פי תקדים וקת אלכסוף או תאכ׳רה אלדי מתל הדא אלכ׳טא או אכתר מנה יקע כתירא׳ ענד אסתכ׳ראג׳ הדה אלאוקאת. ואמא אלצ׳רר אלדי ימכן אן יקע מן קבל דלך פי מקדאר אלכ׳סוף פליס הו ממא יג׳ב אן יעתד בה. פאדא׳ לם יבקי פיהמא אלא אכ׳תיאר [אלמאכ׳ד][278] עלי אנא קד אעלמנאך איהמא ינבגי

[B.2.II.1] אן יכ׳תר.

B.2.III. אלפצל אלתאלת פי מקדאר אכתר הדא אלאכ׳תלאף. פלנבין [אן][279] נסבה נצף קטר פלך אלקמר אלי נצף קטר דאירה אלממר לכי נרשד אלי טריק חסאב אכתר הדא אלאכתלאף. ודלך יכון באלכסופאת באעיאנהא אלדי [][280] בין בטלמיוס נסבה נצף קטר אלחאמל אלי נצף קטר פלך אלתדויר. ודלך אנה אכ׳ד הדה אלכ׳סופאת אלקמריה ואסתכ׳רג׳ אלג׳ז מן אלברוג׳ אלדי כאן פיה חקיקה אלשמס פי וסט אלכ׳סוף אלאול ופי וסט אלתאני ווג׳ד אלבעד בין אלמוצ׳עין, והו בעינה אלבעד בין מוצ׳עי אלקמר לאסתקבאלה אלשמס. אעני אלחרכה אלחקיקיה פי אלזמאן אלדי כאן בין אלכסופין אכתר מן אלחרכה אלוסטי אלתי כאנת ללקמר פי הדא אלזמאן אלמעלום ענדה. פאדא׳ חרכה אלאכ׳תלאף ללקמר אלתי כאנת מעלומה מן קבל אלעלם באלזמאן קד זאדת פי מסיר אלטול [מתל][281] זיאדה אלמסיר אלחקיקי עלי אלוסט פי הדא אלזמאן. ואכ׳ד איצ׳א אלג׳ז מן אלברוג׳ אלדי כאן פיה חקיקה אלשמס פי וסט אלכ׳סוף אלתאלת ווג׳ד אן חרכה חקיקה אלשמס בל חקיקה אלקמר עלי אדואר תאמה פי אלזמאן אלדי בין אלתאני ואלתאלת ינקץ ען אלחרכה אלוסטי ללקמר פי הדה אלמדה באג׳זא מעלומה. פחרכה אכ׳תלאף אלקמר אלמעלומה

76r

פי הדה אלמדה [נקצהא].[282]

B.2.III. פאד אטאנא דלך פלתכן דאירה ממר המרכז בא׳ג׳ה׳, ואלפצל אלמשתרך בינהא ובין סטח אלדאירה אלחאמלה להא כנ׳, ומוצ׳ע אלקמר פי וסט אלכסוף אלאול נקטה א׳. ותחרך אלי אלכ׳סוף אלתאני עלי אדואר תאמה קוס א׳ג׳ב׳, יותר מן אלברוג׳ אג׳זא מעלומה ללזיאדה, והי בחסב חסאב בטלמיוס ג׳ אג׳זא כ׳ד׳ דקי׳. ובין אן קוס ב׳א׳ [ינקץ][283] [][284] מתלהא. ותחרך מן אלכ׳סוף אלתאני אלי אלתאלת קוס ב׳א׳ג׳ עלי אדואר תאמה יותר מן אלברוג׳ ל׳ז׳ דקי׳ ללנקצאן בחסב חסאבה איצ׳א פיבקי קוס א׳ג׳ יזיד גזין מ׳ז׳ דקי׳. ונכ׳רג׳ מן נקטה אלכסוף אלתאני עמוד עלי אלפצל אלמשתרך ינתהי אלי מחיט דאירה אלממר פי אלג׳הה אלמקאבלה והו עמוד ב׳ל׳ה׳. ונכ׳רג׳ מן נקט א׳ג׳ [עמוד][285] עלי כ׳ט ב׳ה׳ [והם][286] א׳מ׳ ג׳ז׳ ונצל א׳ה׳ ג׳ה׳ א׳ג׳. ונכ׳רג׳ מן נקטה ג׳ [אלי][287] א׳ה׳ עמוד ג׳ט׳ ונכ׳רג׳ מן נקטה ד׳ אעני מרכז אלברוג׳ אלי נקט אלכ׳סוף אלאול ואלתאלת כ׳טי ד׳א׳ ד׳ג׳ ואלי נקטתי מואקע עמודי א׳מ׳ ג׳ז׳ כ׳טי ד׳מ׳ ד׳ז׳. ולנאכ׳ד דאירה אלערץ׳

.278 א: -.
.279 א: עתה.
.280 א: בהם.
.281 א: בכדי.
.282 א: היא המחייבת זה הגרעון.
.283 ו: ינקץ׳. א: יחסר.
.284 א: חלקים.
.285 א: שני עמודים.
.286 א: והם. ו: והו.
.287 א: על קו.

פי הדא אלברהאן מקאם דאירה אלברוג̇ אד ליס בינהמא פרק יעתר בה. פלאן כ̇ט א̇מ̇ מעטי ומוצ̇ע אלכ̇סוף אלתאני אעני נקטה ב̇ מן אלברוג̇ הו בעינה מוצ̇ע נקטה מ̇ פתכון זאויה א̇ד̇מ̇ אעני אלתי תנקץ̇ ען אלתקציר בסבב קוס ב̇א̇ הי מעטאה. פנסבה כ̇ט ד̇א̇ והו נצף קטר פלך אלברוג̇ אלמתוהם פי פלך אלקמר אלי אלעמוד אלכ̇ארג̇ מן נקטה א̇ עלי כ̇ט ד̇מ̇ מעטאה. והדא אלעמוד אמא אדא כאן כ̇ט ב̇ה̇ קטר לדאירה אלממר יכון הו בעינה כ̇ט א̇מ̇. ואמא אדא כאן גיר אלקטר פליס הו כ̇ט א̇מ̇, ונחן נעמל פי דלך ופי אסתכ̇ראג̇ אלפצ̇ל אלדי בין אלעמוד ובין כ̇ט א̇מ̇ [עלי קיאס][288] מא עמלנא פי אלשמס. פתכון אדא̇ נסבה ד̇א̇ אלי א̇מ̇ מעטאה ולאן קוס ב̇א̇ בל זאויה ב̇ה̇א̇ מעטאה פנסבה א̇ה̇ אלקטר אלי א̇מ̇ מעטאה פנסבה ד̇א̇ אלי א̇ה̇ מעטאה. ולאן מוצ̇ע נקטה ב̇ הו נקטה ז̇ פזאויה ג̇ד̇ז̇ מעטאה, פנסבה ד̇ג̇ אעני ד̇א̇ אלי אלעמוד אעני ג̇ז̇ מעטאה. ולאן קוס ב̇א̇ג̇ בל זאויה ב̇ה̇ג̇ מעטאה פנסבה ג̇ה̇

76v

אלקטר אלי ג̇ז̇ מעטאה פאדא̇ נסבה ד̇א̇ אלי ג̇ה̇ מעטאה. לאן קוס א̇ג̇ בל זאויה א̇ה̇ג̇ מעטאה פנסבה אלקטר אלי ג̇ט̇ בל אלי ה̇ט̇ תמאמה מעטאה. ויבקי נסבה ג̇ה̇ אלי ט̇א̇ בל אלי ג̇א̇ אלקוי עלי ג̇ט̇ ט̇א̇ מעטאה פנסבה ד̇א̇ אלי א̇ג̇ מעטאה, וקד כאנת נסבה ג̇א̇ אלי קטר דאירה אלממר מעטאה אד קוס א̇ג̇ מעטי. פנסבה ד̇א̇ אעני נצף קטר [פלך][289] אלברוג̇ אלי נצף קטר דאירה אלממר מעטאה.

B.2.III.3 ויכ̇רג̇ בהדא אלטריק אלנסבה אלתי כ̇רג̇ת לבטלמיוס בעד אלאצלאח אלדי יתבין ען קריב והי נסבה סתין אלי כ̇מסה ורבע באלתקריב אלא אלזיאדה או אלנקצאן אלדי מן קבל חסאב אלמדד פי אכ̇תלאף אלאיאם בליאלהא. פאנה ליס יכון בחסב אלאצל אלתאני כמא הו בחסב אצול בטלמיוס לאנה קד תבין אן מן כ̇אצה הדא אלאצל אן יכון אלברג̇ין אלדאן יליאן נקטה אלאסתוא זאידאן פי אלמטאלע. פלדלך ליס יכון מבלג הדא אלאכ̇תלאף במבלג אלדי דכרה. פאן כאן הדא אלאכ̇תלאף כמא דכרה [אעני עלי אלאצל אלאול][290] פיחתאג̇ אן נקול אן אלצ̇רר אלדי יקע מן קבל דלך בחסב הדא אלאצל ליס יצ̇ר לקלתה עלי אנה קד ימכן אן יכון אלדי ילזם ען הדא אלאצל צחיחא̇. והדא אלצ̇רר אלקליל אנמא ילחק אלאצל אלאכ̇ר. ואמא עלי אלראי אלאול פליס יכון פרק אצלא̇. ואללה תעאלי אלעאלם.

B.2.III.4 ואמא כיף נעלם בעד נקטה כל כסוף מן נקטה אקל אלתקציר פהו עלי הדה אלצפה. לאן נסבה ג̇ה̇ אלי ג̇א̇ מעטאה פקוס ג̇ה̇ מעטי בל כל קוס ב̇א̇ג̇ה̇ ויכ̇רג̇ אקל מן נצף דאירה, פנצפה אעני ב̇נ̇ מעטי. פבעד נקטה ב̇ מן נקטה אקל אלסיר או אלתקציר והו נקטה ו̇ מעטי. ואדא עמלנא פי תלאת כ̇סופאת אלאכ̇ירה אלתי אתי בהא בטלמיוס מתל הדא אלעמל ימכננא אצלאח עודה אלאכ̇תלאף בהדא אלוג̇ה בעינה אלדי אצלחהא הו.

B.2.III.5 ומן בעד דלך פאנא נרג̇ע אלי תבין אלפצ̇ול [][291] במתל הדא אלעמל אלדי עמלנאה פי אלשמס. ונסתכ̇רג̇ אלנִסבה מרה תאניה ונסתכ̇רג̇ אלפצ̇ול מרה תאניה בהדה אלנסבה ונכרר אסתכראג̇ אלנסבה בעד אן נזיד או ננקץ̇ עלי אלקסי אלאכ̇תלאף אלתי תוג̇בהא. כמא קיל פי אלשמס אנה כלמא תכררת [][292] כאן אקרב אלי אלצואב [ודמ̇א̇][293]. וט̇אהר איצ̇א אן וג̇ה טריק חסאב אכ̇תלאף

288. א: כמו.
289. א: עגולת.
290. א: -.
291. א: הנזכרות.
292. א: בזה המעשה.
293. א: -.

אלקמר הו בעינה אלטריק אלמסלוך פי אלשמס פננקלה אלי הנא כימא לא נכרר אלמעני אלואחד מרארא̇ כתירה.

77r

B.2.IV.1 אלפצל אלראבע [][294] פי אן אכ̇תלאף מקאדיר אקטאר אלקמר ענד אלבצר ליס יוג̇ב ענה ולא בד אכ̇תלאף אבעאד אלקמר ען אלארץ̇. ומן בעד דלך פאנא נאכ̇ד אכ̇תלאף מא למנט̇ר אלקמר ענד אלבצר פי בעד ואחד מן נקטה סמת אלראס ונסתכ̇רג̇ מן קבלה בעד אלקמר באלקיאס אלי נצף קטר אלארץ̇ כמא פעל דלך בטלמיוס. תם נסתכ̇רג̇ סאיר אכ̇תלאפאת אלמנאט̇ר אלג̇זייה מן קבל אלעלם בבעד אלקמר. ונבין מן קבלה בעד אלשמס ובעד נקטה טרף אלט̇ל, ונסאב אקטאר אלשמס ואלקמר ואלארץ̇ וכדלך נסב אלאג̇ראם בעצ̇הא לבעץ̇.

B.2.IV.2 ונסתכ̇רג̇ מן קבל חדוד אלכ̇סופאת וכדלך מקדאר כל כסוף [ואוקאתה][295] ומקדאר מכת מא לה מכת מנהא באלטריק אלדי פעל דלך בטלמיוס אלא אנא נסתכ̇רג̇ מא נסתכ̇רג̇ מן אכ̇תלאפאת אלמנאט̇ר עלי אן מרכז אלקמר פי בעד ואחד מן מרכז אלארץ̇ דאימא̇ בחסב אלואג̇ב ען הדה אלאצול. ונחן נדכר דלך [פימא בעד][296] בטריק אוג̇ז מא נקדר עליה. וכדלך נעמל עלי אן קטר אלקמר יותר זאויה ואחדה ענד אלבצר פי כל מוצ̇ע פאן תזיד הדה אלזאויה פי בעץ̇ אלמואצ̇ע ופי בעץ̇ אלאוקאת ימכן אן ינסב אלי אסבאב אכ̇רי מתל תגיר אלהוא או תרטיבה יוג̇ב תשכל אלקמר פי וקת דון וקת. פאן דלך ליס ימכן בחסב אלנט̇ר אלטביעי אן ינסב אלי בעד אלקמר אחיאנא̇ וקרבה אחיאנא̇ למרכז אלארץ̇, פאן אלקרב ואלבעד ליס ימכן תוהמה אלא בוצ̇ע פלך כ̇ארג̇ או בוצ̇ע פלך תדויר וקד תבין אמתנאע הדין אלאצלין.

B.2.IV.3 ואלדליל עלי צחה מא קלנאה אן הדא אלאכ̇תלאף אלדי למקאדיר אלאקטאר ליס יוג̇ד ג̇אר אבדא̇ עלי נט̇אם ואחד ענד אלג̇מיע. [פאנא נג̇ד][297] אהל הדא אלעלם אלואצ̇עין הדה אלאצול מכ̇תלפין פימא וג̇דוה מן מקאדיר הדה אלאקטאר פי אלרויה חתי אן אברכ̇ס יקול אנה וג̇ד קטר אלקמר יותר זאויה מסאויה ללדי יותרהא קטר אלשמס ענד אלבצר ענד כונה פי אלמג̇אז אלאוסט מן פלך תדוירה. ובטלמיוס יקול אנה יוג̇דה יותר מתל הדא ענד כונה פי אלבעד אלאבעד ועמל עלי מא וג̇דה מן דלך. וכיף ינסב הדא [אלכ̇טא][298] למתל אברכ̇ס פאן פי הדא אלרצד ליס יחתאג̇ אלי מסאחה אג̇זא אלמסטרה, ובטלמיוס נפסה יקול דלך. פכיף ימכן אן יקע פי דלך שי מן אלכ̇טא. ואלואג̇ב כאן עליה אן יקול אן אברכ̇ס

77v

קד אצאב [פי מא וג̇דה][299] ואן [אלמוג̇ב][300] להדא אלאכ̇תלאף הו מא קלנאה. ולאכן וצ̇עה הדה אלאצול אצ̇טרה אן יצ̇ע קטר אלקמר יותר [זאויה מכ̇תלפה][301] ענד כונה פי מואצ̇ע מכ̇תלפה. תם וג̇ד

294. א: יבאר.
295. ו: ווקאתה. א: עת.
296. א: ב̇ה̇.
297. א: כי.
298. ו: אל. א: הטעות.
299. א: -.
300. ו: אלמואג̇ב. א: המחייב.
301. א: זויות מחולפות.

אכ̇תלאף פי קדר הדה אלזאויה ללראצדין פנסבה אלי הדה אלאצול. וליכון הדא אלמבלג מן קולנא כ̇אפיא̇ פי אכ̇תלאפאת אלקמר אלחקיקיה אלמחתאג̇ה עלמהא פי אלכ̇סופאת.

B.2.V.1 אלפצל אלכ̇אמס פי אלאכ̇תלאפין אלבאקיין ללקמר אעני אלאכ̇תלאף אלמנסוב ענד בטמליוס אלי אלשמס והו אלדי יעמל פיה בכ̇ארג̇ אלמרכז ואלאכ̇תלאף אלמנסוב ענדה אלי אנחראף קטר אלתדויר. אמא אלאצול אלתי עליהא [יג̇רי אמר אלקמר פי][302] אלאכ̇תלאף אלאול והו וחדה אלמחתאג̇ עלמה פי אלכ̇סופאת פקד קלנא פיהא מא פיה כ̇אפיה. ואמא אלנט̇ר פי סאיר אלאכ̇תלאפאת אלתי תט̇הר פיה פי גיר הדה אלאוקאת פבאלואג̇ב כאן לשיין אתנין, אחדהמא אן הדה אלאכ̇תלאפאת תנעדם פי אוקאת אלאתצאלאת בל ופי אלאוקאת אלתי ינעדם פיהא אלאול. פאדא̇ ליס ידכ̇ל צ̇רר מן קבלהא פי אלכ̇סופאת. ואלתאני אן אלאשיא אלתי בהא תתבין הדה אלאכ̇תלאפאת הי אלארצאד באלאלאת וקד תכדב כתירא̇ הדה אלארצאד מן אג̇ל אכ̇תלאף אלמנט̇ר ומן אג̇ל אלראצדין איצ̇א.

B.2.V.2 פאנא נג̇ד אבן אלהיתם קד בין פי אלמקאלה אלראבעה מן כתאבה פי עלם אלמנאט̇ר אן אדראך אלבצר ללכוכב ליס הו באלאסתקאמה ואנמא ידרכה באלאנעטאף. וקאל אן מן ירצד אלכואכב אלתאבתה באלה דאת אלחלק פאנה יוג̇ד בעד אלכוכב אלואחד ען קטב אלעאלם ענד טלועה אקל מן בעדה ענד תוסטה אלסמא, ודלך ליס ימכן אן יכון פי אלכואכב אלתאבתה לו אדרכהא אלבצר עלי אסתקאמה. וקאל איצ̇א אנה אן רצד אלקמר ענד טלועה פאנה יוג̇ד באלאלה בעדה ען סמת אלראס אקל מן אלמקדאר [אלדי יחצל][303] ללבעד פי דלך אלוקת באלחסאב. פיט̇הר מן דלך אן צו̇ אלקמר ליס ימתד אלי תקבי אלאלה עלי אסתקאמה. ואד ליס ידרך אלבצר אלכואכב עלי אסתקאמה ולא באלאנעכאס אד ליס פי אלסמא [][304] ולא פי אלהוא ג̇סם כתיף צקיל תנעכס ענה אלצור פיבקי אן ידרכהא באלאנעטאף, לאן אלבצר ליס ידרך שיא̇ מן אלמבצראת אלא עלי הדה אלוגו̇ה אלתלאתה. וקאל איצ̇א אן ליס פי אלסמא מוצ̇ע ידרך

78r

אלבצר אלכוכב אלדי יכון פיה באלאסתקאמה אלא אדא כאן אלכוכב עלי נקטה סמת אלראס או קריבא̇ ג̇דא̇ ענה. ואדא כאן דלך כדלך פט̇אהר אן מא קאלה בטלמיוס ובינה מן אמר הדה אלאכ̇תלאפאת גיר מותוק בה לאן הדה אלארצאד אנמא הי עלי אן אדראך אלבצר אלכואכב הו עלי אסתקאמה. ולאכן קד אתינא באלאצול אלתי בהא תט̇הר הדה אלאכ̇תלאפאת מן גיר כ̇ארג̇ אלמרכז ומן גיר פלך תדויר מן אג̇ל ג̇לאלה הדה אלאצול אלתי לנא פי דלך ושרף חאלהא והדא חין נבתדי בביאן דלך.

B.2.V.3 אן בטלמיוס וג̇ד ענד אג̇תמאע אלקמר מע אלשמס או אסתקבאלה אכ̇תלאף ואחד ללקמר פקט מרה יזיד אעני אן אלמסיר אלחקיקי יתקדם אלמסיר אלוסט ומרה ינקץ. ווג̇דה יחפט̇ אבדא̇ נט̇אמה בחסב אלג̇ז אלדי הו פיה מן דאירה אכ̇תלאפה. וקד פרגנא מן דכר אלאצול אלתי עליהא יג̇רי אלאמר פי הדה אלאכ̇תלאף. ואמא פי אלאבעאד אלדי בין האדין וכ̇אצה ענד תרביעה אלשמס פאנה אן כאן אלאכ̇תלאף אלאול מוג̇ב זיאדה וג̇ד אזיד ואן מוג̇ב נקצאן וג̇ד אנקץ עלי קדר אלנקצאן ואלזיאדה דאימא̇. ווג̇דוא אכתר הדה אלזיאדה או אלנקצאן ען אלאכ̇תלאף אלאול ג̇זין ונצף באלתקריב.

302. א: נבנה.
303. א: -.
304. א: ובארץ.

B.2.V.4 פלנצ̇ע דאירה אלערץ̇ א̇ד̇ג̇ קטבהא מ̇ ואלדאירה אלחאמלה א̇ב̇ג̇ קטבהא ש̇ ודאירה ממר אלמרכז ח̇ה̇ט̇ס̇ קטבהא ב̇. ונצ̇ע בסבב הדא אלאכ̇תלאף אן מרכז אלקמר ליס עלי הדה אלדאירה נפסהא ידור. ולאכן אנמא ידור עלי דאירה צגירה ממאסה להא ואלקוס אלתי מן קטבהא אלי מחיטהא מסאוי לנצף אעט̇ם הדא אלאכ̇תלאף אלתאני. ולתכן דאירה ה̇ו̇ד̇ קטבהא ז̇ דאירה אבדא̇ עלי דאירה נ̇ז̇מ̇ אלתי קטבהא ב̇ איצ̇א̇ וליכון מיל דאירה א̇ב̇ג̇ עלי דאירה א̇ד̇ג̇ זאידא̇ עלי אלדי קלנאה בקדר קטר הדה אלדאירה. ונקול אמא פי אלאג̇תמאע ואלאסתקבאל פאן מרכז אלקמר יכון עלי אלנקטה מן דאירה ממרה ממאסה לדאירה ח̇ה̇ט̇ס̇ והי נקטה [ה̇][305] אבדא̇ חתי יכון אלאכ̇תלאף אלדי יערץ̇ ענד אלאג̇תמאע ואלאסתקבאל בחסב מא יוג̇בה דאירה ח̇ה̇ט̇ס̇ [לא גיר][306]. ואמא אדא כאן אלקמר פי תרביע אלשמס והו אלמוצ̇ע אלדי פיה הדא אלאכ̇תלאף אלתאני אעט̇ם מא יכון פאן אלקמר יכון עלי נקטה ד̇ מן דאירה ממרה אעני עלי מחיט דאירה [ל̇ד̇כ̇][307] אלתי קטבהא ב̇ איצ̇א̇.

78v

B.2.V.5 פאדא כאן אלקמר חיניד פי נקטה אעט̇ם אלתקציר אעני ה̇ או פי נקטה אקל אלתקציר אעני ס̇ בחית ינעדם אלאכ̇תלאף אלאול פחיניד לם יכון תם אכ̇תלאף אלבתה כמא ילזם ען אצול בטלמיוס לאן מוצ̇ע נקטה ד̇ מן אלברוג̇ הו בעינה מוצ̇ע נקטה ה̇. ואמא אדא כאן אלקמר עלי נקטה ט̇, והי אלנקטה אלתי פיהא אעט̇ם מא יכון מן אלאכ̇תלאף אלאול ללזיאדה פי אלסיר, פחיניד תכון נקטה ד̇ אלתי פיהא אלקמר עלי נקטה כ̇. ויתזיד בדלך אלאכ̇תלאף בקדר קוס ט̇מ̇כ̇ וחיניד תכון הדה אלזיאדה אעט̇ם מא תכון. ואדא כאן פי נקטה ח̇ אלתי פיהא אעט̇ם אלאכ̇תלאף ללנקצאן יכון אלקמר עלי נקטה ל̇ פיתזיד אלאכ̇תלאף איצ̇א̇ ללנקצאן בהדא אלקדר. ואדא כאן פי מא בין נקטתי ה̇ט̇ או ה̇ח̇ פתכון זיאדת אלאכ̇תלאפאת עלי קדר אלאכ̇תלאף אלאול. ואדא כאן אלקמר פי מא בין אלאג̇תמאע ואלתרביע או פימא בין אלאסתקבאל ואלתרביע אעני פימא בין נקטתי ד̇ה̇ מן דאירה ה̇ו̇ד̇ פתכון חיניד אלזיאדאת פי זואיא אלאכ̇תלאף אקל. ויקטע [אלקמר][308] דאירה ממרה אעני דאירה ה̇ו̇ד̇ מרתין פי זמאן אלשהר אלוסט, אעני אן אלפלך אלמתוהם פי הדה אלדאירה יסתכמל דורה עלי קטב ז̇ ואלקמר מרכוזא̇ פיה מרתין פי זמאן אלשהר אלוסט. וקד עלמת אן כל דאירה נדכר הנא הי פי פלך מכ̇צוץ ואלחרכה הי ללאפלאך לא ללדואיר אלמג̇רדה, ואלואחד מנהא דאכ̇ל קרינה. ובאלג̇מלה פאן כל מא יט̇הר מן הדא אלאכ̇תלאף אלתאני בהדא אלאצל יכון מואפקא̇ למא יט̇הר מנה ען אצול בטלמיוס וליס יגאדרה בשי מע כון אלבעד אלדי בין האדין אלאצלין פי אלחקיקה [ופי סהולה אלאחתמאל ללעקל][309] גאיה אלבעד כמא הו ט̇אהר ענהא.

B.2.V.6 ויחתאג̇ אן נצ̇ע אן

79r

קטב ז̇ מנתקל עלי אלדאירתין אלמוצופתין פי אלמקאלה אלאולי בנט̇יר הדא אלמעני פי אלשמס

[B.2.V.6] בסבב כון מרכז אלקמר עלי דאירה אלערץ̇ אבדא̇. פאן אלדי ילזם פי קטב ז̇ ילזם פי מרכז אלקמר חית כאן מן דאירה ה̇ו̇ד̇ת̇. הדא הו אלאצל אלמוצ̇וע ענה להדא אלאכ̇תלאף אלתאני.

305. א: כ̇.
306. א: לא יותה.
307. א: ל̇ד̇מ̇.
308. ו: אלשמס. א: השמש.
309. א: -.

B.2.v.7 ואמא אלאכ̇תלאף אלתאלת פאנה קד יתבין וג̇ודה בהדא אלאצל בעינה אלמוצ̇וע וליס יחתאג̇ פי דלך אלי זיאדה שי אעני זיאדה חרכה ולא אנחראף קטר ולא שי מן הדה אלממתנעאת אלמוצ̇ועה ען בטלמיוס להאדין אלאכ̇תלאפין, פאן [][310] אלשנאעאת אללאחקה לאצולה פי האדין אלאכ̇תלאפין ליס הי מן נוע אלשנאעאת אללאחקה לסאיר אלאצול אלמוצ̇ועה ענה. בל קד ילחקה מן עדם אלאמכאן מא לא יסתטאע בחית אלסכות ענהא כאן לה אג̇וד עלי אנא קד דכרנא ראיה פי מא סלף. ובחסב מא כאן קצדה פליס יסתחק אללום עלי דלך. ונקול אן מן אלדליל אלואצ̇ח עלי צחה הדא אלאצל כון הדא אלאכ̇תלאף אלתאלת לאזמא̇ ען אלתאני עלי אנה צ̇רורי ענה. אעני אנה ליס ימכן וג̇ודה כ̇לוא̇ מנה פי אלמואצ̇ע אלתי פיהא יט̇הר פקט. ואמא כיף ילזם דלך פהו עלי הדא אלוג̇ה.

B.2.v.8 קד קלנא אנה ליס יערץ̇ פי וקת אלאג̇תמאע [ואלאסתקבאל][311] סוי אלאכ̇תלאף אלאול. ואמא פי סאיר אלאשכאל וכ̇אצה פי וקת אלתרביע פאנה יערץ̇ פיה זאידא̇ עלי אלאול אכתר מא יכון מן אלאכ̇תלאף אלתאני אדא כאן אלאול מוג̇ודא̇ כמא דכר. ופי הדא אלוקת איצ̇א פאנה ינעדם אלאכ̇תלאף אלתאלת. ואמא פי סאיר אלאשכאל [סוא הדה אלתלאתה][312] אנה קד יערץ̇ פיה אלאכ̇תלאף אלתאלת והו אלדי נסבוה אלי אנחראף קטר פלך תדויר והו עלי הדא אלוג̇ה.

B.2.v.9 אדא כאן אלקמר פי דאירה אכ̇תלאפה בחית אלזיאדה פי אלמסיר או אלזיאדה פי אלתקציר، וכאן אלקמר פי מא בין אלאסתקבאל ואלתרביע [אלתאני][313] או פימא בין אלאג̇תמאע ואלתרביע [אלאול][314]، פאן אלאכ̇תלאף יתזיד. אעני אן אלתקציר פי אלטול לחקיקה אלקמר יכון אכתר ממא יוג̇בה אלאכ̇תלאפאת אלבאקיה. ואדא כאן אלקמר פי מא בין אלתרביע אלאול ואלאסתקבאל או פימא בין אלתרביע אלתאני ואלאג̇תמאע פאן אלאמר יכון באלעכס. ואמא אדא כאן אלקמר פי דאירה אכ̇תלאפה עלי קוס אלנקצאן מן אלסיר

79v

אעני אלנקצאן מן אלתקציר פאן אלאמר פי כל דלך יכון באלעכס. אעני אנה ינתקץ אלסיר עמא יוג̇בה אלאכ̇תלאפאת אלא̆ול פי אלוצ̇עין אלאולין ויתזיד פי אלתאניין. ואקול אן דלך ילזם מן קבל אנתקאל אלקמר עלי אלדאירה אלפאעלה ללאכ̇תלאף אלתאני אעני דאירה ה̇ו̇ד̇. ודלך אנה אדא כאן אלקמר פי קוס [אלזיאדה או פי][315] אלסיר אעני קוס ס̇ה̇ט̇ וליכן מתלא̇ עלי נקטה ה̇ כמא פי אלצורה וחיניד יכון וקת אג̇תמאע וסט או אסתקבאל. פמא דאמת נקטה ה̇ מנתקלה נחו נקטה ט̇ תאבעה ללאכ̇תלאף אלאול וכאן אלקמר מנתקלא̇ עלי קטב ז̇ מן נקטה ה̇ אלי נקטה ו̇ ואלי ד̇، אעני מן אלאג̇תמאע אלי אלתרביע אלאול או מן אסתקבאל אלי אלתאני، פאן אלאכ̇תלאף יתזיד לאן בהדה אלנקלה יסרע קרבי אלקמר נחו נקטה ט̇. ואכתר הדא אלאכ̇תלאף יכון אדא כאן אלקמר עלי אלוסט בין אלאתצאל ואלתרביע אעני ענד כונה עלי נקטה ו̇. ואדא כאן פי אלוצ̇עין אלתאניין אעני אדא כאן אלקמר עלי קוס ד̇ת̇ה̇ פיכון באלעכס. ואמא אדא כאן אלקמר עלי קוס אלנקצאן אעני עלי קוס ט̇ס̇ח̇ פאן כל דלך ילזם אן יכון באלעכס ודמ̇א̇

310. א: הסברות.
311. ו: ואליסתקבאל. א: וההקבלה.
312. א: ובפרט בעת הרבוע.
313. א: הראשון.
314. א: השני.
315. א: -.

B.2.V.10 ויטׄהר מן דלך אנה ליס יכׄלוא אלאכׄתלאף אלתאני פי הדה אלאוצׄאע מן אכׄתלאף תאלת [כיף מא
כאן][316] יוגׄבה אנתקאל אלקמר עלי הדה אלדאירה. פאדא וצׄע אלאנתקאל עלי מא וצׄענאה כאן
יטׄהר הדא אלאכׄתלאף אלתאלת עלי מא הו עליה. ויטׄהר איצׄא אן מקדאר הדא אלאכׄתלאף עלי
הדא אלוגׄה ליס יכון בינה ובין מקדאר אלאכׄתלאף אלדי וצׄעה בטלמיוס בסבב אלאנחראף פרק יעתר
בה. ויתבין איצׄא אנה ליס יכון פי אלאתצאלאת אלחקיקיה אלמוגׄבה ללכסופאת אכׄתלאף יעתר בה
מן קבל הדא אלאצל אלמוצׄוע להאתין אלאכׄתלאפין, ודלך יתבין במתל אלברהאן אלדי בה בין דלך
בטלמיוס. פהדא מא טׄהר לנא בתופיק אללה תעלי עלינא מן חקיקה אלאצול אלואגׄב וצׄעהא להדה
אלאכׄתלאפאת מן גיר אחתיאגׄ אלי שי מן אלאמתנאעאת אלמוצׄועה ענד בטלמיוס וגירה. והדא שי
ליס יצדר אלא מן לדנה תעלי פאלחמד לה כתיראׄ.

B.2.V.11 וקד בקי עלינא אן נבין כיף יחפטׄ אלקמר אבדאׄ ערץׄ דאירה אלערץׄ למא וצׄע מן אלזואל פי אלערץׄ
בסבב דאירה הׄודׄ אלמוצׄועה ללאכׄתלאף אלתאני ואלתאלת. פנעד אלצורה אלסאלפה ונצׄע דאירה
ממר קטב דאירה

80r

אלערץׄ, אעני אלפלך אלדי פיה דאירה למׄ קטבהא [חׄ][317], והו קטב פלך אלברוגׄ, ודאירה ממר קטב
אלפלך אלדי פיה אלדאירה אלחאמלה דאירה מׄשׄ קטבהא סׄ. ונצׄע קטב סׄ דאיראׄ עלי דאירה סׄמׄ
אלממאסה לדאירה למׄ ותכון מסאויה לדאירה הׄודׄ. [][318] ויכון נקלה קטב סׄ עלי דאירה סׄמׄ מסאויה
לחרכה אלקמר עלי דאירה הׄודׄ בחית אדא כאן מרכז אלקמר עלי נקטה [הׄ][319] אעני עלי מחיט דאירה
הׄטׄ, יכון קטב סׄ עלי נקטה מׄ אעני קטב דאירה אלערץׄ. ויציר אלקמר חיניד עלי דאירה אלערץׄ. ואדא
B.2.V.11/ כאן עלי נקטה דׄ אעני פי אלתרביע כאן קטב סׄ עלי נקטה סׄ מן דאירה מׄסׄ. ובהדא אלוגׄה יכון אלקמר
עלי דאירה אלערץׄ דאימאׄ ויקטע קטב סׄ לדאירה מׄסׄ מרתין פי זמאן אלשהר אלוסט.

B.2.V.12 וינבגי אן תעלם אן הדה אלחרכה אלתי וצׄענאהא לקטב סׄ עלי דאירה מׄסׄ הי בחית ליסת תוגׄב חרכה
ללכואכב פי אלטול ולאכן פי אלערץׄ פקט. פאן הדא ממכן אעני אן ינתקל קטב סׄ עלי דאירה מׄסׄ
תאבעאׄ לאנתקאל אלפלך אלדי פיה תתוהם דאירה מׄסׄ עלי קטב נׄ בחית לא ינתקל אלפלך אלדי קטבה
B.2.V.12/ סׄ [פי אלטול][320] בהדה אלחרכה אלבתה. ואמא וגׄה תבין חסאב [מקדאר][321] אלאכׄתלאפאת אגׄמע
פאנה יכון באלוגׄה אלדי בינא פיה אלאכׄתלאף [אלאול][322] וביאנה קריב. וינבגי אן תכון דאכראׄ
למא אסתתנינאה מראראׄ מן אנקסאם אלפלך אלי עדה אפלאך ללמעני אלמדכור, פאן בדלך תרתפע
B.2.V.12/ אלשכוך אלתי ימכן אשתכאכהא פי הדה אלמוצׄע.

B.2.V.13 [כמלת אלמקאלה אלתאניה מן אלגׄמלה אלתאניה מן כתאב נור אלעאלם][323] ואלחמד ללה כתיראׄ
[אלמקאלה אלתאלתה][324]

316. א: -.
317. א: סׄ.
318. א: והעתקת קטבי (א: קטב) סׄטׄ סביב קטב עגולת הרוחב רלׄ מׄ אחת.
319. א: -.
320. א: -.
321. א: -.
322. א: -
323. א: -
324. ו: 2.

80v

B.3.0.1 אלמקאלה אלתאלתה פי אנתקאל אלכואכב אלתאבתה ופיה פצלאן.

B.3.1.1 אלפצל אלאול פי ראי אלבטרוג̇י פי הדה אלנקלה ואבטאלה. קד יג̇ב עלינא אן נצל בדלך אלכלאם פי אנתקאל אלכואכב אלתאבתה חסב אלתרתיב אלמסלוך ענד אלקדמא וען בטלמיוס. ובחסב מא בינא מן עדם אלתצ̇אד פי אלחרכה אלמסתדירה פאן אצולה פי דלך צחיחה אד ליס יחתאג̇ פיהא אלי כ̇ארג̇ מרכז ולא פלך תדויר. ואנמא [סלכנא][325] פי דלך אלטריק אלתאני למא תקדם מן קולנא. [ואלוג̇ה אלדי עליה אלאמר פי][326] נקלה הדה אלכואכב ענד אלבטרוג̇י הכדא. אן אלכואכב אלתאבתה כלהא מלתחמה [][327] פי כרה ואחדה. ואן הדא אלפלך יתחרך אלי ג̇הה חרכה אלכל תאבעא̇ ללאעלי אלמתחרך אלחרכה אליומיה ועלי קטביה ויקצר בחרכתה הדה ען אלוצול אלי מוצ̇ע אלאעלי ללסבב אלמדכור בעץ̇ אלתקציר. ואד קצדה אלתשבה באלאעלי וטלב הדא אלכמאל לאנה גאיתה، פאנה יתחרך עלי קטביה אלמכ̇צוצה בה חרכה אכ̇רי נחו תלך אלג̇הה. ומע דלך פאנה יבקי לה בעץ̇ אלתקציר פי אלטול עלי מדהב בטמליוס אנה ינתקל אלי כלף קדר מחסוס. או יסתפי בחרכתה חרכה אלאעלי וליס יבקי להא תקציר פי אלטול אלבתה עלי מדהב אלמתאכ̇רון. ואנמא יציר לה אנתקאל פי אלערץ̇ יוג̇בה אנתקאל הדא אלפלך עלי קטביה ללאסתפא.

B.3.1.2 וקאל איצ̇א אן חרכה אלאקבאל ואלאדבאר אלתי וצ̇עהא אלמתאכ̇רון בחסב ארצאדהם ליס הי חרכה באלחקיקה. בל אלחרכה הי מתסאויה אבדא̇ אלי ג̇הה ואחדה לאכן מא דאם אלכוכב מנתקלא̇ עלי אג̇זא אלרבע אלדי פי וסטה נקטה אלאסתוא פאן מא יטלע מע הדא אלרבע [מן][328] מעאדל אלנהאר הו אקל מנהא. פירי לדלך ללכוכב תקהקר אלי כ̇לף ויסמון הדא אלתקציר אקבאלא̇. ומא דאם ינתקל פי אלרבע [אלתאלי לה][329] פאן מא יטלע מעה אכתר מנה וירי לדלך ללכוכב [תטפיף][330] וזיאדה אמאם חרכה אלכל ויסמונהא אדבארא̇. ודלך יט̇הר סוא אנזלנא לה תקציר עלי מדהב בטלמיוס או לא. ועלי הדא אלראי כאן אלכוכב אדא כאן עלי אלנקטה אלנט̇ירה ללמנקלב אלציפי פאנה ירי מנתקלא̇ אלי נקטה אלאעתדאל אלרביעי ומנהא אלי אלאנקלאב אלשתוי ומנהא אלי אל

81r

אעתדאל אלכ̇ריפי עלי כ̇לאף מא וצ̇עוא אלקדמא. הדא הו מדהב הדא אלרג̇ל פי הדה אלנקלה.

B.3.1.3 ואן סלם אן הדא אלפלך יקצר ען אלאעלי עלימא יראה בטלמיוס מן אנה ינתקל אלי כ̇לף סוא כאן אלתקציר מתאל אלאנתקאל אלי כ̇לף אלדי יצ̇עה או אקל מנה או ליס יבקי לה תקציר פי אלטול אלבתה، עלי אנה גיר ואג̇ב פי חק אלאצל אלדי עליה יעמל، פאנה קד סלם ללארצאד אלתי בהא וג̇דת הדה אלנקלה אעני אלנקלה פי אלערץ̇. בל ליס ימכן אן יג̇חדהא וקד וג̇ד באלארצאד אלמתואליה אן אלכואכב מנתקלה עלי כ̇לאף מא וצ̇ע לאנה קד וג̇ד באלרצד אלסמאך אלאעזל אלדי הו קריב מן נקטה אלאסתוא אלכ̇ריפי פי בעדה מן נקטה אלאסתוא נחו אלשמאל באג̇זא מא ווג̇ד הדא אלבעד יקל במרור אלזמאן.

325. א: בארנו.
326. א: ודרך.
327. א: כולם.
328. א: מעגולת.
329. א: הנמשך לו. ו: אלתי לה.
330. א: -.

B.3.1.4 וכדלך אלכואכב אלתי באלקרב מן נקטה אלאסתוא אלרביעי וגׄדת מתבאעדה במרור אלזמאן ען נקטה הדא אלאסתוא נחו אלשמאל. ובאלגׄמלה פאן בטלמיוס וגׄד במקאיסה ארצאדה מע ארצאד אברכׄס וטימוכׄאריס וארסטלס ומילאוש אלמהנדס בקיאס בעץׄ הדה אלארצאד אלי בעץׄ ואלי ארצאד אכׄר מן אלראצדין אלמשהורין פי הדא אלזמאן אלקדים אן אלכואכב אלתי פי נצף אלכרה אלתי פיהא נקטה אלאסתוא אלרביעי [מתבאעדה][331] נחו אלשמאל אבדאׄ במרור אלזמאן ען אלמואצׄע אלתי כאנת פיהא אולאׄ, ואלתי פי אלנצף אלתאני באלעכס. וילזם ען דלך אן ירי אלכוכב מנתקלאׄ עלי חסב מא יגׄב אן ינתקל בחסב אצולה. ובאלגׄמלה פאן כל מא ירי מן אנתקאל כל ואחד מן הדה אלכואכב פי כל ואחד מן אלאצלין יכון עלי [][332] מא ירי פי אלאכׄר סוא כאנת אלכואכב אלתי עלי מנטקה פלך אלברוגׄ או אלכׄארגׄה ענהא אלא פי כוכבין פקט, והמא אלכׄאינה עלי נקטתי אלאנקלאבין פי אבתדא אלנקלה. פאן האדין אלכוכבין יריאן אבדאׄ עלי מוצׄע ואחד פי כל ואחד מן אלאצלין. ולנצׄע לדלך מתלאׄ.

B.3.1.5 דאירה מעאדל אלנהאר גׄטׄדׄ קטבהא אלשמאלי אׄ ודאירה אלברוגׄ גׄהׄדׄ קטבהא אלשמאלי בׄ [ונתוהם][333] הדה אלדאירה [גיר מנתקלה אבדאׄ. ולתכן אלדאירה][334] אלמארה עלי אוסאט אלברוגׄ אלמנתקלה דאירה נׄזׄוׄ קטבהא שׄ עלי אנא נתוהמהא כאנת אולאׄ פי סטח דאירה גׄהׄדׄ ונקטה זׄ עלי נקטה הׄ אלתי הי נקטה אלאנקלאב אלשתוי.

81v

B.3.1.6 וננזל אן אלפלך אלתי פיה אנתקל פי זמאן מא תאבעאׄ לחרכה אלפלך אלאעלי וקצר ען וצול אלי מוצׄעה קוס זׄהׄ מן אלדאירה אלמואזיה למעאדל אלנהאר. ואנתקל הדא אלפלך עלי קטביה אלמכׄצוצה בה טלבאׄ [ללחק][335] באלאעלי קוס זׄכׄ חתי אן אלכוכב אלתי כאן עלי נקטה זׄ, אעני אלנקטה אלנטׄירה ללאנקלאב אלשתוי צאר עלי נקטה כׄ. ואלכוכב אלתי כאן עלי נקטה לׄ מן הדה אלדאירה אעני דאירה נׄזׄוׄ וכאן בעדה מן נקטה נׄ אעני אלנקטה אלנטׄירה ללאסתוא אלכׄריפי קוס לׄנׄ צאר עלי נקטה מׄ, אעני מתבאעדאׄ ענה נחו אלשמאל. ואלכוכב אלתי כאן עלי נקטה סׄ מן הדה אלדאירה וכאן בעדה ען נקטה וׄ, אעני אלנקטה אלרביעיה [אלנטׄירה][336], קוס וׄסׄ צאר עלי נקטה עׄ [אעני מתקארבה][337] אלי גׄהׄה אלגׄנוב. פאדאׄ אלכוכב אלדי עלי נקטה אלאנקלאב פאן מא ירי פיה בחסב כל ואחד מן אלאצלין [שי ואחד לאנה סוא][338] ינתקל אלכוכב אלדי עלי נקטה הׄ אלי כׄלף אלי נקטה כׄ או ינתקל אלי תלך אלגׄהה מן נקטה זׄ מתל הדא אלקוס, פאן באלאצלין גׄמיעאׄ ירי אלכוכב עלי נקטה כׄ פי וקת ואחד. וכדלך ילזם פי אלאנקלאב אלאכׄר. ואמא סאיר אלכואכב אלתי עלי תלך אלדאירה פאנה יכון באלעכס לאן אלכוכב אלדי כאן עלי נקטה לׄ מן דאירה גׄהׄדׄ ונצׄע מתלאׄ אנה אלסמך אלאעזל, והו אלדי בה כאן אלרצד פאנה בעד זמאן וגׄד עלי נקטה מׄ, אעני מתקארבאׄ אלי נקטה גׄ אלתי הי אלנקטה אלכׄריפיה. ובאלואגׄב כאן כדלך עלי ראי בטלמיוס תאבעאׄ לאנתקאל הדא אלפלך אלי כׄלף, אעני מן נקטה הׄ נחו

331. ו: מתעאבדה.
332. א: הפך.
333. א: נחשוב.
334. א: -.
335. ו: לללחק.
336. ו: אלנטׄאירה.
337. א: מנקודת השווי.
338. א: יראה בו כפי העקר השני לפי שבין.

82r

נקטה כֿ, וקד בינא בחסב אצול אלבטרוגי אן אלואגֿב כאן אן יבעד ען תלך אלנקטה נחו אלשמאל אעני מן נקטה לֿ נחו נקטה מֿ מן דאירה נֿזֿו.

B.3.I.7 וכדלך אלכוכב אלדי עלי נקטה סֿ מן דאירה גֿהֿדֿ וכאן בעדה נחו אלשמאל מן נקטה [][339], אעני אלרביעיה, קוס דסֿ, פאנה וגֿד בעד זמאן עלי נקטה עֿ אעני מתבאעדא ענהא נחו תלך אלגֿהה אכתר מן דלך והו לאזם למדהבה וקד תבין בחסב אצול אלבטרוגי אנה כאן יגֿב אן יכון באלעכס. וכדלך יתבין הדא אלאכֿתלאף או שביה בה פי אי מוצֿע וצֿע אלכוכב מן פלכה סוא כאן עלי דאירה נטאקה או כֿארגֿא ענהא. ואדא כאן דלך כדלך פמא הו אדא אלשי אלדי אסתפדנא הדא אלרגֿול בוצֿעה הדא אלאצל אדא לם יכון שי ממא יטֿהר ללחס מטאבק למא יטֿהר ען אצלה. ומן הדא יטֿהר [מא קלנאה גיר מא מרה][340] מן גרץֿ הדא אלרגֿול. ואנא מבתדי מן די בדכר אלאצל אלצחיח להדה אלנקלה עלי אלראי אלתאני אעני אן יכון מא יטֿהר ענה מטאבק למא טֿהר בארצאד אלאקדמין גיר מגאדר לה בשי.

B.3.II.1 אלפצל אלתאני פי וגֿה נקלה הדה אלכואכב. פנקול אן אלשי אלדי מן אגֿלה אצטלח אלאקדמין פי אן הדה אלכואכב אלמסמאה תאבתה כלהא פי כרה ואחדה אנמא הו לאנהם וגֿדוהא חאפטֿה לוצֿעהא בעצֿהא ענד בעץֿ דאימא [כאן][341] אלנקלה ואחדה מתשאבהה פי גֿמיעהא פחכמוא אנהא לכרה ואחדה. וליס האהנא שי ימנע מן אן נקול אן לכל כוכב מנהא פלך מכֿצוץ או לכל אתנין מנהא ויכון לגֿמיעהא נקלה ואחדה ומתסאויה. לאכן מן אגֿל אצטלאחהם פי אלאצל [אלדי עליה יגֿרי אלאמר פי][342] הדה אלנקלה כאן בחסבה פי אלקול אנהא כלהא פי כרה ואחדה כאפיה, ואן כאן קד יבקי עלי אלטביעי אלבחת ען אלסבב אלדי מן אגֿלה כאן הדה אלכתרה מן אלכואכב פי כרה ואחדה וללכואכב אלסבעה אלבאקיה סבעה אכר. וליס הדא פקט בל אן תכון כל כרה מנהא מקסומה אלי אתנין או אכתר כמא תבין. וקד פחץ ארסטו ען דלך ואעטי פיה אסבאבא [ואלבעד ען אליקין טֿאהר פיהא][343]. וקד פרגנא מן דכר מא יערץֿ מן אלמחאל ותכדיב אלארצאד אלכתירה אלמחררה מן קבל וצֿע אלכואכב אלתאבתה כלהא פי

82v

כרה ואחדה מע כון גֿמיע אלחרכאת אלי גֿהה ואחדה. ואן נחן וצֿענא לכל כוכב או כוכבין מנהא כרה ואחדה עלי אן יכון דלך ואגֿבא פיהא או אלא יכון מא יטֿהר ללחס מן נקלתהא עלי מא הו עליה פי גיר הדא אלוגֿה בחסב אלאצול אלחקיקיה אלמדכורה, כנא קד אסתנבטנא אמרא עֿטֿימא פי רפע הדא אלמוצֿע מן אלפחץ והדא יכון במא אצף.

B.3.II.2 ולנדכר אולא אלוגֿה [אלדי עליה יגֿרי אלאמר פי][344] אלכואכב אלתאבתה אלתי פי סטח אלדאירה אלמארה עלי אוסט אלברוגֿ אעני אלדאירה אלתי תרסמהא אלשמס בתקצירה אלאכֿיר. וינבגי אן יפהם הדא אלמעני ענה פי כל מוצֿע נדכר פיה דאירה אלברוגֿ ומנהא יתבין פי אלתי ליסת עלי הדה אלדאירה. ונקול אן אלכוכבין אלתאן כאנתא עלי נקטתי אלאנקלאבין פי אבתדא ואחד מן

339. א: דֿ.
340. א: מה שאמרנו פעמים אחרות.
341. א: וכי.
342. א: -.
343. א: בלתי אמתיות.
344. א: הנכון לתנועת.

אלנקלה קד תבין מן אמרהא אן מא יתבין פיהא בהדא אלאצל מטאבק למא יט̇הר באלרצד, פהאדין אלאתנין פקט ינבגי אן תכון פי כרה ואחדה. ונקול אן אלכוכב אלדי עלי הדה אלדאירה אלדי כאן בעדה פי אלאבתדא ה̇ אג̇זא מן נקטה אלאנקלאב אלשתוי נחו אלנקטה אלרביעיה מתלא̇ פאנה פי כרה אכ̇רי מילהא מתל מיל אלאולי והו פי אלדאירה אלוסטי מנהא, ובעדה מן אלנקטה אלנט̇ירה ללאנקלאב אלשתוי ה̇ אג̇זא איצ̇א פי אלג̇הה אלמקאבלה. אעני אנה אדא וצ̇עת אלנקטה אלנט̇ירה ללאנקלאב אלשתוי מן הדה אלדאירה עלי נקטה אלאנקלאב אלשתוי נפסה כאן בעד אלכוכב ען נקטה אלאנקלאב אלשתוי נחו אלנקטה אלכ̇ריפיה ה̇ אג̇זא. ואלכוכב אלמקאטר לה אדא וצ̇ע פי הדה [][345] נפסהא, פאנה ילזם פיהא מא ילזם פי אלאול.

B.3.II.3 וכדלך יג̇ב אן יכונא האדין אלאתנין [][346] פי כרה ואחדה עלי אלצפה אלמדכורה. ועלי הדא אלמתאל יג̇רי אלאמר פי אלבאקיה סוא כאנת אלתי עלי דאירה אלברוג̇ או כ̇ארג̇ה ענהא באן נצ̇ע כל ואחדה מן אלכ̇ארג̇ה מנהא פי דאירה צגירה מואזיה ללדאירה אלעט̇מי אלתי פי הדה אלכרה אלנט̇ירה לדאירה אלברוג̇, בעדהא ען אלעט̇מי מתל בעדהא ענהא לו כאנת אלכואכב כלהא פי כרה ואחדה. ועלי אן תכון [כל אתנין מתקאטרין][347] פי כרה ואחדה ומיול אלאכר כלהא ואחד. אן קטביהא כלהא דאיראן עלי דאירתין אתנין קטביהמא קטב [][348] מעאדל אלנהאר.

83r

ותכון נקלה הדה אלאכר כלהא [][349] מתסאויה ענד אלחס. ונצ̇ע לדלך מתלא̇ פי אלדואיר ואלחרוף.

B.3.II.4 פלתכון דאריה אלברוג̇ ד̇ט̇ז̇כ̇ קטבהא אלשמאלי ג̇ עלי דאירה ג̇ב̇ אלתי קטבהא קטב [][350] מעאדל אלנהאר והו א̇. ותכון נקטה ד̇ נקטה אלאנקלאב אלשתוי ונקטה [][351] אלציפי ואלנקטה אלרביעיה ש̇ ואלכ̇ריפיה ק̇. ולנצ̇ע כוכבא̇ עלי נקטה ט̇ בעדה מן נקטה ד̇ נחו נקטה ש̇ אעני אלנקטה אלרביעיה ה̇ אג̇זא. וקד ירי הדא אלכוכב במרור אלזמאן בחסב אצול בטמליוס עלי נקטה ז̇ אעני אקרב אלי אלנקטה אלרביעיה בחית יציר שמאליא̇ עמא כאן אולא̇. ונצ̇ע דאירה ה̇ח̇ז̇ל̇ פי פלך אכ̇ר קטבה ב̇ עלי הדא אלמיל ואלנקטה אלנט̇ירה לנקטה אלאנקלאב אלשתוי ה̇ וללציפי ל̇. פאן נחן וצ̇ענא הדא אלכוכב בחסב אצולנא עלי נקטה ח̇, בעדה מן נקטה ה̇ כ̇מסה אג̇זא נחו אלנקטה אלמקאבלה, אעני אלכ̇ריפיה, [אדא כאנת נקטה ה̇ מטאבקה עלי נקטה ד̇][352]. פאנה אדא נקטה ח̇ עלי נקטה ט̇, אעני [מוצ̇ע][353] אלכוכב, פאן באלאצלין ג̇מיעא̇ ירי אלכוכב פי וקת ואחד עלי נקטה ז̇ לאן באלזמאן אלדי ינתקל אלכוכב אלי כ̇לף קוס ט̇ז̇ בחסב אצול בטמליוס יקצר הדא אלפלך בחסב אצולנא קוס ח̇ט̇ מן אלמואזיה. וינתקל אלכוכב עלי קטביה קוס ח̇ז̇ והדא בעינה ילזם פי אלכוכב אלמקאטר לה. וליכון מתלא̇ עלי נקטה נ̇ ופי וקת כון נקטה ח̇ עלי נקטה ט̇ תכון נקטה מ̇ עלי נקטה נ̇, ועלי כל ואחד מן אלאצלין יציראן פי וקת ואחד עלי נקטה ס̇. והאכדא ילזם פי אלבאקיה, ואקטאב הדה אלאפלאך כלהא הי אבדא̇ עלי דאירה ב̇ג̇, וליס יכון פי דלך פרק אלא [][354] אלדי מן

345. א: הכדור.
346. א: בלבד.
347. א: כל אחד מהם.
348. א: גלגל.
349. א: ר̇ל̇ אשר הם בעגולת המזלות.
350. א: גלגל.
351. א: כ̇ היפוך.
352. א: -.
353. א: נקודת.
354. א: החלוף.

83v

[B.3.II.4/X] מן קבל אלמטאלע וקד תבין דלך פי אלמקאלה אלאולי ודלך מא ארדנא.

B.3.II.5 [ואמא חרכה אלאקבאל ואלאדבאר אלמוצ̇ועה ען אלמתאכ̇רין][355] פאן לנא אן נקול אן דלך אנמא הו באלתוהם ואנה מן קבל אזדיאד מטאלע בעץ' אלכואכב עלי בעץ' בחסב קרבהא מן נקטה אלאנקלאב או אלאסתוא. ואיצ̇א פליס ימתנע אן יכון בעץ' הדה אלכואכב אעט̇ם תקצירא̇ מן בעץ' מן אג̇ל אכ̇תלאף אבעאד אלאכר אלתי הי עליהא ען אלאעלי וליס ידרך ללחס הדא אלאכ̇תלאף אלי אלאן לקלתה, פאן אלתקציר נפסה ליס יתחצל אמרה אלי אלאן פצלא̇ ען אכ̇תלאף אלתקציר. וימכן אן הדא [אלתקציר אכ̇תלאף והם][356] אלמתאכ̇רון חתי קאלוא בחרכה אלאדבאר ואלאקבאל. ופימא קלנאה כאפיה פי תחציל אלאצל אלחקיקי להדה אלנקלה. ואלחמד ללה כתירא̇ אלדי אוצלנא אלי הדה אלגאיה.

כמלת אלמקאלה אלתאלתה מן כתאב נור אלעאלם
ואל חמד ללה רב אלעאלמין

B.4.0.1 אלמקאלה אלראבעה פי חרכאת אלכואכב אלכ̇מסה אלמתחיירה והי זחל ומשתרי ומריך וזהרה ועטארד ואכ̇תלאפאתהא ופיהא פצול. אלפצל אלאול פי צדר אלמקאלה. אלפצל אלתאני פי תבין אלאצל אלחקיקי אלמוצ̇וע ענה ללאכ̇תלאף אלאול אלדי יוג̇ד ללכואכב [אלתלאתה][357] אלעלויה והו אלמנסוב אלי אג̇זא אלברוג'. אלפצל אלתאלת פי אלאכ̇תלאף אלתאני אלמוג̇וד להדה [אלתלאתה][358] כואכב והו אלדי יוג̇ד בחסב בעד אלכוכב מן אלשמס וצ̇ע אלאצול אלחקיקיה לה וללאשיא אלגזייה אלתי תרי פיה ותלזם ענה. אלפצל אלראבע פי תבין אכתר אלאכ̇תלאף אלאול אלמנסוב אלי אלברוג' ותבין אלנקטה מן אלברוג' אלתי פיהא אקל אלסיר פי כל ואחד מן אלכואכב אלתלאתה אלעלויה. אלפצל אלכאמס פי תבין קדר אלדאירה אלפאעלה ללאכ̇תלאף אלמנסוב אלי אלשמס פי כל ואחד מן אלכואכב אלתלאתה אלעלויה. אלפצל אלסאדס כיף יסתכ̇רג' מן קבל אלמסיראת אלדוריה אלמסיראת אלחקיקיה.

84r

אלפצל אלסאבע פי [היאה][359] כוכב אלזהרה. אלפצל אלתאמן פי תבין אלנקטה מן אלברוג' אלדי פיהא אקל אלסיר ותבין נסבה נצף קטר פלך אלזהרה אלי נצף קטר אלדאירה אלפאעלה ללאכ̇תלאף אלמנסוב אלי אלברוג' ותבין מקדאר אלדאירה אלפאעלה ללאכ̇תלאף אלמנסוב אלי אלשמס. אלפצל אלתאסע פי [היאה][360] כוכב עטארד. אלפצל אלעאשר

B.4.I.1 אלפצל אלאול פי חרכה אלכואכב אלכ̇מסה אלמתחיירה. קאל יוסף ן̇ נחמיש ומן בעד אנא קד דכרנא ג̇מיע מא ינבגי לנא אן נדכרה פי [היאה][361] אלשמס ואלקמר ואלכואכב אלתאבתה עלי אתם מא ימכן פיהא בחסב אלחק נפסה כמא הו ט̇אהר ממא סלף מן קולנא, פקד חאן לנא אן נדכר אלאצול אלחקיקיה ל[היאה][362] אלכואכב אלכ̇מסה אלמתחיירה. פאן בדלך יתם אלגרץ' פי אסתנבאט אליקין פי הדא אלעלם ורפע אלאקואל ואלאצול אלממתנעה אלמסתנבטה פיה ענד אלקדמא ואלמתאכ̇רין

355. א: -.
356. א: החילוף הטעה.
357. ו: אלתאלתה
358. ו: אלתאלתה
359. ו: האיה.
360. ו: האיה.
361. ו: היה.
362. ו: האיה.

אלי אלאן. פאנא אעלם אן אלנאט̇ר פי כתאבי הדא אמא אן יכון תאם אלפטרה מותר ללחק וחיניד ידפע בעקלה ג̇מיעה אלאקאויל אלמדונה פי הדא אלעלם ען אלקדמא וג̇מיע אלמתאכ̇רין אלי אלאן ויעלם אן ארפע רתבתהא לא יצל אן תכון שכוכ̇א עלי אקאוילנא פיה. ולמתל דלך קאל שלמה המלך ע̇א̇ס̇ אין זכרון לראשונים וגם לאחרונים שיהיו לא יהיה להם זכרון עם שיהיו [לאחרונה][363]. וכלאמנא פי הדא אלכתאב אנמא הו מע הדא אלכאמל. ואמא אן יכון פי פטרתה נאקץ ולא יסע עקלה אלאמור אלגאמצ̇ה לאן אעלי רתבתה אן יפהם אלאצול אלמוצ̇ועה אלי אלאן לא גיר. ואדא לם יפהם אקאוילנא פי דלך יג̇תהד פי נקצ̇הא באקאויל סופסטאניה או שעריה، פאנה ליס ימכן לה גיר דלך. ואלחק יהזא ב̇מתל דלך כמא יהזא באלדין ירידון נקץ̇ אלאמור אלחקיקיה וקדמוא. כמא קאל פי מתל דלך דוד המלך ע̇א̇ס̇ יושב בשמים יצחק ודו. . . .

B.4.1.2 ונרג̇ע אלי גרצ̇נא ולאן פלך אלתדויר בחסב אצול בטלמיוס צ̇רורי פי הדה אל

84v

כואכב אלכ̇מסה אכתר מנה פי אלשמס ואלקמר כמא הו ט̇אהר מנהא וכ̇אצה פי זהרה ועטארד، פאנא קבל אן נשרע פי דכר אלאצול להדה אלכואכב ונדכר [ברהאן][364] ואחד קאטע עלי אמתנאע פלך אלתדויר. פאקול אן ארסטו יבין פי אלמקאלה אלתאניה מן אלסמא ואלעאלם אן אלכואכב ליס להא חרכה דחרג̇ה. וקולה בחסב מא לכ̇צה ענה ן̇ רשד הכדא، קאל פקד ידל עלי דלך ט̇הור אלכ̇יאל אלמערוף בוג̇ה אלקמר עלי חאלה ואחדה. ודלך אנה יתבין מן אמר הדא אלאתר אלט̇אהר פי וג̇ה אלקמר אנה חקיקה לא רויה ואדא פרצ̇נאה חקיקה פלו כאן ללקמר חרכה דחרג̇ה לאכ̇תלפת רויה הדא אלאתר הכדא קולה. ובהדא אלברהאן בעינה ימכננא אן נבין אמתנאע פלך אלתדויר פי אלסמא לאנה אן כאן תם פלך תדויר ללקמר מתלא̇ פאלקמר מרכוז פיה ויג̇ב אן תכ̇תלף רויה הדא אלאתר בחרכה אלקמר פי פלך תדוירה. פאנה אן כאן וצ̇ע אלקמר באלנסבה אלי אבצארנא פי ג̇מיע אג̇זא פלך אלתדויר נסבה ואחדה כאנת חרכתה פי פלך אלתדויר באלתדחרג̇ וליס יכון מרכוזא̇ פיה. ואד ליס כדלך פיג̇ב אן תכ̇תלף רויה הדא אלאתר. ואדא ארתפע אלתאלי ארתפע אלמקדם פאדא̇ וג̇וד פלך אלתדויר פי אלקמר ממתנע.

B.4.1.3 ואד הו ממתנע פי כוכב ואחד פהו ממתנע פי ג̇מיעהא לאן טביעתהא ואחדה באלנוע והדא אלמעני ט̇אהר. ליס יחתאג̇ אלי ברהאן אלא מן אג̇ל אנתשאר הדא אלראי ענד ג̇מיע אלאמם ואקאוילנא פי הדה אלמעאני הי אלאן פי אבתדא אלשהור. ואדא נט̇רוא פיהא דוי אלעקל פאנהם ליס יחתאג̇ אלי ברהאן עלי אמתנאע אלאקאויל אלסאלפה פי דלך. פאנהם ליס יג̇דון פימא נקולה פי דלך אלא חרכאת מסתדירה מתשאבהה מתסאויה פי אזמנה מתסאויה לכל חרכה אקטאב מחדודה ופלך מחדוד מחיט באלכל מרכזה מרכז אלעאלם. ואן אחתג̇נא אן נצ̇ע לכוכב ואחד אפלאך כתירה פאן כל ואחד מנהא הו עלי אלצפה אלמדכורה ואלואחד דאכל אלאכ̇ר ומתחרך בחרכתה ומתחרך איצ̇א חרכה כ̇אצה לה עלי קטביה חתי אנה לו וצ̇ענא עשרה אפלאך לכוכב ואחד אלואחד דאכל אלאכ̇ר. פאן אלפלך אלדאכ̇ל אלאכ̇ר אלאספל מנהא מתחרך באלחרכאת אלתסעה אלתי ללאפלאך אלתסעה אלדי פוקה ומתחרך

85r

איצ̇א בחרכתה אלכ̇אצה לה עלי קטביה. ובתרכיב הדה אלחרכאת בעצ̇הא מע בעץ̇ תחדת אלאכ̇תלאפאת עלי מא הי עליהא. והדא חין אבתדי בדכר מא ועדת לה.

363. ו: באחרונה.
364. ו: בהאן.

B.4.I.4 אעלם אן בטמליוס ומן תקדמה וג̇דוא ללכואכב אלכ̇מסה [אלמתחיירה][365] צנפין מן אלאכ̇תלאף. צנף ואחד מנסוב אלי אלברוג̇ אעני אנה יוג̇ד בחסב אג̇זא אלברוג̇ יזיד או ינקץ עלי אלמסיר אלוסט פי אלטול. ואלאג̇זא אלמנסובה פי אלמסיר אלי אלזיאדה פיהא תכון אלזיאדה אבדא̇ [אלי][366] אלכ̇לאף אלדי נסבה אלי אנתקאל אלאוג̇. ואלאג̇זא אלמנסובה אלי אלנקצאן מן אלמסיר פיהא יכון אלנקצאן אבדא̇ עלי דלך אלמתאל. והדא אלאכ̇תלאף וג̇ד אבדא̇ בחסב אג̇זא אלברוג̇ מעלומה יזיד ובחסב אג̇זא אכ̇ר ינקץ ואן כאן פי אלארצאד ג̇מיעא̇ יוג̇ד פי תשכיל ואחד מן אלשמס לאן בדלך ירתפע אלאכ̇תלאף אלדי מן קבל אלשמס ויבקי הדא אלאכ̇תלאף וחדה. וצנף מנסוב אלי אלשמס אעני אנה יוג̇ד בחסב בעד אלכוכב מן אלשמס יזיד איצ̇א פי אלמסיר אלוסט פי אלטול או ינקץ והדא אלאכ̇תלאף יוג̇ד פי אלאוקאת אלתי יעוד אלאול אלי מא כאן אעני אלי אג̇זא ואחדה מן אלברוג̇. פאנה אדא וג̇דנא פי אג̇זא ואחדה בעינהא מן אלברוג̇ אצנאף מן אלתשכל מן אלכוכב ללשמס מכ̇תלפה וכאן פי אלתשכל אלואחד יזיד ופי אלאכ̇ר ינקץ פאן הדא אלאכ̇תלאף יוג̇ד חיניד כ̇לוא̇ מן אלאול. ווג̇דוא איצ̇א פי אלאכ̇תלאף אלמנסוב אלי אלברוג̇ אן אלזמאן אלדי מן אצגר אלחרכה אלי [אוסטהא][367] אעט̇ם מן אלזמאן אלדי מן [אוסטהא][368] אלי אעט̇מהא, ווצ̇עוא להדא אלאכ̇תלאף אלכ̇ארג̇ אלמרכז. ווג̇דוא פי אלאכ̇תלאף אלמנסוב אלי אלשמס אן אלאמר באלעכס ווצ̇עוא לה פלך אלתדויר. ובעד דלך אסתכ̇רג̇וא קדר אלכ̇רוג̇ ען אלמרכז פי כל ואחד מן אלכואכב אלה̇ ומואצ̇ע אלאוג̇ פי אלברוג̇ וקדר נצף קטר אלתדויר פי כל ואחד מנהא. אמא פי זחל ומשתרי ומריך פאנה בין דלך בתלאתה מן אלאחואל אלתי תסמי טרף אלליל אעני אדא כאן אלכוכב מקאטר לוסט אלשמס לאן חיניד יכון אלכוכב פי אלחציץ̇ אלמרי מן אלתדויר. ואמא פי זהרה ועטארד פאנה בין דלך באלאבעאד אלעט̇מי מן אלשמס אלצבאחיה ואלמסאיה.

B.4.I.5 וקדם לדלך אנה וג̇ד ללתלאתה אלעלויה אן מסיר אלכוכב אלוסט פי אלטול ופי אלאכ̇תלאף אדא ג̇מעה מסאוי

85v

למסיר וסט אלשמס. פאנה וג̇ד לזחל פי נ̇ט̇ סנה, אעני אלתי תבתדי מן אנקלאב או אסתוא ויעוד אליה, ויום ואחד ונצף ורבע באלתקריב, נ̇ז̇ דורה מן אלאכ̇תלאף ומן דוראת אלטול דורתין וג̇ז ואחד ותלתי ג̇ז וג̇ז מן כ̇ מן ג̇ז לתמאם אלזמאן אלמדכור. וללמשתרי פי ע̇א̇ סנה אלא ד̇ איאם ונצף ותלת וג̇ז מן ט̇ו̇ מן יום ס̇ה̇ דורה ללאכ̇תלאף ומן אדואר אלטול סתה אדואר אלא ד̇ אג̇זא ונצף ותלת. וללמריך פי ע̇ט̇ וג̇ איאם וסדס וג̇ז מן כ̇ מן יום ל̇ז̇ דורה ללאכ̇תלאף, מ̇ב̇ דורה ללטול וג̇ אג̇זא וו̇ דקאיק. ווג̇דוא ללזהרה פי ח̇ סנין אלא יומין ורבע וג̇ז מן כ̇ מן יום ה̇ דוראת ללאכ̇תלאף ואדואר אלטול מסאויה לאדואר אלשמס והי ח̇ דוראת אלא ג̇זין ורבע. ולעטארד פי מו̇ סנין ויום ואחד וג̇ז מן ל̇ מן יום קמ̇ה̇ דוראת ללאכ̇תלאף ואדואר אלטול איצ̇א מסאויה לאדואר אלשמס. ודוראת אלטול פי הדה אלכואכב אנמא ארדנא אן תבתדי מן אנקלאב או אסתוא אלי אן תעוד אליה.

B.4.II.1 אלפצל אלתאני פי תבין אלאצל אלחקיקי אלמוצ̇וע ענה ללאכ̇תלאף אלאול אלדי וג̇ד ללכואכב אלתלאתה אלעלויה והו אלמנסוב אלי אג̇זא אלברוג̇. פלננזל פי אצל זחל ומשתרי ומריך דאירה אלברוג̇ אלתי פי פלך כל כוכב מנה אב̇. ולנצ̇ע בסבב אלאכ̇תלאף אלמנסוב אלי אלברוג̇ דאירה

365. ו: אלמתחייה.
366. ו: אלא.
367. ו: אוסאטהא.
368. ו: אוסאטהא.

ג̇מד̇ל̇ חול קטב ב̇ ואלקוס מן דאירה עט̇ימה אלתי מן קטבהא אלי מחיטהא מסאוי לקוס אעט̇ם הדא אלאכ̇תלאף. ואלכוכב ידור עלי הדה אלדאירה חול קטב ב̇ תאבעא̇ לחרכה אלאכ̇תלאף וקטב ב̇ מע פלכה דאיר עלי דאירה אלברוג̇ אעני א̇ב̇ חול קטביהא. ונצ̇ע איצ̇א אן זמאן אלעודתין ואחד، אעני אן זמאן עודה אלכוכב פי דאירה ג̇מד̇ל̇ מסאוי לזמאן עודה קט̇ב ב̇ עלי דאירה אלברוג̇ כמא הו ואג̇ב להדא אלאכ̇תלאף. ולננזל אן אדא כאן קטב דאירה ג̇מד̇ל̇ אעני ב̇ עלי נקטה א̇ כאן אלכוכב עלי נקטה ג̇. ופי אלזמאן אלדי אנתקל קטב ב̇ עלי דאירה אלברוג̇ קוס א̇ב̇ מתלא̇ תאבעא̇ לחרכה אלטול אנתקל אלכוכב אלדי כאן עלי נקטה ג̇ חול קטב ב̇ קוס ג̇ב̇ מן דאירה ג̇מד̇ל̇ אעני דאירה אלממר אלי כ̇לאף חרכה אלקטב שביה בקוס א̇ב̇. וירי לדלך פי מסיר אלטול נקצאן יוג̇בה קוס ג̇ב̇ ומא דאם[369]

86r

B.4.III.1 פי אלאשיא אלג̇זיה אלתי תעלם בעלם אלמטאלע. אדא עלמנא ג̇ז אלשמס פי יום או לילה ערפנא מקדארהא באן נאכ̇ד מטאלע ו̇ ברוג̇ מן ג̇ז אלשמס פי אליום ומן מקאבלהא פי אללילה פי תלך אלמיל ונאכ̇ד ג̇ז מן ט̇ו̇ מנהא פיכון עדה אלסאעאת אלאסתואיה. ואדא אכ̇דנא ג̇ז מן י̇ב̇ מנהא חצל עדה אג̇זא אלסאעה אלזמאניה. וקד יעלם מקדאר אלסאעה אלזמאניה באן נאכ̇ד מן אלג̇דול סדס פצ̇ל אלמג̇מועה לג̇ז אלשמס או מקאבלהא לליל בין אלפלך אלמסתקים ואלמטלוב פיה. פאן כאן ג̇ז אלשמס שמאלי זדנאה עלי ט̇ו̇ ואן כאן ג̇נובי נקצנאה מן ט̇ו̇ חצל אג̇זא אלסאעה אלזמאניה. ולנא אן נרד סאעאת זמאניה אלי אסתואיה באן נצ̇רבהא פי אג̇זאיהא ונאכ̇ד מן אלמג̇תמע ג̇ז מן ט̇ו̇. ובאלעכס באן נצ̇רב אלאסתויה באג̇זאיהא ונקסמהא באזמאן נט̇אירהא מן דלך אלבעד. ואדא אכ̇דנא ג̇ז מן ט̇ו̇ מן צ̇עף אלפצ̇ל אלמדכורה וזדנאה עלי י̇ב̇ או נקצנאה מן י̇ב̇ חצל לנא עדה סאעאת דלך אליום אסתואיא̇. ואיצ̇א אדא עלמנא כם סאעה זמאניה [מצ̇ת][370] מן אלנהאר או אלליל וצ̇רבנאהא פי אג̇זאיהא ואלקינא מא אג̇תמע מן ג̇ז אלשמס עלי תואלי אלברוג̇ או מן מקאבלתהא באלליל במטאלע אלאקלים פחית ינתהי אלעדד פי אלברוג̇ הו אלג̇ז אלטאלע.

B.4.III.2 ואדא צ̇רבנא אלסאעאת מן נצף נהאר [אלמאצ̇י][371] אלי הנא פי אג̇זאיהא כל בנט̇ירה ואלקינא מא אג̇תמע מן ג̇ז אלשמס עלי תואלי אלברוג̇ פי אלמנתצבה פינתהי אלי אלג̇ז מן אלברוג̇ אלדי פי וסט אלסמא פוק אלארץ̇. ואדא אכ̇דנא אלמג̇מוע ללג̇ז אלטאלע פי אלאקלים ונטרח מנה צ̇ פחית ינתהי פי אלמנתצבה מן אלברוג̇ הו אלמתואסט ללסמא פוק אלארץ̇. ובאלעכס נאכ̇ד ג̇מעאת וסט אלסמא פי אלמנתצבה ונזיד עליה צ̇ פחית אנתהי מן אלברוג̇ פי אלאקלים פהו אלג̇ז אלטאלע. ויכ̇תלף נצף אלנהאר פי אלמסאכן בעדד אזמאן אעתדאליה מתל עדד אג̇זא מא בין נצפי נהארהם.

B.4.III.3 פי אלזואיא אלחאדתה מן דאירה אלברוג̇ ודאירה נצף אלנהאר. אעלם אן מערפה תלך אלזואיא וכדלך אלזואיא אלחאדתאת מן אלברוג̇ ואלאפק פי כל מוצ̇ע ומן אלברוג̇ ואלדאירה אלמארה בקטבי אלאפק، אעני אלסמתיה וכדלך קסי תלך אלדאירה אלתי מן סמת אלראס אלי מוצ̇ע תקאטעהא מן אלברוג̇، הי צ̇רורה פי מערפה אכ̇תלאף מנט̇ר אלקמר. ולדלך ראינא אן נבינהא. ונקדם אנא נסמי זואיא קאימה מן דואיר עט̇אם، אן וצ̇ענא קטב ואנפראז בין אלמחיטין באלזאויה רבע אלדאירה אלמרסומה באי בעד כאן אעני צ̇ ג̇זא. ונקצד מן אלד̇ זואיא אלחאדתה פי כל תקאטע אלי ואחדה، והי אלשרקיה אלשמאליה מן אלברוג̇.

369. There are two blank lines at the end of the page in the Vatican MS which could be why the end of 85v does not flow into the beginning of 86r.

370. ו: מצ̇תה.

371. ו: אלמצ̇י.

B.4.III.4 ונבין אולאׄ אן כל נקטתין

86v

בעדהמא מתסאוי ען אסתוא ואחד פזאויתהמא אלמוצופתין מתסאויתין. פלתכן דאירה מעאדל אלנהאר אׄבׄגׄ וקטבהא זׄ, וקוסהא מן אלברוגׄ דׄבׄהׄ וקסי חׄבׄ בׄטׄ בעדהמא ואחד ען בׄ אעני נקטה אלאסתוא. פנרסם זׄכׄחׄ וזׄטׄלׄ פקוסא כׄבׄ בׄלׄ אעני מטאלעהמא מתסאויה, ומיולהמא אעני חׄכׄ טׄלׄ מתסאויתאן. פאלמתלתאת מתסאויאן פזאויא כׄחׄבׄ בׄטׄלׄ בל זׄטׄהׄ מתסאויאן. ונבין איצׄאׄ אן כל נקטתין מן פלך אלברוגׄ כדׄ והׄ בעדהמא סוא ען מנקלב ואחד וליכון בׄ כזאויא זׄדׄבׄ זׄהׄבׄ אלחאדתאת מן נצף אלנהאר מתל קאימתין, לאן זאויתי זׄדׄבׄ זׄהׄבׄ מתסאויתין לתסאויתהא זׄדׄ זׄהׄ מן אגׄל אן בעד דׄהׄ מן בׄ ואחד וזאויה זׄהׄגׄ תעאדל מע כל ואחדה מנהא קאימתין ומׄאׄבׄ.

B.4.III.5 ולתכן נצף אלנהאר אׄבׄגׄדׄ ופלך אלברוגׄ אולאׄ אׄהׄגׄ. ואׄ אלמנקלב קטב ובבעד צׄלע אלמרבע בׄהׄדׄ. פמן אגׄל אן אׄבׄגׄדׄ מארה בקטבי אׄהׄגׄ בׄהׄדׄ, פדׄהׄ רבע, אעני דׄאׄהׄ קאימה פי אי אנקלאב כאן. ונגׄעל אׄהׄגׄ מעאדל אלנהאר ואׄזׄגׄ דאירה אלברוגׄ ואׄ אלאסתוא אלכׄריפי פזׄ אנקלאב שתוי. פקוס דׄהׄ מע גאיה אלמיל אעני הׄזׄ קׄיׄגׄ גׄזא ונׄאׄ דקיקה אעני זאויה דׄאׄזׄ. פאלזאויה אלתי ענד אלאעתדאל אלרביעי מא יבקי לקאימתין והו סׄוׄ גׄזא וטׄ דקאיק.

B.4.III.6 ונגׄעל פי אלדאירה אלאכׄרי מעאדל אלנהאר אׄהׄגׄ ואלברוגׄ בׄזׄדׄ וזׄ אלאסתוא אלכׄריפי וקוס זׄבׄ אלסנבלה. ובבעד צׄלע אלמרבע מן בׄ חׄטׄהׄכׄ ונטלב מקדאר זאויה כׄבׄטׄ. ותעלם בעלם קוס הׄטׄ לאן הׄכׄ רבע פכל ואחד מן קסי חׄבׄ בׄטׄ הׄחׄ רבע מן אגׄל אן אׄבׄגׄדׄ מכׄטוטה עלי קטבי אׄהׄגׄ וחׄהׄכׄ. פלאן נסבה גׄיב קוס בׄאׄ והו מיל ראס אלסנבלה אלי גׄיב קוס תמאמה, אעני אׄחׄ, מולפה מן נסבה גׄיב בׄזׄ אעני ברגׄ אלסנבלה, אלי גׄיב זׄטׄ, ומן גׄיב הׄטׄ אלמגׄהול אלי גׄיב הׄחׄ אלרבע, פיכון אלמגׄהול אלהׄ. פנגׄד קוס הׄטׄ כׄאׄ גׄזא פקוס כׄהׄטׄ אעני זאויה כׄבׄטׄ קׄיׄאׄ. פזאויה ראס אלעקרב הי הדה אלאגׄזא בעינהא ויבקי זאויתא ראס אלתור וראס אלחות תמאם תלך לקאימתין והי....

87r

B.4.III.7 נעלם תגׄזיה אלגׄ ארבאע אלבאקיה. פיכון אלאפק בׄהׄדׄ וקטב כׄ מרתפע ענה לׄוׄ גׄזׄ. ומעאדל אלנהאר אׄהׄגׄ ודאירה אלברוגׄ זׄחׄטׄ ונקטה חׄ רביעיה ונכׄרגׄ ברגׄ אלחמל. וינבגי אן נבין מקדאר חׄהׄ. פנסבה גׄיב קוס כׄדׄ אעני ארתפאע אלקטב אלי גׄיב קוס דׄגׄ, אעני תמאמה לרבע מולפה מן נסבה גׄיב קוס כׄלׄ אעני תמאם מיל לׄ אלי גׄיב קוס לׄמׄ אעני מיל לׄ ומן נסבה גׄיב קוס הׄמׄ אלמגׄהול אלי גׄיב קוס הׄגׄ. פאלמגׄהול אלהׄ ונגׄדה חׄ אגׄזא ולׄחׄ דקיקה. לכן חׄמׄ אעני מטאלע חׄלׄ פי אלמנתצבה כׄזׄ גׄז ונׄ דקיקה יבקי חׄהׄ יׄטׄ גׄזא ויׄבׄ דקיקה. פבין אן אלחות יטלע במתל הדא וכל ואחד מן אלסנבלה ואלמיזאן יטלעאן בתמאם הדה אלאזמאן לצׄעׄף מטאלעהמא פי אלמנתצבה והו לׄוׄ זמאנאׄ וכׄחׄ דקיׄ.

B.4.III.8 ואדא גׄעלנא חׄלׄ קסמין מן אליׄבׄ אעני חמל ותור פנגׄד בהדא אלטריק בעינה קוס מׄהׄ טׄוׄ גׄזא ומׄוׄ דקיקה. פיבקי חׄהׄ מׄאׄ גׄזא ונׄחׄ דקיקה, פאלתור וחדה יטלע מע כׄבׄ זמאנאׄ ומׄוׄ דקיקה. ואלדלו יטלע מע מתל הדא, וכל ואחד מן אלאסד ואלעקרב בתמאם הדה אלאזמאן לצׄעׄף מטאלעהא פי אלמנתצבה והו לׄזׄ זמאן ודקיקתאן. פלמא כאן נצף אלדאירה אלתי מן אלסרטאן אלי אלרמי תטלע פי אלנהאר אלאטול אעני פי יׄדׄ סאעה ונצף בל קׄקׄ יׄזׄ ונצף גׄזׄ ואלנצף יטלע פי אלנהאר אלקציר אעני פי טׄ סאעאת ונצף בל קׄמׄבׄ גׄזא ונצף, יבקי כל ואחד מן אלסרטאן ואלקוס יטלע מע לׄהׄ זמאנה ורבע, וכל ואחד מן אלגׄוׄזא ואלגׄדי פמע כׄטׄ ויׄזׄ דקיקה. ובין אן בהדא אלטריק יעלם מא הו אקל מן תלך אלאגׄזא ומׄאׄבׄ. וכדלך יתבין הדא כלה באלטריק אלאכׄר מן קבל מתלת מׄחׄלׄ.

B.4.III.9 וקד ימכן אן נעטי טריק תאני פי עלם אלמטאלע והו הדא. נכ̇ט אולא̇ פלך נצף אלנהאר א̇ב̇ג̇ד̇ ואפק ב̇ה̇ד̇ ומעאדל אלנהאר א̇ה̇ג̇، ופלך אלברוג̇ [ז̇ה̇ח̇][372] וה̇ אלרביעיה. ונרסם קטעה ט̇כ̇ מואזיה תמר בנקטה ט̇ אלמעלומה אלבעד מן ה̇. ונכ̇רג̇ מן ל̇، אעני אלקטב אלג̇נובי، קסי ל̇ט̇מ̇ ל̇כ̇נ̇ ל̇ה̇. פה̇ט̇ מן אלברוג̇ יטלע מע ט̇כ̇ פי אלמאילה אעני פי ל̇מ̇ נ̇מ̇ לאנהמא מתואזיה מתשאבהה ויטלע פי אלמסתקים מע מ̇ה̇. פאלפצ̇ל בינהמא، אעני נ̇ה̇، יחדה קטעה ל̇נ̇ אלמארה בכ̇.

B.4.III.1 פלנצ̇ע תלך אלצורה פיהא דאירה אלאפק ומעאדל אלנהאר וקטבהא אלג̇נובי

87v

ז̇. ונרסם ז̇ח̇ט̇ ונקטה ח̇ מטאלע ראס אלג̇די וכ̇ מטאלע אי נקטה אכ̇רי שינא מן ג̇די דלו חות אעני אלרבע. פנסבה ג̇יב קוס ח̇ט̇ והי ואחד פי כל אלמיול לאנה גאיה אלמיל אלי ג̇יב תמאמה אלרבע אעני ח̇ז̇ מולפה מן נסבה ג̇יב קוס ט̇ה̇، אעני נצף פצ̇ל אלנהאר אלמעתדל [עלי][373] אלקציר אלי ג̇יב קוס ה̇ל̇ אלמג̇הול، ומן נסבה ג̇יב קוס ל̇כ̇، והו ואחד איצ̇א פי כל אלמיול מן קבל ג̇דול אלמיל، אלי ג̇יב קוס תמאמה אעני כ̇ז̇. פתכון לדלך נסבה ט̇ה̇ אלי ג̇יב ה̇ל̇ ואחדה פי אלמיול כלהא.

B.4.III.1 פנחן נג̇על נקטה כ̇ מרה יטלע פיה אלג̇ז אלמופי כ̇ מן חות ומרה אלג̇ז אלעאשר ומרה אלג̇ז אלא̇، וננתהי אלי אלג̇די בזיאדה עשרה עשרה. ויכון פי כלהא ג̇יוב ל̇כ̇ וכ̇ז̇ מעלומין מן קבל ג̇דול אלמיל، ותבקי פי כל ואחד מנהא נסבה ג̇יב ט̇ה̇ אלי ג̇יב ה̇ל̇ פי כל אלמיול מעלומה ויכון אלמג̇הול אלראבע. ונג̇דה פי אלעשרה אלא̇ אדא כאן ג̇יב ט̇ה̇ מתלא̇ ס̇ ג̇זא ט̇ אג̇זא ול̇ג̇ דקיקה ופי אלב̇ יח̇ ג̇זא ונ̇ז̇ דקיקה ופי אלג̇ כ̇ח̇ ג̇זא ודקיקה. ופי אלד̇ ל̇ו̇ ג̇זא ול̇ג̇ דקיקה، ופי אלה̇ מ̇ג̇ ג̇זא וי̇ג̇ דקיקה ופי אלו̇ נ̇ ג̇זא ומ̇ד̇ דקיקה. ופי אלז̇ נ̇ה̇ ג̇זא ומ̇ה̇ דקיקה، ופי אלח̇ נ̇ח̇ ג̇זא ונ̇ה̇ דקיקה. פאדא כאנת צ̇עף ט̇ה̇ מפרוצ̇ה אד הי אלפצ̇ל בין אלנהארין כאנת נסבה ג̇יב ט̇ה̇ אלי ג̇יב ה̇ל̇ כמא קלנא، פקוס ה̇ל̇ מעלום. פאדא נקצנאה מן מטאלע אלקוס אלמטלוב פי אלמסתקים יבקי מטאלעה פי דלך אלמיל.

B.4.III.1 פנג̇על חית ארתפאע אלקטב ל̇ז̇ ג̇זא וקוס ט̇ה̇ פי הדא אלמוצ̇ע יח̇ ג̇זא ומ̇ה̇ דקיקה. פתנקץ תלך אלאעדאד אלתי דכרנאהא בקדר מא ינקץ ג̇יב יח̇ ג̇זא ומ̇ה̇ דקיקה ען ס̇ ג̇זא. תם נאכ̇ד קוס תלך אלג̇יב ויכון פי אלעשרה אלא̇ ג̇זאין ונ̇ו̇ דקיקה. ופי אלב̇ ה̇ אג̇זא ונ̇ דקיקה، ופי אלתאלתה ח̇ אג̇זא ול̇ח̇ דקיקה ופי אלד̇ י̇א̇ ג̇זא וי̇ז̇ דקיקה، ופי אלה̇ י̇ג̇ ג̇זא ומ̇ב̇ דקיקה ופי אלו̇ ט̇ו̇ ג̇זא ומ̇ו̇ דקיקה ופי אלז̇ י̇ז̇ וכ̇ד̇ דקיקה ופי אלח̇ יח̇ ג̇זא וכ̇ד̇ דקיקה ופי אלט̇ יח̇ ג̇זא ומ̇ה̇ דקיקה והו ה̇ט̇. פננקץ תלך אלאעדאד מן מטאלע אלעשראת פי אלמנתצבה، פיבקי מטאלע אלעשראת פי אלמאילה. ויבקי ללעשרה אלט̇ אעני אלרבע ע̇א̇ ג̇זא ורבע והו מקדאר אלנהאר אלאקצר. ובעלם תג̇זיה הדא אלרבע נעלם תג̇זיה אלארבאע אלבאקיה כמא קד ברהנא קבל. פנעמל ג̇דולא̇ ונבתדי מן מעאדל אלנהאר אלי אלדאירה אלתי אטול נהארהא י̇ז̇ סאעה בזיאדה נצף סאעה. פי אלצף אלאול אלברוג̇ אלי̇ב̇ ופי אלב̇ עשראתהא ופי אלג̇ מטאלעהא ופי אלד̇ ג̇מעתהא.

372. ו: ז̇ה̇ט̇.
373. ו: 2.

2

Translation of the Judeo-Arabic Text of *The Light of the World*

51r

THE BOOK OF *THE LIGHT OF THE WORLD* BY RABBI JOSEPH IBN NAḤMIAS

A book that brings forth the hidden into the light; it made firm the boundaries of the ramparts on high
Look, then, for the thing you want to find; it will be for you a light forever
Show me your way, O Lord, and lead me on a level path; My vindication will come from You, your eyes will behold what is right

When the practitioners of mathematics spoke in conventional, technical language about the hypotheses according to which the heavenly motions proceed, some of them founded the matter upon two hypotheses with regard to which some of the practitioners of physics completely disagreed, without saying anything convincing about the reason for apparent motions and anomalies in the heavens, for their [the physicists'] intention was only to explain the impossibility of these two hypotheses for the celestial bodies with only physical evidence. One of these two hypotheses is their thesis of the eccentric orb and the epicyclic orb. The second is their thesis of allegedly opposing motions in the heavens. They [the physicists] reject the existence of opposing motions in the heavens just as they reject the eccentric orb and the epicyclic orb. That occurred to them only because they thought that a circular motion does [seem to] oppose a circular motion, just 0.1

as a rectilinear motion opposes a rectilinear motion, contrary to what Aristotle demonstrated in *De Caelo*.

Our goal here is to show that what they said about the impossibility of the eccentric and epicyclic orbs is true, so there is no need for another proof. Then we will give, as completely as possible, certain hypotheses from which appear all that used to appear with regard to these two orbs. What was said about the impossibility of opposite motions in the heavens is invalid, for it might be clarified shortly with a demonstration that there is no opposition in circular motion, meaning that a circular motion does not oppose a circular motion, though Aristotle had shown that by inference (*bi-dalāʾil*) only. So we say: the eccentric orb's impossibility, absence of accord, and lack of order are clear from the way it is, as there is no greater monstrosity than the thesis of an orb, revolving about its center, eccentric to the center of the world, with its center revolving about another center. 0.2

51v

It follows from that that there would be between the two orbs either a void, with the impossibility of its existence having already been shown as it is an example of the impossibility of the existence of pure nothingness, or a plenum composed of a foreign body which would change its positions through motion. If these two orbs intersected, what we said would also follow when the center of one moved about the other: the two orbs together would be dispersed and fragmented so that the nature of these orbs would be like the nature of water or air, filling one place and vacating the other, along with the rest of the impossibilities and the violations of truth that follow from these circumstances of which Rabbi Moses, peace be upon him, reminded us in *The Guide of the Perplexed*. 0.3

And more abominable than that is the thesis of an epicyclic orb revolving about its center, with its center rotating also in the width of another orb, for these impossibilities would follow necessarily from it. So how? In many places they were forced to propose both hypotheses together, with the motion of the planet being about the center of its orb, and the planet being carried in this motion upon another orb whose center is removed from it. And what we have said is sufficient to abolish the existence of these two orbs in the heavens. As for what they said about the refutation of the existence of two opposite motions in the heavens because its motions are natural and due to one simple natural mover, one of two things become clear from the essence of what they say. Either there is no opposition in circular motion, I mean that one circular motion does not oppose another circular motion, or that two opposing motions necessarily originate from this one simple mover. If we were to admit that there is opposition [in circular motion] with regard to how the motion 0.4

is natural and due to a natural, single, and simple mover, then that could be explained in this way.

It is a characteristic of the sphere that it can move about any two poles imagined on its surface. Let us assume, for example, that the two poles of the orb of Venus are positioned on the ecliptic, one of them at the point of the head of Cancer and the second at the head of Capricorn. And let us assume that the ecliptic orb moves about its poles from the east to the west, and it carries with it the orb of Venus and its poles. Let us assume also that the orb of Venus moves in this direction about its poles. Then there is no difference between this thesis and the thesis that the ecliptic orb, with its poles, moves with the daily motion 0.5

52r

and moves also about its poles. If so, then it is apparent that as long as one pole, I mean the north pole of the orb of Venus, moves from the beginning of Cancer to the beginning of Aries with the motion of the ecliptic orb, and the south pole through the opposite quadrant, then the two motions, I mean the motion of the equinoctial and the motion of the orb of Venus, will be in one direction.

At the time when the two poles abide at the equinox points, the two motions will intersect at a right angle. As long as the north pole moves from the beginning of Aries to the beginning of Capricorn, and the second through the opposite quadrant, then the two motions would be opposite according to their opinion. So I wish I knew how uniform, natural motions proceeding according to a single order could be for half the time uniform in one direction and opposite in the other half? Rather, in essence, one motion opposes itself at different times although it is forever uniform. That would be extremely hideous, though this demonstration was taken from the perceived directions of the motions. Thus, it has become apparent from that that a circular motion does not oppose a circular motion. And if we were to assume that [one circular motion can oppose another], for it is impossible, then it would become evident from that that these two motions which are opposed according to their school of thought, are from this one natural simple mover necessarily forever maintaining its order. So were it the part of this mover that there never originate from it opposite motions, then it would be necessary, when the two poles arrive at the two points of the equinoxes, that the motion stop, and that is impossible. 0.6

Since that has been explained, then the impossibilities that follow from them are only from their thesis of the eccentric orb and the epicyclic orb, and of this impossibility is this measure which you see. How amazing is the magnitude of what the chief of this discipline in particular, Ptolemy, deduced from these hypotheses and how he availed those who came after him with this majestic science that facilitates, for whoever wishes, finding the times of the returns of the 0.7

planets, and knowing their positions in their orbs in longitude and latitude, and knowing the measures of their size and the ratio of some to others, and their ratio to the Earth, and the knowledge of the times of the eclipses of those [planets] that are eclipsed, and the measure of the eclipsed portion of its body at each time, all this with the hypotheses upon which this is based being imagined so far from the truth. It is not necessary to presume like Ptolemy,

52v

and propose these hypotheses being of the opinion that things are themselves in that manner because that which appears to the senses follows that which proceeds from those hypotheses.

I am greatly amazed by what the great philosopher Rabbi Moses, upon him be 0.8
peace, said in the 24th chapter of the second part of *The Guide of the Perplexed*, when he enumerated the impossibilities that follow from these two hypotheses. He said: "How can the various motions of the stars come about? Is it in any way possible that circular motion should be on the one hand equal and perfect, and that on the other hand the things that are observable should be observed in consequence of it, unless this be accounted for by making use of one of the two principles or both of them?"[1] until he said that God, may He be exalted, has "enabled man to have knowledge only of what is beneath the heavens."[2] As for the forms of the heavens and its motions, He alone, may he be exalted, knows. Because of that King David, upon him be peace, said, "The heavens are the heavens of the Lord."[3] And let us return to where we were.

Because of the possibility of these things appearing to the senses according 0.9
to how he [Ptolemy] structured them, these hypotheses became accepted by all. Therefore, the gate of investigation into the true hypotheses came to be closed for a long time. That hindered the perfection of this majestic wisdom until al-Biṭrūjī came and became aware of these impossibilities which follow from these hypotheses. He claimed to have invented a model for these motions and anomalies, without the thesis of an eccentric orb or an epicyclic orb. He mentioned that he became aware of that when he heard that Abū Bakr Ibn al-Ṭufayl had come across a model and hypotheses other than those two that had been proposed, except that

1. For the preceding quotation from the *Guide*, I have followed Maimonides (trans. Shlomo Pines, intro. Leo Strauss): *The Guide of the Perplexed* (Chicago and London: University of Chicago Press, 1963): vol. 2, p. 325.

2. From "enabled" until the end of the sentence, I have followed Maimonides (trans. Pines), *Guide*, vol. 2, p. 327.

3. All translations of biblical passages follow the 1917 Jewish Publication Society translation. This quotation is from Psalm 115:16.

he followed the faction of physicists in rejecting opposite motions. He [Biṭrūjī] began by mentioning the manner of the motions and how all that which is in the heavens moves in one direction, with all that follows from these [motions] agreeing with that which appears to the senses. This extent of what he has said is correct; I mean that it is possible that it be as he has said.

As for when he gave the reasons that necessitate the anomalies, nothing that follows from those proposed hypotheses corresponds to that which appears to the senses, until now, unless his intent was to teach us how there is some anomaly in these motions, other than the one which appears to the senses. So I consider this to have been his school of thought, and if not, then why would he need to renew observations as he said? It might be clear to one looking in his book, in many other places, that his intent was precisely that 0.10

53r

unnecessary intention, for he has caused us to enter into a perplexity and impossibilities greater and more abominable than that which we were in. Despite what we have said about the deficiency of his hypotheses and their many deviations from the truth, as will be clear subsequently when we arrive at these places with God's assistance, he [Biṭrūjī] merits no small thanks for being the first one to become aware of that and to write a book about it.

So from the time I saw his book, I continued to contemplate for a long time until I arrived at the truth of the hypotheses necessitating the anomalies for these motions as will be shown afterward if God, may He be exalted, wills. Know that our objective in this book is to propose hypotheses for the anomalies instead of the two aforementioned hypotheses, I mean the eccentric and the epicyclic orbs. As for the matter of opposite motions, it would have been necessary to leave them as they are according to Ptolemy due to how he showed the lack of opposition in the motions. But due to the diffusion of this opinion among the masses, I mean that there is opposition in circular motion, we have seen fit to propose hypotheses for these motions so that the motion will even be in the single direction in which it was according to them, except when matters compel the thesis of a motion like that which was our thesis in the preceding example, I mean motions which are imagined to be together at one time and opposite at another time, though they are forever uniform proceeding in a single order. 0.11

Thus we say that the thing that compelled them to propose the two motions which, according to them, are opposite, is the movement of the Sun, Moon, and the rest of the planets toward the north and south. Were it not for that, there would be nothing necessitating, according to them, that their movement toward the east be a motion in reverse any more than it be a lag, with Ptolemy having 0.12

clarified that in the first treatise of his book. So, what makes it out of the question to say that each one of the orbs that is beneath the uppermost moves in the direction of the uppermost orb following its motion and about its poles, and that it lags through this motion behind the motion of the uppermost because of the loss of the power reaching it from the uppermost due to its distance from the first mover? It is not in its nature to accept more than that since it is not carried in this motion with the uppermost orb according to how the two [orbs] would then be one thing both moving with a uniform motion.

53v

It moves also about its own two poles in that direction as if its intention were that 0.13
it make up for its lag behind the uppermost through the first motion, since its goal and intention is toward the uppermost and to resemble it, since it is its goal. Through that motion it moves toward the north and the south by the measure of the distance of its poles from the poles of the uppermost. With that motion also, each orb falls short of arriving at the position of the uppermost orb, and resembling it, by a limited amount, namely the measure of the distance of each orb from the first mover that imparts the movement to it.

Therefore, the closer its position to the uppermost [orb], the less its final lag. 0.14
Through this doctrine, the disagreement among practitioners of this science regarding the order of the orbs disappears, because it is clear that their order is the cause of their lag. So, the poles particular to each orb beneath the uppermost lag in each revolution by a limited amount in the two circles of their paths, and that is on the measure of the distance of each of them from the first mover. But the lags of the planets that are above these orbs are less than that by the measure of the motion of each one about its poles, until the orb of the fixed stars makes up through its particular motion what its orb lagged, according to some. Or a small lag remains according to the school of thought of others. And that which is beneath it lags more than that leading ultimately to the lowest orb of the Moon, the lag of which is greater than that of every orb above it due to its great distance from the uppermost.

Therefore, it is apparent from the case of the orb that is the mover for the 0.15
daily motion, it being perceived by the intellect since it has no sensible precursor because it is completely simple, that it is above all [the other orbs] since its motion is one, simple, and all-encompassing. As for the rest of the orbs that are beneath it, their motions are not simple, and even the orb of the fixed stars has two motions as there are for the rest of the orbs that are beneath it. Also, our greatest goal is to explain that a heavenly orb must necessarily have its concave surface be wholly tangent to the convex surface of that beneath it with it being completely spherical on the outside and inside, and with its convex surface being tangent to the concave surface of the other orb above it, if there is another orb above it, with the number of the orbs being as it was. And

54r

there is no harm in this. It is necessary for one who looks into this book of ours, that if he seeks from this book this matter that intellects fell short of perceiving to such an extent, he find it explained within, so that there is no doubt. We say that even if it were necessary, according to this perspective, that we begin by talking about the uppermost orb and conclude by looking into the lowermost, we have made the motion of the Sun the beginning, owing to the system of the hypotheses followed until now, so that the distinction between these hypotheses and those be facilitated for one looking into this, and so that there will be facilitated for him a distinction between the paths taken, through the ease of their conceivability and their proximity to the truth.

Our discourse in this book is with one who has already looked into the books 0.16
set forth in this science until now even though he did not need to read these books. I opine that one who has already looked into the ideas set forth, the hypotheses and the reasons of this science from Ptolemy and others, has wasted his time with them. When he looks into this book of ours there occurs to him what occurs to one who lingered for a long time in darkness unable to exit suddenly into the light. But when he did exit into it he took pleasure in it, and he perceived the advantage of the light over the darkness, and about the likes of that the prophet said: "The people that walked in darkness have seen a great light; they that dwelt in the land of the shadow of death, upon them hath the light shined."[4]

It is necessary that you know that if we said that the center of such and such 0.17
a star moves on such and such a circle, whose pole is on such and such a circle, then each circle is traced on the surface of a particular orb from among the orbs which are according to the aforementioned description. Likewise, if we named the motion of the particular planet the lag, then we do this so we can affirm the hypotheses even according to the school of thought of those who hold there to be opposition in circular motion, and you already know our opinion about that. According to our opinion it is necessary that it be named direct motion and not lag and you take what you wish from the two schools of thought, for our hypotheses for the anomalies agree with both.

We have divided this book into two parts. I amass in the first part all of the 0.18
geometrical preliminaries and other things that should be presented to communicate the manner of the motions, with all of them being shared between Ptolemy and us since the manner of a motion

54v

does not ever sully them. We mention them in the most succinct way and the most proximate to the comprehension and in this part are five chapters. The

4. Isaiah 9:1.

first chapter is about the assumed preliminaries that need to be presented for this science, and what follows from them. The second chapter is about all of the geometrical preliminaries that need to be presented before communicating the motions of the stars. The third chapter is about the instruments required for the observations through which this science is perfected. The fourth chapter is about the partial declinations and the rising times at *sphaera recta* and at *sphaera obliqua* and other particular things that follow from them. The fifth chapter is about the measure of the solar year. I mention in the second part the manner of the motions and the hypotheses on the basis of which it is desirable to proceed regarding the sensible variations for the motions and positions of the celestial bodies, that being why I first intended this book. This is the time at which I begin to mention that which I promised.

The first chapter from the first part—the assumed premises whose presentation is necessary for this science. There are 24 premises. (A) The Sun and the Moon and the rest of the stars move forever from east to west on parallel circles. They rise, then culminate over us, then set and are hidden for a time; then they return to their points of ascent from the beginning and that recurs always, uniformly. (B) The north pole is elevated above the horizon, and the south pole is depressed below it in each of the latitudes in the northern hemisphere, that being the latitude of each horizon from the Earth's equator toward the north except at *sphaera recta*. The stars whose distance from the north pole is less than its local elevation do not ever set but rather revolve about the pole upon parallel circles on the measure of the distance of the stars from the pole about which they revolve. (C) The stars near to the forever visible [stars] linger less in darkness than those that are far from them. (D) The magnitude of the stars and their light is forever in a single state except that they are magnified when they set and that is due to the vapors of humidity that enter between them and the sight. (E) They intersect the plane A.I.1

55r

of the Earth when they set. (F) The distances from the Earth upward are equal. (G) It is desirable that the form of the body follows from its motion. (H) The motion of the heavens is the most perfect motion and the most facile. (I) The body with the most perfect motion is the spherical. (J) Of various shapes with equal circumferences, that which subtends larger angles has a larger surface area. (K) The body of the heavens must be greater than anything else. (L) The body of the orb has, of all of the bodies, the strongest resemblance between one part of it to another. (M) Heavenly bodies like the planets appear round from different directions at the same time for anyone who sees them. (N) Measurements through instruments agree except in this way [which follows]. (O) The rising and setting of all the stars for those in the east precedes the stars' rising and setting for those in the west. (P) Whoever in the east who recorded the time of a lunar eclipse, which is at the same time for all, is farther beyond the meridian than whoever in the west recorded it, and that [difference in the recorded times] is on the measure of the difference of distance between them. (Q) Every one who travels by sea toward the mountains or any elevated place, however much they come closer, they see it increasing [in size] as if rising forth from the water. (R) In each part of the Earth night and day are equal two times a year, and this equality is at the [Sun's] mean passage between the solstices, I mean between the shortest and longest days. (S) Midday is always when the Sun culminates, I mean when it is in the zenith circle perpendicular to the equinoctial. (T) Each horizon, I mean the surface extending from the sight, bisects the celestial sphere, and that is because we always see six zodiacal signs appearing, then the six hidden ones rise and those set. (U) At midday, the shadow of the eastern gnomon falls together with the shadow of the western gnomon on a single straight line. (V) Lunar eclipses, at every point in the heavens, occur at the time of diametrical opposition. (W) The judgment of the gnomon in any given part of the surface of the Earth is like its judgment were it assumed to be at the center of the Earth, and likewise for the centers of the instruments employed in this science. (X) If a man

55v

moved with a motion that is faster than the rest of the existing motions, then it is necessary that he not see anything moving in the direction of his motion; rather, he sees everything as if it is moving opposite his motion.

So if these things are premised, then it is clear from what we said that the heavens are spherical and their motion is also spherical, and that is shown with premises A, B, C, D, E, F, G, H, I, J, K, L, M, and N. And the syllogism, however, is complete either through a single premise of these on the condition that the second is known or through two premises together, such as H and I and J with Q. From some of these premises what is sought, that the motion is spherical, is shown from accidents that follow from them; and with some of them it is shown from how the form is spherical due to the manner of the motion. From some of them what is sought is shown through a regular syllogism and with some of them through a reductio ad absurdum and all of that is apparent in it. It is shown that the Earth, with all its parts, is spherical according to the senses, with analogy to the universe, through premises B, O, P, and V. And through some of these premises it is shown that it does not depart [from the center of the universe] either up or down. And through some of them it is shown that it does not depart [from the center of the universe] to the west or the east, and through some of them that it does not depart [from the center of the universe] toward one of the poles. Through that, the opinion compounded from these becomes null, and it is clear that the Earth does not have a transpositional motion through the premises A.I.2

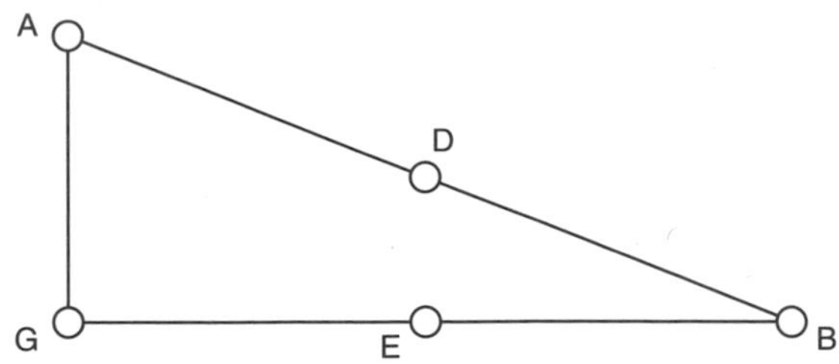

FIGURE 2.1

from which it was shown that the Earth is in the center. And it is shown through premise X that the Earth does not move about its center with the heavens being fixed, as a group of people had presumed.

It is shown that the classes of the primary motions in the heavens are two because we find that all that is in the heavens rises and traverses the heavens and sets each day upon parallel circles, the greatest of which is the equinoctial since the horizon bisects it, and this is the first motion that causes the universe to revolve equally. We find the Sun and the Moon and the five planets moving toward the east, I mean opposite the first motion, with varying motions or lags, according to one of the two schools of thought as we mentioned, and by their movement toward the east, or its lag, they move also toward the north and south with a uniform movement. So we have judged that this second motion is about the poles of a circle inclined A.I.3

56r

to the equinoctial and it is named the ecliptic, with the motion of the Sun being upon it and tracing it. On its two sides, toward the north and the south, is the path of the Moon and the five planets. We imagine a circle passing through the poles of the universe and through its [the ecliptic's] poles, intersecting it on right angles at two points, with the [point] north of the equinoctial being called the summer solstice, and the second the winter [solstice]. The point of the intersection of the ecliptic with the equinoctial with the passage through it to the north is called the vernal equinox, and the second is autumnal, with the poles of the ecliptic adhering to their positions on the colure, although they move with the first motion. The circle of the meridian is a circle forever perpendicular to the horizon passing through the poles of the universe.

Now that we have finished that, let us begin with the individual demonstrations and we construct our investigation on the basis of the division of the circumference of the circle into 360°, which everyone has taken as a convention, and the division of the diameter into 120 parts owing to the facility of calculations A.I.4

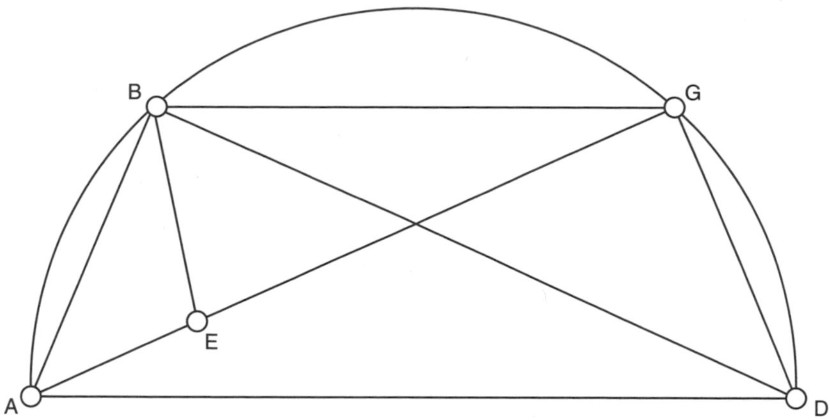

FIGURE 2.2

with these numbers; likewise [a convention is] the division of the degree into 60 minutes and the minute into 60 seconds and likewise ad infinitum until further division can be set aside without harm.

(See figure 2.1.) The second chapter about all of the geometric preliminaries that need to be presented before communicating the motions of the stars. Let us first present the method for knowing the measures of the chords of the degrees of the circle. Thus, let us set forth as a lemma that if we wanted to divide a known number into mean and extreme ratio,[5] then we take a square and the square of its half and combine them. We take the root of the sum and we subtract from it half of the number, with the remainder being the greater portion. Let us propose an example with lines. Let the known number be *BG*, and let *AG* be equal to half of *BG*, and it encompasses, with line *GB*, a right angle. Then we connect *AB* and subtract from *BA* half of *BG* and let it be *BD*. We separate from *BG* [segment] *GE*, equal to *DA*, so I say that *GE* is the greater portion. Its demonstration is because the product of *BG* with itself, rather with *BE* and *EG*, and the product of *GA* with itself equals [the product of] *BA* with itself, rather *BD* with itself, and *DA* with itself, and twice the product of *BD* with *DA*. We subtract [the product of] *BD* with itself, I mean [the product of] *GA* with itself. There remains the product of *BG* with *BE* and *EG*, equal to *DA*, I mean the product of *EG* with itself and twice the product of *BD* with *DA*, I mean the product of *BG* with *EG*. There remains the product of *BG* with *BE* being equal to the product of *EG* with itself, and that is what we desired. Euclid had demonstrated precisely that in the second part of his book with a demonstration potentially the same as this one. A.II.1

5. Literally: a ratio with a middle and two extremities.

56v

(See figure 2.2.) Since we have presented this, if we therefore know the measure of the side of the inscribed hexagon, that is, the chord of an arc of 60°, it being known because it is half the diameter of that by analogy with which we wish to know the measures of the chords, then we might know the measure of the side of the inscribed decagon, namely the chord of 36° since it is the greatest part of it, as was shown in the 15th part of Euclid. Let us present as a lemma also that if there were [inscribed] in a circle a quadrilateral like *GDBA*, then the product of *BD* and *GA* is equal to the product of *DA* and *GB* together with *BA* and *GD*. So we make angle *ABE* equal to [angle] *DAG* if *ABD* is greater than it, so *ABD* equals *EBG* and *BDA* equals *BGE*, so triangles *ABD BGE* are similar. Thus the ratio of *BG* to *GE* equals *BD* to *DA*, and the product of *BG* and *AD* equals the product of *BD* and *GE*. Likewise the triangles *ABE* and *BGD* are similar, so the product of *BA* and *DG* equals the product of *BD* and *EA*. So the product of *BD* and *AG*, all of it, is equal to the product of *AB* and *DG* and *AD* and *GB* together. If *AB* and *AG* were given, then *BG*, which is between them, is given because *BD* and *GD* are given due to the complementarity of the two halves of two circles. As *AD* the diameter is given, then *GB* is given. Therefore we know the chord of 24° from the surplus of the side of the hexagon over that of the decagon. A.II.2

(See figure 2.3.) I say also that if we know the chord of arc *AG*, then we know the chord of half of it, let it be *AD*, since because *AE*, half of it [line *AG*], is given, and *AZ* is given, so *EZ* is given. *ED* comes to be given, and *AD* also. Therefore, we might know the chord of 12° and the chord of 6 [degrees] and the chord of 3 [degrees] and the chord of 1½°, and the chord of ¾° through bisection. In this way, it emerges that the chord of 1;30° is 1;34;15°, and the chord of 0;45° is 0;47,8°. We know, by this, all of the remaining chords by [increments of] a degree and a half in that we begin with the difference between a degree and a half and a half circle. A.II.3

We try now to find the chord of 1° from 1;30° and from 0;45°, since we are not able to demonstrate that geometrically.[6] It is in this way. I say that if we extended in a circle two unequal lines, the ratio of the longer to the shorter is smaller than the ratio of the arc upon the longer to the arc upon the shorter. Euclid had demonstrated that. So let us propose A.II.4

57r

that if *AG* is the chord of a degree, and *AD* is the chord of 0;45°,[7] then line *AG* is less than one and one-third of line *AD*, since arc *AG* is one and one-third of

6. Judeo-Arabic: we are not able to demonstrate that by lines. Cf. Toomer, *Ptolemy's Almagest*, p. 54: "cannot be found by geometrical methods."

7. Both MSS have 1;45°, but it should be 0;45°. See Toomer, Ptolemy's "*Almagest*," pp. 55–6 (as Ibn Naḥmias follows Ptolemy's method for approximating the chord of 1°).

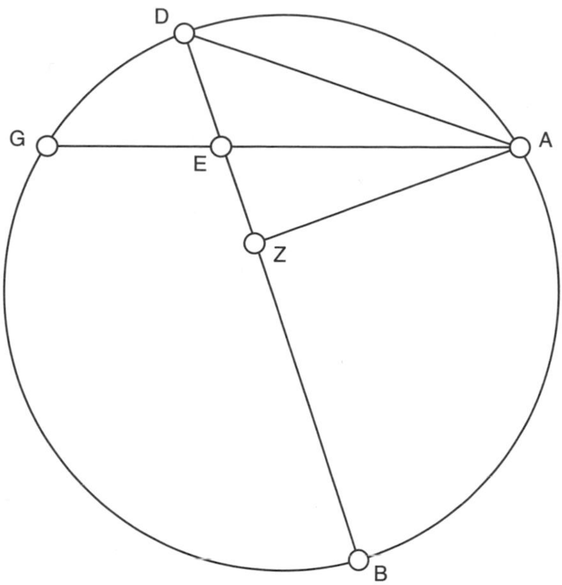

FIGURE 2.3

[arc] *AD*. So line *AG* is less than 1;2,50,40°. Also, we assume that *AD* is the chord of 1°, with *AG* being the chord of 1½°, so line *GA* is less than one and one-half [times] line *AD* and line *AD* is more than 1;2,50°, I mean two-thirds of *AG*, so the lesser and the greater divide the third. The chord of a degree of arc is 1;2,50,20°. Therefore, we know all of the [remaining] arcs by increments of ½°, because we know the chord of half of the circle, I mean the diameter. And if we know the chord of ½°, then we might know the chord of 179;30° from the difference between them. Likewise we know the rest by decreasing [increments of] half of a degree. We might know also the sine of each arc because the sine is half the chord of twice the arc, and that is what we wanted [to demonstrate].

(See figure 2.4.) And we also present as a lemma that if there occurred in a circle, such as circle *AB*, three points, such as *ABG*, provided that each one of arcs *GB* and *BA* is smaller than a half circle, and we connect *AEG* and *DEB* with the center, then I say that the ratio of the sine of arc *AB*, I mean *AS* the perpendicular on *DB*, to the sine of arc *BG*, I mean *GH* the perpendicular, is also equal to the ratio of *AE* to *EG* because the triangles are similar. A.II.5

(See figure 2.5.) Let us present also what needs to be presented for demonstrations of spherical matters. For every two great circles on the surface of a sphere such as *ADBG* and *ALB*, there can be separated from one of them, from one of the two points of intersection, two arcs such as *AD* and *AG*, with each of them A.II.6

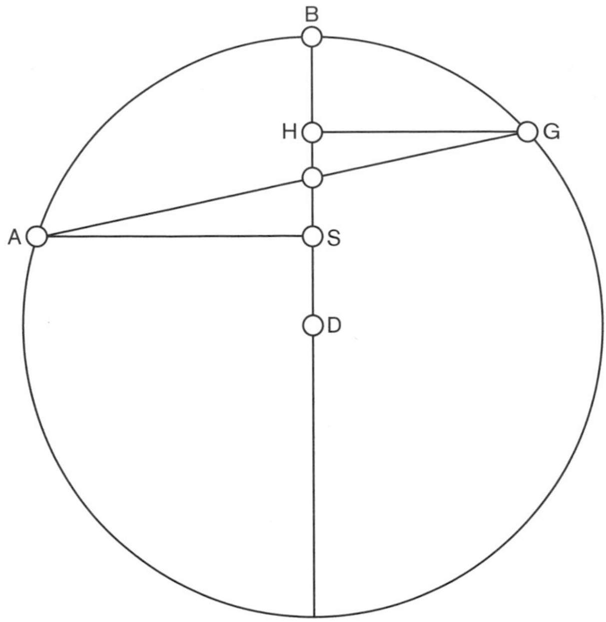

FIGURE 2.4

being smaller than a half circle. We extend from points *DG* two perpendiculars upon the plane of the other circle. I say that the ratio of the sine of [arc *AD* to the sine of][8] arc *AG* is like the ratio of the perpendicular extending from point *D* to the other [perpendicular]. Its demonstration is: let the common portion of the two circles, I mean their diameters, be line *AB*, and we extend upon it the two perpendiculars *DZ* and *GE*, they being the sines of the two arcs. If they were also perpendicular to the surface of circle *ALB*, then we have shown what we wanted. And if they were not like that, we extend *DT GH* from points *DG* perpendicular to the surface of circle *ALB*. We connect *EH TZ*, and because lines *DZ GE* are parallel, and likewise *DT GH*, then angles *ZDT EGH*

57v

are equal with *H* and *T*, being right angles. So the two triangles are similar; and the ratio of *DZ* to *GE*, I mean the sines, is like the ratio of *DT* to *GH*, I mean the perpendiculars, and that is what we wanted [to prove]. It would, likewise, be evident were the arcs separated off on a single side from the point of intersection.

(See figure 2.6.) If that has been presented, let great circle arcs *GD* and *BE* intersect great circle arcs *AB* and *AG* on the surface of a sphere also upon point A.II.7

8. Following the insertion in the Hebrew.

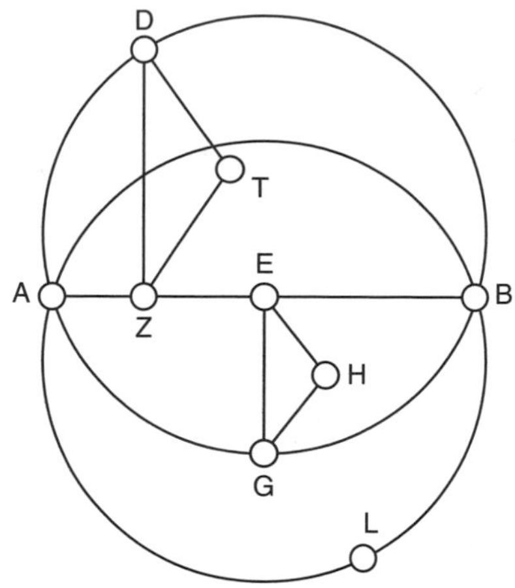

FIGURE 2.5

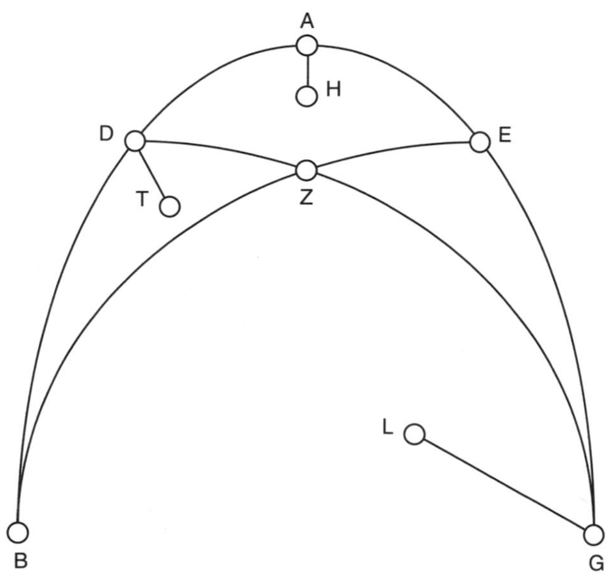

FIGURE 2.6

Z. I say that the ratio of the sine of arc *GE* to the sine of arc *EA* is [compounded][9] of the ratio of the sine of arc *GZ* to the sine of arc *ZD* and from the ratio of arc *BD* to the sine of arc *BA*. Let us extend from points *GAD* perpendiculars to the surface of circle *BZE*, and they are *GL AH DT*. Because for every three lines, the ratio of the first to the third is compounded of the ratio of the first to the second and of the ratio of the second to the third, then let us make perpendicular *DT* the middle term of the ratio between *GL* and *AH*. Then the ratio of perpendicular *GL* to perpendicular *AH*, I mean the ratio of the sine of arc *BE* to the sine of arc *EA*, is compounded of the ratio of *GL* to *DT*, I mean the ratio of the sine of arc *GZ* to the sine of arc *GD* and from the ratio of *DT* to *AH*, I mean the sine of arc *BD* to the sine of arc *BA*. Likewise, it also becomes clear that the ratio of the sine of arc *GZ* to the sine of arc ZD is compounded of the ratio of the sine of arc *GE* to the sine of arc *EA* and from the ratio of the sine of arc *AB* to the sine of arc *BD*, if we made perpendicular *AH* the middle [term] between *GL* and *DT*. Likewise, it becomes clear through compounding that the ratio of the sine of arc *GA* to the sine of arc *AE* is compounded of the ratio of the sine of arc *GD* to the sine of arc *DZ* and from the ratio of the sine of arc *ZB* to the sine of arc *BE* in that we extend the perpendiculars from points *AEZ* on the surface of circle *BDA*. Hence, the compounding is explained for each one of the sides. This demonstration is, however, provided that each side, with its counterpart, be in one arc. And the rest of the ratios that are compounded from these six sines become clear from the demonstration that Ibn Muʿādh al-Jayyānī brought, and it is like this:

(See figures 2.7 and 2.8.) I say that the product of the first sine and the fourth A.II.8 combined with the sixth, I mean the solid body that is encompassed by these three sines is equal to the product of the second and the third combined with the fifth. So we propose that line *GB* is equal to the fourth sine and that *BH* equals the sixth, enclosing with *GB* a right angle. We extend line *BA* from point *B* perpendicular to the plane of the two lines, and it is equal to the first sine, and we complete the two surfaces

58r

GH GA with parallel sides that encompass solid *DH*. Let us also make solid *MS*[10] provided that line *ML* is equal to the third sine, and *LS* equal to the fifth sine

9. The distinction between the compounding and the composition of ratios is explained in Euclid (intro., trans., and comm. Thomas L. Heath): *The Thirteen Books of the Elements* (Cambridge: Cambridge University Press, 1908; repr., Mineola, NY: Dover Publications, 1956), vol. 2: pp. 132–5. Here and elsewhere, Ibn Naḥmias used *taʾlīf* (lit. composition) sometimes to denote what Euclid would call compounding.

10. Following the Hebrew recension. Ar. ME, but there is no E pictured in the MS figure.

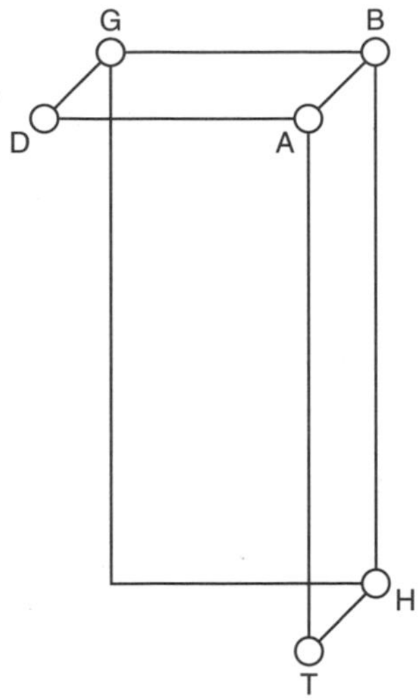

FIGURE 2.7

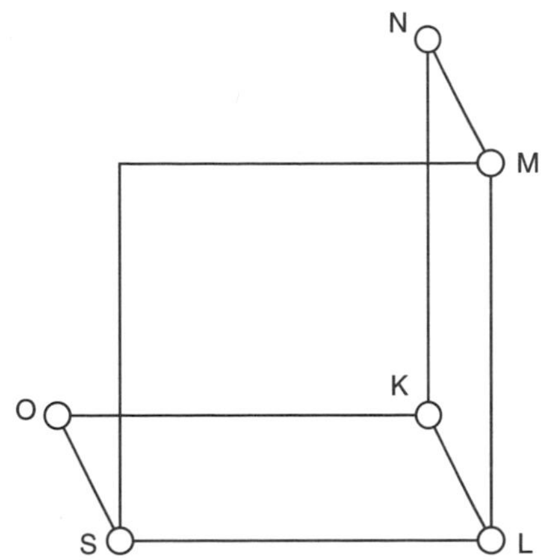

FIGURE 2.8

and *LK* equal to the second. Because the ratio of surface *MS* to surface *GH* is compounded from [the ratio of] *LM* to *GB* and from the [ratio of] [*LS*][11] to *BH*, so the bases *MS GH* are interchangeable in ratio with the heights, and the solid is equal to the solid.

And since we have presented [this],[12] we say that the ratio of any side taken from one solid to any other [side][13] from another solid is compounded from the four remaining [sides], like how the ratio of *BA*, namely the first sine, to *LM*, namely the third sine, I mean the sine of arc *GZ* in the previous figure, is compounded from the ratio of line *LK*, namely the second, to sine *BG*, namely the fourth, and from the ratio of [*LS*],[14] namely the fifth, to *BH*, namely the sixth. That is because the ratio of surface *KS* to surface *GH*, rather the ratio of line *BA* to line *LM*, owing to the interchangeability of the ratios, is compounded from these aforementioned four. Thus, the rest of it is clear. If it was taken with its counterpart from that solid, then it would not be compounded from what is left, and that is what we desired [to prove]. A.II.9

(See figure 2.9.) We say that for every four quantities, the ratio compounded A.II.10

11. Following the Hebrew recension. Ar. LE, but there is no E pictured in the MS figure.
12. Following the Hebrew recension. Ar. has nothing.
13. Following the Hebrew recension. Ar. position (*waḍʿ*).
14. Following the Hebrew recension. Ar. LE, but there is no E pictured in the MS figure.

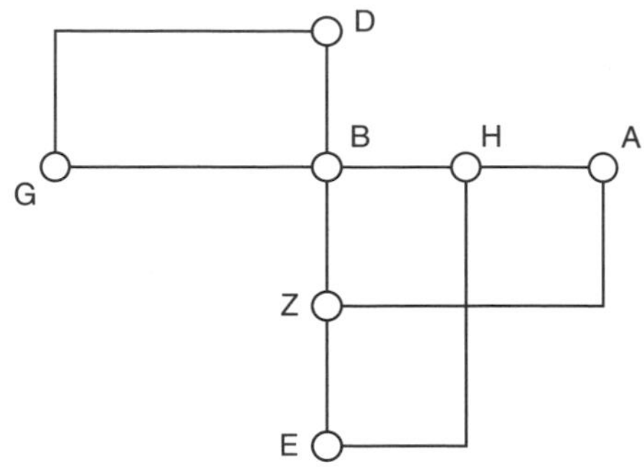

FIGURE 2.9

from the first to the second and from the third to the fourth is precisely the ratio that is compounded from the first to the fourth and from the third to the second. An example: let there be two surfaces with parallel sides, surfaces *DG AZ*, with angle *B* shared between them. The ratio of surface *AZ* to surface *DG* is compounded from the ratio of the first line *AB* to the second [line] *BG* and from the ratio of the third [line] *ZB* to the fourth [line] *BD*. I say that it is also compounded of the ratio of the first [line] *AB* to the fourth [line] *BD* and from the ratio of the third [line] *ZB* to the second [line] *BG*. Its demonstration: let us extend *BZ* on a perpendicular until it equals *BA*, and let it be *BE*, and we make *BH* equal to *BZ*, and we complete surface *HE*. So the ratio of surface *HE*, I mean *AZ*, to the surface *DG* is compounded from the ratio of side *EB*, I mean *AB*, to *DB*, and from the ratio of *HB*, I mean *BZ*, to *BG*, and that is what we wanted [to prove].

It results from that that each ratio compounded from these four quantities is A.II.11
compounded from them

58v

in two ways. For that, we make a small table, and we number in the first column the compounded ratios that can be compounded from the six quantities. In the second and third columns are the ratios that can be compounded from the rest and those [ratios] which cannot be [compounded]. And in the four remaining columns, we put across from every compounded ratio, for which compounding is possible, the ratios that are compounded from it. And those for which that is not possible we leave aside, and we do not write a letter for them. Every ratio compounded from them might be inverted so that the number of the compounded ratios would be twice that which is written in the table. And this is how to draw up the table:

TABLE 5.1

The Number	The compounded ratio		The ratios that are compounded from it			
1	A	B	G	D	E	W
2	A	B	G	W	E	D
3	A	G	B	D	E	W
4	A	G	B	W	E	D
x	A	D	x	x	x	x
5	A	E	B	W	G	D
6	A	E	B	D	G	W
x	A	W	x	x	x	x
x	B	G	x	x	x	x
7	B	D	A	G	W	E
8	B	D	A	E	W	G
x	B	E	x	x	x	x
9	B	W	A	G	D	E
10	B	W	A	E	G	B
11	G	D	A	B	W	E
12	G	D	A	E	W	B
x	G	E	x	x	x	x
13	G	W	A	B	D	E
14	G	W	A	E	D	B
15	D	E	B	A	G	W
16	D	E	B	W	G	A
x	D	W	x	x	x	x
17	E	W	A	B	D	G
18	E	W	A	G	D	B

(See table 5.1.) If one of these six were unknown we could extract it, since we knew that the first multiplied with the fourth, then with the sixth is like the second multiplied with the third, then with the fifth. So if, for example, the sixth were the unknown, the second could be multiplied with the third, then with the [fifth],[15] and we divide the product by the product of the first with the fourth. The sixth emerges, and according to this example we work the rest. A.II.12

The third chapter about the instruments needed for the observations through A.III.1

15. My emendation. Both MSS read "fourth."

which this science is perfected. In order to show the measure of the inclination of the ecliptic from the equinoctial, I mean the inclination of each of the solstices from it, that being equal to that which is between the poles, we make two circles of brass

59r

with a squared-off surface, and we place the smaller in the hollow of the larger so that it revolves in its interior toward the north and the south. We divide the larger into 360°, and we place on the diameter of the smaller two plates facing each other with two pointers in their middles [touching][16] the face of the outer [ring]. We erect the two rings on a pillar perpendicular to the horizon and on the local meridian so that they are both in the plane of the local meridian. If we rotate the inner ring so that the lower plate is shaded by the upper one when the Sun is in the meridian, [the tip of the measuring instrument][17] shows us the measure of the distance of the center of the Sun from the zenith point in the meridian circle. Through this instrument we know the measure of the arc that is between the zenith and between each of the solstices; therefore, that which is between the solstices is known. For this in particular they have made another instrument in that they have traced on a [block][18] of stone or wood a quadrant through which they extract what emerges from the aforementioned instrument and nothing else. Thus, we have not mentioned it since our intention is concision insofar as it does not lead to a shortcoming in this science.

The second instrument is that which is known as the armillary sphere (*dhāt* A.III.2
al-ḥalaq). We take two equal rings with squared-off surfaces, and we attach one to the other at right angles so that one of them is the ecliptic. We place two pegs on its poles, and we attach to them a ring touching them from outside, with it revolving fluidly in longitude, and another [ring] inside revolving also in longitude. We divide this inner ring and the orb of the ecliptic into their degrees. We attach to the inner ring a delicate ring revolving in its plane toward [either of] the poles for observing the latitude, and in it are two protruding, diametrically opposite apertures. We attach to the poles of the circle of the equinoctial a large ring outside of the ring of measurement so that, if we erect it parallel to the true local meridian at the altitude of the pole of the locale in which we are, the revolution of the inner ring about these poles follows the first motion of the universe.

So when we wish to observe with this instrument, we put the outside ring that A.III.3
is [revolving] upon the pole of the ecliptic at the degree of the Sun, and we rotate it so that the intersection of the two rings becomes across from the Sun, I mean

16. Cf. Toomer, Ptolemy's "*Almagest*," p. 62: grazing.
17. Cf. Toomer, Ptolemy's "*Almagest*," p. 62: tips of the pointers.
18. Cf. Toomer, Ptolemy's "*Almagest*," p. 62: plaque.

when they are shaded by themselves. And if we put, in place of the Sun, a fixed star, the degree of which we know,

59v

we would also make the intersection [of the two fixed rings] across from it by placing the sight in one of the two sides of the ring until the star is seen as if it adheres to its surfaces.[19] Then we rotate the inner ring toward the Moon, or any star we wish to measure, until we see it through the two diametrically opposite apertures. Therefore, we know the position of the star in longitude by the degrees of the ecliptic and in latitude by the degrees of the inner circle.

The third instrument, which is named the parallactic rods (*dhāt al-masāṭir*). A.III.4
We take two square rods whose length is not less than four cubits and with a thickness so that they do not twist, and we pierce them at the two ends in the middle of two lines that we trace in the middle of their [the rods'] widths. We connect them with a pin through the two holes so that it is possible for one of them to rotate about the pin with the second fixed at a base, and we erect it, fixed like that, perpendicular to the plane of the horizon parallel to the line of the local meridian so that we picture one inserted inside the other, facing south. And we attach to the rod that is not fixed [in the base] two markers, with two apertures in their middles, on the middle of the line running through the middle of the alidade with the aperture for the eye of the observer being smaller than the other. We make two points, in the middles of the two lines that run through the middle, whose distance from the center of the pin is the same and the most possible. We divide the line separated off in the rod with the base into 60°. We attach at the point on the rod that is fixed [in the base] a thin rod, easy to rotate, whose length is such that it reaches the end of the other when it comes apart from the one fixed [in the base]. So when we wish to observe the visible distance of the Moon from the zenith point we rotate the rod with the [two markers][20] in the plane of the local meridian, when the Moon was in the circle of the local meridian, until the Moon is seen in its entirety. We learn, through the thin rod, the distance that obtained between the two ends of the lines of the two equal rods and we place it on the line divided into 60. We take the arc that is upon it, and it is the distance of the visible center of the Moon from the zenith point in the orb that is traced about the two poles of the horizon, and all of this can be perceived with the armillary sphere.

19. Toomer, Ptolemy's "*Almagest*," p. 219, note 5, commented on how this phrase was corrupted in the original Greek and in the Arabic tradition, communicating mistakenly the opposite faces of the ring. *The Light of the World* has a corrected reading, referring to the two faces as the nearer and farther distance of the same side of the ring.

20. The Arabic MS is illegible; I am following the Hebrew MS.

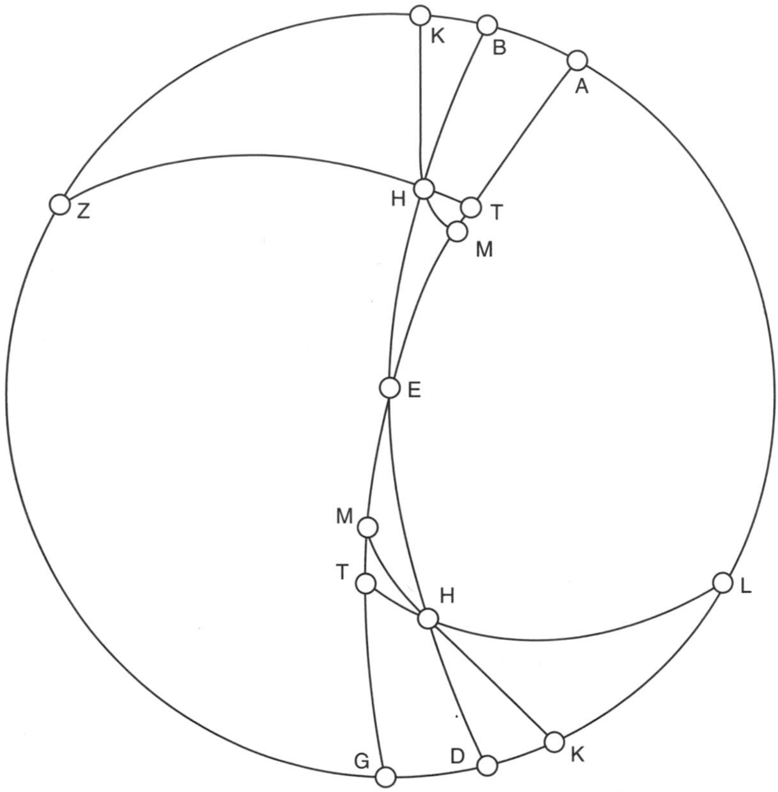

FIGURE 2.10

The fourth instrument, which is named the [dioptra].[21] We make a rod with a length of four cubits, and we draw in its middle a line and we put at its end a small peg. Then we attach to this rod a plate so that it can move upon the length of the A.III.5

60r

rod with it being always perpendicular to the surface of the rod. We make in the middle of the plate a round aperture, with the distance of its lower edge from the surface of the rod being equal to the distance of the top of the peg from it. Then we divide the line that is in the middle of the rod into the parts that we wish so that, through them, the diameter of the circle of the aperture is known. With that instrument we may know the measure of the diameter of the Moon or any planet that we wish, if its distance was known, in that we make the point of the sight at the top of the peg and we move the block along the length of the rod until the star,

21. Literally: "the one with the moving plate" (*dhāt al-shuṭba al-sayyāra*).

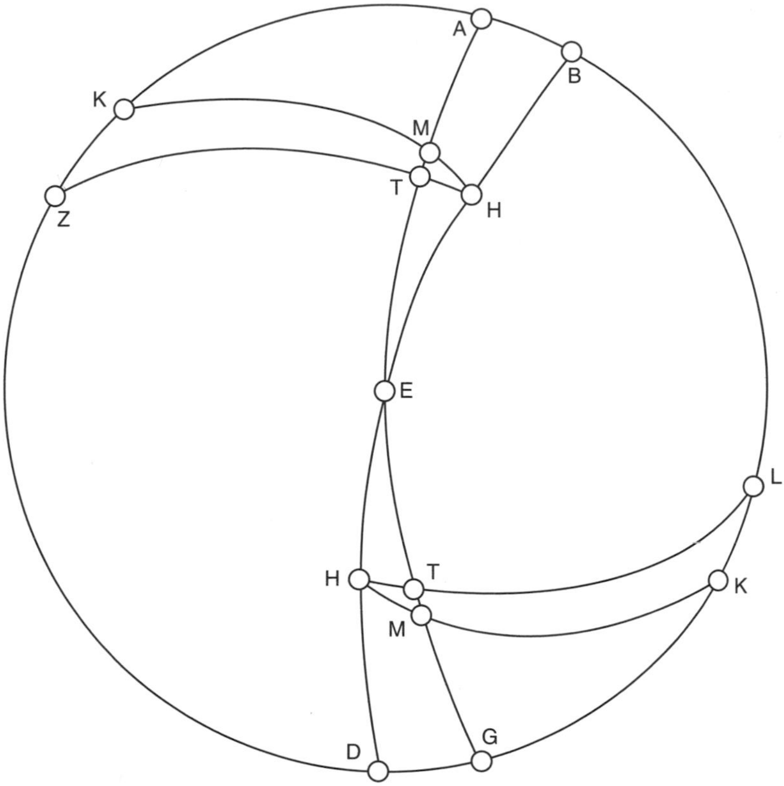

FIGURE 2.11

all of it, is seen. This is because the ratio of the line that is between the sight and the aperture to the radius of the aperture is like the ratio of the distance of the planet to its radius. We also know the planet whose visible body is greater than or equal to another planet.

The fourth chapter, about the knowledge of partical declinations and the rising times at *sphaera recta*, and at *sphaera obliqua*, and other particular things that follow from them. We know the arc that is between the solstices, through the instrument which was made for that, to be 47°43' of the colure according to the opinion of Ptolemy. And likewise we know through this instrument the latitude of the climes, that is, the distance of the zenith from the equinoctial, rather, the altitude of the pole. A.IV.1

(See figures 2.10 and 2.11.) Let the colure be *ABGD*, and the equinoctial is *AEG*, and the half circle from the ecliptic that is from Capricorn to Cancer is *BED* in the first figure and the other half [circle] is *BED* in the second [figure]. The south pole of the equinoctial is *Z*, and we take arc *EH* from the ecliptic from the side of A.IV.2

the equinoctial to be assumed. We extend arc *ZHT* in the two figures according to how this demonstration is made for both figures. So because the ratio of the sine of arc *ZA* the quadrant to the sine of arc *AB*, namely the maximum inclination, is compounded of the ratio of the sine of arc *ZT* to the sine of the unknown arc *TH* and from the ratio of the sine of the assumed arc *HE* to the sine of arc *EB*. The unknown is the fourth [quantity], and it is given in what has preceded in the second chapter. And if the sine was given, then the arc is given.

We wish to know the measure of what rises from the equinoctial with assumed parts from the ecliptic at *sphaera recta* and at *sphaera obliqua*, that is, the measure of arc *ET* rising in the right horizon, and let it be *ZHT* with [known] arc *EH*. That is according to this way, because the ratio of the sine of arc *ZB* to the sine of arc *BA*, A.IV.3

60v

both of them being known, is compounded of the ratio of the sine of arc *ZH* to the sine of arc *HT*, both of them being known, and from the ratio of the sine of the unknown arc *TE* to the sine of arc *EA*. Thus the unknown is the fifth [quantity], and we extract it through what has preceded. Also, we extend arc *KHM*, I mean the horizon at *sphaera obliqua*, and we intend to derive arc *EM* that is rising in this horizon with arc *EH* from the ecliptic. That is by deriving the measure of arc *TM*, I mean the difference of the rising times between *sphaera recta* and *sphaera obliqua*. It is according to this way: because the ratio of the sine of arc *ZK*, I mean the depression of the pole, to the sine of arc *KA*, its complement, both known, is compounded of the ratio of the sine of arc *ZH* to the sine of arc *HT*, both known, and from the ratio of the sine of the unknown arc *TM* to the sine of arc *MA*, the quadrant. So arc *TM* is given, and we subtract it from the rising times of the assumed arc at *sphaera recta* in the half that is in the first figure [2.10], or we add it to it in the half that is in the second [2.11], obtaining the rising times of that arc at *sphaera obliqua*. Likewise, we produce this ratio on the other side of the Earth in that we extend arc *LHT* from the north pole, with arc *KHM* being a segment from the inclined horizon and, like that, we extract the rising times of each degree of the circle in [each] latitude, and that is what we wanted.

In this way it is possible to draw up tables for the rising times at any assumed latitude. It is clear from this that if the degree of the Sun on some day were known, then the measure of that day would be known in that we take the rising times of the six zodiacal signs that are rising on that day or night in that clime and we take from it $\frac{1}{15}$, then the number of the equinoctial hours obtains. And if we took $\frac{1}{12}$, the [time-]degrees of the seasonal hours would obtain. If we knew how many seasonal hours passed from the day or night, then we would multiply them by their degrees and add them to all of the rising times that are from the beginning of Aries to the degree of the Sun. And [if] we seek the degree of the A.IV.4

ecliptic all of whose rising times in the clime, in the direction of the zodiacal signs from the beginning of Aries, are this

61r

sum, then it is the degree that is rising.

If we multiplied the seasonal hours from the past day's noon to now by their degrees, each with its counterpart, and [add it to][22] the cumulative rising times of the degree of the Sun in the direction of the signs at *sphaera recta*, then it [the Sun] reaches the degree of the ecliptic that is culminating at this time. If we take the sum [of the rising times] for the rising degree from the beginning of Aries in the clime and we subtract from it 90, and if it were less, then after we add to it 360°, the place it reaches in the ecliptic at *sphaera recta* is the degree that is culminating. Conversely, we take the sum [of the rising times] for [the degree of][23] mid-heaven at *sphaera recta* and we add to them 90. Wherever it reaches the ecliptic in the clime, that is the rising degree at that time. Whoever wants to extract something like that needs to have the tables of the rising times that Ptolemy drew up, they being derived in these ways mentioned in this chapter, as all of that is easy to grasp. A.IV.5

The fifth chapter, about the measure of the solar year. It is necessary that the measurement of the return of the Sun be according to the ecliptic, if it began from a stationary point so that it returns to it, and the most suitable of the points for that are the points of the equinoxes and the solstices because at them the separation and the change between the times of the year are more than at any other. And it is not necessary that the measurement of the return be with reference to a fixed star, due to its slow movement and also [because] it might be possible that one errs and measures with a planet. Hipparchus and many of the ancients agreed about the measure of the time period of the year in that they observed the time of the spring and autumnal equinoxes with a brass ring erected in the plane of the equinoctial. That [the time of the year] is known from it in that it is illuminated on both of its sides when the Sun abides at the point of the equinox. They said that this time period is 365¼ less $\frac{1}{300}$ of a day in that they observed a vernal equinox, and they observed a second time after 300 Egyptian years, with each of those years being 365 days. They found it 74 days after the completion of the 300 years in the place of the 75th day that makes up for the additional quarter in the 300 years. They divided this time period by the degrees of the circle, I mean 360°, so the mean daily motion emerges, I mean the mean lag according to one of the two [hypotheses]. And they made a table for the mean motion for A.V.1

22. Following Toomer, Ptolemy's "*Almagest*," p. 104. Both MSS read "we subtract it from."
23. Following the Hebrew. The Judeo-Arabic MS omits the word.

61v

the days and the hours and the months and the years in that they placed in the first column the number of the time period, from whatever type it was, and across from it in the rest of the columns the numerical value (*'adad*) of the motion that is proper to it.

Ibn Sīnā said that this is not its meaning. If the time of the equinox was at night then we can know at which time of the night it was in that we look at the measure of the difference between the altitudes of the two local noons, the one that came before and the one that came after the equinox. The ratio of this difference to the difference between the altitude of the later local noon and the altitude of the noon of the equinox is equal to the ratio of the time period that is between the two local noons to the time period that is between the time of the equinox and the later local noon. A.V.2

The first part is complete and praise to the one God. There is no Lord and no Creator other than him. A.V.3

The second part about the manner of the motions and the hypotheses that need to be our thesis for the sensible anomalies for the heavenly bodies, with this part being divided into treatises. The first treatise is about the motions of the Sun and its anomaly and in it are three chapters. The first chapter is about the hypothesis that is our thesis for the mean motion. The second chapter is about the hypothesis that is our thesis for the anomaly. The third chapter is about demonstrating the amount of the greatest anomaly that is seen with the Sun and the way to calculate the individual anomalies and showing the point on the ecliptic in which there is the least lag or speed. B.O.1

62r

The second treatise is about explaining the hypotheses that should be used for the motions of the Moon, and for the anomalies that are seen in it. In it are five chapters. The first chapter is about the hypotheses given for the Moon's motions in revolution. The second chapter is about the explanation of the first true anomaly that appears with the Moon. The third chapter is about the greatest amount of this anomaly. The fourth chapter is about how the variations in the measures of the visible diameters of the Moon do not entail the variation of the distances of the Moon from the Earth. The fifth chapter is about the two remaining anomalies for the Moon, I mean the anomaly related, for Ptolemy, to the Sun, namely that for which he used an eccentric, and the anomaly related, for him, to the inclination of the diameter of the epicycle. B.O.2

The third treatise is about the movement of the fixed stars, and in it are two chapters. The first chapter is about the opinion of Biṭrūjī regarding this movement and its refutation. The second chapter is about the way in which the stars move. B.O.3

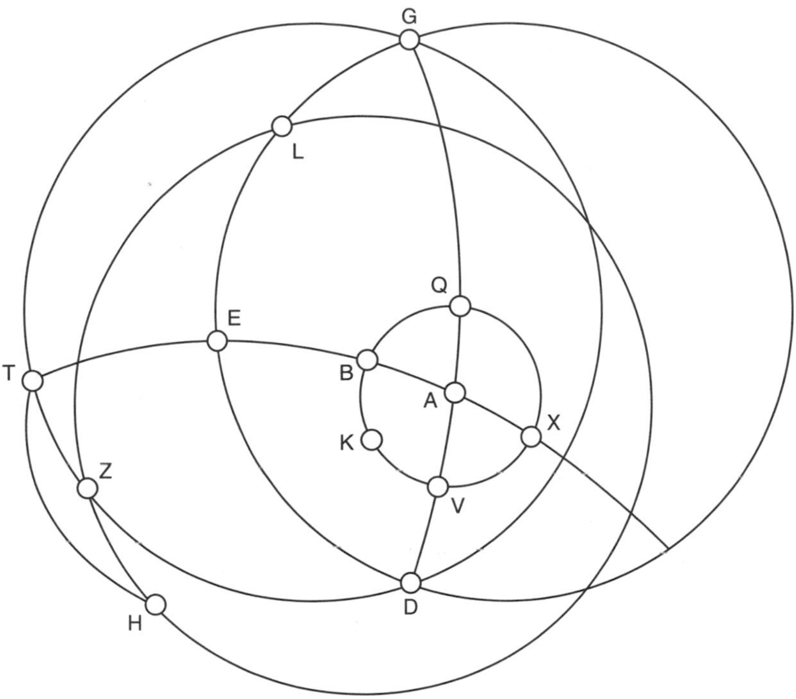

FIGURE 2.12

The first chapter from the first treatise, about the hypothesis that is our thesis for the mean motion. We say that the Sun was found lagging behind the first, uppermost orb, I mean behind the daily motion, by such an amount so that it completes a revolution in its lag one time each year. Although its motion is on circles that the senses tell us are parallel to the circle of the equinoctial about whose poles is the first motion, the Sun is found moving through its lag to the north and south [of the equinoctial]. The amount of the Sun's separation from the equinoctial is limited to the inclination of the ecliptic from the equinoctial. B.1.I.1

(See figure 2.12.) That is imagined according to what I describe. We assume that the circle of the equinoctial is *GED* with its pole at *A*, and the circle of the inclination of the Sun is *GTD* with its pole at *B*. The circle passing through the two points of the solstices and through the two poles of the equinoctial is *TEBA*, and the circle passing through the two equinoxes and through the two poles of the equinoctial is *GAD*. We suppose that in one day the movement of pole *B* on the circle of its path, which is circle *BQXV*, following B.1.I.2

62v

the motion of the universe from the east to the west, is through arc *BXK* so it lags through arc *KB*. If we grant that the Sun, for example, is on point *T*, I mean the point of the solstice, then the Sun had moved, for example, with its maximum inclination toward the west, through the motion of the poles, and had fallen short of a complete revolution by an arc from the circle parallel to the equinoctial, passing through the solstice, similar to arc *KB*; let it be arc *TH*.

The position of the circle of the Sun at that time, when its pole is at *K*, is like B.1.1.3 the position of circle *LZH*. The maximum inclination will be on point *H* from circle *TH*, so it [the Sun] will have fallen short of completing a revolution by arc *HT* from the [circle] parallel [to the circle of the celestial equinoctial]. We grant also that in this period of time in which the pole moves through arc *BXK*, along with the extreme inclination in which the Sun was, through the complement of arc *HT* from a complete circle, the circle of the Sun moves about its poles. The Sun moves with it also in this direction through arc *HZ* striving for perfection and in imitation of the uppermost [orb], longing for it in order to make up for some of what its [the Sun's circle's] poles fell short of, from the uppermost, due to the rupture of the power descending to it. Through this motion, the Sun falls at one time north of the equinoctial and at one time south of it. Therefore this motion is called the motion in latitude. The lag of the Sun, after these two motions together, fell short of arriving at the place from which began the measure that Ptolemy perceived, namely approximately 59'. Its motion in latitude on that day is the amount of the difference between the latitude of point *Z* and the latitude of point *H*.

And because the motion of longitude with the Sun, I mean the final lag of the B.1.1.4 longitude, is equal to the motion of latitude, it is therefore necessary that arc *HZ*, I mean the motion of the orb about its poles, be equal to the final lag of the Sun, I mean 59'. Thus the lag of the pole, I mean arc *KB*, is equal to the motion in latitude and the lag in longitude, both of which are equal, together. The sum total of all that is seen with the Sun is that it lags each day behind a complete revolution by the aforementioned amount, that it is observed moving always upon

63r

circles parallel to the equinoctial, and that it also moves to the north and to the south by the perceived amount. With this alone it is complete, without any need for the thesis of two opposite motions, because [the Sun] would reach the solstice, I mean point *T*, in one revolution in one day with the motion in latitude from point *T* to point *Z* being either in reverse, or in one direction. So through both hypotheses together there is completed all of that which is seen of the [Sun's] motions being in the same way, with the Sun forever being in the ecliptic. There is no difference between the two [opinions] except with the matter of rising times, for it [the motion of the Sun] is, according to these hypotheses, the contrary of what it is with those because it is necessary, according to this opinion, that the Sun, when it is in the ecliptic at the solstice, move toward decreasing rising times. And if the Sun is at the equinox, then the opposite, until this anomaly is completed for the quadrant, I mean an increase to equal the decrease.

They do not have evidence for why their doctrine about that is worthier than B.1.I.5 this doctrine in this respect. So this is the hypothesis that ought to be our thesis for the mean motion by those whose thesis is that there is opposition in circular motion. You know from what we said previously that the first hypothesis is worthier due to what we demonstrated regarding the lack of opposition in circular motion.

The second chapter about the hypothesis that is our thesis for the anomaly. I B.1.II.1 say that the anomaly that is seen with the Sun is that it is seen moving in divisions of the ecliptic at times faster and at times slower and at times in the middle between these two. This observed variation forever preserves its order according to what they found; I mean the slow motion that is found for it is always at certain degrees, and likewise with the fastest motion and the mean. So it is desirable that we make our thesis for this anomaly true hypotheses that agree with our school of thought.

We say that the center of the Sun, were it to move with the motion par- B.1.II.2 ticular to it upon the ecliptic, or upon its circle with an inclination equal to the inclination of the ecliptic, it would not be necessary that there be observed any anomaly at all in longitude. This is because it would be necessary that the Sun move in equal periods of time through equal arcs from the circle of its inclination because it is carried upon it in the motion. I mean it would delimit equal angles about the poles of the ecliptic in equal time periods. But the Sun does not traverse equal arcs from the circle of its inclination in equal time periods, so the opposite of the premise is produced, and it is that the center of the Sun is not carried in

63v

its particular motion upon the inclined circle, although it is not found outside of it.

Thus, it results from that, that the center of the Sun moves on a small circle, B.1.II.3 whose pole is forever carried on a great circle inclined to the circle of the Sun's declination by the amount of the radius of that small circle. Let that small circle be called the circle of the path of the center, and it takes the place of the orb of the epicycle that Ptolemy used as his thesis. Let it be according to this description:

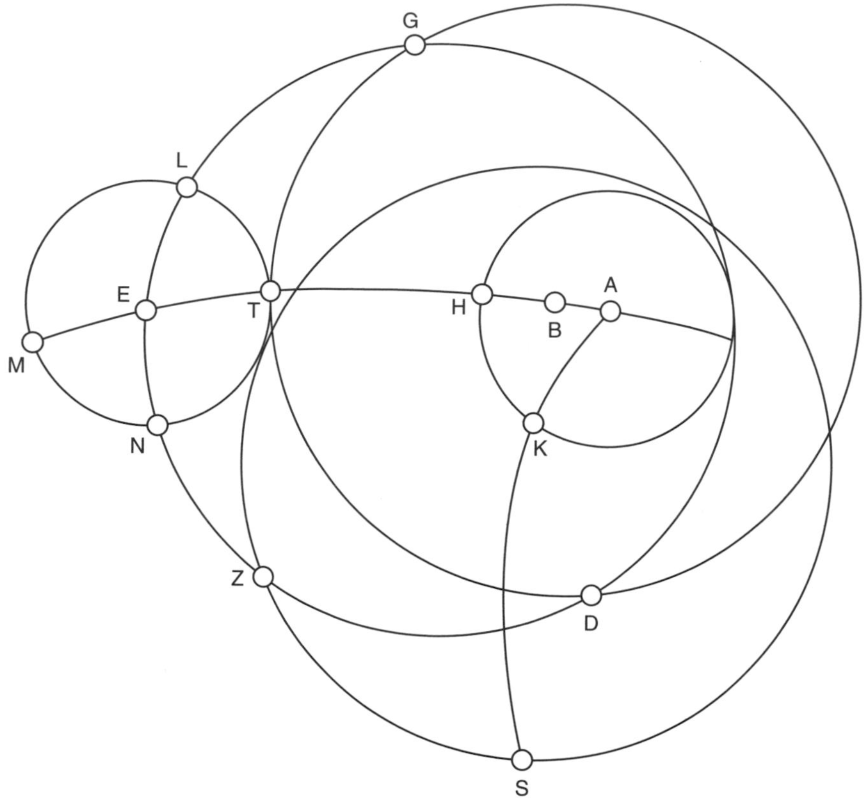

FIGURE 2.13

(See figure 2.13.) Let the ecliptic be *GTD*, its pole *B*,[24] and the pole of the equinoctial is *A*. The inclined circle, the deferent for the pole of the circle of the path, is *GED* with its pole at *H*. The circle of the path of this pole is *HK*, whose pole is the pole of the equinoctial. Let us grant that the circle of the path of the center is *LMNT*, whose pole is *E*, and it is tangent to the ecliptic so that the arc emerging from its pole to its circumference is equal to the maximum inclination of circle *GED* to circle *GTD*. Let us propose, for example, that point *T* is the point of the solstice, and let us propose also that when the pole of the circle of the path of the center is on point *E*, according to what is in the picture, then the center of the Sun is on point *T*, touching the ecliptic. B.1.II.4

When pole *H* moves in one day, for example, following the motion of the uppermost orb, and lags behind a complete revolution by the aforementioned B.1.II.5

24. The position of pole B is not indicated precisely in the MSS figure.

measure, and let it be equal to arc *HK*, then pole *H* is on point *K*. Also, point *E*, I mean the pole of the circle of the path of the center, will be on point *S*, lagging also by an arc resembling arc *HK*. By this same amount the Sun lags also, since it is carried on the circle of its path. In this period of time the pole of the circle of the path of the center moves about point *K* and carries the Sun with it from point *S* to point *Z*, for example, with the measure of arc *SZ* being half of the first lag. So there remains the lag of the pole, rather the Sun's final lag, which is half of the first lag, I mean approximately 59'.

This would be the case were the B.1.II.6

64r

center of the Sun forever on point *T* from the circle of its path and were the center of the Sun to have no motion about point *E*. But if we granted that, it would be necessary that the Sun never retain the latitude of the ecliptic. Namely, if the pole of the circle of the path of the center were on point *Z*, the Sun would have a latitude from the ecliptic because the latitude of point *Z* from the corresponding point on the ecliptic is smaller than arc *ET*. And as it [that pole] moves toward the node, it [the latitude of point *Z* from the point corresponding to it on the ecliptic] is less, so the latitude of the Sun from the ecliptic becomes greater.

And because the matter of the Sun is the opposite of that, namely that the B.1.II.7
Sun is never found outside the latitude of the ecliptic, it is accordingly necessary that the center of the Sun have a motion about the poles of the circle of its path. Therefore, our thesis is that this motion be from point *T* toward point *N*, for this motion can be proposed in any direction we wish because it is from among the motions that can be imagined at one time in the first direction and at one time in the opposite. Our thesis is that the measure of this motion be equal to the final lag, I mean 59' per day. The Sun in that manner retains the latitude of the ecliptic, because when the Sun moves from point *T* toward point *N* through an arc resembling *SZ*, the ratio of the decrease of the latitude of point *Z* from the latitude of point *S* to the latitude of point *S* is like the ratio of the decrease of the latitude of the point which the center of the Sun reaches on this day, beginning from point *T* toward *N*, from the latitude of point *E* to the latitude of point *E*. When *E* is on point *Z*, the Sun also retains the latitude of the ecliptic due to it moving on the circle of its path.

Thus, there occurs for the Sun, on account of this movement, an increase in its B.1.II.8
final lag, for if it [the Sun] were always on point *T*, then its lag would be equal to the lag of the pole. As long as the Sun moves on arc *LTN*, the final lag increases. On account of that, its motion, I mean its lag, in the direction of the signs will be observed to be faster, and as long as it [the Sun] is conveyed on arc *LMN* its final lag starts to decrease; therefore, its lag will observed to be slower. The amount of increase and decrease is the measure of great circle arc *LN*. It falls to you to

propose this anomaly so that the motion of the pole of the circle of the path of the center will be in reverse, for example, from point *E* to point *Z*, with circle *SZ* removed from the picture. Let point *T* from the circle of the path be named the point of greatest speed or lag, and point *M* be named the point of least lag or speed. They are the points that bisect halves *NTL NML*.

And it might have been possible that this anomaly B.1.II.9

64v

be imagined in the one half in a way other than this, I mean even if we had not made our thesis for the Sun a motion about the poles of the circle of its path with the hypothesis in which the motions are in a single direction. That is because it appears at first that the two points of the intersection of the circle of the path of the center with the inclined circle, I mean points *LN*, are forever on circle *SZT* because the movement of the poles of the circle of the path is upon this circle. And if pole *E* were upon point *Z*, with points *LN* being upon points *HK*[25] from circle *SZ* in the figure that is after this, the lag of the Sun would be more than the lag of the pole, even if the Sun never separated from point *T*, because point *T* is at that time slanted toward point *D* from the pole that is upon point *Z*. This slant always increases until the pole is at the point of intersection. The greatest slant comes about at that time, and therefore an increase in its final lag appears. And when it moves through the quadrant following it, this slant starts to revert and decrease until it completes the quadrant, so point *T* returns to its position, and the position of point *T* and point *E* from the ecliptic become one.

Therefore, in this quadrant, and likewise in the quadrant that follows it, there B.1.II.1
is seen a decrease in the lag. In the other quadrant an increase in the lag equal to what there was in quadrant *EZD* also appears, so the anomaly would be imagined to be what it was in the other thesis. Except, if the anomaly were proceeding along these lines, the Sun would not forever retain the latitude of the ecliptic; rather, it would retain this latitude only in half [circle] *GTD*. And in the second half [circle], it might depart from this latitude by the measure of arc *MET*, and that occurs due to our thesis that the Sun not be moving about the poles of the circle of its path. Thus, our thesis must be that the anomaly is from the motion of the Sun about the two poles of the circle of its path, because only through that will the Sun be found on each degree of the circle of its path and come to retain the aforementioned latitude. I mentioned to you all of this by way of mathematics and to shore up the mind of one who looks into these things so as to firm up his thinking about them, and because points *LN* adhere to circle *SZ* according to this hypothesis, then according to that, the anomaly is two anomalies.

One of them is that which is on account of the movement of the orb, in which B.1.II.1

25. I don't know the figure to which Ibn Naḥmias has referred.

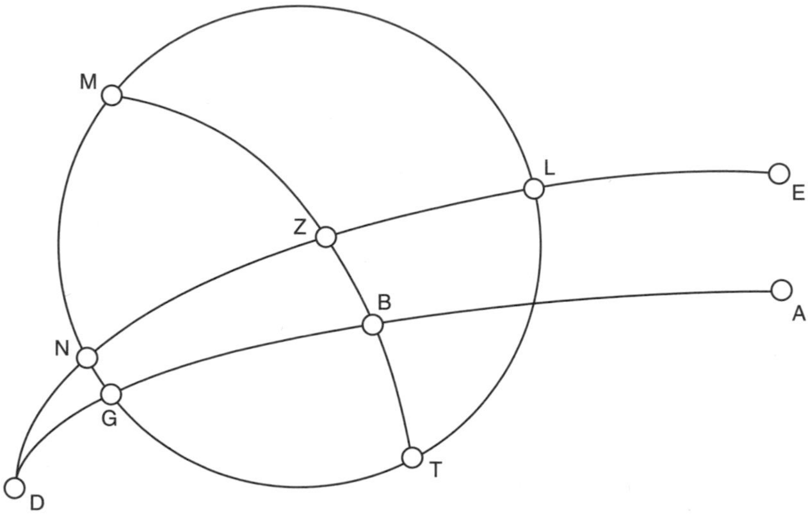

FIGURE 2.14

there is the circle of the path, about its poles. The second is the aforementioned [the one due to the slant of *SZ* to *ED*], and we

65r

explain later how it occurs with the help of God, may He be exalted. If our thesis is that the hypothesis for this anomaly is the first hypothesis, I mean according to the hypotheses of Ptolemy for opposite motions, in that the motion of the pole of the circle of the path will be in reverse, from point *E* toward point *Z*, then on account of that the two points of intersection *LN* will stay on circle *GED*, since we have no need at this time to propose circle *SZ*.

We say also, as is necessary according to this hypothesis, that if our thesis is that points *LN* never depart from circle *DZE*, not even by a little bit, then it follows necessarily that the Sun never preserves the latitude of the circle of its inclination even though it moves on the circle of its path in the aforementioned way. The deviation that occurs due to that is great because when the departure from the latitude is conjoined with the point of the equinox, it makes an impression on the observation of the time of the equinox, showing that the discrepancy is a mistake of no small measure. B.1.II.12

(See figure 2.14.) And as for how that appears, it is in this way. We repeat the preceding figure and we grant that the pole of the circle of the path of the center is on point *Z*. It is our thesis that the center of the Sun moves on the circle of its path from point *T* toward point N through an arc similar to *EZ*. We say that the Sun does not reach point *G*, for arc *TG* is not equal to *EZ*. The demonstration of that is that the ratio of the decrease of the latitude of point *Z* on the circle inclined from B.1.II.13

the latitude of point *E* to the latitude of point *E* is like the ratio of the decrease of the latitude of the point which the Sun reaches on the circle of its path, beginning at point *T* on arc *ZT*, to arc *ZT*, with arc *ZT* being equal to the latitude of point *E*. So the decrease of the latitude of point *Z* from the latitude of point *E* is great circle arc *TB*. It is therefore necessary that the decrease of the latitude of the point on the circle of its path that the Sun reaches from arc *ZT* is arc *TB* also.

But the decrease of the latitude of point *G* on the circle of the Sun's declina- B.1.II.1
tion from the latitude of point *E* is more than arc *TB* because the decrease of the latitude of point *G* is more than the decrease of the latitude of point *Z*. And if the Sun, on the circle of its path in that period of time, had reached point *G*, I mean

65v

the point of its intersection with the circle of the Sun, then it would have decreased in latitude by more than arc *TB*. This is a contradiction. Therefore, the arc similar to arc *EZ* that the Sun reaches is smaller than arc *TG*, so the Sun therefore does not reach the circle of its inclination, I mean circle *DBA*. In general, the Sun might reach the latitude of its mean position in the ecliptic, I mean point *B*, not the latitude of its true position.

Since that has become clear, it is incumbent on us to show how the center B.1.II.1
of the Sun retains forever the circumferences of these two circles, I mean the circumference of the ecliptic and the circumference of the circle of the path according to each of the two hypotheses, I mean according to the thesis of opposite motions following the opinion of Ptolemy and according to the thesis of their being in one direction. Through that, what the ancients found and what Ptolemy worked with becomes clear, that the period of time with this anomaly that is from the greatest motion to its mean is less than the period of time that is from its mean to its least. And we explain this concept first according to the opinion of Ptolemy, I mean by our thesis of opposite motions, then we follow by explaining it according to the second opinion.

(See figure 2.15.) Let us grant that the figure is the way it is, with it having been B.1.II.1
shown that if the center of the Sun was carried on the circle of its path, I mean circle *LTNM*, and if pole *Z* moved through arc *ZE* and the Sun moved from point *T* toward point *N* through an arc similar to it, then the center of the Sun does not reach, through this movement in this quadrant, the circumference of the ecliptic, I mean the intersection of the circle of the path with the ecliptic, namely point *X* in this figure. And were it to move like this, then at the completion of the quadrant any departure from the latitude might add up to be a lot. Because the two points of the intersection, according to this hypothesis, forever retain their positions on circle *GED* and do not separate from it, then it would appear that the center of the Sun is not carried on the circle of its path, I mean circle *LTNM* itself, although it is not found departing from it. As for how that can be, it is in this way.

We imagine that when point *Z* is on point *E*, with point *T* from circle *LTNM* B.1.II.1

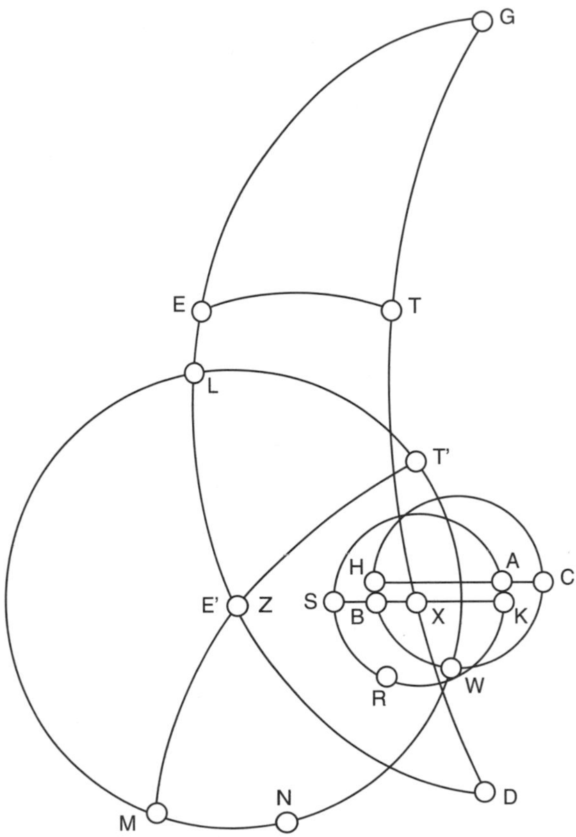

FIGURE 2.15

on point *T* from the ecliptic, pole *A* from circle *HC* is superimposed on point *T*, and great circle arc *AH* is also on great circle arc *TE*. We imagine also about pole *H* a circle *SRK* equal to it, with great circle arc *KS* being superimposed on arc *TE*. The measure of the great circle arc extending from

66r

the pole of each of these two [circles] to their circumference is equal to half of the arc of the greatest anomaly necessitated by the arc from the circle of the path of the center, which is similar to the arc of the greatest solar anomaly, whose measure is 2°, I mean half of the arc of the circle of the path similar to the arc of the greatest anomaly. And we grant that the center of the Sun is carried in its motion on circle *SRK* and that the pole of circle *SRK* is also carried in its motion on circle *HWC*. When pole *A* is on point *T* from circle *LTNM* and on point *T* from the ecliptic, at that time the pole of circle *SRK*, I mean *B*, is on point *H*. And at that time the center of the Sun would be on point *K* from circle *SRK* which was at that time superimposed on pole *A*, which is on point *T* from the ecliptic. When point *Z* moves through arc *ZE*, then pole *A* moves on circle *LTNM* about pole *Z*, from point *T* toward point *A*, through an arc equal to arc *ZE*. Also, the pole of circle *SRK*, I mean *B*, moves about pole *A* from point *H* toward point *B* through an arc also similar to *ZE*.

At that time point *K*, in which there is the center of the Sun, separates from point *A* on the way toward point *N*, and through that, the Sun began to have a latitude from circle *LTNM*, for arc *KBS* is forever parallel to arc *HAC* through this movement. The center of the Sun also moved about pole *B* from point *K* to point *X* from circle *SRK* through an arc also similar to arc *EZ*. The center of the Sun by that means reaches the circumference of circle *LTNM* and also the circumference of the ecliptic, so that when pole *Z* moves through the degrees of the quadrant less the degrees of the greatest anomaly, then at that time arc *TA* will be less than a quadrant by an arc similar to the arc of the greatest anomaly. Pole *B* will be near point *W* on circle *HWC*, with the center of the Sun being near point *K*, I mean close to a quadrant away from point *K*. B.1.II.1

If we made each of these two motions not equal to the movement of pole *Z* but more than it by the amount B.1.II.1

66v

of the greatest deviation in the entire quadrant, then the center of the Sun at that time would be on point [*R*],[26] I mean [a quadrant from][27] point *K*. In that manner, the center of the Sun has become an entire quadrant away, beginning at point *T* on circle *LTNM*, I mean on point *N*, with point *N* at that time being at the point of the intersection of the deferent circle with the ecliptic, except that this does not follow in the rest of the quadrants. Therefore, our thesis is that the motions are uniform because the deviation which occurs due to that is not sensible, so then the center of the Sun, through these two motions, might forever adhere to the ecliptic and also to the circumference of the circle of its path. Likewise, that itself is necessary in each of the remaining quadrants, I mean that through these two movements it [the Sun] is found retaining forever the circumferences of these two circles, and that is what we intended to show.

And that, precisely, results, even if we granted that arc *SBK* were never parallel through this motion to arc *HAC* in that point *K* from circle *SRK* is forever superimposed upon point *A* as is necessary according to what we said about pole *B* being moved around pole *A*, for the two points of intersection of circle *SRK* with circle *HWC* would always adhere to their positions on circle *HWC*, with B.1.II.2

26. Following the Hebrew recension. Judeo-Arabic: K.

27. Following the Hebrew recension. Judeo-Arabic: In the quadrant of point.

circle *HWC* having moved about its poles, and circle *SRK* and its pole having moved with it by the aforementioned measure. In that period of time, the center of the Sun moved from point *K* toward point *R* by twice the aforementioned amount because if pole *B* moves on circle *HWC* through a quarter circle toward point *W*, then point *S* comes to be on the circumference of circle *LTNM*. So the motion of the center of the Sun about point B in this time period needs to be twice the aforementioned amount. Thus, it should be found forever retaining the circumferences of the two circles as was previously shown, and that is what is best.

Second, let us assume according to the second opinion that the two points of intersection [*LN*] forever retain their positions on circle *SZ*, the circle about whose two poles the pole of the circle of the path of the center forever revolves, with this being the only requirement according to this hypothesis. We say that B.1.II.21

67r

when point *S*, I mean the southern extremity, is on point *E*, then the two points of intersection *LN* are also on circle *ED* with point *T* from the circle of the path, which is the point of greatest lag, is on point *T* from the ecliptic. So when point *S* lags behind point *E* by half the parallel circle which passes through it and point *E*, and [if] the pole of the circle of the path moved through a quarter of the deferent circle [i.e., the inclined circle carrying the pole of the circle of the path of the center of the Sun], then it [the pole] comes to be on point *D*, the point of the intersection of *ED* with the ecliptic.

(See figure 2.16.) Let us grant that *GZ* is equal to *ZK*, so the point of greatest lag, which would have had to be on point *Z* had the two points of intersection maintained their positions on circle *ED*, has come to be on point *T*, which bisects arc *LTN*. The measure of arc *ZT* is equal to arc *GL* with it [arc *GL*] equaling twice the inclination of point *E* from the equinoctial, which is known. Then the center of the Sun, which was at first on point *T*, the point of greatest lag, will need to move in this period of time, about the pole of the circle of its path, I mean *D*, only from point *T* on the circle of the path to point *A* on the ecliptic. The mean motion in this period of time about this pole is a quarter of the circle of the path. Thus, we need to grant that the center of the Sun was moved in this period of time toward point *Z* by the two orbs described above through the arc of the difference of *TA* from a quadrant. B.1.II.22

Let it be, according to this description, that the center of the Sun moves always on circle *TWX* whose pole is *B*, with pole *B* forever moving on circle *HBW* whose pole is *T*, which is carried on the circle of the path. The measure of the arc from the circle of the path, which goes from the pole of each of these two circles to their circumference, is equal to half of the arc of twice the aforementioned inclination less the arc of the greatest deviation, I mean the difference of *TE* over *KA*. But *KA* B.1.II.23

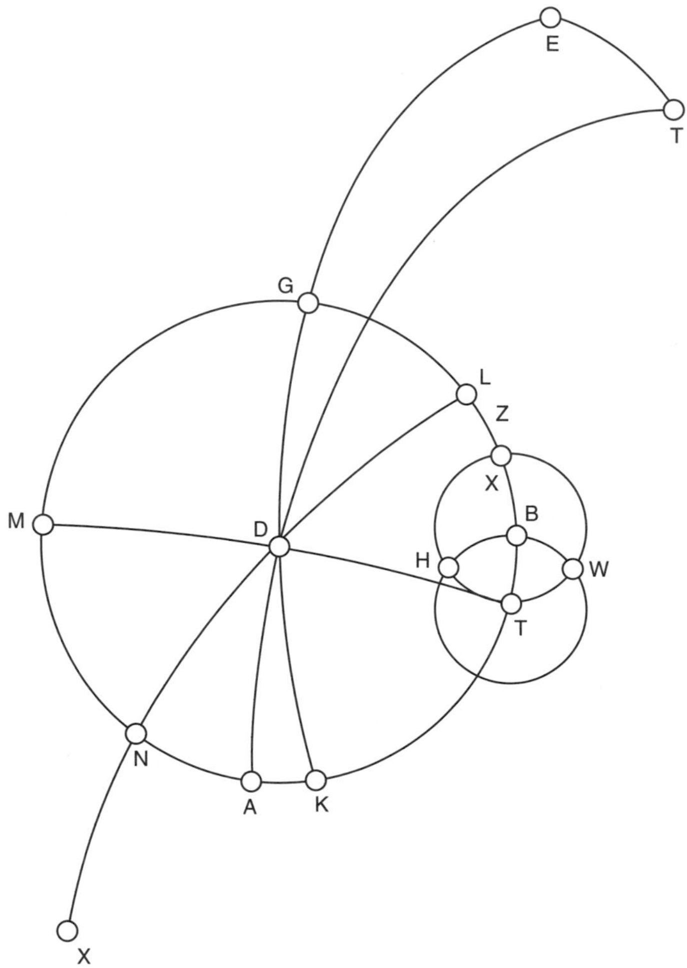

FIGURE 2.16

is similar to the arc of the greatest deviation; accordingly, arc *XZ* is equal to *KA*. And if points *S* and *D*, which is the pole of the circle of the path, are on point *E*, then point *T*, I mean the point of greatest lag, will be on

67v

point *T* from the ecliptic. In it would be the center of the Sun, and point *B* would be on point *H* from circle *HBW*.

And in the period of time in which point *S* lagged through half of the parallel B.1.II. circle and the pole of the circle of the path moved through a quadrant, pole *B* moved about pole *T* through a quadrant also, I mean from point *H* to point *B*. The center of the Sun which was on point *T* moved about pole *B* through a half circle, I mean from point *T* to point *W* and [then] to point *X* with the center of the Sun coming to point *X*. Also in this period of time, it [the center of the Sun] moved about the pole of the circle of the path through a quarter of a circle and came to point *A* because arc *XTA* is a quadrant.

And in the quadrant following it, the center of the Sun returned, through these motions, to point *T*, I mean from circle *TWX*, not to the point of maximum lag. Point *T* returned to point *Z*, I mean that point *T* bisected at that time the half which circle *ED* comprised, because the points of intersection also returned to *ED*. The Sun also moved about the pole of the circle of the path through a second quadrant and came to point *M*, with point *M* being at that time the point on the ecliptic opposite point *T*. The Sun forever preserves the circumference of the ecliptic, and that is what we desired. B.1.II.25

Through that precisely it results that the time period from the greatest lag to its mean is smaller than the time period from its mean to its smallest. Because of that, we needed to make it our thesis that, in each of the two hypotheses, the orb in which there is the circle of the path has these two orbs in which there are these two circles. Were it not for that, we would be satisfied, for the anomaly, with having our thesis be only the two circles on the ecliptic, and without the circle of the path and that which carries it. So you have seen the truth of each of these two hypotheses, and there remains only to choose the approach, though I think the first hypothesis is more appropriate due to the ease of working with it, and that is what we desired. B.1.II.26

So we have produced all of the true hypotheses for the motion of the Sun and the anomalies that appear with it, while following the opinions of Ptolemy about all that. By no means should those hypotheses be sullied by the impossible because these hypotheses are established upon all of the motions being upon fully spherical orbs. There is no motion except that which is equal and uniform in equal time periods, and circular about defined poles, or following from and mediated by a circular motion also about B.1.II.27

68r

its poles, with no need either for an eccentric or an epicycle. It also results from what we have said about the Sun that the tables which Ptolemy used for the motions and the anomalies are themselves the tables needed for calculations with these hypotheses, except that that which is called motion in reverse with him [Ptolemy] is called lag with us in the second hypothesis. And this shall become clear in the future.

As for al-Biṭrūjī, his opinion about this anomaly is an opinion far from the truth, for he opines that the center of the Sun moves always on the inclined circle B.1.II.28

provided that it [i.e., the center of the Sun] is carried on it in motion. The cause of the visible anomaly is the movement of the poles of the circle of its [i.e., the Sun's] inclination on the circle of the path of the poles of the ecliptic. He alleged that if he positioned the poles of the Sun's inclined circle so that they lag behind equal segments of the circle of the path of the poles of the ecliptic in unequal times, then the Sun would likewise be shown, in the ecliptic, to lag equal arcs in unequal times. It would be as he said if we had meant by anomaly that the Sun lags through the degrees of the ecliptic in longitude in varying time periods, even if the Sun left it [the ecliptic] in latitude. It likewise follows necessarily from his thesis that even if the Sun lags in longitude through the quadrant which is from the winter solstice to the vernal equinox, for example, in less than one-fourth of the days of the year, it does not traverse a complete quadrant in this period of time from its inclined circle until it reaches the node of this circle with the celestial equinoctial as it reached the ecliptic, since there is nothing here [in Biṭrūjī's models] necessitating that it [the Sun] traverse equal arcs of the ecliptic in unequal times.

So if the Sun were never on the ecliptic, and if it were not at the time of the equinox on the ecliptic, I mean at the time when the Sun is at the beginning of Aries, as follows necessarily from his [al-Biṭrūjī's] thesis, then how can the observations which Ptolemy put in place, and on the basis of which he worked, be accepted for the times of the equinoxes which he observed with the brass ring erected in the plane of the equinoctial in that the Sun illuminated the ring from both sides at this time? It is not possible with its [the instrument's] placement that the times between when the Sun is at the points of the equinoxes and solstices be found, through this instrument, to be essentially any different, B.1.II.

68v

because it does not illuminate them on both sides, except at the time when the Sun is at the point of the intersection of the Sun's inclined circle with the equinoctial, and it [the Sun] does not reach these points except at equal times.

It is also his thesis, because of the lag of the pole on the circle of its path in different periods of time, that it moves on a circle whose pole is removed [from the pole of the equinoctial]. On account of that, the Sun is not always found on the ecliptic, while he affirms that the Sun does not leave the ecliptic. It is unbelievable how he neglected all that! I think his goal, indeed, was to give hypotheses for some anomaly which he created himself, one which does not follow necessarily from the opinions of the ancients. In spite of that, he constructed his demonstrations on the basis of the observations of the ancients, so because of that he fell, clearly lapsing and unknowingly, into confusion! And we had derived for these hypotheses, which this man made his theses, ways in which what is sought may appear, except that these are not devoid of impossibilities that are stronger and B.1.II.

more repugnant than the hypotheses that have been followed until now. Therefore, we have cast aside all that from this book, and we return to where we were and we show the amount of this anomaly.

The third chapter, about showing the measure of the greatest visible solar anomaly, and an approach to calculating the particular anomalies, and showing the point from the ecliptic in which there is the least lag or speed. To prepare for that, we show first the ratio of the radius of the orb of the Sun to the radius of the circle of the path of the center. Likewise, it is necessary that we show the points from the ecliptic in which there is the greatest and the least speed, I mean the point on the ecliptic in which the center of the Sun is if it were on the point of the least speed. That becomes clear with the observations through which Ptolemy showed the measure of the eccentricity and the position of the apogee on the ecliptic. He showed that by finding the time period that is from the time of the Sun's arrival at the point of the spring equinox to the point of its alighting at the point of the summer solstice to be 94½ days. And from the summer [solstice] to the autumnal [equinox] to be 92½ days, with the Sun traversing on average in 94½ days 93;9° and traversing in 92½ days, [on average] 91;11°. B.1.III.1

(See figure 2.17.) So let the circle of the path of the center be *ABG* and the portion common to it and the plane of its deferent circle B.1.III.2

69r

be line *NLK*. We position the Sun at the time of the vernal equinox at point *A*, and it moves to the summer solstice through arc *AB*, the measure of which is 93;9°, transecting from the ecliptic an arc of 3;9° in lag. And it moves from the summer solstice to the fall equinox through arc *BG* the measure of which is 91;11° transecting from the ecliptic an arc of 1;11° in lag also. We extend from point *B* a perpendicular to the common portion that reaches the circumference of the circle of the path on the other side, and it is perpendicular *BLE*. We extend from points *AG* upon line *BE* two perpendiculars *AM GZ*, and we connect *AE GE AG*, and we extend from point *G* line *GT* upon line *AE*, if it was extended perpendicularly, because angle *AEG* is obtuse. We extend from point *D*, I mean the center of the ecliptic, to points *AG* lines *DG* and *DA* and to the points *MZ* lines *DM* and *DZ*.

So because the position of point *B* on the ecliptic is point *M*, let angle *ADM*, I mean that which is less than the lag according to the second opinion or that which is less than the speed according to the first opinion, be given according to arc *BA*. So the ratio of line *DA*, I mean the radius of the ecliptic imagined in the orb of the Sun, to the perpendicular extending from point *A* upon line *DM* is given. As for when line *BE* is a diameter of the circle of the path, then this perpendicular is itself line *AM*. As for when it [*BE*] was something other than the diameter, then it is not line *AM*. We work on extracting the ratio on the condition B.1.III.3

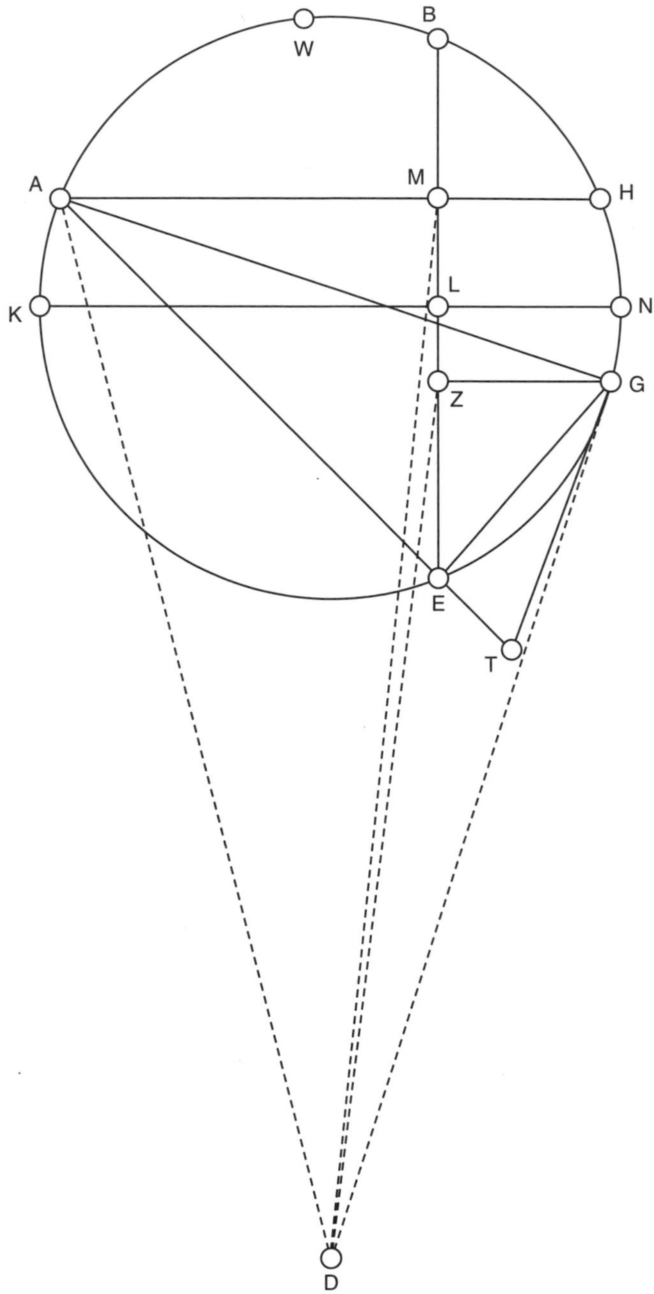

FIGURE 2.17

that it is line *AM*, and we pay no attention right now to the difference between them since the difference is little. Then, let us return to showing this excess and through it we correct the ratio.

Therefore, the ratio of *DA* to *AM* is given, and because arc *BA*, rather angle *BEA*, is given, then the ratio of *AE*, the diameter, to *AM* is given, so the ratio of *DA* to *AE* is given. Also, because the position of point *B* with respect to the ecliptic is point *Z*, then angle *GDZ* is given. So the ratio of *DG*, I mean *DA*, to the perpendicular, I mean *GZ*, according to the aforementioned condition, is given. And because arc *BG*, rather angle *BEG*, is given, then the ratio of *GE* the diameter to *GZ* is given. Therefore, the ratio of *DA* to *GE* is given. And because angle *AEG*, rather *GET*, is given, then the ratio of *GE* the diameter to *GT*, rather to *ET*, is given, so the ratio of *GE* to *TA*, rather to B.1.III.4

69v

GA, that is potentially equivalent to *GT* [to] *TA*, is given. So the ratio of *DA* to *AG* is given, with the ratio of *GA* to the diameter of the circle of the path having been given, because arc *ABG* was given. The ratio of *DA*, I mean the radius of the orb of the Sun, to the radius of the circle of the path, is given. That the radius of the circle of the path is approximately one twenty-fourth the radius of the circle of the Sun emerges through the ratio that Ptolemy derived. That is after the correction that we are going to mention shortly.

As for how we know the degree of the ecliptic in which the Sun is, if it were in the point of the least lag, it is that we know the distance of point of *B* from the point of the least lag which is known because the ratio of *GE* to *GA* is given. Thus arc *GE*, rather arc *BGE*, is given, so half of it, I mean *BN*, is given. Then the distance of point *B* from the point of least lag, I mean point *W*, is given. This is the distance in the direction of the signs of the ecliptic of the point of the summer solstice from the point in which the Sun is if it were at the point of least lag, and let this point be approximately 5° from Gemini, as Ptolemy derived it. Let the point of the greatest lag be at the degree of the ecliptic opposite it. B.1.III.5

If we have shown that, we return and show the measure of the difference that is between the two lines *AM* and *GZ* and between the two perpendiculars emerging from points *AG* upon lines *DM* and *DZ* so we can correct the ratio in that manner. We present for that the lemma that we proved in the first treatise, and it is this. If there were in a circle three points so that each one of the arcs separated by these points were smaller than a half circle, and a line connected the first point and the third, I mean the chord of the arc that is the sum of the two arcs, then a line connected the center of the circle and the middle point of the three, dividing the aforementioned line in two. Then the ratio of the sine of one of the separated arcs to the sine of the other is like the ratio of the arc from the aforementioned chord which follows this arc to its other part. B.1.III.6

If we have presented that, let us imagine that line *AM* is the portion common to the circle of the path and to the circle of the ecliptic. Let us extend it [*AM*] to point *H*, so it becomes line *AMH*, the chord of the arc from the ecliptic cut off by arc *ABH* from the circle of the path. And if line *DM* were extended perpendicularly, it would come to subtend this arc [*ABH*] from the circle of the ecliptic, there having occurred in this arc three points, two of them being points *AH* and the middle [point] being where line *DM* cuts this arc. So the two arcs separated off by these points are given because the arc from point *A* to the B.1.III

70r

middle point is on the measure of angle *ADM*, with the second arc being on the measure of the angle that increases or decreases because of arc *BH*. It is given because arc *BW* is given, and according to the ratio, the arc from the ecliptic cut off by *BW* is given, and likewise the arc cut off by arc *WH*. Thus, the arc cut off by *BH* becomes given. According to the lemma, the ratio of the sine of the arc that is from point *A* to the middle point, namely the perpendicular extending from point *A* upon line *DM*, which is given, to the sine of the second arc, namely the perpendicular extending from point *H* upon this line, which is also given, is equal to the ratio of the line *AM* to the line *MH*. And since all of line *AH* is given, then line *AM* is given. Similarly, the measure of line *GZ* has been shown, and that is what we desired.

If we repeat the derivation of the ratio with these quantities, and the ratio between the radii emerges, then we derive the measure of the differences a second time. We repeat the derivation of the ratio a second time, after increasing or decreasing the arcs by the amount of the variation these arcs necessitated, due to the Sun moving about the poles of the two aforementioned circles, and the correct ratio emerges. It appears that it is possible to propose a hypothesis to tie the mean motions to the times of the observations that he mentioned for the equinoxes or the solstices so that we can derive at any time we wish the mean positions of the Sun in the ecliptic in the circle of the path. B.1.III

(See figure 2.18.) In order to explain a way to calculate the individual anomalies, let the circle of the path of the center be *AEGM* whose center is *Z* and the portion common to it and to the plane of the deferent circle, when it is tangent to the ecliptic, is line *BZM*. Let line *AZG* be perpendicular to it, then of points *AG*, one of them is the point of the greatest lag, and the second is the point of least lag. Let the point of least lag be *G*, and let us grant that the point of mean passage is *B*, and we connect *BZ* and *DB* and *DZ*. Because the ratio of *DB* to *BZ* is given, with angle *Z* being a right angle, so angle *BDZ*, namely the angle of greatest anomaly, is given. And let us propose that the center of the Sun is on point *H*, provided that its distance from point *G*, I mean the point of greatest lag, is given. We extend B.1.III

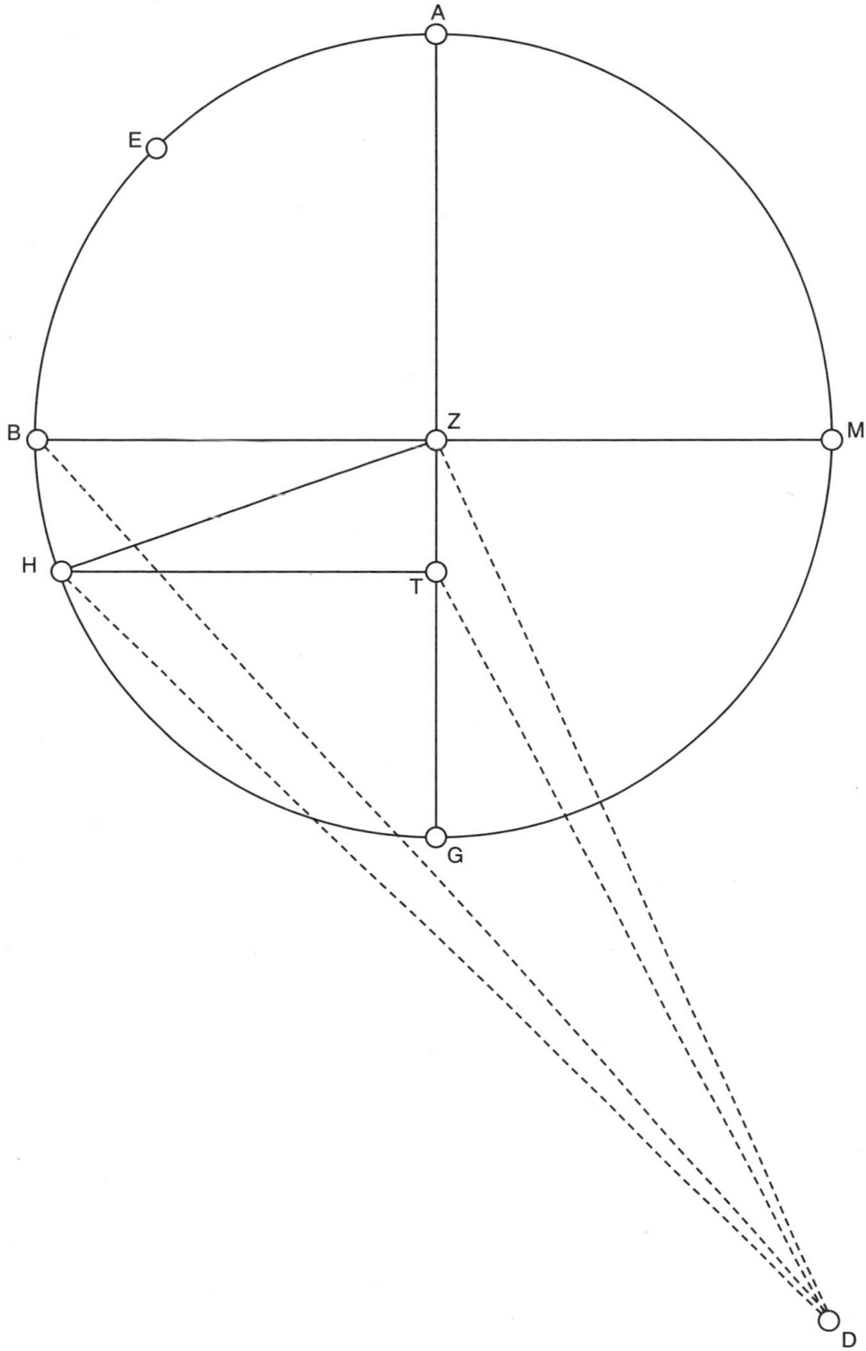

FIGURE 2.18

from point *H* perpendicular *HT* upon line *AG*, and we connect *DH* and *DT*. Because arc *HG*, rather angle *HZG*, I mean the angle of the mean

70v

motion is given, the ratio of *ZH* to *HT* is given. Therefore, the ratio of *DH*, I mean the radius, to *HT* is given, with angle *HTD* being a right angle. Then angle *HDT*, I mean the angle of the anomaly necessitated by arc *HG*, is given, and likewise it is shown for the rest of the positions. Through the likes of that, it is shown that were the angle of the anomaly given, then the angle of the uniform motion is given. From that, the angle of the true motion, I mean the visible motion, is given in that we add the angle of the anomaly to the mean motion if the Sun were upon arc *GEA*, and subtract it from it were it [the Sun] upon arc *ABG*.

Because if the Sun were upon the point of the greatest lag, I mean point *G*, and B.1.III. moved to point *H* by the measure of the mean motion, then the center of the Sun would also move in this direction, increasing [in speed] due to the aforementioned concept, through its motion about the poles of the two aforementioned circles. The amount of this movement is the measure of the arc of the anomaly. Thus, we add to arc *HG* the equal of the arc of the anomaly which it necessitates, and we calculate a second time the arc of the anomaly that this combined arc entails, it being the arc of the true anomaly. We continue to add this arc to the mean motion to derive the anomaly as long as the Sun is on the arc that is increasing in lag, I mean the arc that is between the two mean passages from the side of the greatest speed. And we continue to subtract it in the rest of the circle, and we calculate the anomaly that the rest necessitates, it being the true anomaly. So we have produced all upon which it is desirable to establish the motion of the Sun and its anomaly, with all of it corresponding to that which emerges from the hypotheses of Ptolemy, aside from the exception particular to these hypotheses on whose account there is no harm, with no need for the thesis of an eccentric or epicyclic orb. And God, exalted, is all-knowing. The first book of the book *The Light of World* is complete. Praise to God, exalted.

71r

The second treatise about explaining the hypotheses according to which it is B.2.0.1 desirable to work with the motions of the Moon and its visible anomalies, and in it are five chapters.

The first chapter is about the hypotheses given for the Moon's motions in B.2.I.1 revolution. As for the movement of the Moon according to the second opinion, this movement resembles the movement of the Sun except that its lag is greater than the lag of the Sun due to the rupture of power due to its distance from the prime mover. A multitude of anomalies appear in it also, so it is necessary that our given hypotheses about that be based on the observations of Ptolemy, as with the Sun. We say that Ptolemy found the Moon moving away from the ecliptic toward the north and toward the south by equal degrees. Therefore, it is necessary that the Moon move on a circle inclined to the ecliptic by the measure of the Moon's distance in latitude from the ecliptic, which was found to be around 5°. The return in latitude was also found to be faster than the return in longitude. That is, through his observations, it was found to complete the return in latitude in a period of time less than that in which it completes the return in longitude. That is not possible according to their hypotheses unless the node moves through the degrees of the orb of the ecliptic toward the west, I mean in the opposite direction of the signs. He inferred that the node moves also since the maximum latitude is found toward the north at one time and toward the south in the same degree of the ecliptic with the Moon being at both times either north or south of the equinoctial. It was found at one time in that same degree [of the ecliptic] with no latitude.

He also inferred, because he found that the Sun and Moon might be eclipsed B.2.1.2
in each degree of the ecliptic preserving this order, that there are therefore two motions in latitude. One of them is the motion of the Moon about the poles of its particular orbs, I mean about the poles of the circle of the Moon's latitude. Through this motion the Moon is seen at one time south of the ecliptic and at one time north, not just in a single degree of the ecliptic so that, if the greatest southern latitude were at one degree, the greatest northern latitude would be in the diametrically opposite degree and from the other extreme of the equinoctial. This would be the case even if the two points of the nodes were at rest, maintaining single degrees of the ecliptic. Through that motion, a latitude would occur for the Moon with respect to both the ecliptic and the equinoctial.

The second motion is the movement of the circle of latitude with the two nodal B.2.1.3
points through the degrees of the ecliptic in

71v

the opposite direction of the flow [of the signs], preserving its inclination to the ecliptic. Through this [second] motion mixed with the first a latitude occurs for the Moon. The Moon comes to be at one time at its extreme distance in the south and one time at its extreme distance in the north in one of the zodiacal signs with the Moon being at the two points on a single side of the equinoctial. That is because the movement of the node, retaining its extreme inclination to the ecliptic, is nothing but the movement of the two poles of the orb of the Moon, I mean the two poles of the circle of the latitude of the Moon upon a circle of the path whose pole is the pole of the orb of the ecliptic, with the arc from its pole to its circumference being equal to the extreme inclination of the circle of the latitude to the ecliptic. If it is agreed that the pole of the orb of the Moon is at its extreme distance from the pole of the universe and [if] the Moon is at the southern extremity, for example, on the circle of its inclination, then the Moon at that time is at the most extreme possible inclination to the ecliptic and the circle of the equinoctial that it can have. At that time the inclination of the Moon to the equinoctial is the two inclinations together toward the south, and when it is in the northern extreme, then it would be toward the north at the diametrically opposite degree.

As for if it were agreed that the pole of the orb of the Moon is at its extreme proximity to the pole of the universe in the circle of its path and [if] the Moon were at the northern extremity, then the Moon at that time is at its extreme inclination to the ecliptic to the north at that precise degree from which it had inclined toward the south at first. The latitude of the Moon from the equinoctial at that time was in the direction of the south by the amount by which the greater of the two inclinations exceeded the smaller. And if it were in the southern extremity, then the opposite. For that we named this motion the motion of the latitude that is with reference to the ecliptic and we named the first [motion] the motion of the total latitude. Ptolemy ascertained these motions through his successive observations of lunar eclipses, which were thus not affected by parallax. He found the daily mean motion in longitude to be 13;10,35° approximately and the motion in latitude, I mean the total latitude, to be 13;13,46° approximately. He found the motion in anomaly, I mean the motion of the Moon in the orb of the epicycle that was his thesis, to be 13;3,54° per day approximately. So the increase of the motion in latitude over the motion in longitude is 0;3° per day. The movement of the node in reverse is by this amount, so the motion that is upon the circle of latitude itself produces the motion in longitude except for these 3' that are subtracted from the longitude because of the movement of the node in reverse. That is because the inclination of this circle to the ecliptic is so little that it entails neither an increase nor a decrease in the motion in longitude. B.2.1.4

72r

(See figure 2.19.) Let us propose first the hypothesis for the mean motions in longitude and latitude. Let the parecliptic be circle *GTD* whose pole *B* revolves upon circle *LBN*, whose pole is the pole of the equinoctial, and it is *A*. Let the inclined circle for the Moon also imagined in this orb be circle *GED* with pole *M*, and this pole is forever revolving upon the circle of the path *LMN* whose pole is the pole of the ecliptic, I mean *B*. The solstitial colure is circle *ETBA*. According to the first opinion, there is no need for anything except to imagine these two circles, with this description, in the orb of the Moon, provided that it is divided as we said. As for according to the second opinion, our thesis is that in the orb of the Moon are two other circles in place of the ecliptic and the inclined circle, each of them in the plane of their counterpart from those in the uppermost orb, with the pole of B.2.1.5

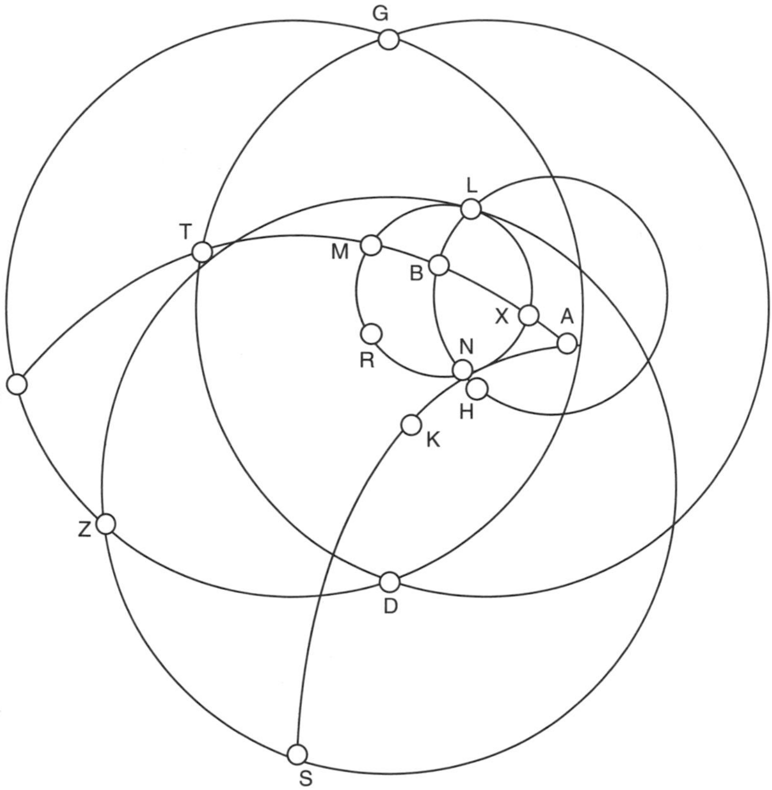

FIGURE 2.19

the circle of the latitude revolving on a circle of the path whose pole is the pole of the parecliptic.

We propose, for example, that the Moon is in its southern extremity, and let it be point *E*, with its position in the ecliptic being the point of the winter solstice. And let, at that time, the circles of the ecliptic and latitude imagined in the orb of the Moon be in the planes of their imagined counterparts in the uppermost orb. Likewise, their corresponding poles are on one axis, with the latitude of the Moon from the equinoctial being the two inclinations combined. Our thesis is that in one day, for example, the orb of the Moon moves in its entirety following the daily motion, with the poles of both the ecliptic and the circle of the latitude imagined in it lagging behind the pole of the uppermost sphere by an arc from the circle of the path of the pole of the ecliptic, whose measure is twice the final lag. B.2.1.6

72v

The pole of the ecliptic came to be upon point *H* from the circle of the path of the pole, with the pole of the circle of the latitude being on point *K*. An example of that is that the point of the winter solstice imagined in the orb of the Moon and the point of the southern extremity on the circle of latitude lagged through two arcs from the two circles parallel to the equinoctial with the measure of each [arc] being twice the measure of the final lag, I mean equal to arc *BH*. For example, the circle of the latitude that was first imagined in the plane of the circle *GED* became circle *SZ*, and the southern extremity that was on point *E* came to be on point S.

It is desirable that we explain how the center of the Moon traverses the quadrant from the circle of its latitude that is from point *S* to the intersection of the circle of the latitude with the ecliptic, I mean the node, in the period of time in which it traverses quadrant *ED*, beginning from point *E*, opposite the motion of the universe according to the opinion of Ptolemy, and through our opinion according to the first opinion. Despite that, it does not traverse a complete quadrant from the parecliptic in this period of time just as it does not traverse quadrant *TD* from the ecliptic in this time period according to the first opinion. That, however, follows from the first thesis because we propose in it that node *D* moves in the direction of the motion of the universe, I mean toward point *T*, so the Moon traverses quadrant *ED* in the time period in which it traverses less than quadrant *TD*. B.2.1.7

According to this second hypothesis, it is by this description. The center of the Moon moves about the pole of the circle of the latitude, I mean pole *K*, in one day through arc *SZ* whose measure is the degrees that Ptolemy says are the motion of the latitude, they being 13;13,46° approximately. This motion is not this amount in the ecliptic imagined in the orb of the Moon, but rather less than it by 3'. That is, if our thesis was that point *S*, I mean the southern extremity of the circle of the latitude imagined in the orb of the Moon, lagged in one day behind its arrival at its position by 3' more than the lag of the point of the winter solstice imagined in this orb, then the alignment that was between point *S* and between the point of the solstice deviated. The position of point *S* became 3' away from the beginning of Capricorn while preserving its maximum inclination. I mean that the inclination of point *S* in B.2.1.8

73r

latitude from this point that it is upon is equal to the inclination that it had at first from the beginning of Capricorn.

Thus, if the pole of the ecliptic imagined in the orb of the Moon, I mean *B*, lagged through arc *HB*, it is necessary that the pole of the circle of the latitude, I mean *M*, not arrive at point *K*, in that the three poles would be on a single great circle arc like arc *AHKS*, for were it like that, the lag of the southern extremity would be like the lag of the solstice. Thus, it follows that pole *M* does not move, B.2.1.9

preserving its position with respect to it [point *B*]; rather it is as if the pole of the ecliptic precedes and follows it on the circumference of circle *LMN* from point *M* toward point *N*. Therefore, if pole *B* arrived at point *H*, and we imagine circle *LMN* on pole *H*, with point *M* being upon pole *K*, then the pole of the circle of the latitude that was on pole *M* does not arrive at point *K* but rather at point *R*, for example. This pole continues to move on this circle in that direction, following the motion of the poles of the ecliptic imagined in it, so that the lag of the extremity becomes forever greater than the lag of the solstice by that amount each day.

The center of the Moon moves through arc *SZ* from the circle of the latitude, B.2.I.10
with it being seen to have traveled, through the ecliptic, through an arc less than that by the aforementioned amount. The Moon reaches the latitude of the degree of the ecliptic that it reaches according to the hypotheses of Ptolemy. And it also reaches the node in the time period during which it reaches both it and the latitude of that very same degree upon which is the node according to his hypotheses. After that, the final lag in longitude remains the aforementioned amount due to the weakening of the power reaching it, with the motion of the two nodes preserving forever the extreme inclination. That is because the movement of the two poles of the circle of latitude is about a circle whose pole is the pole of the ecliptic, and it is circle *LMN*. And this movement is not in reverse because this movement, indeed, follows the motion of the poles of the ecliptic since its intent is that its lag be less than the lag of the circle of the latitude, I mean the lag of the orb in which it is, for the motion belongs to the orb, not to the abstract circles. It is desirable that you bear in mind what we said in the first treatise about the division of the orb into two orbs or three or more as this is

73v

needed when our thesis for the one orb is two motions or three about different poles.

The second chapter, about the explanation of the first, true anomaly that B.2.II.1
appears with the Moon. Since we have brought hypotheses through which the mean motions in longitude and latitude appear according to the way in which they are, let us explain now the [][28] way in which things are with the first anomaly, with it alone being the one necessary to know for eclipses. That corresponds to what we did with the Sun according to this description.

(See figure 2.20.) Let the inclined circle which for the Moon is imagined in the B.2.II.2
uppermost starred orb, I mean the circle of the latitude, be circle *GED*, with its pole *M* revolving upon a circle whose pole is the pole of the orb of the ecliptic. Our thesis is that in this orb is a circle inclined to it by the measure of the radius

28. Following the Hebrew. Judeo-Arabic adds: first.

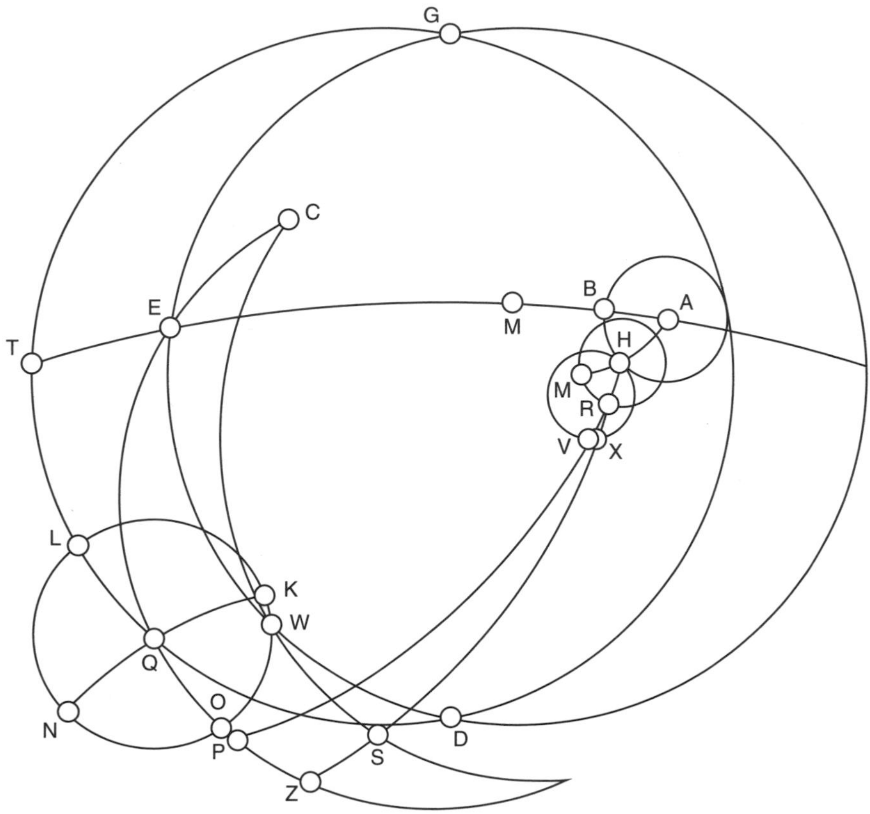

FIGURE 2.20

of the circle of the path that we will propose, namely circle *GTD*, whose pole also revolves about the circle of the path whose pole is the pole of the circle of the latitude. The pole of the ecliptic is *B* revolving on circle *BH*, whose pole is the pole of the equinoctial, and let it be *A*. Let us imagine, in the orb of the Moon, two circles in place of the circle of the latitude and the circle inclined to it, with each one of them being in the plane of its counterpart from the [circles] imagined in the starred orb and with the poles of the counterparts being on a single axis.

We propose, for example, that the positions of point E, the southern extremity B.2.II.3
from the circle of the latitude,

74r

and likewise point *T* from the deferent circle, are the point of the winter solstice on the ecliptic. Our thesis is that the center of the Moon is revolving on a small circle whose pole is carried upon and is revolving about the circle imagined in the orb of the Moon in the plane of circle *GTD*. Let this circle be circle *KLNO*, with its pole being *Q*, provided that we grant that pole *Q* is on point *T*, with the arc that is from its pole to its circumference being equal to the arc resembling arc *ET*, I mean the extreme inclination of the deferent circle to the circle of the latitude. So when we place the pole of the circle of the path of the center as if it were on point *T*, with the center of the Moon in the circle of its path being at the extreme that follows the ecliptic, I mean on point *K*, then the center of the Moon would be in the southern extremity of the circle of latitude. Its position on the ecliptic would be the point of the winter solstice.

Let us first explain this matter according to the second opinion. We propose, B.2.II.4
for example, that on one day the pole of the ecliptic imagined in the orb of the Moon, which was on point *B*, lagged behind arriving at its position by arc *HB* from the circle of the path of the pole, whose measure is twice the final lag in longitude. It moved, along with itself, the circle of the path of the poles of the circle of the latitude and the deferent circle, I mean circle *MR* and circle *XV*. In this period of time the pole of the circle of the latitude moved in reverse through arc *MR*, whose measure is the 3'. So the pole of the circle of the latitude came to be on point *R*, and the pole of the deferent circle on point *X*. Therefore, the circle of latitude that was at first in the plane of circle *GED*, is circle *CS*, whose southern extremity, which was first on point *E*, came to be upon point *S*. The southern extremity of the deferent circle that had been on point *T* came to be on point *Z*, and the pole of the circle of the path of the center that had been on point *T* came to be on point *Z* also. Therefore, the two southern extremities, which belonged to the circle of the latitude and the deferent circle, that were on points *SZ*, along with the pole of the circle of the path of the center, lagged behind arriving at their positions by 3' more than the point of the winter solstice imagined in the orb of the Moon lagged, retaining their two latitudes in the ecliptic. Our thesis is also that on that day the pole of the circle of the path of the center moved around the pole of its deferent circle, following the motion of the universe, from point *Z* to point *Q*, by the measure of arc *ZQ*, equal to the daily motion of the latitude, even if points *LO*, I mean the points of the intersection, forever retained the circumference of circle *CZ*.

Then, on account of the two B.2.II.5

74v

deferent orbs for the two aforementioned circles with the Sun, the likes of which need to be our thesis for the Moon as we said previously, the two points of intersection will be as if they retain the circumference of circle *GTD* with respect to the center of the Moon. Therefore, the lag of the Moon becomes the amount that we said was the final lag in longitude. Our thesis is also on that day that the center of the Moon, which had been on *K*, moves about the pole of the circle of the path in the direction of the motion of the universe also through arc *KW* whose measure is the degrees that we said belonged to the anomaly. But if we grant the hypothesis in this way, it would be necessary that the center of the Moon not arrive at the circumference of the circle of latitude imagined in the starred orb. That is because the degrees of the latitude are more than the degrees of the anomaly, for arc *ZQ* is 13;13,46° and arc *KW* is 13;3,54°.

But the matter of the Moon is not like this, as it shall appear with respect to the eclipses and their magnitude that the Moon never leaves the latitude of the circle of the latitude. So it results from that, that the motion of the pole of the circle of the path of the center about the pole of its deferent circle is equal to the motion in anomaly and that the pole of the deferent circle and its southern extremity move also in this direction about the pole of the circle of latitude, retaining forever their inclination to the circle of latitude, by the measure of the surplus of the motion in latitude over the motion in anomaly, and it is approximately 10'. In this way, we imagine the pole of the circle of the path of the center and the southern [extremity] of the deferent circle to be upon point *Z*, with the pole of the deferent circle at that time on point *X* from the circle of its path. On this day, the pole of the deferent circle moves, about the pole of the circle of latitude, I mean *R*, in the direction of the motion of the universe from point *X* to point *V* by the measure of this arc equal to the surplus of the motion in latitude over the motion in anomaly, namely 10'. The southern extremity that was on point *Z* moved by this amount and came to be on point *P*. B.2.II.6

And by this amount the Moon increased in longitudinal motion. So on this day, the pole of the circle of the path of the center moved about the pole of the deferent circle in that direction, with a motion equal to the anomaly, and let it be arc *PQ*. Through equal degrees, I mean by the measure of the degrees in anomaly, the center of the Moon also moved in the direction of the motion of the universe, from point *K* on the circle of its path to point *W*. The decrease of the latitude of point *W* from the latitude of point *K* becomes equal to the decrease B.2.II.7

75r

of the latitude of point *Q* on the deferent circle from the latitude of point *P*. Through that, the center of the Moon arrives at the latitude of the degree on the circle of latitude imagined in the starred orb that corresponds to the degree on the circle of latitude imagined in the orb of the Moon in which it is. Still, we need to add here the concept of the two circles that we mentioned for the Sun that corresponds to this.

(See figure 2.21.) We explain this matter according to the first opinion. We say that we imagine, according to the first opinion, that the southern extremity of the deferent circle, I mean point *T*, moves opposite to the motion of the universe B.2.II.8

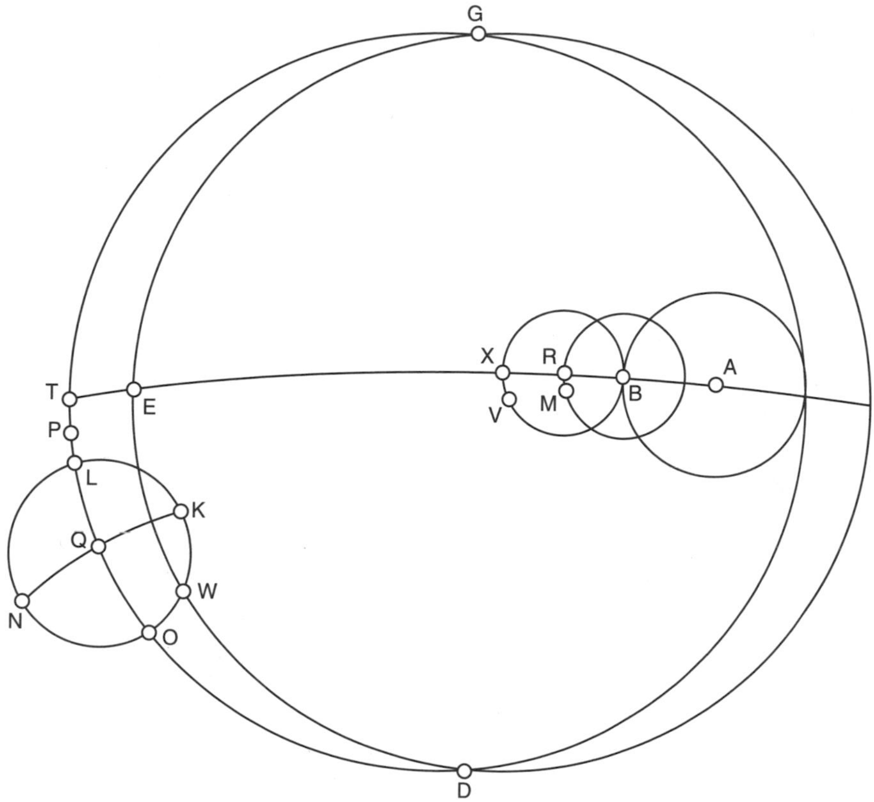

FIGURE 2.21

through the degrees of the circle of latitude, forever retaining its latitude to it, by the measure of the surplus of the motion of the latitude over the motion in anomaly. For example, let this motion in one day be from *T* to *P*, in the second figure, through the movement of its pole from *X* to *V*, with the pole of the circle of the path moving in that direction upon the deferent circle by the measure of the motion in anomaly. The two nodes move in the opposite direction, also by the aforementioned measure, from *M* to *R*, in this aforementioned picture. The Moon moves in the circle of its path and upon the two aforementioned circles by the measure of the motion in anomaly.

So what follows from this thesis is like that which follows from our other thesis, as this is because whether our thesis was that the southern extremity, which was on point *Z* with its circle, moves in the direction of the motion of the universe by the aforementioned amount, retaining its latitude from the circle of B.2.II.9

the latitude, through the movement of its pole about the two circles of their path about the two poles of the circle of the latitude, or [whether our thesis was that] the pole of the circle of the path of the center moves about the two poles of the deferent circle by the measure of the motion in anomaly

75v

in this direction, with the two nodes moving through the movement of the two poles on the circle of the path *MR*, whose pole is the pole of the ecliptic, by the opposite of the aforementioned measure, the pole of the circle of the path of the center reaches one point through both hypotheses together. There is no distinction by any means between them in latitude. This is what necessitates the occurrence of an eclipse and also that which necessitates the magnitudes of the eclipses according to the distance from the node.

As for the difference that is between them, it is with regard to the increase B.2.II.10
or decrease in longitude that is due to the rising times as we showed in the first treatise. The harm that can befall one from that occurs mostly with the advance of the time of the eclipse or its delay, which is equal to this error, or a greater error that occurs frequently with the derivation of these times. As for the harm that can befall that with regard to the measure of the eclipse, it is not necessarily sensed. Nothing remains except to choose the starting point, according to how we taught you which of them it is necessary to choose.

The third chapter, about the measure of the maximum of this anomaly. Let us B.2.III.1
show [now][29] the ratio of the radius of the orb of the Moon to the radius of the circle of the path so that we can point out the way to compute the maximum of this anomaly. That is, through the eclipses themselves, Ptolemy showed the ratio of the radius of the deferent to the radius of the orb of the epicycle. Namely, he took these lunar eclipses and derived the part of the ecliptic in which the true Sun was in the middle of the first eclipse, and in the middle of the second, and he found the distance between the two positions, it itself being the distance between the two positions of the Moon due to its being in opposition with the Sun. I mean, the true motion in the period of time that was between the two eclipses is more than the mean motion of the Moon in this known period of time. Thus, the lunar motion in anomaly, that was known by knowing the period of time, increased with the speed in longitude equal to the increase of the true speed over the mean in this period of time. He took also the degree of the ecliptic in which the true Sun was in the middle of the third eclipse and found that the motion of the true Sun, rather the true [motion] of the Moon through complete revolutions, in

29. Following the Hebrew.

the period of time that is between the second and third [eclipses], is less than the mean motion of the Moon in this period by a known number of degrees. So the known lunar motion in anomaly

76r

in this period is its decrease. B.2.III.2

(See figure 2.22.) Since we have set that out, let the circle of the path of the center be *BAGE* and the portion common to it and the plane of its deferent circle is *KN*, with point A being the position of the Moon in the middle of the first eclipse. It moved toward the second eclipse through complete revolutions through arc *AGB*, subtending from the ecliptic known degrees for the increase [in speed], they being, according to the calculation of Ptolemy, 3°24'. It is clear that arc *BA* is less than it by an equal amount. It [the Moon] moves from the second eclipse to the third through arc *BAG* by complete revolutions subtending from the ecliptic 37' in decrease, also according to his calculation, so arc *AG* comes to increase by 2°47'. We extend from the point of the second eclipse a perpendicular to the common portion, namely perpendicular *BLE*, reaching the circumference of the circle of the path on the opposite side. We extend from points *A* and *G* [two perpendiculars][30] upon line *BE*, namely *AM GZ*, and we connect *AE GE AG*. We extend from point *G* to *AE* perpendicular *GT* and we extend from point *D*, I mean the center of the ecliptic, to the points of the first and third eclipses two lines *DA DG* and to the points of the positions of perpendiculars *AM GZ*, lines *DM DZ*. Let us take the circle of latitude in this demonstration to be in the place of the ecliptic since there is no sensible difference between the two of them. Because line *AM* is given and the position of the second eclipse, I mean point *B* on the ecliptic, is itself the position of point *M*, then angle *ADM*, I mean that which is less than the lag because of arc *BA*, is given. So the ratio of line *DA*, namely the radius of the parecliptic, to the perpendicular going out from point *A* on line *DM* is given. And this perpendicular, if line *BE* were a diameter of the circle of the path, is itself line *AM*. And as for if it [line *BE*] were not the diameter, then it is not line *AM*, and we shall treat that and the derivation of the difference between the perpendicular and line *AM* by analogy with what we did with the Sun. Therefore, the ratio of *DA* to *AM* is given, and because arc *BA*, rather angle *BEA*, is given, then the ratio of *AE* the diameter to *AM* is given, and the ratio of *DA* to *AE* is given. Because the position of point *B* is point *Z* so angle *GDZ* is given, and then the ratio of *DG*, I mean *DA*, to the perpendicular, I mean *GZ*, is given. Because arc *BAG*, rather angle *BEG*, is given, then the ratio of *GE*

30. Following the Hebrew. Judeo-Arabic: a perpendicular.

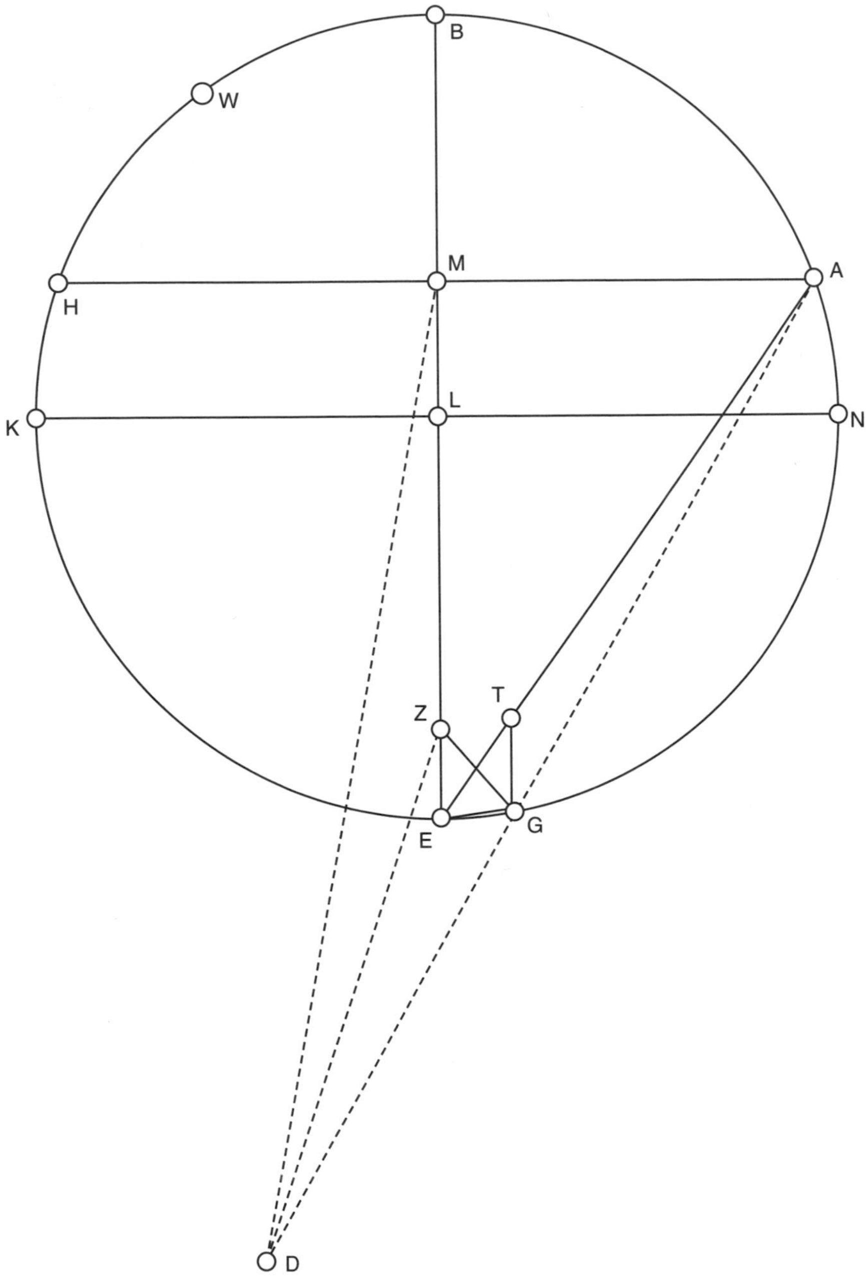

FIGURE 2.22

76v

the diameter to *GZ* is given, so therefore the ratio of *DA* to *GE* is given. Because arc *AG*, rather angle *AEG*, is given, so the ratio of the diameter to *GT*, rather to *ET* its complement, is given. The ratio of *GE* to *TA*, rather to *GA* which is potentially equal to *GT* [to] *TA*, comes to be given, so the ratio of *DA* to *AG* is given, with the ratio of arc *GA* to the diameter of the circle of the path being given, as arc *AG* is given. Thus, the ratio of *DA*, I mean the radius of the orb of the ecliptic, to the radius of the circle of the path is given.

In this way, the ratio that emerged for Ptolemy after the correction that will soon be explained emerged, and it is approximately the ratio of 60 to five and a quarter, less the increase or decrease stemming from the calculation of the time intervals regarding the variation of the nychthemerons. According to the second hypothesis, it is not as it is according to Ptolemy's hypotheses because it has been shown that a particular feature of this hypothesis is that the two zodiacal signs that follow the point of the equinox increase in rising times. Thus, the amount of this anomaly is not the amount that he mentioned. And if this anomaly was as he mentioned, I mean according to the first hypothesis, then we need to say that the harm that occurs because of that according to this hypothesis is of no consequence owing to its paucity, such that it might be possible that what follows from this hypothesis be correct. This little harm does, however, attach itself to the other hypothesis. But as for according to the first opinion, there is no difference at all. And God, exalted, is all-knowing. B.2.III.3

As for how we know the distance of the point of each eclipse from the point of least lag, it is in this way. Because the ratio of *GE* to *GA* is given, arc *GE* is given, rather all of arc *BAGE*, and it comes out to be less than half a circle. So half of it, I mean *BN*, is given. Then the distance of point *B* from the point of least path or lag, point *W*, is given. And if we performed this operation on the last three eclipses that Ptolemy brought, it is possible to correct the return of the anomaly in this same way in which he himself corrected it. B.2.III.4

After that we return to explain the differences in the same way as we did with the Sun. We derive the ratio a second time, and we derive the differences a second time through this ratio, and we repeat the derivation of the ratio after we increase or decrease the arcs by the anomaly that necessitates them. As was said with the Sun, the more it is repeated, the closer to the truth it is, and that is what we wanted to prove. It is clear also that the way of calculating the lunar anomaly is precisely the path taken with the Sun, and we transpose it so as not to repeat one concept many times. B.2.III.5

77r

The fourth chapter, about how the visible variation of the measures of the diameters of the Moon does not necessarily entail the variation of the distances of the B.2.IV.1

Moon from the Earth. Then, we take a certain visible lunar parallax at a single distance from the zenith point, and we derive from it the distance of the Moon by analogy to the radius of the Earth as Ptolemy did. Then we derive the rest of the individual parallaxes by knowing the distance of the Moon. From it we explain the distance of the Sun, the distance of the point of the edge of the shadow, the ratios of the diameters of the Sun and the Moon and the Earth, and likewise the ratios of the bodies to each other.

And we derive from this the limits of the eclipses and similarly the measure B.2.IV.2
of each eclipse, its times, and the amount of the lingering of that which lingers in the way in which Ptolemy did, except we derive what we derive from the parallaxes provided that the center of the Moon is always at one distance from the center of the Earth, as is necessitated by these hypotheses. We mention that afterward in the most concise way of which we are capable. Likewise, we work on the condition that the diameter of the Moon subtends one angle at the sight in each position, and if this angle increases in some locales and at some times, then it can be attributed to other reasons, such as the alteration of the air or its dampening, that necessitate the variegation of the Moon at one time or another. It is not possible from the standpoint of physics that it be attributed sometimes to the Moon's distance from and proximity to the center of the Earth, for proximity and distance can be imagined only with the thesis of an eccentric or an epicycle, with the impossibility of these two hypotheses having been shown.

The evidence for the rectitude of what we have said is that this variation of the B.2.IV.3
measures of the diameters is not found to occur forever in a single way with all [observers]. For we find that the practitioners of this science whose thesis is these hypotheses [the epicycle and eccentric] disagree about what they found regarding the visible measures of these diameters, so that Hipparchus says that he found the diameter of the Moon subtending an angle equal to that subtended by the diameter of the Sun at the sight when it is at the mean passage from its epicyclic orb. And Ptolemy says that he found it subtending its equal when it is at its greatest distance, and he worked according to what he found regarding that. How can this error be attributed to the like of Hipparchus, for in this observation there is no need to measure the degrees of the rod, as Ptolemy himself has said. So how is it possible that any error occurs? It was necessary for him to say that Hipparchus

77v

was correct in what he found and that the cause of this variation is what we said. But his proposing these hypotheses led him to the thesis that the diameter of the Moon subtends a different angle when it is in different positions. Then he found a variation in the observed measure of this angle, so he attributed it to these hypotheses. Let this measure of our speech be sufficient for the true anomalies of the Moon, the knowledge of which is needed for eclipses.

The fifth chapter, about the two anomalies remaining for the Moon, I mean the anomaly Ptolemy related to the Sun, namely that which he treats with the eccentric, and the anomaly Ptolemy attributed to the inclination of the diameter of the epicycle. As for the hypotheses according to which the matter of the Moon proceeds with the first anomaly, with knowledge of it alone being necessary for eclipses, we have already spoken sufficiently about it. And as for looking into the rest of the variations that appear with it at times other than these, it was necessary due to two things, one of them being that these anomalies disappear at the times of the conjunctions, rather in the times that the first [anomaly] disappears. Therefore, no harm comes to the eclipses because of them. The second [thing] is that the things through which these variations are shown are observations performed with instruments, and these observations might often lie due to parallax and also due to the observers. B.2.V.1

We find Ibn al-Haytham having explained in the fourth treatise of his book on optics that the sight's perception of a star is not on a straight line but rather on a bent one. He said that whoever observes the fixed stars with an armillary sphere finds the distance of a single star from the pole of the cosmos at its rising to be less than its distance at culmination, which would not be possible for the fixed stars if the sight perceived them on a straight line. He said also that if the Moon is observed at its rising, then its distance from the zenith, found through the instrument, is less than the measure obtained by calculation for the distance at that time. It results, then, that the light of the Moon does not extend to the two apertures of the instrument in a straight line. Since the sight perceives the stars neither on the perpendicular nor through reflection, since there is neither in the heavens nor in the atmosphere a dense polished body off of which the images reflect, it remains [only] that it [the sight] perceives them on a curved line because the sight does not perceive anything visible except in these three ways. And he said also that there is no place in the heavens where the sight B.2.V.2

78r

might perceive a star on the perpendicular, except if the star was on the point of the zenith or very near to it. And if that was like that, then it appears that what Ptolemy said and showed about these anomalies is not dependable because these observations are predicated upon the sight's perception of stars being on the perpendicular. But we have brought hypotheses through which these anomalies appear without an eccentric and without an epicyclic orb due to our hypotheses' majesty and noble circumstances, and this is when we begin to explain that.

Ptolemy found, at the time of the Moon's conjunction with the Sun or its opposition, only a single lunar anomaly at one time increasing, I mean that the true speed is ahead of the mean speed, and at one time decreasing. He found it forever preserving its order according to the degree in which it is on the circle B.2.V.3

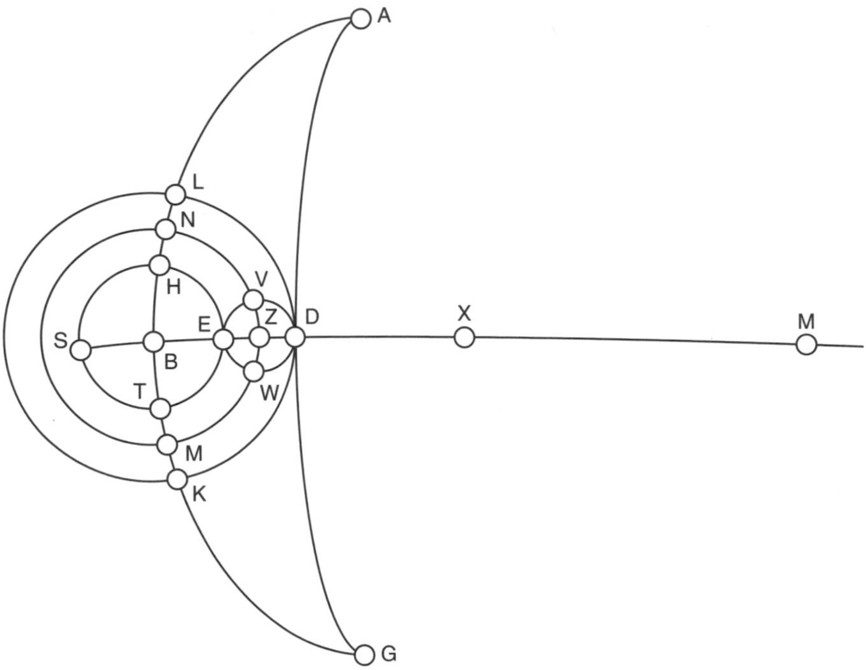

FIGURE 2.23

of its anomaly. And we are through with mentioning the hypotheses according to which things proceed with this anomaly. As for at the distances between these two [opposition and conjunction], and especially when it [the Moon] is at quadratures with the Sun, if the first anomaly necessitated an increase, then the speed was found to be increasing even more, and if it necessitated a decrease, it was found to be decreasing even more, always over the measure of decrease and increase. They found the greatest increase or decrease over the first anomaly to be 2½° approximately.

(See figure 2.23.) Our thesis is that the circle of the latitude is *ADG*, whose pole B.2.V.4
is *M*, with the deferent circle being *ABG*, whose pole is *X*, and the circle of the path of the center is *HETS*, whose pole is *B*. Our thesis is, because of this anomaly, that the center of the Moon does not revolve upon this circle itself. Rather, it revolves on a small circle tangent to it, with the arc that is from its pole to its circumference being equal to half of the maximum of the second anomaly. Let circle *EWD* with its pole *Z* be revolving forever about circle *NZM* whose pole is also *B*, and let the inclination of circle *ABG* to circle *ADG* be increased over that which we had said by the measure of the diameter of this circle. As for conjunction and

opposition, we say that the center of the Moon is upon the point from the circle of its path tangent to circle *HETS*, with it being forever point *E*, so that the anomaly that appears at conjunction and opposition is that which is entailed by circle *HETS*, nothing else. And as for if the Moon is at quadratures with the Sun, that being the position in which this second anomaly is the greatest it can be, then the Moon is on point *D* from the circle of its path, I mean on the circumference of circle *LDK* whose pole is also *B*.

78v

So if the Moon were at that time at the point of the greatest lag, I mean *E*, or at the point of least lag, I mean *S*, where the first anomaly disappears, then at that time there is no anomaly at all as follows from Ptolemy's hypotheses because the position of point *D* with respect to the ecliptic is itself the position of point *E*. As for when the Moon was on point *T*, namely the point at which there is the greatest increase in the speed from the first anomaly, so at that time point *D*, in which the Moon was, was on point *K*. Through that the anomaly increases by the measure of arc *TMK*, and at that time this increase is at its greatest. If it were on point *H*, in which the anomaly was at its greatest decrease, the Moon is on point *L*, and the anomaly also increases in decrease by this amount. When it was between points *ET* or *EH*, then the increases in the anomaly are over the amount of the first anomaly. And when the Moon was between conjunction and quadrature or between opposition and quadrature, I mean between points *DE* from circle *EWD*, then at that time the increases in the angles of the anomaly are less. And the [Moon][31] traverses the circle of its path, I mean circle *EWD*, twice during the time of a mean month. I mean that the orb imagined for this circle, with the Moon fixed in it, completes a revolution about pole *Z* twice in the period of the mean month. You have learned that each circle that we mention here is in a particular orb, with the motion belonging to the orbs not to the abstract circles, with each one of them being inside its companion. On the whole, all that appears from this second anomaly through this hypothesis agrees with what appears from the hypotheses of Ptolemy and does not depart from them at all, although the distance that is between these two hypotheses in reality and in the ease of intellectual conceivability is most extreme, as is apparent. B.2.V.5

And we need to make our thesis that B.2.V.6

79r

pole *Z* moves on the two circles described in the first treatise that correspond with this concept for the Sun, because the center of the Moon is forever on the

31. Both MSS read "Sun," but see the commentary.

circle of the latitude. That which is necessary for pole *Z* is necessary for the center of the Moon wherever it was on circle *EWDT*. This is the hypothesis that is our thesis for this second anomaly.

As for the third anomaly, its existence is explained through this hypothesis itself that has been our thesis, and there is no need to add anything, I mean neither an increase of a motion nor an inclination of a diameter nor anything of these impossibilities proposed by Ptolemy for these two anomalies, for the repugnancies attached to his [Ptolemy's] hypotheses for these two anomalies are not of the type of repugnancies attached to the rest of the hypotheses that were his thesis. Rather, the impossibility of that which is not feasible befalls it so that silence about it is best, since we have previously mentioned his opinion. And given what his intention was, he does not deserve blame. We say that the third anomaly necessarily following from the second is clear evidence for the rectitude of this hypothesis [of ours]. I mean that its [the third anomaly's] existence is not possible without it [the second anomaly] only in the positions in which it appears. And as for how this is necessary, it is in this way. B.2.V.7

We have said that in the times of conjunction and opposition there appears only the first anomaly. As for other configurations, particularly at the time of quadrature, [the Moon] appears to increase over the first [anomaly] as much as can be due to the second anomaly, if the first anomaly existed as he mentioned. And in this time, also, the third anomaly vanishes. In other configurations, the third anomaly might appear, and it is that which they attributed to the inclination of the diameter of the epicycle, and it is in this way. B.2.V.8

If the Moon was in the circle of its anomaly in the place of an increase in speed or lag, and the Moon is between opposition and the second quadrature or between conjunction and the first quadrature, then the anomaly increases. I mean that the lag in the longitude for the true position of the Moon is more than that necessitated by the remaining anomalies. When the Moon is between the first quadrature and opposition or between the second quadrature and conjunction, then the matter is reversed. And as for if the Moon was upon the arc of decreasing speed in the circle of its anomaly, B.2.V.9

79v

I mean the decrease in lag, then everything concerned with this is reversed. I mean that the speed decreases from what the first anomalies require in the first two positions, and it increases in the second two. I say that that follows from the movement of the Moon on the circle that produces the second anomaly, I mean circle *EWD*. Namely, if the Moon was in the arc of increase or speed, I mean arc *SET*, and let it be, for example, upon point *E* as in the figure, then at that time it is the time of a mean conjunction or opposition. Point *E* continues to move toward point *T*, following the first anomaly, with the Moon moving about point *Z* from point *E* to point *W* and to *D*, I mean from conjunction to the first quadrature or from opposition to the second. The anomaly increases because through this movement it speeds up as the Moon is closer to point *T*. The maximum of this anomaly is when the Moon is at the mean between conjunction and quadrature, I mean when it is on point *W*. If it were in the second two positions, I mean if the Moon were on arc *DVE*, then it is the reverse. As for if the Moon was on the arc of decrease, I mean on arc *TSH*, then it follows that all of that is the reverse, and that is what we desired to prove.

From that it appears that the second anomaly in these positions is not devoid of a third anomaly, however it may be, entailed by the movement of the Moon on this circle. If the movement was proposed the way it was in our thesis, this third anomaly would appear in the way it is. It also appears that there is no sensible distinction between the measure of the anomaly, in this way, and between the measure of the anomaly that was Ptolemy's thesis because of the inclination. It is also clear that there is not, in the true conjunctions necessitating eclipses, a sensible anomaly with this hypothesis that is our thesis for these two anomalies, with that being shown by the same demonstration with which Ptolemy showed it. This is what resulted for us through success granted by God exalted, about the truth of the hypotheses that were necessarily our thesis for these anomalies, with no need of any of the impossibilities proposed by Ptolemy and others. This is something that originates only from God, exalted, much praise to him. B.2.V.10

(See figure 2.24.) There remains for us to explain how the Moon forever retains the latitude of the circle of latitude due to the departure in latitude due to circle *EWD* that is our thesis for the second and third anomalies. So we reiterate the previous figure, and we propose the circle of the path of the pole of the circle B.2.V.11

80r

of the latitude, I mean the orb in which there is circle *LM* with pole *H*, namely the pole of the ecliptic orb, and the circle of the path of the pole of the orb in which there is the deferent circle, circle *MX* whose pole is *S*. Our thesis is that pole *S* is revolving upon circle *SM*, which is equal to circle *EWD*, tangent to circle *LM*. The movement of pole *S* upon circle *SM* is equal to the motion of the Moon upon circle *EWD*, so that if the center of the Moon were on point *E*, I mean upon the circumference of circle *ET*, then pole *S* is upon point *M*, I mean the pole of the circle of the latitude. The Moon comes to be at that time upon the circle of the latitude. If it were upon point *D*, I mean at quadrature, pole *S* would be on point *S* from circle *MS*. In this way the Moon is always upon the circle of the latitude, and pole *S* traverses circle *MS* twice in the period of a mean month.

And you should know that this motion that was our thesis for pole *S* on circle B.2.V.12

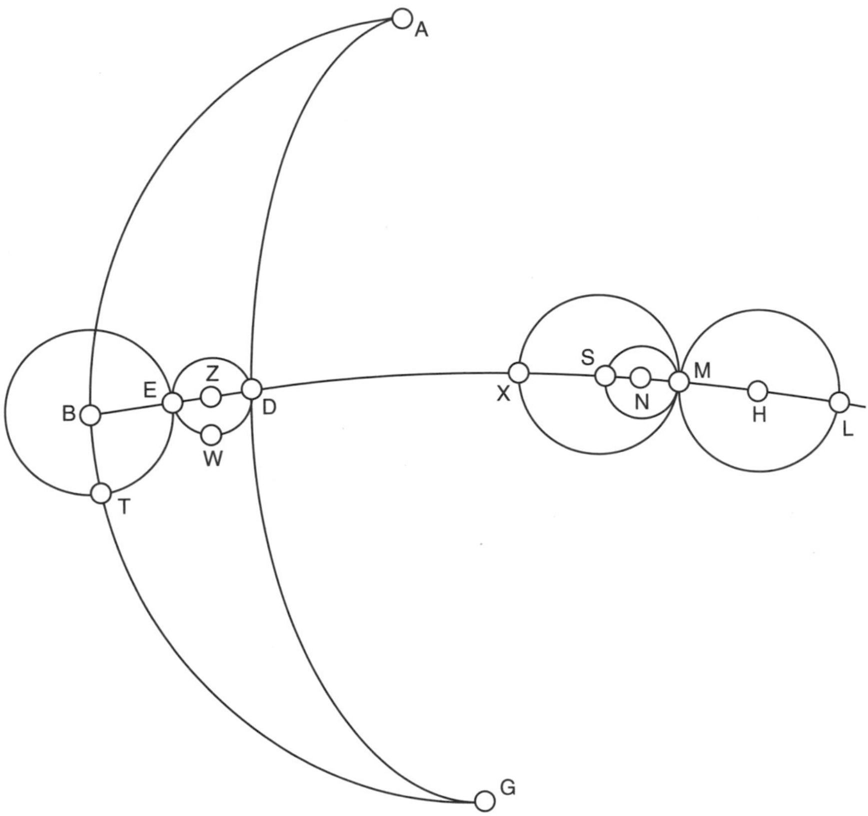

FIGURE 2.24

MS is so that it does not necessitate a motion for the planet in longitude but only in latitude. So, this, I mean that pole *S* moves on circle *MS* following the movement of the orb in which circle *MS* is imagined about pole *N*, is possible so that the orb whose pole is *S* does not ever move in longitude through this motion. As for the way of explaining the calculation of the measure of the anomalies together, it is through the way we explained with the first anomaly, and its explanation is near. It is necessary that you recall what we mentioned many times of the division of the orb into a number of orbs for the aforementioned concept. Through that, any possible doubts disappear in this place.

The second treatise of the second division of the book *The Light of the World* B.2.V.13
has been completed and much praise to God.

80v

[The third treatise][32] about the movement of the fixed stars; it has two chapters. B.3.0.1

The first chapter, about al-Biṭrūjī's opinion about this movement, and its refutation. It is incumbent upon us that we connect that to a discussion of the movement of the fixed stars according to the arrangement followed by the ancients and Ptolemy. Since there is no need for an eccentric or an epicyclic orb with them, his hypotheses about that are correct according to what we demonstrated regarding the lack of opposition in circular motion. We, however, have followed the second path with that because of what we have said previously. And the way according to which things are, for al-Biṭrūjī, with the movement of these stars is like this: All of the fixed stars adhere to a single sphere. This orb moves in the direction of the motion of the universe, following the uppermost [orb] that moves with the daily motion, and about its poles, and it lags, through this motion, behind arriving at the position of the uppermost, for the aforementioned reason, by some of the lag. Since its [the orb of the fixed stars'] intention was to resemble the uppermost orb and seek this perfection, for that is its goal, then it moves about its particular poles with another motion in that direction. Still, some lag in longitude remains for it, which according to Ptolemy's school of thought is that it moves backward by a sensible amount. Or it completes with its motion the motion of the uppermost, so there absolutely does not remain for it a lag in longitude according to the school of thought of the recent astronomers. There would occur for it, however, a movement in latitude necessitated by the movement of this orb about its poles while completing [the motion of the uppermost]. B.3.1.1

He [Biṭrūjī] said also that the motion of trepidation that was the thesis of recent astronomers according to their observations is not a true motion. Rather, the motion is forever equal in one direction, but as long as the star moves through the degrees of the quadrant in whose middle is the point of the equinox, then what rises with this quadrant from the equinoctial orb is less than it. Thus, there is seen for this star a retreat, and they name this lag accession. As long as it moves in the quadrant that [follows it],[33] then that which rises with it is more than it, and there is observed, due to that, for this star, a hastening (*taṭfīf*) and an increase over the motion of the universe, and they name it recession. That appears whether or not we assumed it to be a lag according to the school of thought of Ptolemy. According to this opinion the star, if it were upon the point corresponding to the summer solstice, then it would be observed moving toward the point of the spring equinox and from it to the winter solstice and from it to the B.3.1.2

32. These words were repeated at the end of the previous folio.

33. Following the Hebrew. Judeo-Arabic: "That belongs to it."

81r

fall equinox according to the contrary of the thesis of the ancients. This is the school of thought of this man for this movement.

If it were assumed that this orb lags behind the uppermost according to what Ptolemy saw, in that it moved in reverse whether the lag was equal to the movement in reverse that was his thesis or less than it, or that absolutely no lag in longitude remained for it, as it is not necessary for the truth of the hypothesis according to which we work, then it [the hypothesis] was accepted for the observations through which this movement was found, I mean the movement in latitude. Rather, it is not possible to deny it, it having been found through successive observations that the stars move according to the contrary of his thesis, because it had been found through observations that the distance in degrees of Spica Virginis toward the north, which is near the point of the autumn equinox, from the point of the equinox at which this distance was found, decreases with the passage of time. B.3.1.3

Likewise the stars that are near the point of the spring equinox were found to become distant with the passage of time from the point of this equinox toward the north. In sum, Ptolemy found by comparing his observations with the observations of Hipparchus and Timocharus and Aristyllus and Mileus the geometer, comparing some of these observations to the others and to the observations of other famous observers in this ancient time, that the stars that are in the hemisphere in which there is the point of the spring equinox forever become distant toward the north, with the passage of time, from the positions in which they were at first, and those [fixed stars] that are in the other half, the opposite. From that it follows that the star is seen moving according to how it necessarily moves according to his hypotheses. In sum, all that is seen of the movement of each of these stars with one of the hypotheses is according to what is seen with the other [hypothesis], whether they were the stars upon the equator of the ecliptic orb, or outside of it except for only two stars, namely those that were upon the two points of the two solstices at the beginning of the movement. These two stars are seen forever in one place with each of the two hypotheses. Let us propose an example for that.[34] B.3.1.4

(See figure 2.25.) The circle of the equinoctial is *GTD* whose north pole is *A*, and the ecliptic is *GED* whose north pole is *B*, and we imagine that this circle never moves. Let the circle passing through the middles of the moved zodiacal signs be circle *NZW* whose pole is *X* such that we imagine it was at first in the plane of circle *GED* with point *Z* being upon point *E* which is the winter solstice. B.3.1.5

34. In the Judeo-Arabic MS, there are no labeled points in the figure. The labeled points come from the Hebrew recension.

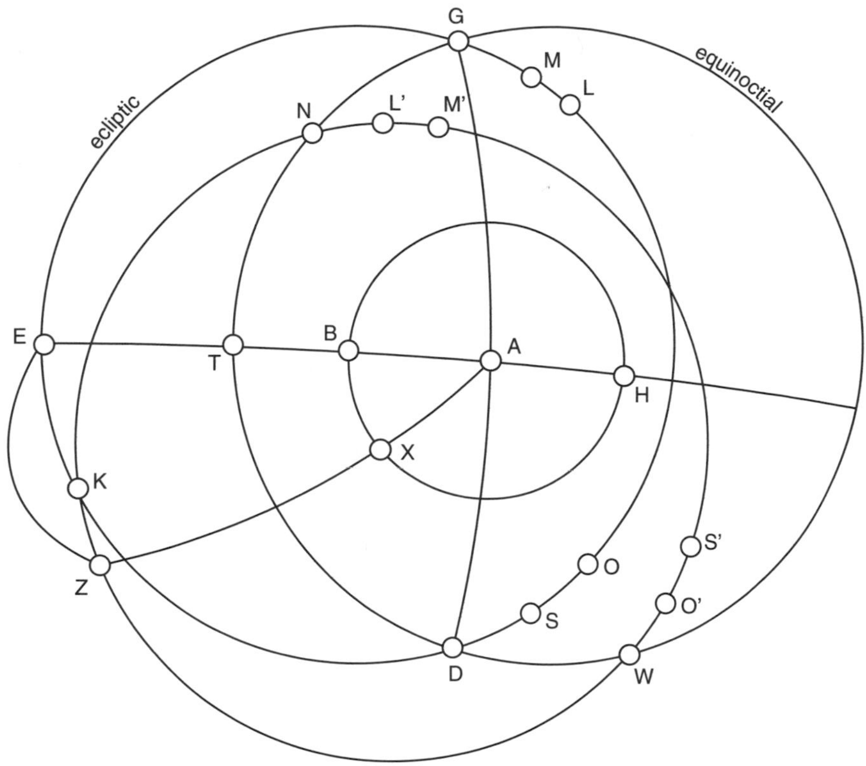

FIGURE 2.25

81v

We grant that the orb in which it is moved in some period of time, following the motion of the uppermost orb, and that it lagged behind arriving at its position by arc *ZE* from the circle parallel to the equinoctial orb. This orb moved about poles particular to it through arc *ZK*, desiring to catch up to the uppermost, until the star that had been on point *Z*, I mean the point corresponding to the winter solstice, came to be on point *K*. The star that was upon point *L* from this circle, I mean circle *NZW*, with its distance from point *N*, I mean the point corresponding to the fall equinox, being arc *LN*, came to be upon point *M*, that is to say distancing itself from it toward the north. The star that was upon point S from this circle, whose distance from point *W*, I mean the point corresponding to the spring point, was arc *WS*, came to be upon point *O*, I mean coming closer in the direction of the south. Thus, the star that is upon the point of the solstice is seen in the same way, according to each of the hypotheses, because whether the star that was on point *E* moves in reverse to point *K*, or if it moves in that direction B.3.1.6

from point *Z* through the likes of that arc, through the two hypotheses together the star is seen on point *K* at a single time. Likewise this follows for the other solstice. As for the rest of the stars that are upon that circle, it is the contrary because the star that was upon point *L* from circle *GED*, and we propose for example that it is Spica Virginis, namely that through which the observation occurred, is, after a period of time, found upon point *M*, I mean coming closer to point *G*, the fall point. And according to the opinion of Ptolemy, it would necessarily be like that, following the motion of this orb in reverse, I mean from point *E* toward

82r

point *K*, with us having shown that, according to the hypotheses of al-Biṭrūjī that it would be necessary that it become distant from that point toward the north, I mean from point *L* to point *M* from circle *NZW*.

Likewise, the star that is upon point *S* from circle *GED*, with its distance B.3.I.7
toward the north from a point, namely the spring [point], being arc *DS*, is thus found after a period of time on point *O*, I mean it becomes distant from it in that direction by more than that, as follows from his school of thought, with it having been shown according to the hypotheses of Biṭrūjī that it would be necessary that it be the opposite. Likewise, this anomaly, or something resembling it, becomes evident in whatever position the star was proposed on its orb, whether it was on the orb's equator or outside it. And if that is like that, then what is the benefit to us of this man making his thesis this hypothesis, if nothing of what appears to the senses corresponded to what results from his hypothesis? Therefore, [what we said at other times][35] about this man's goal appears to be the case. I begin now to mention the correct hypotheses for this movement according to the second opinion, I mean so that what appears from it will correspond with what appears from the observations of the ancients, not departing from it at all.

The second chapter about the manner of the movement of these stars. We say B.3.II.1
that the reason why the ancients spoke conventionally about these so-named fixed stars being all in one sphere is that they found them retaining their positions with regard to each other always, as if the movement was one, wholly uniform, so they judged that it was due to a single sphere. There is nothing that prevents us from saying that each star, or every two of them, has its own particular orb, and that all of them have a single and equal motion. But it is due to how they spoke conventionally about the hypothesis through which things happen with this motion that the doctrine that all of the fixed stars are in one sphere is sufficient, even if it used to remain incumbent on the natural philosopher to look into the reason why this multitude of stars is in one sphere and why, for the seven remaining planets, there are seven spheres. Not only this but also that each sphere of them is divided into

35. Following the Hebrew. Judeo-Arabic: מא קלנאה גיר מא מרה.

two or more as was shown. Aristotle had investigated that and gave reasons for it, with their distance from certitude being apparent. We have finished mentioning the impossibilities and the denial of recorded observations that occur from the thesis of all of the fixed stars being in

82v

one sphere with all of the motions in one direction. If our thesis was that for each star or two of them there was a single sphere, provided that what follows from it, or that what appears to the senses of their movement not be in some other way, according to the aforementioned true hypotheses, then we have deduced a great matter by removing this subject from investigation, and it is as I describe.

Let us mention first the way in which this occurs with the fixed stars that are B.3.II.2
in the plane of the circle passing through the middle of the zodiacal signs, I mean the circle that the Sun traces through its final lag. It is desirable that this meaning be understood wherever we mention the ecliptic, and from it, [the matter] will be explained for that which is not on this circle. We say that the two stars that were upon the two solstice points at a single beginning of the movement, with it having been shown from that which concerns them that what follows from them with this hypothesis corresponds to what appears through observation, that only these two [stars] are necessarily alone in one sphere. We say that the star that is upon this circle, whose distance in the beginning was 5° from the point of the winter solstice in the direction of the spring point, for example, is in another sphere, in the middle circle between them, whose inclination equals the inclination of the first sphere, with its distance from the point corresponding to the point of the winter solstice also being 5° in the opposite direction. I mean that if the point corresponding to the winter solstice in this circle were positioned upon the winter solstice itself, the distance of the star from the point of the winter solstice toward the fall point would be 5°. And for the star diametrically opposite it, if it were positioned in this [sphere],[36] then that which followed for the first follows there.

Likewise, it follows that these two be in a single sphere according to the afore- B.3.II.3
mentioned description. The matter proceeds with the remainder according to this example, whether they were upon the ecliptic or outside of it, in that our thesis is that each one of those outside of it is in a small circle parallel to the great circle corresponding to the ecliptic in this sphere, with its distance from the great [circle] being like its distance from it [the ecliptic], were all the stars in a single sphere. Provided that every two [stars] are diametrically opposite in one sphere and the inclinations of the spheres, all of them, are one, then the two poles of

36. Following the Hebrew.

each orb are revolving upon two circles whose two poles are the pole[s] of the equinoctial.

83r

The movement of all of these spheres is equal according to the senses. We propose for that an example with circles and letters.[37]

(See figure 2.26.) Let the ecliptic be *DTZK*, whose north pole *G* is on circle *GB*, whose pole is the pole of the equinoctial, namely *A*. Point *D* is the point of the winter solstice and [*K*][38] the point of the summer [solstice][39] with the spring [equinox][40] point being *X* and the fall point being *Q*. Let us position a star upon point *T* whose distance from point *D* in the direction of point *X*, I mean the spring point, is 5°. That star might be seen, with the passage of time, according to the hypotheses of Ptolemy, to be upon point *Z*, I mean nearer to the spring point so that it comes to be to the north of what it had been at first. Our thesis is that circle *EHZL* is in another orb, whose pole is *B*, at this inclination, with the point corresponding to the point of the winter solstice being *E* and [that corresponding to] the summer [solstice] being *L*. If we propose this star, according to our hypotheses, to be on point *H*, its distance from point *E* is 5° toward the opposite point, I mean the fall point, were point *E* superimposed on point *D*. If point *H* were on point *T*, I mean the position of the star, then through the two hypotheses together the star would be seen at a single time on point *Z*, because in the time period in which the star moves in reverse through arc *TZ*, according to the hypotheses of Ptolemy, this orb lags according to our hypotheses through arc *HT* from the parallel [circle]. The star moves about its poles through arc *HZ*, with this itself following for the star diametrically opposite it. Let it be, for example, on point *N* and at the time when point *H* is on point *T*, point *M* is on point *N*, and according to each one of the two hypotheses they come to be at a single time on point *S*. And this follows likewise for the remainder, with the poles of these orbs, all of them, being forever upon circle *BG*. There is no difference with that except B.3.II.4

83v

with regard to the rising times, this having been shown in the first treatise. And that is what we desired.

As for the motion in accession and recession which was the thesis of the recent astronomers, it is for us to say that it is imagined, with it being due to the increase B.3.II.5

37. There are no labeled points on the figure in the Judeo-Arabic MS. The labeled points come from the Hebrew MS.

38. The Hebrew has a K here, which would make sense. There is nothing in the Judeo-Arabic.

39. The Hebrew adds *hippuḵ* (solstice).

40. The Hebrew adds *hashwa'ah* (equinox).

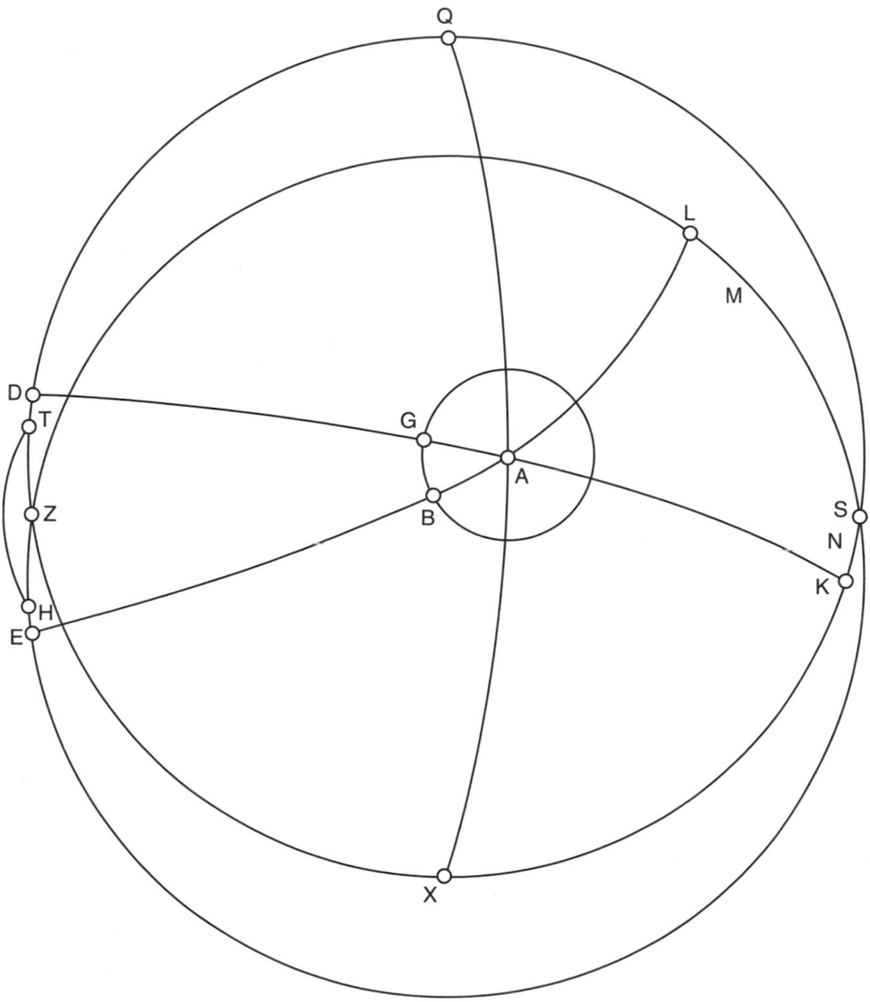

FIGURE 2.26

of the rising times of some of the stars over others according to their proximity from the point of the equinox or the solstice. Also it is not impossible that some of these stars have a lag greater than others due to the varying distance of the spheres upon which they are from the uppermost, with the senses not having yet perceived this variation due to its paucity, for the lag itself has not turned out to be any different, to now, from the variation of the lag. It is possible that this lag is an anomaly of the recent astronomers' imagination, to the extent that they propounded the motion in accession and recession. What we have said is sufficient to obtain the correct hypothesis for this movement. Much praise to God, who has caused us to arrive at this end. The third treatise of the book *The Light of the World* has been completed. Praise to God, master of the worlds.

The fourth treatise is about the motions of the five planets, namely Saturn, Jupiter, Mars, Venus, and Mercury, and their anomalies, and there are chapters. The first chapter is the preliminaries of the treatise. The second chapter is the explanation of the true hypothesis that is our thesis for the first anomaly found for the three upper planets, and it is the one related to the degrees of the ecliptic. The third chapter is about the second anomaly found for these three planets, it being that which is found to be according to the distance of the planet from the Sun, and [about] proposing the true hypotheses for it and for the particular matters that are seen in it and follow from it. The fourth chapter is about showing the maximum of the first anomaly, that which is related to the ecliptic, and showing the point from the ecliptic in which there is the least speed for each one of the three upper planets. The fifth chapter is about showing the measure of the circle that produces the anomaly that is related to the Sun for each one of the three upper planets. The sixth chapter is about how the true speeds are derived from the direct motions in revolution. B.4.0.1

84r

The seventh chapter is about the model for the planet Venus. The eighth chapter is about showing the point on the ecliptic in which there is the least speed, and explaining the ratio of the radius of the orb of Venus to the radius of the circle that produces the anomaly related to the ecliptic, and showing the measure of the circle producing the anomaly related to the Sun. The ninth chapter is about the model for the planet Mercury. The 10th chapter is [on the latitude of the five planets].[41]

The first chapter about the motion of the five planets. Joseph ibn Naḥmias said: and after we have mentioned all that is desirable to mention with the model of the Sun and the Moon and the fixed stars in the most complete way possible according to the truth itself, as is apparent from what we said previously, so it has come time for us to mention the true hypotheses for the model[s] of the five planets. Through that, the goal of deriving the certain of this science and of eliminating the impossible declarations and hypotheses that the ancient and recent astronomers until now had advanced is completed. I know that one who looks into this book of mine is either of perfect constitution, preferring the truth, when he pushes away with his intellect all of the doctrines recorded from the ancient and all the recent scholars until now in this science, knowing that those of its highest rank get to the point where they have no doubts about our doctrine B.4.1.1

41. Following the Hebrew recension

in it [this science].[42] As an example of that, King Solomon said, "There is no remembrance of them of former times; neither shall there be any remembrance of them of latter times that are to come, among those that shall come after."[43] And what we say in this book is for this perfect one. Or he [the reader] is defective in constitution and his intellect cannot comprehend obscure matters because his highest rank is that he understands the hypotheses that have been proposed until now and nothing else. And if he does not understand what we said, he will try to refute it with sophistry or poetry; for it is impossible for him to do otherwise. Truth mocks the likes of that as it mocks those who desired the refutation of true things who have come before. As King David said about the likes of them: "He that sitteth in heaven laugheth.[44]

And we return to our goal. Because the orb of the epicycle is more necessary, B.4.1.2
according to the hypotheses of Ptolemy, for these

84v

five planets than for the Sun and the Moon, as is apparent with them, and especially for Venus and Mercury, then before we begin to mention the hypotheses for these planets, we mention one decisive demonstration for the impossibility of the orb of the epicycle. I say that Aristotle shows in the second book of *De Caelo* that the planets do not have a rolling motion. What he said according to what Ibn Rushd summarized is as follows: he said that the known specter in the face of the Moon appearing in a single way might be evidence of that [that the planets do not roll] for the Moon. Namely, as it is clear from the case of this apparent trace in the face of the Moon that it is true and not a vision, so if we supposed it to be true, then if the Moon had a rolling motion, the view of this trace would likewise vary, he says. Through this demonstration in particular we can show the impossibility of the orb of the epicycle in the heavens, because if there were an epicyclic orb for the Moon, for example, then the Moon would be fixed in it, and it would be necessary that the view of this trace vary with the motion of the Moon in the orb of its epicycle. So if the position of the Moon with respect to our sight in all parts of the orb of the epicycle is one, then its motion in the orb of the epicycle would have to be through rolling and would not be fixed in it. And if it is not like that, then it would be necessary that the view of this trace vary. If the consequent is eliminated, then the antecedent is eliminated; therefore the existence of the orb of the epicycle for the Moon is impossible.

Since it is impossible for one planet, then it is impossible for all of them since B.4.1.3
their nature is one in species (*nawʿ*), and this meaning is apparent. There is no

42. Literally: they do not get to the point where they have doubts.

43. Ecclesiastes 1:11.

44. Psalm 2:4.

need for a demonstration except on account of the spread of this opinion among all peoples, but our doctrines about these concepts are beginning to become widespread. If those possessed of intellect were to look into it, they would not need a demonstration of the impossibility of the preceding doctrines. And they would find, in what we say about that, only uniform, equal, circular motions in equal times, with each motion having defined poles, with a defined orb surrounding the universe whose center is the center of the world. If we need our thesis for one planet to be many orbs, then each of them is according to the aforementioned description, with one being inside the other moving through its motion, and moving also with a motion particular to it about its poles, even if our thesis were ten orbs for a single planet, one inside the other. The lowest orb inside the others moves with the nine motions that belong to the nine orbs that are above it, moving

85r

also with its particular motion about its poles. Through the compounding of these motions with each other, the anomalies occur in the way that they are. This is when I begin to mention what I promised.

Know that Ptolemy and those who came before him found for the five planets B.4.1.4
two types of anomalies. One type is related to the ecliptic, I mean that it is found, according to the degrees of the ecliptic, increasing or decreasing over the mean speed in longitude. For parts of the path related to the increase in speed, the increase is always in the reverse [direction] of that which he related to the movement of the apogee. And for parts related to the decrease in speed, the decrease is always according to that example. This anomaly is found forever increasing, according to known parts of the ecliptic, and decreasing according to other parts, even though in all the observations it was found in a single configuration with respect to the Sun, so that the anomaly that is with respect to the Sun disappears, and this anomaly alone remains. There is a second type [of anomaly] related to the Sun, I mean that it [the planet] is found, according to the distance of the planet from the Sun, also increasing over the mean speed in longitude, or decreasing, with this anomaly being found at the times, I mean in certain parts of the ecliptic, when the first returns to what it was. If we found in certain parts of the ecliptic various configurations of the planets with the Sun, and in one configuration it [the second anomaly] is increasing and in the other it is decreasing, then this anomaly exists independently from the first. They found also with the anomaly related to the ecliptic that the time period from the smallest motion to its mean is greater than the time period that is from its mean to its greatest, and their thesis for this anomaly was the eccentric. They found for the anomaly related to the Sun that the matter is reversed, and their thesis for it was the orb of the epicycle. After that they derived the measure of the eccentricity for each of the five planets, the positions of the apogee in the ecliptic, and the measure of

the radius of the epicycle for each of them. As for Saturn, Jupiter, and Mars, he showed that with three cases named the extreme of night, I mean when the planet was diametrically opposite to the mean Sun, because at that time the planet is at the visible nadir of the epicycle. As for Venus and Mercury, he demonstrated that with the greatest morning and evening distances from the Sun.

He prefaced that by finding for the three upper planets that the mean speed in longitude and anomaly, if he summed them, would be equal to B.4.I.5

85v

the speed of the mean Sun. So he found for Saturn in 59 years, I mean that which begins from a solstice or equinox and returns to it, and 1½ days and ¼ approximately, 57 revolutions in anomaly and two revolutions from the revolutions in longitude and 1⅔° and $\frac{1}{20}$° to complete the aforementioned period of time. For Jupiter [he found] in 71 years less 4½ days and ½ and ⅓ and $\frac{1}{16}$ of a day, 65 revolutions in anomaly and six revolutions in longitude less 4½° and ⅓°. For Mars in 79 years and 3⅙ days and $\frac{1}{20}$ of a day are 37 revolutions in anomaly, 42 revolutions in longitude and 3°6'. They found five revolutions in anomaly for Venus in 8 years less 2¼ days and $\frac{1}{20}$ day, with the revolutions in longitude being equal to the revolutions of the Sun, and they are eight revolutions less 2¼°. And for Mercury [they found] 145 revolutions in anomaly in 46 years and $1\frac{1}{30}$ days, with the revolutions in longitude also being equal to the revolutions of the Sun. We want the revolutions in longitude with this planet, however, to begin at a solstice or an equinox so that they will return to it.

Chapter two, about explaining the true hypothesis proposed for the first anomaly that is found for the three upper planets, namely that related to parts of the ecliptic. Let us grant, for the hypothesis of Saturn and Jupiter and Mars, that the ecliptic that is in the orb of each of them is *AB*[45]. Let us propose, because of the anomaly related to the ecliptic, circle *GMDL* to be about pole *B*, with the great circle arc that is from its pole to its circumference being equal to the arc of the greatest of this anomaly. The planet revolves on this circle about pole *B* following the motion of the anomaly, with pole *B*, along with its orb, revolving about the ecliptic, I mean *AB*, about its poles. Our thesis is also that the time period of the two returns is one. I mean that the time period of the return of the planet in circle *GMDL* is equal to the time period of the return of pole *B* about the ecliptic as is necessary for this anomaly. Let us grant that if the pole of the circle *GMDL*, I mean *B*, were on point *A*, the planet would be on point *G*. In the time period in which pole *B* moved upon the ecliptic, through arc *AB* for example, following the motion in longitude, the planet that was on point *G* would move about pole *B* through arc *GB* from circle *GMDL*, I mean the circle of the path, in the opposite B.4.II.1

45. There is no figure in the MS.

of the motion of the pole, [through an arc] similar to arc *AB*. There is observed, therefore, in the path in longitude, a decrease that arc *GB* entails and as long . . .

86r

About the particular things that are known through the knowledge of rising B.4.III.
times. If we know the degree of the Sun on a day or night, we know its measure [of time] in that we take the rising times of the six zodiacal signs from the degree of the Sun on the day or its opposite in the night at that latitude, and we take $\frac{1}{15}$ of it, and it is the number of the equinoctial hours. And if we took $\frac{1}{12}$ of it, the number of the [time-]degrees of the seasonal hours would obtain. The measure of the seasonal hour will be known in that we take from the table $\frac{1}{6}$ of the difference of the sums of the rising times of the degree of the Sun, or its opposite for the night, between *sphaera recta* and the sought [latitude]. So if the degree of the Sun were to the north [of the equinoctial], we add it to 15 [degrees], and if it were to the south, we subtract it from 15, and the [time-]degrees of the seasonal hour obtain. It is for us to turn the seasonal hours into equinoctial hours by multiplying them by their parts,[46] and we take $\frac{1}{15}$ of the product. Vice versa, we multiply the equinoctial hours by their [time-]degrees and divide it by the time periods of their counterparts from that distance.[47] If we take $\frac{1}{15}$ of twice the aforementioned difference and we add it to 12 or we subtract it from 12, we have the number of the equinoctial hours of that day. Also, if we knew how many seasonal hours had elapsed from the day or the night, and we multiplied it by its [time-]degrees, and we add to it (*alqaynā*) the cumulative rising time in the clime for the degree of the Sun in the direction of the signs or for its opposite at night, and wherever the number arrives in the ecliptic, it is the ascendant.

And if we multiplied the hours from the previous noon to here by their degrees, B.4.III
each with what corresponds to it, and add to it the summed [rising times] of the degree of the Sun in the direction of the signs at *sphaera recta*, then it ends up at the degree of the ecliptic that is culminating. If we take the summed [rising times] for the degree that is rising in the clime and we subtract 90, then wherever it ends up at *sphaera recta* is that [degree] which is culminating. Vice versa, we take the cumulative [rising times of the traversal of] mid-heaven at *sphaera recta*, and we add 90, then wherever it ends up on the ecliptic, that is rising for that clime, it is the ascendant. The meridian varies in the climes by the number of time periods of equinoctial hours equal to the number of degrees between its two noons.

46. By multiplying them by their parts, he meant that one should multiply the number of seasonal hours, twelve, by the length of one equinoctial hour in time-degrees at that latitude. Cf. Toomer, *Ptolemy's "Almagest,"* p. 104.

47. The time periods of their counterparts are the length of a seasonal hour at that latitude in time-degrees. Cf. Toomer, Ptolemy's "*Almagest*," p. 104.

On the angles that occur between the ecliptic and the meridian circle. Know that the knowledge of these angles and likewise the angles occurring between the ecliptic and the horizon in each position, and from the ecliptic and the circle passing through the poles of the horizon, I mean the zenith, and likewise the arcs of that circle that is from the zenith to the position of its intersection with the ecliptic is necessary for knowledge of the lunar parallax. Therefore, we see fit to explain them. We preface this by naming as right the angles from great circles, were we to position a pole with the separation between the two circumferences enclosing the angle being the quarter of the circle traced at any distance, I mean 90°. And from the four angles occurring at each intersection we intend one, and it is the northeast angle from the ecliptic.[48] B.4.III.3

(See figures 2.27 and 2.28.) We demonstrate first that every two points [on the ecliptic] B.4.III.4

86v

equidistant from one equinox describe equal angles. Let the equinoctial circle be *ABG*, with its pole being *Z*, and its arc from the ecliptic is *DBE*, with arcs *HB BT* being of one distance from *B*, I mean the point of the equinox. We trace *ZKH* and *ZTL* so arcs *KB BL*, I mean their rising times, are equal, with their latitudes, I mean *HK TL*, being equal. So the triangles are equal, and angles *KHB BTL*, rather *ZTE*, are equal. We explain also that for every two points from the orb of the ecliptic, like *D* and *E*, with their distances together from a single solstice, and let it be *B*, are like angles *ZDB* and *ZEB*, occurring on the meridian, equal to two right angles. This is because angles *ZDB ZEB* are equal because they are both equal to *ZD ZE*, due to the distance of *DE* from *B* being one and angle *ZEG* equaling, with each one of them, two right angles. And that is what we wanted to explain.

(See figure 2.29.) Let the meridian be *ABGD*, with the orb of the ecliptic being at first *AEG*. With *A*, the solstice, as a pole, on the distance of the side of the [inscribed] square is [arc] *BED*. So due to how *ABGD* passes through the poles of *AEG BED*, then *DE* is a quadrant, I mean *DAE* is a right angle at any solstice. We make *AEG* the equinoctial and *AZG* the circle of the ecliptic, with *A* being the fall equinox, so *Z* is a winter solstice. So arc *DE* with the extreme inclination, I mean *EZ*, is 113°51′, I mean angle *DAZ*. Thus the angle that is at the spring equinox is the supplement, and it is 66°9′. B.4.III.5

(See figure 2.30.) We make in the other circle the equinoctial *AEG* and the ecliptic is *BZD*, with *Z* being the fall equinox, and arc *ZB* being Virgo. On the distance of the side of the [inscribed] square from [pole] *B* is [half-circle] *HTEK*, B.4.III.6

48. Cf. Toomer, Ptolemy's "*Almagest*," p. 105. Ptolemy defined this angle as the one "to the rear of the intersection of the circles and to the north of the ecliptic."

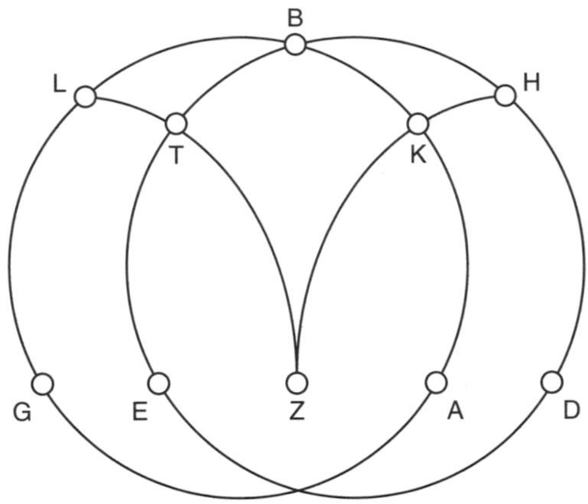

FIGURE 2.27

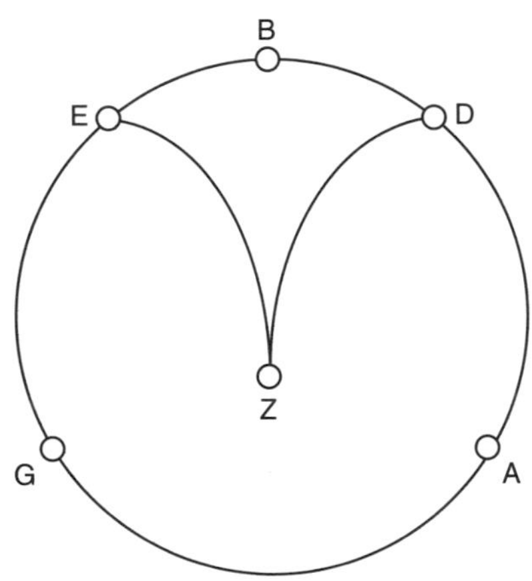

FIGURE 2.28

and we seek the measure of angle *KBT*. It is known through the knowledge of arc *ET*, because *EK* is a quadrant, so each of arcs *HB BT EH* is a quadrant on account of how *ABGD* is traced about the poles of *AEG* and *HEK*. Because the ratio of the sine of arc *BA*, it being the latitude of the first of Virgo to the sine of the arc of its complement, I mean *AH*, is compounded from the ratio of the sine of *BZ*, I mean the zodiacal sign Virgo, to the sine of *ZT*, and from the sine of *ET*, the unknown, to the sine of *HE*, the quadrant, then the unknown is the fifth. We find that arc *ET* is 21°, and arc *KET*, I mean angle *KBT*, is 111°. The angle at the first of Scorpio,

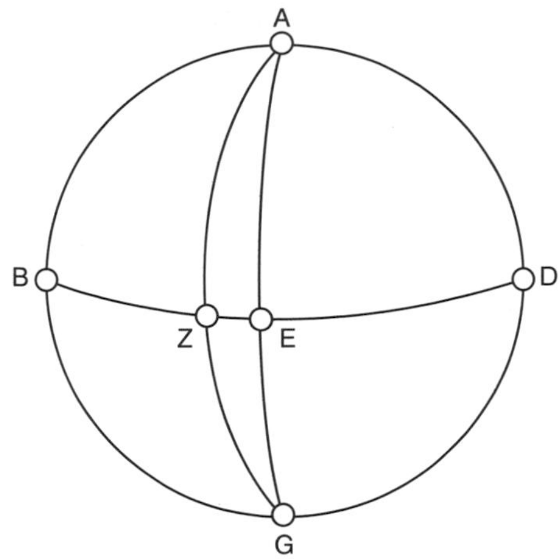

FIGURE 2.29

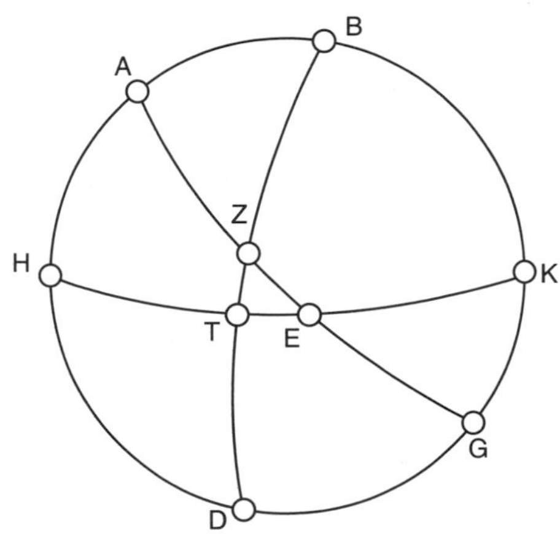

FIGURE 2.30

it is these degrees themselves, and the angles at the first of Taurus and the first of Pisces are the supplement of that and it . . . [49]

87r

(See figures 2.31 and 2.32.) We know the division of the three remaining quadrants. Let the horizon be *BED* with pole *K* being raised above it by 36°. The equi- B.4.III.7

49. The MS contains but does not refer to figure 2.31, perhaps because 86v does not lead into 87r.

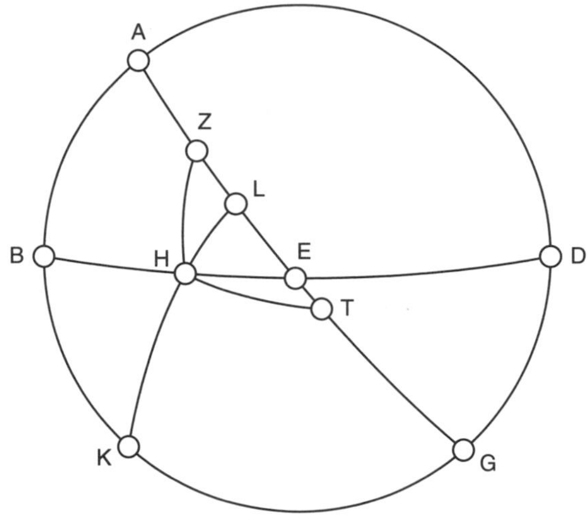

FIGURE 2.31

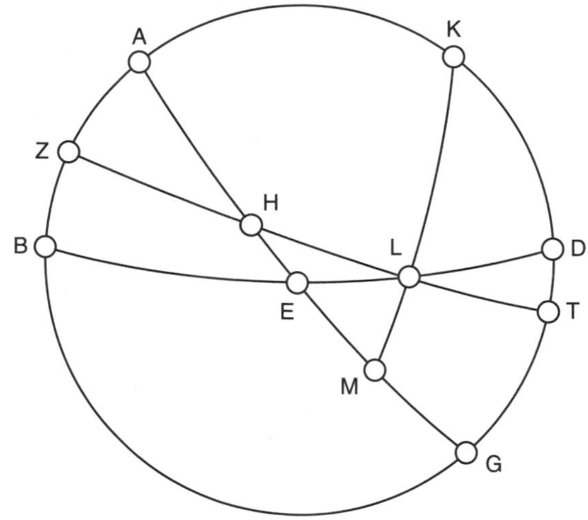

FIGURE 2.32

noctial is *AEG*, and the ecliptic is *ZHT*, with point *H* being the spring point, and we extend the sign of Aries. It is desirable that we show the measure of *HE*. So the ratio of the sine of arc *KD*, I mean the altitude of the pole, to the sine of arc *DG*, I mean its complement, is compounded of the ratio of the sine of arc *KL*, I mean the complement of the latitude of *L*, to the sine of arc *LM*, I mean the latitude of

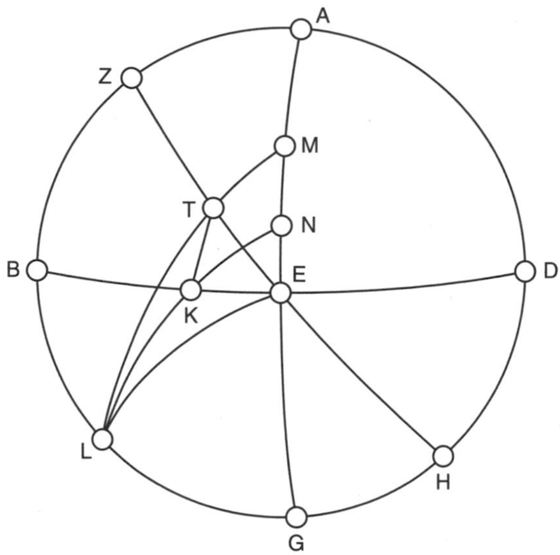

FIGURE 2.33

L, and from the ratio of the sine of the unknown arc *EM* to the sine of arc *EG*. So the unknown is the fifth quantity, and we find it to be 8°38'. But *HM*, I mean the right ascension of *HL*, is 27°50', and *HE* becomes 19°12'. So he demonstrated that the rising time of Pisces was equal to this, with each of Virgo and Libra rising with the complement of these time periods for twice their right ascensions, it being 36 time-degrees and 28'. B.4.III.8

If we made *HL* into two divisions of the 12 signs, I mean Aries and Taurus, so we find in precisely that way arc *ME* to be 15°46'. *HE* becoms 41°58', so Taurus alone rises with 22 time-[degrees] and 46 minutes. Aquarius's rising time equals that, and each of Leo and Scorpio [rising] with the complement of those time periods for twice their right ascensions, it being 37 time-[degrees] and 2'. So when the half circle from Cancer to Sagittarius rises on the longest day, I mean in 14½ hours, rather 217;30°, and on the shortest day, rising in nine hours and a half, rather 142;30°, then each of Cancer and Sagittarius come to rise with 35 [time-]degrees and ¼ and each of Gemini and Capricorn with 29 [time-degrees] and 17'. It is clear in this way that one might know [rising times of arcs] less than these degrees, and that is what we wanted to explain. Likewise this is all demonstrated another way with triangle *MHL*.

(See figure 2.33.) It shall be possible for us to give a second way to know the rising times, and it is this. We trace first the orb of the meridian *ABGD*, the horizon *BED*, the equinoctial *AEG*, and the ecliptic orb *ZEH*, with *E* being the spring point. We trace a parallel segment *TK* passing through point *T* of known distance B.4.III.9

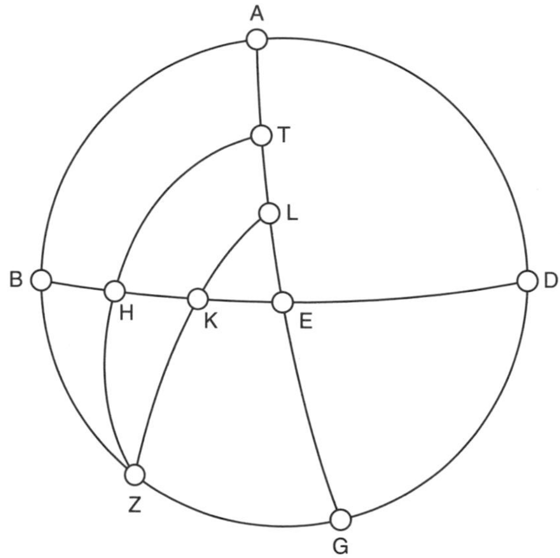

FIGURE 2.34

from *E*. We extend from *L*, I mean from the south pole, arcs *LTM LKN LE*. So *ET* from the ecliptic rises at *sphaera obliqua* with *TK*, I mean *LM* and *NM*, because they are parallel and uniform, with it rising at *sphaera recta* with *ME*. The difference between them, I mean *NE*, is bordered by segment *LN* passing through *K*.

(See figure 2.34.) Let us propose this figure, with the circle of the horizon and B.4.III.1
the equinoctial, whose south pole is

87v

Z. We trace *ZHT*, with point *H* being the rising time of the beginning of Capricorn, and *K* is the rising time of any other point that we wish from the quadrant of Capricorn, Aquarius, and Pisces. So the ratio of the sine of arc *HT*, and it is the same at each latitude because it is the extreme inclination, to the sine of its complement I mean *HZ*, is compounded from the ratio of the sine of arc *TE*, I mean half of the surplus of the mean day over the shortest, to the sine of the unknown arc *EL*, and from the ratio of the sine of arc *LK*, and it is also the same at every inclination as given in the table of the inclination, to the sine of the arc of its complement, I mean *KZ*. Therefore, the ratio of *TE* to the sine of *EL* is one at all latitudes.

We designate point *K*, with there at one time rising with it the degree comple- B.4.III.1
menting *K* from Pisces, and at one time the 10th degree, and at one time the first degree, and we end up at Capricorn by increasing 10 by 10. And there is with all of

them the sines of *LK* and *KZ* being known through the table of the inclination, so the ratio of the sine *TE* to the sine of *EL* at all latitudes becomes known for each interval, it being the fourth unknown. We find [the ratio] in the first 10 if the sine of *TE* were, for example, 60° to 9°33' and in the second [60° to] 18°57', and in the third [60° to] 28°1'. And in the fourth [60° to] 36°33', and in the fifth [60° to] 43°13', and in the sixth [60° to] 50°44'. And in the seventh [60° to] 55°45', and in the eighth [60° to] 58°55'. And if the double of [arc] *TE* were assumed, since it is the difference between the two days, the ratio of the sine of *TE* to the sine of *EL* would be as we said, so arc *EL* is known. So if we subtracted it [arc *EL*] from the rising times of the sought arc at *sphaera recta*, its rising times would remain at that latitude.

So we take [the case] where the altitude of the pole is 37°, with arc *TE* at this B.4.III.12 position being 18°45'. These numbers that we mentioned are decreased by the amount that the sine of 18°45' is less than 60°. Then we take the arc of that sine and it is, in the first 10 [degrees], 2°56', and in the second ten 5°50', and in the third 8°38', and in the fourth 11°17'. And in the fifth, 13°42', and in the sixth, 15°46', and in the seventh 17°24', and in the eighth 18°24', and in the ninth 18°45', and it is ET. We subtract these numbers from the rising times of the 10 [degree interval]s at *sphaera recta*, so there remains the rising times of the 10 [degree interval]s at *sphaera obliqua*. There comes to be for the nine 10 [degree interval]s, I mean the quadrant, 71¼°, and it is the measure of the shortest day. By knowing the division of this quadrant we know the division of the remaining quadrants as we demonstrated previously. So we make a table, and we begin from the equinoctial to the circle whose longest day is 17 hours by increments of a half hour. In the first column are the 12 zodiacal signs, and in the second are their 10 [degree interval]s, in the third are the rising times, and in the fourth their accumulated time-degrees.

3

Hebrew Recension of *The Light of the World*

127b

ספר אור עולם [בתכונה חברו][1] יוסף נחמיש

ספר אשר הוציא לאור נעלם
יצב גבולות רום לכל חילם
משם דבר חפץ מצוא בקש
יהיה לך תמיד לאור עולם
הורני יי דרכך ונחני באורח מישור
מלפניך משפטי יצא עיניך תחזינה מישרים

[אומר][2] כי בעלי החכמה הלמודית כשדברו בעקרי [][3] תנועות השמימיות [][4] הסכימו לבנות הענין
0.1 על שני עקרים כבר חלקו עליהם בשניהם בכלל קצת בעלי החכמה הטבעית מבלתי שיאמרו בסבת מה שיראה מן התנועות והחלופים בשמים מאמר [מספיק][5], לפי שכוונתם היתה לבאר מניעות אלו השני עקרים בחק הגרמים השמימים בהוראות טבעיות לבד. האחד משני העקרים הוא הנחתם גלגל יוצא המרכז וגלגל ההקפה. והשני הנחתם בשמים תנועות הפכיות לפי סברתם. והם מרחיקים היות תנועות הפכיות בשמים כמו שירחיקו הגלגל יוצא המרכז וגלגל ההקפה. ואמנם קרה להם זה לפי שהם סוברים כי תנועה סבובית תהיה הפך תנועה סבובית, כמו שתנועה ישרה הפך תנועה ישרה, בהפך מה שביאר ארסטו בספר השמים ועולם.

1. ו: ל.
2. ו: -.
3. ו: אלתי עליהא יג׳רי אלאמר פי.
4. ו: מנהם.
5. ו: מקנע.

0.2 וכוונתינו בזה []‏[6] לבאר כי מה שאמרו במניעות הגלגל יוצא המרכז וגלגל ההקפה אמת עד שאין צרך למופת אחר [על זה].[7] ואחר זה נתן עקרים אמתיים יראה מהם מה שהיה יראה [בשמים][8] מצד [הנחת][9] אלו השני גלגלים על הצד היותר שלם שאיפשר. ואמנם מה שאמרו במניעות תנועות הפכיות בשמים הוא בטל לפי שכבר יתבאר []‏[10] כי אין בתנועה הסבובית הפך במאמר מפתי, ר״ל כי תנועה סבובית לא תהיה הפך תנועה סבובית, כי ארסטו אמנם ביאר זה בהוראות לבד. ונאמר אמנם הגלגל יוצא המרכז גלוי מענינו המניעות והריחוק מן ההזדמנות ועזיבת הסדר, לפי שאין דבר יותר זר מהנחת גלגל יוצא המרכז ממרכז העולם סובב סביב מרכזו ומרכזו סובב סביב מרכז אחר.

0.3 ויתחייב מזה שיהיה בין שני הגלגלים אם רקות וכבר נתבאר מניעות מציאותו וכי הוא כמו מניעות מציאות ההעדר המוחלט, או מלוי מגשם נכרי ימיר מקומותיו בתנועה. ואם היו אלו השני גלגלים מתחתכים, יתחייב [עם][11] מה שאמרנו גם כן כשיתנועע המרכז האחד סביב האחר שיתפוצצו השני גלגלים []‏[12] וישתנו עד שיהיה טבע אלו הגלגלים כמו טבע המיים או האויר יריקו מקום וימלאו מקום, ושאר מה שיתחייב מאלו ההנחות מן המניעיות והענינים החולקים על האמת אשר זכר אותם רבינו משה ע״ה בספר מורה הנבוכים.

127a

0.4 ויותר זר מזה הנחת גלגל ההקפה סובב סביב מרכזו ומרכזו ג״כ סובב בעצמות עובי גלגל אחר, כי אלו המניעיות יתחייבו ממנו [כל שכן],[13] שכבר הכריח אותם ברוב המקומות לשום השני עקרים [],[14] ושתנועת הכוכב סביב מרכז גלגלו והכוכב נשוא בזאת התנועה בגלגל אחר יוצא המרכז ממנו. וזה המאמר יספיק בבטול []‏[15] אלו השני גלגלים בשמים. אבל מה שאמרו בבטול היות שתי תנועות הפכיות בשמים לפי שתנועתם טבעית ולמניע אחד []‏[16] פשוט, מעצם מאמרם זה יתבאר אחד משני דברים. אם שלא יהיה בתנועה הסביבית הפך ר״ל שלא תהיה תנועה סבובית הפך תנועה סבובית, או שיצאו בהכרח שתי תנועות הפכיות מזה המניע האחד הפשוט. אם נודה שיש בה הפך מצד שהתנועות טבעיות ולמניע טבעי אחד פשוט וזה יתבאר בזה [המופת].[17]

0.5 כי מסגולת הכדור שאיפשר לו להתנועע סביב כל שני קטבים יצויירו בשטחו. ונשים דרך משל כי קטבי גלגל נוגה מונחים על עגולת חגורת המזלות, האחד מהם בנקודת ראש סרטן והשני בנקודת ראש גדי. ונשים כי גלגל המזלות נעתק סביב קטביו ממזרח למערב, יועתק עמו גלגל נוגה וקטביו. ונשים כי גלגל המזלות נעתק סביב קטביו ממזרח למערב, ויעתק עמו גלגל נוגה וקטביו. ונשים ג״כ שגלגל נוגה נעתק לצד זה סביב קטביו. ואין הפרש בין זאת ההנחה להנחת גלגל המזלות עם קטבין

6. ו: אלמוצׄע.
7. ו: -.
8. ו: כל מא.
9. ו: -.
10. ו: בעד קליל.
11. ו: -.
12. ו: גׄמיעאׄ.
13. ו: פכיף.
14. ו: גׄמיעאׄ.
15. ו: כון.
16. ו: טביעי.
17. ו: אלוגׄה.

נעתק בתנועת היומית ונעתק ג̇כ̇ סביב קטבין. [א̇כ̇ ידוע][18] שכל זמן שהקטב האחד ר̇ל̇ הצפוני מגלגל נוגה נעתק מראש סרטן עד ראש טלה בהעתקת גלגל המזלות, והקטב הדרומי הרביע המקביל, יהיו שתי התנועות ר̇ל̇ תנועת גלגל המישור ותנועת גלגל נוגה לצד אחד.

0.6 ובעת חנות הקטבים בנקודות השווי, יהיו שתי התנועות מתחתכות על זויות נצבות. וכל זמן שהקטב הצפוני נעתק מראש טלה עד ראש גדי, והשני הרביע המקביל, יהיו התנועות הפכיות לפי דעתם. מי יתן אדע איך יהיו התנועות המתדמות הטבעיות ההולכות על סדר אחד בחצי הזמן מתדמות לצד אחד ובחציו הפכיות? אבל התנועה האחת בעצמה תהיה הפך עצמה בזמנים מחולפים עם היותה מתדמה תמיד? זה בתכלית הזרות, וזה המופת נלקח מצד משיגי התנועה. א̇כ̇ כבר נתבאר מזה כי תנועת סבובית לא תהיה הפך תנועה סבובית. ואפילו נודה זה, על שהוא נמנע כבר, נתבאר מזה כי אלו השתי תנועות שהם הפכיות לפי סברתם הם מצד הכרח שמירת זה המניע הטבעי האחד הפשוט סדרו תמיד. ואם היה חק זה המניע שלא יצאו ממנו תנועות הפכיות, יחויב כשיגיעו הקטבים לנקודות השיווי שתעמד התנועה וזה נמנע.

0.7 אחר שהתבאר זה א̇כ̇ הנמנעות הנמשכות להם אמנם הוא מצד שהם מניחים גלגל יוצא המרכז וגלגל ההקפה בלבד, ובזה מן המניעות זה השיעור שתראה. והתמה מרוב מה שהמציא מאלו העקרים בפרט ראש [][19] החכמה בטלמיוס ומה שהרויח לבאים אחריו מזאת החכמה היקרה עד היות נקל למי שירצה מציאות זמני חזרות הכוכבים וידיעת מקומותם בגלגליהם באורך וברוחב, וידיעת שעורי גדלם וערך קצתם אצל קצת וערכם אצל הארץ, ובעתי לקות מה שילקה מהם ושיעור הנלקה מגופם בכל זמן, עם היות היסודות אשר עליהם הבנין בזה הרחוק מהציור בחקם, עם כל זה אין ראוי לחשוב [כאיש][20] כמו בטלמיוס בשמו אלו העקרים שהוא סובר כי הענין הוא כך, בעבור היות מה שיראה

126b

לחוש נמשך למה שיצא מאלו העקרים.

0.8 ואני תמה מאד ממאמר הפלוסוף הגדול רבי משה ע̇ה̇ בפר̇ כ̇ד̇ מהחלק השני ממורה הנבוכים כשהביא מספר הנמנעות המתחייבות לאלו השני עקרים. אמר [ואיך ימצאו לכוכבים אלו התנועות המחולפות [ואם יש][21] שם צד [יהיה איפשר בהנחתו][22] שתהיה התנועה הסבובית שווה שלמה ויראה בה מה שיראה [זולתי][23] באחד משני העקרים או בשניהם יחד],[24] עד שאמר כי האל יתעלה נתן יכולת לאדם לדעת מה שתחת השמים בלבד. ואמנם צורות השמים ותנועותם הוא לבדו היודע. [ועל דומה לזה][25] אמר דוד המלך ע̇ה̇ השמים שמים [להש̇][26] וגו̇. ונשוב למה שהיינו בו.

18. ו: פאדא כאן דאלך כדלך פט̇אהר אנה.
19. ו: הדא.
20. ו: -.
21. תרגום אבן תיבון: או שיש.
22. תרגום אבן תיבון: שיתכן עמו.
23. תרגום אבן תיבון: אלא.
24. תרגום אבן תיבון: 50b–51a.
25. ו: ולדלך.
26. ו: לש̇.

0.9 ולהיות הדברים נראים לחוש כפי מה שחבר אותם [הודו הכל באלו העקרים].[27] ולזה היה שער החקירה על העקרים האמתיים סגור זמן רב. והיה זה מונע מהשלים זאת החכמה הנכבדה עד שבא אלבטרוגי המתעורר על אלו הנמנעות המתחייבות מאלו העקרים. וחשב כי חדש תכונה לאלו התנועות והחלופים מבלתי הנחת גלגל יוצא המרכז ולא גלגל הקפה. וזכר כי הוא נתעורר לזה מעט שמעו לאבי בכר בן אלטופיל שהוא פגש תכונה ועקרים מבלתי אלו המונחים, אלא שהוא נמשך בזה לכת הטבעיים בהרחקת התנועות ההפכיות. והתחיל בהזכרת איכות התנועות ואיך יהיה כל אשר בשמים מתנועע לצד אחד ויהיה מה שיתחייב ממנו דבק אל הנראה לחוש. ועד זה השיעור מדבריו הוא אמת, ר״ל שאיפשר להיות כמו שאמר.

0.10 ואמנם כשהביא הסבות המחייבות לחלופים לא ידובק דבר ממה שיתחייב [מעקריו לדבר מה שיתחייב][28] מאלו העקרים המונחים עד עתה ר״ל למה שיראה לחוש, זולתי אם היתה כונתו להודיענו איך נמצא חלוף מה באלו התנועות מבלתי החלוף הנראה לחוש. ואני חושב כי זאת היתה דעתו ואם לא לאיזה דבר היה צריך לחדש עיון כמו שאמר. וכבר יתבאר למעיין בספרו ממקומות אחרים רבים כי זאת כונתו [][29] וזו כונה בלתי ראויה, לפי שהכניס אותנו במבוכה ונמנעות יותר גדולות ויותר זרות מאשר היינו בהם. ועם מה שאמרנו מחסרון עקריו בזה ורוב רחוקם מן האמת כמו שיתבאר [][30] בהגיענו לאותם המקומות בעזרת השם, ראוי לשבחו הרבה להיותו ראשון מי שהתעורר על זה וחבר בו ספר.

0.11 ואני מעת ראותי ספרו לא הסירותי זמן גדול מחשבתי בזה עד שהגעתי לאמיתת העקרים המחייבים החלופים באלו התנועות כמו שיתבאר כל זה [בה׳][31] יתעלה. ודע כי כונתינו בזה הספר אמנם הוא לשום עקרים לחלופים במקום העקרים הנזכרים, ר״ל היוצא המרכז וגלגל ההקפה. ואמנם ענין התנועות ההפכיות ראוי היה להניחם כמו שהם על דעת בטלמיוס למה שנתבאר מהעדר ההפכות בהם. אבל מצד התפשטות זה הדעת אצל ההמון, ר״ל כי בתנועה הסבובית הפך, ראינו לשום העקרים לתנועות האלה ואם על שיהיו אצל הצד אשר הוא אצלם אחד, זולתי במקום שיכריח הענין להניח תנועה כמו התנועה אשר שמנו במשל הקודם, ר״ל התנועות אשר יחושב שהן אחת בזמן והפכיות בזמן אחר עם היותן תמיד מתדמות הולכות על סדר אחד.

0.12 ונאמר כי הדבר אשר הכריחם לשום התנועות אשר הם אצלם הפכיות אמנם הוא העתקת השמש והירח ושאר הכוכבים לצד הצפון והדרום. ולולי זה לא יהיה להם דבר יחייב שתהיה העתקתם לצד

126a

המזרח תנועה להפך יותר מהיותה קצור, וכבר זכר בטלמיוס זה במאמר הראשון מספרו. א״כ איזה דבר ימנע [מלהיות][32] כל אחד מהגלגלים אשר תחת העליון מתנועע לצד תנועת העליון נמשך לתנועתו ועל קטביו, ויהיה מקצר בזאת התנועה [][33] העליון לחולשת הכח היורד עליו מהעליון בעבור

27. ו: צארת הדה אלאצול מתסלמה מן אלג׳מיע.
28. ו: -.
29. ו: בעינה.
30. ו: בעד.
31. ו: פימא בעד אן שא אללה.
32. ו: מן אן נקול אן.
33. ו: ען חרכה.

רחוקו מן המניע הראשון? ואין בטבעו לקבל יותר מזה לפי שאינו נשוא בזאת התנועה בעליון על ששניהם דבר אחד ומתנועעים תנועה שוה.

0.13 ויתנועע [][34] על קטביו המיוחדים בו לזה הצד כאלו כונתו להשלים מה שקצר מן העליון בתנועה הראשנה לפי שכונתו ותכליתו אמנם הוא כונת העליון והתדמנותו בו לפי שהוא תכליתו. ובזאת התנועה יעתק לצד הצפון ולצד הדרום בשעור מרחק קטביו מקטבי העליון. ועם זאת התנועה גֿכֿ יחסר לכל גלגל מהם שעור מוגבל מהגיע למקום העליון וההתדמות בו, וזה בשעור מרחק כל אחד מהם מן המניע הראשון המקנה לו התנועה.

0.14 לזה כל מה שיהיה הנחתו יותר קרוב לעליון יתמעט קצורו האחרון. ובזה המאמר יסתלק המחלוקת אשר בין בעלי זאת החכמה בסדר הגלגלים לפי שהוא ידוע כי סדורם סבת קצורם. אֿכֿ כל אחד מהגלגלים אשר תחת העליון יקצרו קטביו המיוחדים בו בכל סבוב שעור מוגבל בעגולת עברתם וזה בשעור מרחק כל אחד מהם מן המניע הראשון. אבל הכוכבים שהם על אלו הגלגלים קצורם פחות מזה בשעור תנועת כל אחד [מהם][35] על קטביו עד שגלגל הכוכבים הקיימים ישלים בתנועתו המיוחדת לו מה שקצר גלגלו אצל כת, או ישאר לו קצור מעט על דעת אחרים. ואשר תחתיו יקצר יותר מזה עד שיגיע לגלגל הירח התחתון אשר קצרו יותר מקצור כל אחד ממה שלמעלה ממנו לרוב רחוקו מהעליון.

0.15 אֿכֿ הגלגל המניע התנועה היומית והוא המושג בשכל לפי שאין לו מקדים יושג בחוש מחמתו לפי שהוא פשוט שלם הפשיטות גלוי מעניינו שהוא יותר עליון מכלם לפי שתנועתו אחת פשוטה כוללת הכל. ואמנם שאר הגלגלים אשר תחתיו אין תנועתם פשוטה עד שגלגל כוכבי שבת יש לו שתי תנועות כמו לשאר הגלגלים אשר תחתיו. וגֿכֿ רוב כונתינו לבאר כי אין בכאן דבר יחייב להיות בשמים גלגל זולתי אשר ימשש עקמימותו לגבנונית אשר תחתיו משוש שלם, עם היותו כדורי שלם העגול מבחוץ ומבפנים וגבנויתו ממשש לעקמימות אשר למעלה ממנו, אם יש [שם][36] גלגל אחר יהיה מספר הגלגלים מה שיהיה. כי זה לא יזיק. וזה הראוי למעיין בספרנו זה לבקש אותו מזה הספר כי הוא ימצא אותו מבואר בענין שאין בו ספק, וזה הענין אשר נלאו השכלים מהשיג אותו [עד עתה].[37] ונאמר כי ואם היה ראוי כפי זה העיון להתחיל במאמר בגלגל העליון ולהגיע בעיון עד התחתון, שמנו ההתחלה מתנועת השמש [להמשך][38] לסדר העקרים הנהלכים עד עתה לבעבור יתישר למעיין [השמש היו שני מיני העקרים][39] ויראה לו החלוף בין הדרכים הנהלכים בקלות סבול השכל אותם ובקרובם מהאמת.

0.16 לפי [כונתינו][40] בזה הספר לדבר עם מי שקדם לו העיון בספרים המונחים בזאת החכמה עד עתה ואם אינו צריך בקריאתו לאלו הספרים. ואני יודע כי מי שקדם לו העיון בדעות המוצעות יסודות וסבות לזאת החכמה מבטלמיוס וזולתו והוציא זמנו בהם. כאשר יעיין בספרינו זה יקרה לו במקרה המתמיד לישב בחשך שאינו יכול לצאת פתאום לאור. ואמנם כשידריג בו יהנה ממנו

34. ו: איצֿא.
35. ו: -.
36. ו: פוקה.
37. ו: אלי הדה אלגאיה.
38. ו: -.
39. ו: אלתמיז בין הדה אלאצול ובין תלך.
40. ו: כלאמנא.

125b

וישיג יתרון האור מן החשך ועל [היוצא][41] בזה [אמר][42] הנביא עה העם ההולכים בחשך ראו אור גדול, יושבי בארץ צלמות אור נוגה עליהם.

0.17 ודע כי כשנאמר כי מרכז כוכב פלוני מתנועע על עגולה פלונית קטבה על עגולה פלונית, כי כל עגולה מהם נרשמת בשטח גלגל מיוחד מן הגלגלים אשר על התאר הנזכר. וגכ כשנקרא לתנועת הכוכב המיוחדת לו קצור כי זה [][43] לישב העקרים אפילו על דעת הסוברים כי בתנועה הסבובית הפך, וכבר ידעת דעתינו בזה. והראוי על דעתינו שתקרא מהלך לא קצור ואתה קח לך הטוב בעיניך משני הדעות ואחר שעקרינו בחלופים יסכימו עם שני הדעות.

0.18 וחלקתי זה הספר לשני כללים. הכלל הראשון אכלול בו כל ההקדמות הלמודיות וזולתם הראויות להקדימם על הספור באיכות התנועות וכולם משותפות בין בטלמיוס ובינינו לפי שלא יתערב עמהם איכות תנועה [].[44] נזכור אותם בדרך היותר קצר והיותר קרוב אל השכל ובזה הכלל חמשה פרקים. הפרק הראשון בהקדמות המושלמות והראיות להקדימם לזאת החכמה והנולד מהם. הפרק השני לכל הקדמות [הלמודיות][45] אשר יאות להקדימם קודם הספור בתנועת הכוכבים. הפרק השלישי [בכלים][46] אשר אנו צריכים אליהם לעיונים אשר בהם תשלם זאת החכמה. הפרק הרביעי בידיעת הנטיות החלקיות והמצעדים בכדור הישר והנוטה ודברים אחרים חלקיים ימשכו מהם. הפרק החמישי בשיעור זמן שנת השמש. הכלל השני אזכור בו איכות התנועות והעקרים אשר עליהם יאות לעשות בחלופים המורגשים לגרמים השמימיים בתנועותיהם ומקומותם, ולזה היתה תחלת המחשבה בזה הספר. ומכאן אתחיל לזכור מה שיעדתי אליו.

A.I.1 הפרק הראשון מן הכלל הראשון ההקדמות המושלמות הראויות להקדימם לזאת החכמה. הם כד הקדמות. א כי השמש והירח ושאר הכוכבים יתנועעו תמיד ממזרח [][47] על עגולות הנכוחיות, זורחים ועולים עלינו ויורדים ונסתרים ממנו זמן מה ושובים למקום זריחתם כבראשונה וזה בהתדמות תמיד. ב כי הקטב הצפוני גבוה על האופק והדרומי עמוק ממנו בכל אחד מהנטיות בחצי הצפוני מן הארץ, וזה [בשיעור][48] נטיית כל אפק מקו השוה לצד הצפון זולתי במקום שהכדור ישר. וכי הככבים אשר מרחקם מהקטב הצפוני פחות מהגובה לא ישקעו תמיד אבל יסובבו תמיד הקטב על עגולות נכוחיות שהן בשיעור מרחק הככבים הסובבים עליהם מהקטב. ג כי הככבים הקרובים אל הנגלים תמיד ישהו בהסתרה פחות מן הרחוקים מהם. ד כי גודל הככבים ואורם תמיד על עניין אחד אלא שהם גדלים אצל השקיעה וזה בסבת הבל לחות [הארץ][49] הנכנס אז בין הראות ובינם. ה שהם מתחתכים בשטח הארץ כשהם שוקעים. ו כי מרחקי הארץ למקומות העליונים הם שווים [וזאת תתבאר בהקדמה הרביעית וכאן הן הקדמה אחת].[50] ז צורת הגשם יאות שתהיה נמשכת לתנועתו. ח כי תנועת השמים

41. ו: מתל.
42. ו: -.
43. ו: מנא.
44. ו: אלבתה.
45. ו: אלהנדסיה.
46. א: בכל יום. ו: פי אלאת.
47. ו: אלי אלמגרב.
48. ו: -.
49. ו: -.
50. ו: -.

היא היותר קלה מכל התנועות והיותר מהירה. טֹ כי הגשם היותר קל התנועה הוא הכדורי. יֹ כי התמונות המחולפות אשר הקפם שוים

125a

מי שיש לה רב זויות תכיל יותר. יאֹ כי גשם השמים ראוי להיות יותר גדול ממה שזולתו. יבֹ כי גשם הגלגל יותר דומה קצתו לקצתו מכל הגשמים. יגֹ כי הגרמים השמימיים כמו הכוכבים יראו מעוגלים מצדדים מחולפים זמן אחד אצל כל הרואה אותם [וטבע המקיף אותם דומה לטבעם].[51] ידֹ כי ההקשים בכלים [לא][52] יסכימו זולתי בזה הצד, [רצוני לומר בשום התמונה כדורית.][53] טוֹ כי זריחת כל הככבים ושקיעתם אצל המזרחיים קודם לזריחתם ושקיעתם אצל המערביים [][54] וזה כי כתב [מי שכתב][55] מן המזרחיים זמן לקות ירח שהוא בזמן אחד אצל הכל [ימצא][56] יותר מתאחר מחצי היום מן הנכתב מן המערביים. [יוֹ][57] כי זה בשעור חלוף המרחק אשר ביניהם. יזֹ כל ההולך בים לנוכח הר או מקום גבוה כל מה שיתקרב אליו יראה אותו נוסף [בגובהו][58] וכאלו הוא צומח מן המים. יחֹ כי בכל מקומות הארץ ישתוו היום והלילה שתי פעמים בכל שנה וזה השווי הוא אצל המעבר האמצעי בין שני ההפוכים רלֹ בין היום היותר קצר והיותר ארוך. יטֹ כי חצית היום הוא תמיד בהיות השמש באמצע השמים רלֹ בעגולה הקדקדית הנצבת על גלגל המישור. כֹ כי כל אחד מן האפקים רלֹ השטח היוצא מן הראות יחתוך הכדור לשני חצאין וזה לפי שאנחנו נראה תמיד ששה מזלות נגלות ואחר כן יעלו הששה הנסתרות ויסתרו הן. כאֹ כי צל העמוד המזרחי עם צל העמוד המערבי יפלו שניהם על קו אחד ישר בהשואת היום [והלילה].[59] כבֹ כי הלקיות הירחיים בכל קצות השמים אמנם יהיו אצל הקבלה על הקטר. כגֹ כי משפט עמוד הצל באיזה צד הושם משטח הארץ כמשפטו אם הושם במרכז הארץ וגֹכֹ מרכזי הכלים אשר ישתמשו בהם בזאת החכמה. כדֹ כשיתנועע האדם בתנועה יותר מהירה משאר התנועות הנמצאות יתחייב שלא יראה דבר מתנועע לצד תנועתו, אבל יראה כל דבר כמו מתנועע להפך תנועתו.

A.I.2 ואחר הנחת אלו הדברים, מבואר ממה שאמרנו כי השמים כדוריים ותנועתם גֹכֹ כדוריית וזה יתבאר בהקדמה הראשונה ובבֹ ובגֹ ובדֹ ובהֹ ובוֹ ובזֹ ובחֹ ובטֹ וביֹ וביאֹ וביבֹ וביגֹ ובידֹ. וההקש אמנם ישלם אם בהקדמה אחת מאלה על שהשנית ידועה ואם בשתי הקדמות יחד כחֹ עם הטֹ והיֹ עם הי[א].[60] ומקצת אלה ההקדמות יתבאר הדרוש כי התנועה כדורית מצד מקרים ישיגו אותם ומקצתם יתבאר זה מצד שהצורה כדורית מצד איכות התנועה. ומקצתם יתבאר הדרוש בהקש ישר ומקצתם בהקש החלוף וכל זה מבואר בהן. ויתבאר כי הארץ בכל חלקיה כדורית אצל החוש בהקש אל הכל בהקדמה הבֹ ובטוֹ וביוֹ וב[יזֹ].[61] [ויתבאר כי הארץ באמצע השמים בהקדמה הדֹ ובהֹ ובכֹ ובכאֹ ובכבֹ][62] ובקצת אלו

51. ו: -.
52. ו: -.
53. ו: -.
54. ו: יוֹ.
55. ו: -.
56. ו: -.
57. ו: -.
58. ו: -.
59. ו: -.
60. ו: זֹ.
61. ו: כבֹ.
62. ו: -.

ההקדמות יתבאר שאינה מוסרת לא למעלה ולא למטה. ובקצתם שאינה מוסרת לצד מערב או למזרח ובקצתם שאינה מוסרת לצד אחד מן הקטבים. ובזה יתבטל דעת המורכב מאלה, ויתבאר כי הארץ אין לה תנועת העתקה בהקדמות אשר נתבאר מהם כי הארץ באמצע. ויתבאר כי אינה מתנועעת סביב מרכזה

124b

והשמים [שוקטים][63] כמו שיחשבו אחרים בהקדמה הכדׄ.

A.I.3 ויתבאר כי מיני התנועות הראשונות בשמים שתים לפי שאנו מוצאים כל מה שבשמים יזרח וימצע השמים וישקע בכל יום על עגולות נכוחיות אשר היותר גדולה מהם [גלגל][64] המישור לפי שהאופק יחתוך אותה לשני חצאין וזוהי התנועה הראשונה המסבבת הכל בשוה. ונמצא השמש והירח והחמשה הנבוכים מתנועעים לצד המזרח, רׄלׄ הפך הראשונה, תנועות מחולפות או מקצרים על אחת מהדעות כמו שזכרנו, ובהעתקתם לצד מזרח או קצורים יעתקו גׄכׄ לצד צפון ולצד דרום העתקה מתדמה. [ולפי זה][65] גזרנו כי זאת התנועה השניה היא על קטבי עגולה נוטה מגלגל המשור ותקרא חגורת המזלות ותנועת השמש רושמת אותה ועליה. ומשני צדיה מעבר הירח וחמשה כוכבים הנבוכים לצד צפון ולצד דרום. ונחשוב עגולה עוברת בקוטבי הכל ובקטביה, והיא תחתוך אותה על זויות נצבות על שתי נקודות, הצפוניות מגלגל המישור תקרא הפוך החום והשנית הפוך הקור. ונקודת חתוך הנוטה עם גלגל המישור אשר המעבר בה לצפון תקרא השואת הקיץ והשנית [השואת][66] החורף, וקטבי הנוטה קיימים במקומם מן העוברת בקטבים עם שהם מתנועעים בתנועה הראשונה. ועגולת חצי היום היא עגולה נצבת על האפק תמיד עוברת בקוטבי הכל.

A.I.4 ואחר שהשלמנו זה נתחיל במופתים החלקיים ונבנה הענין על מה שהסכימו עליו הכל והוא חלוק מקיף העגולה לשׄסׄ חלקים וחלוק האלכסון לקׄכׄ חלקים לקלות אלו המספרים אצל החשבון, וגׄכׄ חלוק החלק לסׄ דקים והדקה לסׄ שנים וגׄכׄ לאין תכלית עד שלא יזיק הנחת החלוקה.

A.II.1 הפרק השני בכל ההקדמות הלימודיות הראויות הקדימם קודם הספור בתנועות הככׄבים. ונקדים ראשונה המאמר בידיעת שיעורי יתרי חלקי העגולה. ונקדים לזה כשרצינו לחלק מספר ידוע על הערך הנקרא בעל אמצע וקצוות נקח מרובע ומרובע חציו ונקבצם יחד. ונקח גדר הנקבץ ונגרע ממנו חצי המספר והנשאר הוא החלק הגדול. ונשים המשל בקוים. ויהיה המספר הידוע בׄגׄ ויהיה אׄגׄ שוה לחצי גׄבׄ ויקיף עם קו גׄבׄ בנצבה. ונדביק אׄבׄ ונגרע מן בׄאׄ חצי בׄגׄ ויהיה בׄדׄ. ונחתוך מן בׄגׄ כמו דׄאׄ ויהיה גׄהׄ, אומר כי גׄהׄ החלק הגדול. המופת לפי שהכאת בׄגׄ בעצמו אבל בבׄהׄ ובהׄגׄ וגׄאׄ בעצמו כמו בׄאׄ בעצמו אבל בׄדׄ בעצמו ודׄאׄ בעצמו וכפל בׄדׄ בדׄאׄ. נפיל בׄדׄ בעצמו רׄלׄ גׄאׄ בעצמו. ישאר בׄגׄ בבׄהׄ ובהׄגׄ כמו דׄאׄ רׄלׄ הׄגׄ בעצמו וכפל בׄדׄ בדׄאׄ רׄלׄ בׄגׄ בהׄגׄ. ישאר בׄגׄ בבׄהׄ כמו הׄגׄ בעצמו ומׄשׄלׄ. וזה בעצמו כבר ביארו אקלידס בשני מספרו במופת כחו כח זה המופת.

A.II.2 ואחר שהקדמנו זה אׄכׄ כאשר נדע שעור צלע המשושה הנופל בעגולה אשר הוא יתר קשת סׄ חלקים, והוא ידוע לפי שהוא חצי האלכסון אשר בהקש אליו נרצה לדעת שעורי היתרים, נדע [גׄכׄ][67] שעור צלע המעושר אשר בזאת העגולה, והוא יתר קשת לׄוׄ חלקים לפי שהוא החלק הגדול ממנו כמו

63. ו: -.
64. ו: -.
65. ו: -.
66. ו: -.
67. ו: -.

שנתבאר בטו מאקלידס. ונקדים גכ כשיהיה בעגולה תבנית בעל ארבע צלעות כגדבא יהיה הכאת בד בגא כמו הכאת דא בגב ובא בגד יחד. ונשים זוית אבה כמו דאג אם יהיה אבד גדלה ממנה, אם כן אבד כמו הבג ובדא כמו [גדה][68] אכ משולשי אבד בגה מתדמים.

124a

וערך בג אל [גו][69] כבד אצל דא אכ הכאת בג באד כמו הכאת בד בגה. וגכ משולשי אהב בגד מתדמים תהיה הכאת בא בדג כמו הכאת בד בהא. אכ הכאת בד באג כלו כמו הכאת אב בדג ואד בגב יחד. אכ כשיהיו אב אג ידועים יהיה בג אשר ביניהם ידוע לפי שבד גד ידועים לתשלום חצאי העגולות. ואד האלכסון ידוע יהיה גב ידוע. אכ כבר נדע יתר כד חלקים מן המותר אשר בין צלע המשושה וצלע המעושר.

A.II.3 ואומר גכ כי כאשר נדע יתר קשת אג נדע יתר חציה ויהיה אד לפי שאה חציו ידוע ואז ידוע יהיה הז ידוע. וישאר הד הידוע ואד גכ. אכ כבר נדע יתר יב חלקים ויתר ששה ויתר שלשה ויתר חלק וחצי ויתר חצי ורביע [].[70] ויצא בזה הדרך כי יתר חלק וחצי יהיה חלק ולד דקים וטו שניים ויתר חצי ורביע [מח][71] דקים וח שניים. ונדע בזה כל היתרים העודפים בחלק וחצי בשנתחיל מן המותר בין חלק וחצי ובין חצי העגולה.

A.II.4 [ואחר זה][72] נשתדל להמציא יתר חלק אחד מצד [ידיעת][73] חלק וחצי [וחצי][74] ורביע לפי שאי אפשר להביא מופת על זה על דרך הנדסי. ויהיה בזה התאר. אומר כשנוציא בעגולה שני קוים בלתי שוים כי ערך [היותר ארוך][75] אצל היותר קצר קטן מערך הקשת אשר על הארוך אצל הקשת אשר על הקצר, וזה כבר נתבאר [בספר][76] אקלידס. ונשים אג יתר חלק ואד יתר חלק וחצי ורביע. יהיה קו אג פחות מקו אד ושלישית [יחד][77] לפי שקשת אג שוה לאד ושלישיתו. אכ קו אג פחות מחלק ושני דקי ונ שניים ומ שלישיים. וגכ נשים אד יתר חלק ואג יתר חלק וחציו, יהיה קו אג פחות מקו אד עם חציו, אכ קו אד יותר מחלק ושני דקי ונ שניים רל שני שלישי אג והפחות והיותר יחלקו [המ][78] שלישיים. אכ יתר חלק מקשת הוא חלק ושני דקים ונ שניים וכ שלישיים. ולפי זה כבר ידענו כל היתרים העודפות בחצי חלק לפי שידענו יתר חצי העגולה רל האלכסון. וידענו יתר חצי חלק, אכ כבר נדע יתר קעט חלקים וחצי מן המותר אשר ביניהם, וגכ נדע השאר בגרוע חצי חלק. ונדע גכ בקע כל קשת לפי שהבקע הוא חצי יתר כפל הקשת ומשל.

A.II.5 ונקדים גכ כשיפלו בעגולה כעגולה אב שלש נקודות כאבג על שתהיה אחת מקשתי גב בא פחות מחצי עגולה, ונדבק אהג ודהב מן המרכז, אומר כי ערך בקע קשת אב רל אס הנצב [וגם][79] דב אצל

68. א: בגה.
69. ו: גז. Euclid, *The Elements*, vol. 2, p. 225, reads GE.
70. ו: בתנציף.
71. ו: מז.
72. ו: אלאן.
73. ו: -.
74. ו: ומן קבל גז.
75. א: יותר הארוך.
76. ו: -.
77. ו: -.
78. ו: -.
79. ו: עלי.

בקע קשת בג רל גח הנצב גכ כערך אה אצל הג לפי שהמשולשים מתדומים. [שתי קשתות זכ כל מעגולת זכל הקטנה דומות לשתי קשתות אה הד מעגולת אהד הגדולה כל אחת לחברתה. אומר כי ערך בקע קשת אה אל בקע קשת הד כערך בקע קשת זכ אל בקע קשת כל, וכי ערך בקע קשת אה אל בקע קשת אד כלו כערך בקע קשת זכ אל בקע קשת זכל. ונשים מרכז השתי עגולות מ ונוציא קוי מזא מכה מלד ונוציא בקעי אב דג זח לט אז זכ. ערך אב אל זח כערך אמ אל מז רל דמ אל למ אבל דג אל לט. ובחלוף אב אל דג כזח אל לט וגכ ערך אב אל זח כערך אמ אל זמ

123b

אבל או אל זנ. ובחלוף מכ אל מו כזח אל זנ וזה מה שרצינו.][80]

A.II.6 ונקדים גכ מה צריך להקדימו למופתים על העניינים הכדוריים. כל שתי עגולות גדולות בשטח כדור כאדבג ואלב יחתך מן אחת מהן שני קשתות כאד אג מן אחת מנקודות החתוך, וכל אחת מהן פחות מחצי עגולה. ויצאו מנקודות דג שני עמודים על שטח העגולה האחרת. אומר כי ערך [בקע קשת אד אצל][81] בקע קשת אג כערך העמוד היוצא מנקודת ד אצל האחר. המופת יהיה החתוך המשותף לשתי העגולות רל אלכסוניהם קו אב ונוציא עליו עמודי דז גה והם בקעי הקשתות. ואם היו גכ עמודים על שטח עגולת אלב כבר בארנו מה שרצינו. ואם לא [][82] נוציא מנקודות דג עמודי דט גח על שטח עגולת אלב. ונדביק הח טז ולפי שקוי דז גה נכוחיים, וגכ דט גח, יהיו זויות זדט הגח שוות וח וט נצבות. אכ המשולשים מתדמים וערך דז אצל גה רל הבקעים כערך דט אצל גח רל העמודים ומשל. וגם כן יתבאר אם נחתכו הקשתות מצד אחד מנקודת החתוך.

A.II.7 ואחר שקדם זה יתחתכו בין שתי קשתות אב אג מעגולות גדולות והם בשטח כדור קשתי גד בה מעגולות גדולות גכ על נקודת ז. אומר כי ערך בקע קשת גה אצל בקע קשת הא מחובר מערך בקע קשת גז אצל בקע קשת זד ומערך בקע קשת בד אצל בקע קשת בא. ונוציא מנקודות גאד עמודים על שטח עגולת בזה והן גל אח דט. ולפי שכל שלשה קוים, ערך הראשון אצל השלישי מחובר מערך הראשון אצל השני ומערך השני אצל השלישי, ונשים עמוד דט אמצעי בערך בין גל אח. יהיה ערך עמוד גל אצל עמוד אח [][83] ערך בקע קשת בה אצל בקע קשת הא מחובר מערך גל אצל דט [][84] ערך בקע קשת גז אצל בקע קשת [זד][85] ומערך דט אצל אח רל ערך בקע קשת בד אצל בקע קשת בא. וגכ יתבאר כי ערך בקע קשת גז אצל בקע קשת זד מחובר מערך בקע קשת גה אצל בקע קשת הא ומערך בקע קשת אב אצל בקע קשת בד כאשר נשים עמוד אח אמצעי בין גל ודט. וגכ יתבאר בהרכבה כי ערך בקע קשת גא אצל בקע קשת אה מחובר מערך בקע קשת גד אצל בקע קשת דז ומערך בקע קשת זב אצל בקע קשת בה בשנוציא העמודים מנקודות אהז על שטח עגולת בדא. וגם כן יתבאר ההרכבה בכל אחת מן הצלעות. [ומן כאן נתבאר כי בערך ההרכבה אם היתה קשת בזה מעגולה קטנה קטבה על עגולת אדב, אשר עליה נפילת העמודים, כי זה בעצמו יתחייב. רל כי הערך יהיה מחובר כפי הנזכר כמו שהיה לו היתה קשת בזה מעגולה גדלה. וענין זה נצטרך אליו לעתיד. וגכ בערך הפירוק לא יתחייב בהכרח שתהיינה אלו

80. ו: -.
81. ו: -.
82. ו: יכונא כדלך.
83. ו: אעני.
84. ו: אעני.
85. ו: גד.

123a

הארבע קשתות כולם מעגולות גדולות כי זה בעצמו יתחייב כשנשים קשת מאלו הארבעה מעגולה קטנה אם לא תהיה זאת הקשת מן העגולה אשר עליה נפילת העמודים. אבל קטבה יהיה על הקשת מהעגולה אשר עליה נפילת העמודים בהכרח.][86] וזה המופת אמנם הוא על שיהיה כל צלע עם חברו בקשת אחד. ויתבארו שאר הערכים אשר יתחברו מאלה הששה בקעים במופת אשר הביא לזה אבן מעאד אלגיאני והוא זה.

A.II.8 אומר כי הכאת הבקע הראשון ברביעי והנקבץ בששי רל המוגשם אשר יקיפו בו אלו השלשה בקעים שוה להכאת השני בשלישי והנקבץ בחמישי. ונשים קו גב שוה אל הבקע הרביעי ובח שוה לששי, יקיף עם גב בנצבת. ונוציא מנקודת ב קו בא נצב על שטח השני קוים, ויהיה שוה אל הבקע הראשון, ונשלים שטחי גח גא [אח][87] הנכוחיות הצלעות יקיפו במוגשם דח. ונעשה מוגשם [מס][88] גכ על שיהיה קו מל שוה לבקע השלישי, ולס שוה לחמישי ולכ שוה לשני. ולפי שערך שטח מס אצל שטח גח מחובר מן למ אצל גב ומן [לס][89] אצל בח יהיו תושבות מס גח [מתהפכות בערך][90] עם הגובהים אכ המוגשם שוה למוגשם.

A.II.9 ואחר שהקדמנו [זה][91] נאמר כי ערך איזה צלע לוקח מן המוגשם האחד אצל איזה צלע לוקח מן האחר מחובר מן הארבעה הנשארים, המשל כי ערך בא והוא הבקע הראשון, אצל למ והוא הבקע השלישי רל בקע קשת גז בצורה שעברה מחובר מערך קו לכ והוא השני, אצל [][92] בג, והוא הרביעי, ומערך [לס][93] והוא החמישי אצל בח והוא הששי. וזה לפי שערך שטח כס אצל שטח גח, אבל מערך קו בא אצל קו למ, [להפוך הערך],[94] מחובר מאלו הארבעה ערכים הנזכרים. וכזה יתבאר השאר. ואמנם כשילקח עם חברו מזה המוגשם לא יתחבר מן השאר ומשל.

A.II.10 ונאמר כי כל ארבעה שיעורים הערך המחובר מן הראשון אצל השני מן השלישי אצל הרביעי הוא בעצמו הערך אשר יתחבר מן הראשון אצל הרביעי ומן השלישי אצל השני. המשל יהיו שני שטחים נכוחיי הצלעות, והם שטחי [כג][95] אז, זוית ב משותפת להם. וערך שטח אז אצל שטח דג מחובר מערך קו אב הראשון אצל בג השני ומערך זב השלישי אצל בד הרביעי. אומר שהוא מחובר גכ מערך אב הראשון אצל בד הרביעי ומערך זב השלישי אצל בג השני. המופת נוציא בז על ישר עד שיהיה שוה לבא ויהיה בה ונשים בח שוה לבז ונשלים שטח חה. יהיה ערך שטח חה רל אז אצל שטח דג מחובר מערך צלע הב רל אב אצל דב ומערך חב רל בז אצל בג ומשל.

A.II.11 ויראה מזה כי כל ערך יחובר מד מאלה השיעורים יתחבר

86. ו: -.
87. ו: -.
88. ו: מה.
89. ו: לה.
90. ו: מכאפיתין.
91. ו: -.
92. ו: גיב.
93. ו: לה.
94. ו: ללתכאפו.
95. ו: דג.

122b

מהם משני פנים. ונעשה לזה לוח קטן ונשים בטור הראשון מספר הערכים החבוריים אשר איפשר לחברם בששה שעורים. ובטור השני והשלישי הערכים אשר יתכן התחברם מהשאר ואשר לא יתכן. ובארבעה טורים הנשארים נשים לנוכח כל ערך מחובר הערכים אשר מהם התחבר באשר איפשר בהם התחברות. ואשר לא יתכן בהם זה נניח אותם ולא נכתוב בהם אות כלל. וכל ערך מחובר מאלו אפשר שיתהפך עד שיהיה מספר הערכים המחוברים כפל הכתובים בלוח. וזה רושם הלוח.

A.II.12 ואם היה אחד מאלו הששה נעלם נוציא אותו לפי שכבר ידענו כי הכאת הראשון ברביעי והנקבץ בששי שווה להכאת השני בשלישי והנקבץ [בחמישי],[96] ונחלק הנקבץ על הכאת הראשון ברביעי, יצא הששי ועל זה המשל יאות לעשות בנשאר.

A.III.1 הפרק השלישי בכלים אשר אנו צריכים אליהם לעיונים אשר בהם תשלם זאת החכמה. בסבת ידיעת שיעור נטיית עגולת המזלות מגלגל המישר רל׳ נטיית כל אחד מן ההפוכים ממנו, והוא שוה למה שבין הקטבים, נעשה שתי טבעות נחשת מרובעות השטח ונכניס הקטנה בחלל הגדולה בכדי שתסובב בחלל הגדולה לצד צפון ולדרום. ונחלק הגדולה בשס׳ חלקים ונשים על קטר הקטנה שתי ידות [קטנות][97] באמצעותם שתי יתידות מגיעות לשטח החצונה. ונקים השתי טבעות על עמוד נצב על האפק ועל קו חצי היום כדי שיהיו בשטח עגולת חצי היום. וכשנסבב הטבעת התיכונה עד שיפול צל היד העליונה על התחתונה בהיות השמש בחצי היום, יורנו קצת היתד על שיעור מרחק מרכז השמש מנקודת נכח הראש בעגולת חצי היום. ובזה הכלי ידענו שיעור הקשת אשר בין נכח הראש [][98] וכל אחד מההפוכים. אם כן מה שבין ההפוכים ידוע. וכבר עשו לזה עצמו כלי אחר בשהרשימו בלבנה מאבן או עץ רובע עגולה והוציאו בה כל מה שיצא מזה הכלי הנזכר [לא פחות ולא יתר].[99] ולזה לא זכרנו אותו לפי שכוונתינו הקצור במה שלא יביא לידי חסרון בזאת החכמה.

A.III.2 הכלי השני והוא הנקרא בעל הטבעות. נקח שתי טבעות מרובעות השטחים שוות ונרכיב אחת מהן עם חבירתה על זויות נצבות על שתהיה אחת מהם עגולת המזלות. ונועץ על שתי קטביה שתי יתידות ונרכיב עליהן טבעת מבחוץ ימשש אותה, ותהיה קלת הסבוב באורך, ואחרת מבפנים תסובב ג״כ באורך. ונחלק זאת הפנימית [ועגולת][100] המזלות לחלקיהן. ונרכיב בפנימית טבעת דקה תסובב לצד הקטבים בשטחה לעיון הרוחב ובה שני נקבים מקבילים בולטים. ונרכיב בקטבי עגולת גלגל המישור טבעת גדולה מחוץ לטבעת ההקש לכשנקים אותה נכוחית [לעגולת][101] חצי היום האמיתית על גובה קטב המקום שאנחנו בו, יהיה סבוב [הטבעות הפנימיות][102] סביב אלו הקטבים נמשך לתנועת הכל הראשונה.

A.III.3 וכשרצינו לעיין בזה הכלי נשים הטבעת החצונה אשר על קטבי המזלות על חלק השמש ונסבב אותה עד שיהיה חתוך השתי טבעות נכח השמש רל׳ עד שיאפילו עצמן. ואם שמנו במקום השמש כוכב

96. א: ברביעי. ו: אלכאמס. פאדא כאן מתלא׳ אלמג׳הול אלסאדס צ׳רב אלתאני פי אלתאלת תם פי אלכ׳אמס.

97. ו: מתואג׳התין.

98. ו: ובין.

99. ו: לא גיר.

100. ו: ופלך

101. ו: -.

102. ו: אלחלקה אלדאכ׳לה.

[קים][103] אשר חלק מקומו ידוע נשים גֹכֹ החתוך נכחו בשנשים הראות באחד מצדדי [הטבעות][104] עד שיראה הכוכב כאלו הוא מדובק בשטחיה. ואחר כן נסובב

122a

הטבעת הפנימית לצד הירח או איזה כוכב נרצה [לדעת מקומו][105] עד שנראה אותו בשני הנקבים המקבילים. ובזה נדע מקום הכוכב באורך בחלקי עגולת המזלות וברוחב בחלקי הפנימית.

A.III.4 הכלי השלישי והוא הנקרא בעל השרביטים [בערבי דאת אלמסאטר].[106] נקח שני שרביטים מרובעים ארכם לא פחות מארבע אמות ובעובי שלא יתעותו, ונקוב בהם חורים אצל הקצוות באמצע שני קווים נשרטט אותם באמצע רחבם. ונקשר אותם בבריח יכנס בנקבים בכדי שיוכל לסובב האחד על הבריח כשיהיה השני נעוץ בתושבת, ונקים אותו [][107] נצב על שטח האופק נכוחי לקו חצי היום [נעוץ בתושבת][108] על [שיהיה דבוק][109] האחד בחברו פונה לדרום. ונרכיב בשרביט הבלתי נעוץ שתי ידות באמצעותם שנים נקבים באמצע הקו הממצע השרביט, והנקב הקרוב לעין הרואה יותר קטן מן השנית. ונרשום באמצא השני קוים האמצעיים שתי נקודות מרחקן ממרכז הבריח השווה ויֹתר מה שנוכל. ונחלק הקו הנפרש בבעל התושבת לששים חלקים. ונרכיב בנקודה אשר בשרביט הנעוץ שרביט דק קל הסבוב ארכו בכדי שיגיע לקצה האחר כשיתפרק מן הנעוץ. וכשנרצה לעיין מרחק הירח הנראה מנקודת נֹכח הראש נסובב השרביט בעל השתי ידות בשטח עגולת חצי היום, כשיהיה הירח בעגולת חצי היום, עד שיראה הירח כלו. [ונרשום][110] בשרביט הדקה מרחק [הנפרש][111] בין שני קצות קוי השרביטים השווים ונשים אותו על הקו הנחלק בסֹ. ונקח הקשת אשר עליו והוא מרחק מרכז הירח הנראה מנקודת נֹכח הראש [בעגולה][112] הנרשמת על קטבי האפק וכל זה אפשר להשיגו בבעל הטבעות.

A.III.5 הכלי הרביעי הנקרא בעל היד המתנועעת. ונעשה שרביט ארכו דֹ אמות ונתאר באמצעו קו ונשים בקצו יתד קטן. אחר כן נרכיב בזה השרביט יד בכדי שתוכל להתנועע על אורך השרביט עם היותה נצבת תמיד על שטח השרביט. ונשים באמצע היד נקב עגול יהיה מרחק זרו התחתון משטח השרביט כמו מרחק ראש היתד ממנו. ונחלק הקו אשר באמצע השרביט בחלקים שנרצה בכדי שיהיה אלכסון עגולת הנקב ידוע באלו החלקים. ובזה הכלי נדע מדידת קטר הירח או איזה כוכב נרצה, כשיהיה מרחקו ידוע, בשנשים נקודת הראות אצל ראש היתד ונניע היד על אורך השרביט עד שיראה הכוכב כלו, לפי שערך הקו אשר בין הראות והנקב אצל חצי אלכסון הנקב כערך מרחק הכוכב אצל חצי אלכסונו. ונדע גֹכֹ הכוכב אשר [גופו יותר גדול][113] מכוכב אחר אצל הראות או שוה לו.

103. א: קיום. ו: תאבת.
104. ו: אלחלקה.
105. ו: קיאסה.
106. ו: -.
107. ו: מרכוזה כדלך.
108. ו: -.
109. ו: נצור אנצֹמאם.
110. ו: ונתעלם.
111. ו: אלדי חצל.
112. ו: פי אלפלך.
113. א: יותר גופו גדול.

A.IV.1 הפרק הרביעי בידיעת הנטיות החלקיות והמצעדים בכדור הישר והנוטה ודברים אחרים חלקיים ימשכו מהם. כבר ידענו הקשת אשר בין ההפוכים בכלי אשר הושם לזה ונמצא זה הקשת מז חלקים מג דקים מן העגולה העוברת בקטבים לדעת בטלמיוס. וגכ נדע בזה הכלי נטיות מקומות הישוב רל מרחק נכח הראש [מגלגל][114] המישור אבל גובה הקטב.

A.IV.2 ותהיה העגולה העוברת בקטבים אבגד [וגלגל][115] המישור אהג וחצי עגולת המזלות אשר מראש גדי עד סרטן בהד בצורה הראשונה והחצי השני בהד בשנית. וקטב [גלגל][116] המישור הדרומי

121b

ז ונשים קשת הח מעגולת המזלות מצד השווי מונח. ונוציא קשת זחט בשתי הצורות על שזה המופת הוא עשוי לשני הצורות. ולפי שערך בקע קשת זא הרביע אצל בקע קשת אב, והוא תכלית הנטייה, מחובר מערך בקע קשת זט אצל בקע קשת טח הנעלם ומערך בקע קשת חה המונח אצל בקע קשת הב. והנעלם הרביעי ויהיה ידוע במה שקדם בפרק השני. ואחר שהבקע ידוע הקשת ידוע. [ויתבאר ממה שנתבאר בהקדמות כי כאשר נשים נקודת א מעגולת משוה היום בצורה הראשונה קטב ורשמנו במרחק קשת אב רביע עגולה קטנה והיא קשת בס, ולקחנו קשת סו ממנה ידוע כי נטיית נקודת ו ממשוה היום והיא קשת ונ תהיה ידועה גכ בזה הדרך בעצמו. וכאשר נשים קשת סו, דומה לקשת הח, אל בקע קשת אב מחובר מערך בקע קשת חה אל בקע קשת הב ומערך בקע קשת זט אל בקע קשת טח הנעלם, ומחובר גכ מערך בקע קשת סו אל בקע קשת סב רל ערך בקע קשת הח אל בקע קשת הב ומערך בקע קשת וט אבל זט אל בקע קשת נו הנעלם, [אכ ישאר בקע קשת טח שוה לבקע קשת ונ][117] וזה מה שרצינו.][118]

A.IV.3 וגכ נרצה לדעת שיעור העולה [מגלגל][119] המישור עם חלקים מונחים מעגולת המזלות בכדור הישר ובכדור הנוטה רל שיעור קשת הט העולה באפק הישר ויהיה זחט עם קשת הח. וזה יהיה על זה הדרך לפי שערך בקע קשת זב אצל בקע קשת בא הידועים מחובר מערך בקע קשת זח אצל בקע קשת חט הידועים ומערך בקע קשת טה הנעלם אצל בקע קשת הא. והנעלם הוא החמישי ונוציא אותו במה שקדם. וגכ נוציא קשת כחמ רל האופק בכדור הנוטה ונכוון להוציא קשת המ העולה בזה האופק עם קשת הח מהמזלות. וזה בשנוציא שיעור קשת טמ רל מותר המצעדים אשר בין הכדור הנוטה והישר. וזה יהיה על זה הצד, לפי שערך בקע קשת זכ רל שפלות הקטב אצל בקע קשת כא תשלומו הידועים מחובר מערך בקע קשת זח אצל בקע קשת חט הידועים, ומערך בקע קשת טמ הנעלם אצל בקע קשת מא הרביע. אכ קשת טמ ידועה [נגרע][120] אותה ממצעדי הקשת המונח בכדור הישר בחצי אשר בצורה הראשונה, או נוסיף אותה עליה בחצי אשר בשנית, ויצאו מצעדי זאת הקשת בכדור הנוטה. וגכ נעשה זה הערך בצד האחר מהשווי בשנוציא קשת לחט מן הקטב הצפוני, וקשת כחמ חתיכה מהאופק הנוטה, ונוציא כזה מצעדי כל חלק מחלקי העגולה [בכל][121] נטייה ומשל.

A.IV.4 ובזה הדרך אפשר לרשום לוחות

114. ו: -.
115. ו: -.
116. ו: -.
117. א: 2.
118. ו: -.
119. ו: -.
120. א: נדע. ו: וננקצה.
121. ו: פי.

121a

למצעדים [בלא][122] נטייה מונחת. ויתבאר מזה כי אם היה חלק השמש ביום מה ידוע יהיה שעור זה היום ידוע בשנקח מצעדי הששה מזלות העולים בזה היום או הלילה בזה האקלים ונקח מהם חלק מטו̇, יעלה מספר השעות השוות. ואם לקחנו חלק מי̇ב̇ יעלו חלקי השעה הזמנית. ואם ידענו מספר השעות הזמניות שעברו מן היום או הלילה, נכפול אותן בחלקיהן ונוסיף אותם על כל המצעדים אשר מראש טלה עד חלק השמש. ונבקש החלק מן המזלות אשר כל מצעדיו באקלים על סדר המזלות מראש טלה [שוה לנקבץ][123] והוא החלק הזורח.

A.IV.5 וג̇כ̇ נכה השעות הזמניות מחצי היום שעבר עד עתה בחלקיהן כל אחת בדומה לה ונפיל הנקבץ ממצעדי חלק השמש על סדר המזלות בכדור הישר, יגיע החלק מן המזלות אשר באמצע השמים בעת הזאת. ואם לקחנו הנקבץ לחלק הזורח מראש טלה באקלים וחסרנו ממנו תשעים [][124] אחר שנוסיף עליו חלקי עגולה [שלמה אם לא היו שם תשעים][125] יהיה המקום אשר יגיע מהמזלות [בכדור][126] הישר הוא החלק אשר באמצע השמים. וג̇כ̇ בהפך נקח הנקבץ [לחלק][127] אמצע השמים [בכדור][128] הישר ונוסיף עליו תשעים, והמקום אשר יגיע מהמזלות באקלים הוא החלק הזורח בעת הזאת. והרוצה להוציא דבר מזה צריך להיות אצלו לוחות המצעדים אשר רשם אותם בטלמיוס והם המוצאות באלו הדרכים הנזכרים בזה הפרק ואז יהיה כל זה נקל להעשות.

A.V.1 הפרק החמישי בשעור זמן שנת השמש. מהראוי להיות הקש חזרת השמש בעגולת המזלות כשיתחיל מנקודה בלתי נעתקת עד שיחזור אליה, והנקודות היותר ראויות לזה הן נקודות השווים וההפוכים לפי שבהן יראה ההפרש והשנוי בין עתי השנה יותר מזולתם. ואין ראוי להיות הקש החזרה בהקש אל כוכב קיים בעבור העתקתו המתונה וג̇כ̇ אפשר שיטעה האדם ויקיש בכוכב נבוך. ואברכס ורוב הקדמונים הסכימו בשעור זמן השנה באשר עיינו עת השואת הקיץ והשואת החורף בטבעת נחשת היתה יושבת בשטח [גלגל][129] המישור. ויודע זה ממנה כי תאיר בעת חנות השמש בנקודת ההשואה משני צדדיה. ואמרו כי זה הזמן הוא ש̇ס̇ה̇ יום ורביע זולתי חלק מש̇ מיום כשעיינו השואת קיץ [אחת][130] ועיינו אותה פעם אחרת אחר שלש מאות שנה מצריות וכל שנה מהם ש̇ס̇ה̇ יום. ומצאו אותו אחר ארבעה ושבעים יום [יותר][131] מהש̇ שנה במקום החמשה ושבעים יום המגיע לרביע הנוסף בשלש מאות שנה. וחלקו זה הזמן על חלקי עגולה ר̇ל̇ ש̇ס̇ חלקים תצא התנועה האמצעית ביום ר̇ל̇ הקצור האמצעי כפי אחת [הדעות].[132] ועשו לוחות לתנועות האמצעיות בימים ובשעות ובחדשים ובשנים ושמו בטור הראשון מספר הזמן מאיזה מין שהיה ובנכחו בשאר הטורים מספר התנועה המגיע לו.

122. ו: פי כל.
123. ו: הו הדה אלג̇מלה.
124. ו: ואן כאן אקל.
125. ו: -.
126. ו: פי.
127. ו: -.
128. ו: פי.
129. ו: -.
130. ו: -.
131. ו: מן תמאם.
132. ו: אלנתין (ר̇ל̇ אלאתנתין).

A.V.2 ואמר ן׳ סינא מה שזה ענינו. ואם היה עת ההשואה בלילה נדע באיזה עת היה מן הלילה בשנראה שעור [המרחק][133] בין גבהי חצאי שני הימים [][134] והמתאחר להשואה. וערך זה [המרחק][135] אצל [המרחק][136] אשר בין גובה חצי היום המתאחר ובין חצי יום ההשואה כערך הזמן אשר בין חצאי שני הימים אצל הזמן אשר בין עת ההשואה וחצי היום המתאחר.

A.V.3 נשלם הכלל הראשון והשבח לאל האחד [אין זולתו].[137]

120b

B.O.1 הכלל השני באיכות התנועות והעקרים הראוי להניחם לחלופים המורגשים לגרמים [השמימיים][138] וחלקתי זה הכלל [לארבעה][139] מאמרים. המאמר הראשון בתנועות השמש וחלופו ובו שלשה פרקים. הפרק הראשון ביסוד המוסד לתנועת המשתוה. הפרק השני בעקר המונח לחלוף. הפרק השלישי בבאור [שעור][140] החלוף היותר גדול הנראה לשמש ודרך חשוב החלופים החלקיים ובאור הנקודה מהמזלות אשר בה הקצור או המהלך היותר קטן.

B.O.2 המאמר השני בבאור היסודות אשר עליהם יאות לעשות בתנועות הירח והחלופים הנראים בו, ובו חמשה פרקים. הפרק הראשון בעקרים הנתונים לתנועות הירח הסבוביים. הפרק השני בבאור החלוף הראשון האמיתי [][141] בירח. הפרק השלישי בשעור היותר גדול מזה החלוף. הפרק הרביעי [יבאר][142] כי חלוף שעורי קטרי הירח אצל ההבטה לא יחוייב בהכרח חלוף מרחקי הירח מן הארץ. הפרק החמישי בשני החלופים הנשארים לירח ר״ל החלוף הנערך אצל בטלמיוס לשמש, והוא אשר ישתמש בו ביוצא המרכז, והחלוף הנערך אצלו לעקום קטר גלגל ההקפה.

B.O.3 המאמר השלישי בהעתקת הכוכבים הקיימים ובו שני פרקים. הפרק הראשון בדעת הבטרוגי בזאת ההעתקה וביטולו. הפרק השני בדרך העתקת [אלו][143] הכוכבים.

[המאמר הרביעי בתנועות החמשה כוכבים הנבוכים שבתאי צדק מאדים נוגה כוכב וחלופיהן ובו עשרה פרקים. הפרק הראשון בהקדמות המאמר. הפרק השני בביאור העקר האמיתי אשר יסדנו אותו לחלוף הראשון הנמצא לשלשת הכוכבים העליונים והוא הנערך לחלקי המזלות. הפרק השלישי בחלוף השני הנמצא לשלשת הכוכבים האלה והוא אשר ימצא כפי מרחק הכוכב מהשמש, והצעת העקרים האמִתיים לו ולדברים החלקיים הנראים בו ותמשכי היותר גדול ממנו. הפרק הרביעי בביאור החלוף הראשון המתייחס למזלות וביאור הנקודה מהמזלות אשר בה המהלך היותר קטן בכל אחד מהשלשה כוכבים העליונים. הפרק החמישי בביאור שעור העגולה אשר היא פועלת החלוף המתייחס לשמש בכל אחד מהשלשה כוכבים העליונים. הפרק הששי איך נוציא מצד המהלכים הסבוביים המהלכים האמִתיים. הפרק השביעי בתכונת כוכב נוגה. הפרק השמיני בביאור הנקודה מהמזלות

133. ו: אלתפאות.
134. ו: אלמתקדם.
135. ו: אלתפאות.
136. ו: אלתפאות.
137. ו: לא רב גירה ולא כאלק סואה.
138. א: שמיימיים.
139. ו: אלי.
140. ו: קדר מקדאר.
141. ו: אלט֒אהר.
142. ו: פי.
143. ו: -.

אשר בה המהלך היותר קטן וביאור ערך חצי קטר גלגל נוגה אל חצי קטר העגולה הפועלת החלוף הנערך למזלות, וביאור שעור העגולה אשר היא פועלת החלוף הנערך אל השמש. הפרק התשיעי בתכונת כוכב הנקרא עטארד. הפרק העשירי במרחב החמשה כוכבים הנבוכים.][144]

B.1.1.1 הפרק הראשון מן המאמר הראשון ביסוד המוסד לתנועה המשתוה. נאמר כי השמש נמצא מקצר מהגלגל העליון, רׄלׄ מהתנועה היומית, בשיעור שישלים סבובו בקצורו בכל שנה פעם אחת. ועם שתנועתו על עגולות נכוחיות בחוש לעגולת [גלגל][145] המישור אשר על קטביה התנועה הראשונה, ימצא בקצורו נעתק לצד הצפון ולדרום. ושיעור רחוקו ממנה מוגבל והוא כמו נטיית עגולת חגורת המזלות ממנה.

B.1.1.2 וזה יצוייר על מה שאתאר. והוא שנשים

120a

עגולת [גלגל][146] המשור גׄהׄדׄ קטבה אׄ ועגולת נטיית השמש גׄטׄדׄ קטבה בׄ. והעגולה העוברת בנקודׄתׄ ההפוכים ובקטבי [גלגל][147] המישור טׄהׄבׄאׄ והעוברת בשוויים ובקטבי [גלגל][148] המישור גׄאׄדׄ. ונשים כי ביום אחד יתנועע קטב בׄ על עגולת מעברו והיא עגולת בׄקׄשׄתׄ נמשך לתנועת הכל ממזרח למערב קשת בׄשׄכׄ, ויקצר קשת כׄבׄ. וכאשר נשים השמש דרך משל על נקודת טׄ, רׄלׄ נקודת ההפוך, אׄכׄ כבר התנועע השמש [][149] עם תכלית נטייתו לצד המערב בתנועת הקטבים וחסר מעגולה שלימה קשת מן העגולה הנכוחית [לגלגל][150] המישור העוברת בנקודת ההפוך דומה לקשת כׄבׄ, ותהיה קשת טׄחׄ.

B.1.1.3 ואז תהיה הנחת עגולת השמש כשיהיה קטבה על כׄ כהנחת עגולת לׄזׄחׄ. ותהיה תכלית הנטייה על נקודת חׄ מעגולת טׄחׄ ואז יקצר מהשלים עגולה שלימה קשת חׄטׄ מן הנכוחית. ונשים גׄכׄ כי בזה הזמן אשר יתנועע הקטב קשת בׄשׄכׄ ותכלית הנטייה אשר היה בו השמש תשלום קשת חׄטׄ לעגולה שלמה, תתנועע עגולת השמש סביב קטביה. ותעתיק השמש עמה גׄכׄ לזה הצד קשת חׄזׄ בבקשו השלמות וההתדמות בעליון, כי נכסוף נכסף להשלים קצת מה שקצרו קטביו מן העליון בסבת שבירת הכח היורד עליו. ובזאת התנועה יקרה לו שיהיה פעם צפוני [לעגולת][151] המשור ופעם דרומי. ולזאת תקרא זאת התנועה תנועת הרוחב. וישאר קצור השמש אחר אלה שתי תנועות נקבצות גורע מהגיע אל המקום אשר ממנו התחיל השיעור אשר השיג אותו בטלמיוס והוא נׄטׄ דקים בקירוב. ותהיה תנועתו ברוחב בזה היום בשעור חסרון רוחב נקודת זׄ מרוחב נקודת חׄ.

B.1.1.4 ולפי שתנועת האורך בשמש רׄלׄ קצור האורך האחרון שוה לתנועת הרוחב, יתחייב מזה שתהיה קשת חׄזׄ, רׄלׄ תנועת הגלגל על קטביו, שוה לקצור השמש האחרון רׄלׄ [נׄטׄ][152] דקיׄ. אׄכׄ קצור הקטב, רׄלׄ קשת כׄבׄ, שוה לתנועת הרוחב וקצור האורך השוים יחד. אׄכׄ כל הנראה בשמש שהוא גורע כל יום מהקפה

144. ו: -.
145. ו: -.
146. ו: -.
147. ו: -.
148. ו: -.
149. ו: מתלא.
150. ו: -.
151. ו: ען.
152. ו: נׄטׄ. א: זׄטׄ.

שלמה השעור הנזכר, ושהוא נראה מתנועע תמיד על עגולת נכוחית [לעגולת][153] המישור, ויתנועע גֹּכֹ לצפון ולדרום השעור המושג. בזה בלבד ישלם ואין צורך להנחת שתי תנועות הפכיות, וזה כי אין הפרש בין שישלים ההפוך רֹלֹ נקודת טֹ הקפה אחת ביום אחד ותהיה תנועת הרוחב מנקודת טֹ עד נקודת זֹ להפך ובין שיהיו התנועות לצד אחד כי בכל אחד מהעקרים ישלם כל הנראה בו מן התנועות על ענין אחד עם היות השמש תמיד על עגולת המזלות. ואין ביניהם הפרש אלא בענין המצעדים, כי באלה העקרים יהיה הפך אלו לפי שיתחייב לזה הדעת שתהיה תנועת השמש, בהיותו במזלות אשר אצל ההפוך, לגרוע מצד המצעדים. ובהיותו אצל השווי יהיה בהפך עד שזה החלוף ישלם בכל רביע רֹלֹ התוספת

119b

במגרעת.

B.1.I.5 ואין להם ראיה על שאמרם בזה יותר אמִתי מזה המאמר מזה הצד. וזהו היסוד אשר יאות להניחו לתנועה המשתוה אצל האומרים כי בתנועה הסבובית הפך. וככה לדעת ממה שנזכר ממאמרינו כי העקר הראשון יותר אמִתי למה שבארנו מהעדר ההפכיות בתנועה הסבובית.

B.1.II.1 הפרק השני בעקר המונח לחלוף. ואומר כי החלוף הנראה לשמש רֹלֹ שהוא נראה מתנועע בחלקי המזלות פעם במהירות ופעם במתינות ופעם במיצוע ביניהם. וזה החלוף הנראה בו ישמור תמיד סדורו כפי מה שמצאו, רֹלֹ כי התנועה המתונה אמנם תמצא לו בחלקים [ידועים][154] בעצמם תמיד וגֹכֹ התנועה המהירה והאמצעית. ויאות לשום לזה החלוף עקרים אמִתיים יסכימו לסברתינו.

B.1.II.2 ונאמר כי מרכז השמש אם היה מתנועע תנועתו המיוחדת לו על עגולת המזלות או על עגולתו אשר נטייתו שוה לנטיית עגולת המזלות, לא יתחייב שיראה בכאן חלוף באורך כלל, לפי שהשמש בהכרח יתנועע בזמנים שוים קשתות שוות מעגולת נטייתו לפי שהוא נשוא עליה בתנועה. רֹלֹ שהוא מגביל זויות שוות אצל קטבי המזלות בזמנים שוים. אבל השמש לא יחתוך הקשתות השוות מעגולת נטייתו בזמנים שווים, יוליד מקביל הקודם והוא כי מרכז השמש אינו נשוא בתנועתו המיוחדת לו על העגולה הנוטה עם שאינו נמצא חוץ ממנה.

B.1.II.3 יראה מזה כי מרכז השמש מתנועע על עגולה קטנה קטב זאת העגולה נשוא תמיד על עגולה גדולה נוטה על עגולת נטיית השמש בשעור חצי קטר זאת העגולה הקטנה. ותקרא זאת העגולה הקטנה עגולה מעבר המרכז ותעמוד זאת העגולה במקום גלגל ההקפה אשר שם אותו בטלמיוס ונשתמש בו, ויהיה על זה התֹאר.

B.1.II.4 תהיה עגולת המזלות גֹטֹדֹ קטבה בֹ וקטב גלגל המישור אֹ. והעגולה הנוטה הנושאת לקטב עגולת המעבר גֹהֹדֹ קטבה חֹ. ועגולת מעבר זה הקטב חֹכֹ קטבה קטב גלגל המישור. ונשים עגולת מעבר המרכז לֹמֹנֹטֹ קטבה הֹ ותהיה ממששת לעגולת המזלות עד שתהיה הקשת היוצאת מקטבה להקיפה שוה לתכלית נטיית עגולת גֹהֹדֹ על עגולת גֹטֹדֹ. ונשים דרך משל כי נקודת טֹ היא נקודת ההפוך, ונשים גֹכֹ כי כשיהיה קטב עגולת מעבר המרכז על נקודת הֹ כמו שהוא בצורה, יהיה מרכז השמש על נקודת טֹ הממששת לעגולת המזלות.

153. ו: ל.

154. ו: ואחדה.

B.1.II.5 וכשיעתק קטב חׄ ביום אחד המשל נמשך לתנועת העליון וקצר מהשלים הקפה שלמה השעור הנזכר ויהיה משל קשת חכׄ והנה קטב חׄ על נקודת כׄ. תהיה גכׄ נקודת הׄ רלׄ קטב עגולת מעבר המרכז על נקודת סׄ מקצר גכׄ קשת דומה לקשת חכׄ. וכמו זה השעור יקצר השמש גכׄ לפי שהוא נשוא על עגולת מעברו. ובזה הזמן יתנועע קטב עגולת מעבר המרכז על נקודת כׄ ויעתיק השמש עמה מנקודת סׄ עד נקודת זׄ דרך משל, ויהיה שעור קשת סזׄ חצי הקצור הראשון. וישאר קצור הקטב אבל קצור השמש האחרון חצי הקצור הראשון רלׄ נטׄ דקים בקירוב.

B.1.II.6 וזה אם היה מרכז השמש על נקודת טׄ מעגולת מעברו תמיד, ולא תהיה למרכז השמש תנועה על קטב הׄ. אבל אם שמנוה יתחייב שלא ישמור השמש רוחב עגולת המזלות תמיד. וזה כי אם יהיה קטב עגולת מעבר המרכז על נקודת זׄ היה מתחדש לשמש רוחב מעגולת המזלות לפי

119a

שרוחב נקודת זׄ מן הנקודה הנכחית לה מעגולת המזלות יותר קטן מקשת הטׄ. וכל מה שנעתק לצד החתוך יתמעט ויתגדל רוחב השמש מעגולת המזלות.

B.1.II.7 ולפי שהענין בשמש בהפך זה, וזה כי לא ימצא חוץ מרוחב עגולת המזלות תמיד, יתחייב מזה שיהיה למרכז השמש תנועה על קטבי עגולת מעברו. ואם נשים זאת התנועה מצד נקודת טׄ לצד נקודת נׄ, כי זאת התנועה איפשר לה להניחה לאיזה צד שנרצה לפי שהיא מן התנועות אשר ידומה בהן שהן פעם לצד הראשונה ופעם בהפך. ושמנו שעור זאת התנועה שוה לקצור האחרון, רלׄ נטׄ דקים ביום. יהיה השמש בזה שומר לרוחב עגולת המזלות לפי שכשיעתק השמש מנקודת טׄ לצד נקודת נׄ קשת דומה לקשת סזׄ, יהיה ערך חסרון רוחב נקודת זׄ מרוחב נקודת סׄ אצל רוחב נקודת סׄ כערך חסרון רוחב הנקודה אשר יגיע אליה מרכז השמש בזה היום, בשהוא מתחיל מנקודת טׄ לצד נקודת נׄ מרוחב נקודת הׄ אצל רוחב נקודת הׄ. וכשתהיה נקודת הׄ על נקודת זׄ יהיה השמש גכׄ שומר לרוחב עגולת המזלות בסבת העתקתו על עגולת מעברו.

B.1.II.8 ויתחדש לשמש בסבת זאת ההעתקה תוספת בקצורו האחרון לפי שאם היה על נקודת טׄ תמיד היה קצורו בשוה על מה שהוא עליו קצור הקטב. וכל מה שיתמיד השמש נעתק על קשת לנטׄ יתגדל הקצור האחרון. ותראה בסבת זה [][155] רלׄ קצורו לצד סדר המזלות יותר מהירה, וכל מה שיתמיד בהעתקתו על קשת למנׄ יהיה קצורו האחרון מתמעט [והולך],[156] ויראה לזה קצורו יותר מתון. ויהיה שעור התוספת והמגרעת בשיעור קשת לנׄ מן העגולה הגדולה. והרשות בידך לשום זה החלוף בשתהיה תנועת קטב עגולת מעבר המרכז להפך, המשל מנקודת הׄ לצד נקודת זׄ, ואז תפיל עגולת סזׄ מהצורה. ותקרא נקודת טׄ מעגולת המעבר [][157] הקצור הגדול או המהלך, ונקודת מׄ הקצור הקטן או המהלך. והן הנקודות החולקות לחצאי נטלׄ נמלׄ לחצאין.

B.1.II.9 ואיפשר שידומה זה החלוף במחצית האחת בזולת זה הדרך, רלׄ אפילו שלא נשים לשמש תנועה על קטבי עגולת מעברו בעקר אשר יונחו בו התנועות לצד אחד. וזה לפי שנקודות חתוך עגולת מעבר המרכז עם העגולה הנוטה, רלׄ נקודות לנׄ, יראה בתחלת המחשבה שהן יושבות תמיד על עגולת סזטׄ לפי שהעתקת קטבי עגולת המעבר אמנם הוא על זאת העגולה. וכאשר יהיה קטב הׄ על נקודת זׄ, ושתי נקודות לנׄ על שתי נקודות חכׄ מעגולת סזׄ בצורה אשר אחרי זאת, יהיה קצור השמש יותר מקצור

155. ו: חרכתה.
156. ו: -.
157. ו: נקטה.

הקטב עם שהשמש לא יתפרק תמיד מנקודת ט֗, לפי שנקודת ט֗ תהיה אז נוטה מהקטב אשר הוא על נקודת ז֗ לצד נקודת ד֗. ותוסיף תמיד זאת

118b

הנטייה עד שיהיה הקטב בנקודת החתוך. ותשלם אז רוב הנטייה ולזה יראה בקצורו האחרון תוספת. וכשהוא נעתק ברביע הנמשך לו תחזור זאת הנטייה [לאחוריה][158] ותתחסר עד שישלים הרביע, ותשוב נקודת ט֗ למקומה ויהיה מקום נקודת ט֗ ונקודת ה֗ מהמזלות אחד.

B.1.II.1 ולזה יראה בזה הרביע חסרון בקצור וג֗כ֗ ברביע הנמשך לו. וברביע האחר יראה ג֗כ֗ תוספת בקצור כמו שהיה ברביע הז֗ד֗ והיה החלוף מדומה על מה שהוא עליו בהנחה האחרת. לולא זאת היה החלוף הולך על זה הסדר היה השמש בלתי שומר תמיד רוחב עגולת המזלות ואמנם ישמור זה הרוחב בחצי ג֗ט֗ד֗ בלבד. ואמנם בחצי השני יסור מזה הרוחב בשעור קשת מ֗ה֗ט֗ וזה יקרה בסבת הנחת השמש בלתי מתנועע על קטבי עגולת מעברו. ולזה יתחייב להניח כי החלוף אמנם הוא מצד תנועת השמש סביב קטבי עגולת מעברו, לפי שבזה בלבד ימצא השמש על כל חלק מעגולת מעברו, ויהיה בזה שומר לרוחב הנזכר. ואמנם זכרתי כל זה על דרך השתדלות בלִמוד ולחדד שכל המעיין באלו הדברים כדי שיתיישב דעתו בהם, ששתי נקודות ל֗נ֗ נמשכת לעגולת ס֗ז֗ על זה העקר ולפי זה יהיה החלוף שני חלופים.

B.1.II.1 האחד מהם אשר יהיה כפי העתקת הגלגל אשר בו עגולת המעבר סביב קטביו. והשני זה הנזכר, אנחנו נבאר איך ילך העניין בזה [ב֗ה֗].[159] ואם שמנו עקר זה החלוף העקר הראשון, ר֗ל֗ כפי עקרי בטלמיוס בתנועות ההפכיות בשתהיה תנועת קטב עגולת המעבר להפך מנקודת [][160] לצד נקודת ז֗, כפי זה יהיו [ג֗כ֗][161] נקודות חתוך ל֗נ֗ שומרות לעגולת ג֗ה֗ד֗ לפי שאין בנו אז צורך לשום עגולת ס֗ז֗.

B.1.II.1 ונאמר ג֗כ֗ אם שמנו כי נקודות ל֗נ֗ לא יתפרקו מעגולת ד֗ז֗ה֗ תמיד ולא בדבר מועט כלל כמו שהוא מחוייב כפי זה העקר, יתחייב שלא ישמור השמש רוחב עגולת נטייתו תמיד עם היותו מתנועע על עגולת מעברו על הצד הנזכר. והחלוף הנופל מזה הצד גדול לפי שכשיתקבץ ההסרה מן הרוחב אצל נקודת השווי יפול בסבתו בעיון עת השווי, והוא המורה על זה החלוף טעות בלתי קטן.

B.1.II.1 ואמנם הוראת זה יהיה על זה הצד. נחזור הצורה שעברה ונשים קטב עגולת מעבר המרכז על נקודת ז֗. ונשים כי מרכז השמש התנועע על עגולת מעברו מנקודת ט֗ לצד נקודת נ֗ קשת דומה לקשת ה֗ז֗. ואומר כי לא יגיע השמש לנקודת ג֗ וכי קשת ט֗ג֗ בלתי דומה לקשת ה֗ז֗. מופת זה כי ערך חסרון רוחב נקודת ז֗ מהעגולה הנוטה מרוחב נקודת ה֗ אצל רוחב נקודת ה֗ כערך חסרון רוחב הנקודה אשר אליה יגיע השמש מעגולת מעברו בהתחילו מנקודת ט֗ מקשת ז֗ט֗, אצל קשת ז֗ט֗, וקשת ז֗ט֗ שוה לרוחב נקודת ה֗. א֗כ֗ חסרון רוחב נקודת ז֗ מרוחב נקודת ה֗ הוא קשת ט֗ב֗ מן העגולה הגדולה. ויתחייב מזה שיהיה חסרון רוחב הנקודה אשר אליה יגיע השמש מעגולת מעברו מקשת ז֗ט֗ הוא קשת ט֗ב֗ ג֗כ֗.

B.1.II.1 אבל חסרון רוחב נקודת ג֗ מעגולת נטיית השמש מרוחב נקודת ה֗ יותר גדול מקשת ט֗ב֗ לפי שחסרון רוחב נקודת ג֗ גדול מחסרון רוחב נקודת ז֗. ואם הגיע השמש בזה הזמן בעגולת מעברו לנקודת ג֗, ר֗ל֗ נקודת חתוכו עם

158. ו: -.
159. ו: פימא בעד בעון אללה תעלי.
160. ו: ה֗.
161. ו: -.

118a

עגולת [נטיית][162] השמש, היה החסרון מן הרוחב יותר גדול מקשת טב̇. זה הפך. א̇כ̇ הקשת הדומה לקשת הז̇ אשר אליו יגיע השמש הוא קטן מקשת טג̇, והשמש א̇כ̇ לא יגיע לעגולת נטייתו, ר̇ל̇ עגולת דב̇א̇. ודרך כלל הנה יגיע השמש אל רוחב מקומו האמצעי מהמזלות, ר̇ל̇ נקודת ב̇, לא אל רוחב מקומו האמיתי.

B.1.II.15 ואחר שנתבאר זה יאות לנו לבאר איך ישמור מרכז השמש מקיפי אלו השתי עגולות, ר̇ל̇ מקיף עגולת המזלות ומקיף עגולת המעבר תמיד על דעת כל אחד מהעקרים, ר̇ל̇ כפי הנחת התנועות הפכיות על דעת בטלמיוס וכפי הנחתם לצד אחד. ובזה יתבאר מה שמצאו הקדמונים ובנה עליו בטלמיוס בזה החלוף [כי][163] הזמן אשר מן התנועה אשר יותר גדולה עד האמצעית פחות מן הזמן אשר מן האמצעית לקטנה. ונבאר זה הענין ראשונה על דעת בטלמיוס, ר̇ל̇ בהנחת התנועות הפכיות, ואחר נמשיך הביאור על הדעת השני.

B.1.II.16 ונשים הצורה כמו שהיא וכבר נתבאר כי אם היה מרכז השמש נשוא על עגולת מעברו, ר̇ל̇ עגולת ל̇ט̇נ̇מ̇, והתנועע קטב ז̇ קשת זה̇ ונעתק השמש מנקודת ט̇ לצד נקודת נ̇ קשת דומה לה, כי לא יגיע מרכז השמש בזאת ההעתקה בזה הרביע למקיף עגולת המזלות, ר̇ל̇ לחתוך עגולת המעבר עם עגולת המזלות והוא נקודת ש̇ בזאת הצורה. [ואם היה כן הנה][164] יתקבץ בהשלימו הרביע הסרה ברוחב הרבה. ולפי שנקודת החתוך בזה העקר שומרות תמיד מקומן מעגולת ג̇ה̇ד̇ ולא יתפרקו ממנה, יראה מזה כי מרכז השמש אינו נשוא על עגולת מעברו, ר̇ל̇ עגולת ל̇ט̇נ̇מ̇ עצמה עם שאינו נמצא חוץ ממנה. ואמנם איכות הענין בזה הוא על זה הדרך.

B.1.II.17 נצייר כי כשתהיה נקודת ז̇ על נקודת ה̇ ונקודת ט̇ מעגולת ל̇ט̇מ̇נ̇ על נקודת ט̇ מעגולת המזלות, כי קטב א̇ מעגולת ח̇צ̇ מדובק על נקודת ט̇, וקשת א̇ח̇ אשר מעגולה גדולה יהיה אז על קשת ט̇ה̇ אשר מעגולה גדולה ג̇כ̇. ונצייר על קטב ח̇ ג̇כ̇ עגולת ס̇ר̇כ̇ שוה לה ויהיה [אז][165] קשת כ̇ס̇ אשר מעגולה גדולה ג̇כ̇ מדובק על קשת ט̇ה̇. ויהיה שעור הקשת מן העגולה הגדולה היוצאת מקטב כל עגולה מאלו למקיפה שוה לחצי קשת החלוף היותר גדול אשר תחייב אותו הקשת מעגולת מעבר המרכז הדומה לקשת החלוף הגדול לשמש אשר שעורו שני חלקים, ר̇ל̇ חצי קשת עגולת המעבר הדומה לקשת החלוף הגדול. ונשים כי מרכז השמש נשוא בתנועתו על עגולת ס̇ר̇כ̇ וקטב עגולת ס̇ר̇כ̇ נשוא בתנועתו ג̇כ̇ על עגולת ח̇ו̇צ̇. וכשיהיה קטב א̇ על נקודת ט̇ מעגולת ל̇ט̇נ̇מ̇ ועל נקודת ט̇ מעגולת המזלות, יהיה אז קטב עגולת ס̇ר̇כ̇ ר̇ל̇ ב̇ על נקודת ח̇. והיה אז מרכז השמש על נקודת כ̇ מעגולת ס̇ר̇כ̇ שהיתה אז מדובקת על קטב א̇ אשר על

117b

נקודת ט̇ מעגולת המזלות. וכשיתנועע קטב ז̇ קשת זה̇ יעתק קטב א̇ על עגולת ל̇ט̇נ̇מ̇ סביב קטב ז̇ מנקודת ט̇ עד נקודת א̇, קשת דומה לקשת זה̇. ויעתק ג̇כ̇ קטב עגולת ס̇ר̇כ̇, ר̇ל̇ ב̇, סביב קטב א̇ מנקודת ח̇ עד נקודת ב̇ קשת דומה לקשת זה̇ ג̇כ̇.

162. ו: -.
163. ו: מן.
164. ו: ואנה לו אנתקל כדלך.
165. ו: -.

B.1.II.1 ואז התפרק נקודת כׄ אשר בה מרכז השמש מנקודת אׄ לצד נקודת נׄ, ויתחדש בזה לשמש רוחב מעגולת לׄטׄנׄמׄ, לפי שקשת כׄבׄסׄ נכׄחי תמיד לקשת חׄאׄצׄ בהעתקה הזאת. ויעתק מרכז השמש גׄכׄ סביב קטב בׄ מנקודת כׄ עד נקודת שׄ מעגולת סׄרׄכׄ קשת דומה לקשת הׄזׄ גׄכׄ. ובזה יגיע מרכז השמש למקיף עגולת לׄטׄנׄמׄ ולמקיף עגולת המזלות גׄכׄ, עד כי כשיתנועע קטב זׄ חלקי הרביע חוץ מחלקי החלוף הגדול יהיה אז קשת טׄאׄ גורע מרביע קשת דומה לקשת החלוף הגדול. ויהיה קטב בׄ קרוב מנקודת וׄ מעגולת חׄוׄצׄ, ומרכז השמש קרוב לנקודת [רׄ],[166] רׄלׄ קרוב מרביע כשיתחיל מאצל נקודת כׄ.

B.1.II. ואם שמנו כל אחת מאלו ההעתקות בלתי דומה להעתקת קטב זׄ אבל יותר ממנה בשעור החלוף הגדול ברביע כלו, היה אז מרכז השמש על נקודת [רׄ][167] רׄלׄ [על רביע שלם מנקודת][168] כׄ. והיה בזה מרכז השמש על רביע שלם מעגולת לׄטׄנׄמׄ כשהוא מתחיל מאצל נקודת טׄ, רׄלׄ על נקודת נׄ, ונקודת נׄ תהיה אז על נקודת חתוך העגולה הנושאת עם עגולת המזלות, אלא שלא יתחייב זה בשאר הרובעים. ולזה שמנו התנועות דומות לפי שההסרה המתחדשת מזה בלתי מורגשת, אׄכׄ מרכז השמש באלו השתי העתקות ישמור עגולת המזלות תמיד, וישמור מקיף עגולת מעברו גׄכׄ. וזה בעצמו יתחייב בכל רביע [][169] מהנשארים, רׄלׄ כי הוא ימצא בשתי ההעתקות האלה שומר למקיפי השתי העגולות האלו תמיד וזה היתה כונתינו לבארו.

B.1.II. וזה בעצמו יראה בהניח קשת סׄבׄכׄ בלתי נכוחי לזאת התנועה תמיד לקשת חׄאׄצׄ, בשתהיה נקודת כׄ מעגולת סׄרׄכׄ מדובקת תמיד על קטב אׄ, כמו שהוא מחוייב כפי מאמרנו כי קטב בׄ נעתק סביב קטב אׄ, לפי שנקודות חתוך עגולת סׄרׄכׄ עם עגולת חׄוׄצׄ שומרות תמיד מקומן מעגולת חׄוׄצׄ, ועגולת חׄוׄצׄ נעתקה סביב קטביה, והעתיקה עמה עגולת סׄרׄכׄ וקטבה השעור הנזכר. ומרכז השמש נעתק בזה הזמן מנקודת כׄ לצד נקודת רׄ כפל השעור הנזכר לפי שכשנעתק קטב בׄ על עגולת חׄוׄצׄ רובע עגולה לצד נקודת וׄ, היתה נקודת סׄ על מקיף עגולת לׄטׄנׄמׄ. וצריך שתהיה תנועת מרכז השמש סביב קטב בׄ בזה הזמן כפל השעור הנזכר, ובזה ימצא שומר תמיד למקיפי השתי עגולות כמו שנתבאר וזהו יותר טוב. [ואנחנו בארנו זה הענין בכאן ונעתיק אותו לתכונת הירח כדי שלא נחזר הענין האחד פעמים רבות. ובידך לחלק גלגל השמש למספר גלגלים על הדרך הנזכר בסבת אלו השתי עגולות לפי שכל אחת מהן צריך להניח אותה בגלגל מיוחד האחד מקיף בחברו. ויראה מזה גׄכׄ כי הזמן אשר מן התנועה היותר גדולה עד האמצעית יותר קטן מן הזמן אשר מן האמצעית עד הקטנה בשעור אשר ביאר אותו בטלמיוס. וזה מבואר כאשר נשים שעור שני חצאי הקשתות יותר מאשר הושם בשעור אשר יתבאר בעקר העקום.][170]

B.1.II. ונשים שנית על פי הדעת השני כי נקודות החתוך שומרות תמיד מקומן מעגולת סׄזׄ והיא העגולה אשר על קטביה יסובב תמיד קטב עגולת מעבר המרכז, וזה

117a

בלבד הוא הראוי לפי זה העקר. ונאמר כי כשתהיה נקודת סׄ, רׄלׄ תכלית הדרום, על נקודת הׄ, יהיו נקודות חתוך לׄנׄ על עגולת הׄדׄ גׄכׄ, ותהיה נקודת טׄ מעגולת המעבר אשר היא נקודת הקיצור הגדול על נקודת טׄ מעגולת המזלות. וכאשר תקצר נקודת סׄ מנקודת הׄ חצי העגולה הנכוחית אשר תעבור

166. ו: כׄ.
167. ו: כׄ.
168. ו: פי רבע נקטה.
169. ו: אלארבאע.
170. ו: -.

בה ובנקודת הֹ, ונעתק [קטר][171] עגולת המעבר רביע העגולה הנושאת יהיה על נקודת דֹ, והיא נקודת חתוך עגולת הדֹ עם עגולת המזלות.

B.1.II.22 ונשים גזֹ שוה לזכֹ אכֹ נקודת הקצור הגדול אשר היה מן הראוי שתהיה על נקודת זֹ אם היו נקודות החתוך שומרות מקומן מעגולת הדֹ, היא על נקודת טֹ והיא החותכת קשת לנֹטֹ לשני חצאיו, שעור קשת זטֹ שוה לקשת גלֹ, והוא שוה לכפל נטיית נקודת הֹ מעגולת [גלגל][172] המישור אשר הוא ידוע. אכֹ צריך שיתנועע מרכז השמש שהיה ראשונה על נקודת טֹ, רלֹ נקודת הקצור הגדול, סביב קטב עגולת מעברו רלֹ דֹ בזה הזמן מנקודת טֹ מעגולת המעבר עד נקודת אֹ מעגולת המזלות בלבד. ותנועתו האמצעית בזה הזמן סביב זה הקטב אמנם היא רביע עגולת המעבר. ולזה צריך לשום כי מרכז השמש נעתק בשני הגלגלים המתוארים [][173] בזה הזמן לצד נקודת זֹ שעור מה שיחסר טאֹ מרביע.

B.1.II.23 ויהיה על זה התאר כי מרכז השמש נעתק תמיד על עגולת טוֹשֹ קטבה בֹ, וקטב בֹ נעתק תמיד על עגולת חבוֹ שקטבה טֹ אשר הוא נשוא על עגולת המעבר. ושעור הקשת מעגולת המעבר היוצא מקטב כל אחת מאלו השתי עגולות למקיפה שוה לחצי מותר קשת כפל הנטייה הנזכרת על קשת החלוף הגדול, רלֹ מותר [טזֹ על טוֹ][174] [לפי ש][175]כאֹ דומה לקשת החלוף הגדול, ויהיה לפי זה קשת שזֹ שוה לכאֹ. וכאשר יהיו נקודת סֹ ודֹ אשר הוא קטב עגולת המעבר על נקודת הֹ תהיה נקודת טֹ, רלֹ נקודת הקצור הגדול, על נקודת טֹ מעגולת המזלות. ובה היה מרכז השמש ויהיה קטב בֹ על נקודת חֹ מעגולת חבוֹ.

B.1.II.24 ובזמן שקצר נקודת סֹ חצי העגולה הנכוחית ונעתק קטב עגולת המעבר רביע עגולה, יעתק קטב בֹ סביב קטב טֹ רביע עגולה גכֹ, רלֹ מנקודת חֹ עד נקודת בֹ. ויעתק מרכז השמש שעל נקודת טֹ סביב קטב בֹ חצי עגולה, רלֹ מנקודת טֹ עד נקודת וֹ ועד נקודת שֹ, וחנה מרכז השמש על נקודת שֹ. ונעתק גכֹ בזה הזמן סביב קטב עגולת המעבר רביע עגולה וחנה על נקודת אֹ לפי שקשת שטאֹ רביע.

B.1.II.25 וברביע הסמוך לו [ונמשך לו][176] חזר מרכז השמש באלו התנועות לנקודת טֹ, רלֹ מעגולת טוֹשֹ לא לנקודת הקצור הגדול.

116b

וחזרה נקודת טֹ לנקודת זֹ, רלֹ כי נקודת טֹ תחלוק אז החצי אשר תגבול אותו עגולת הדֹ לשני חצאין לפי שנקודות החתוך חזרו גכֹ אל עגולת הדֹ. ונעתק השמש גכֹ סביב קטב עגולת המעבר רובע שני וחנה על נקודת מֹ, ונקודת מֹ תהיה אז על הנקודה מעגולת המזלות המקבלת לנקודת טֹ. ויהיה השמש שומר תמיד למקיף עגולת המזלות. וזה מה שרצינו.

B.1.II.26 ומזה [][177] יתבאר כי הזמן אשר מן הקצור הגדול עד האמצעי פחות מהזמן אשר מן האמצעי עד הקטן. ובסבת זה הוצרכנו לשום הגלגל אשר בו עגולת המעבר עם אלו שני הגלגלים אשר בהם אלו השתי עגולות בכל אחד משני העקרים. ולולא זה היה מספיק בסבת החלוף כשנשים אלו השתי עגולות

171. ו: קטב.
172. ו: -.
173. ו: אל.
174. ו: טהֹ עלי כאֹ.
175. ו: לאכן.
176. ו: -.
177. ו: בעינה.

בלבד על עגולת המזלות מבלי שום עגולת המעבר והנושאת אותה. אכׄ כבר ראית אמיתת כל אחד מאלו השני עקרים וכי לא ישאר בהם זולת הבחירה [][178] בלבד, עם כל זה מדעתו כי העקר הראשון יותר ראוי לסמוך עליו מצד שהוא יותר נקל המעשה [].[179]

B.1.II.2 [וגׄכׄ השמש ימצא בזה העקר השני שומר עגולת המזלות בדרך אשר בו נלקח המשל בלבד והוא
X.1 כאשר תהיינה שתי נקודות המהלך היותר גדול והיותר קטן בשתי נקודות ההפוכים. וגׄכׄ כשיהיו בשאר חלקי המזלות כשנשים כי כשתהיה המתנועעת בשטח השוקטת קטב עגולת המעבר אז הוא בתכלית היותר גדולה אשר למתנועעת מגלגל המישור לא מהמזלות. ומשם תתחיל התנועה והתחלת תנועת השתי עגולות הוא כשתהיה נטיית המתנועעת מעגולת המזלות פחות מה שאיפשר ומנקודת החתוך אשר לעגולת המעבר עם המזלות אז אשר בה השמש. וידוע כי שעור השתי עגולות ישתנה כפי מקום נקודת המהלך הגדול מעגולת המזלות.

B.1.II.2 ואמנם כשתהיה נקודת המהלך הגדול מחלפת מקומותיה בחלקי המזלות כמו שהוא בירח בסבת
X.2 שמירת הרוחב, נצטרך להוסיף בזה דברים אחרים קשה לציירם ואין כונתינו להגדיל קובץ הספר בזה. ולזה בחרנו העקר הראשון לפי שאין הפרש בו בין שתהיינה אלו הנקודות נקודות ההפוכים או זולתם, או שתהיינה מתחלפות בחלקי המזלות כי המופת הנזכר בו ידובק עם איזו הנחה תונח וזה מה שרצינו.

B.1.II.2 וכבר ימצא כל זה בדרך אחר יותר קצר ויותר שלם מבלתי הנחת אלו השתי עגולות הקטנות ושיהיה
X.3 [מרכז][180] השמש נשוא על עגולת המעבר עצמה. ונקדים לזה זאת ההקדמה.

B.1.II.2 נשים נקודה מה או קטב או מרכז כוכב בשטח גלגל ותהיה נקודת כׄ. אומר כי איפשר שתעתק זאת
X.4 הנקודה על קשת דׄכׄ מעגולה גדולה שיעורה החלקים שנרצה, ותגיע מדׄ לכׄ מעגולת דׄגׄכׄ הגדולה בתנועות סביביות. תשוב מכׄ לדׄ מבלתי השלמת עגולה, לא עגולת דׄגׄכׄ ולא זולתה, ולא תסור תמיד מעשות זה. ואם נרצה שתהיה זאת ההעתקה על עגולות קטנות כבר קדם ביאור זה בעקר החלוף כשנשים שתי עגולות קטנות שוות על קשת דׄכׄ כל אחת מהם בגלגל מיוחד, שעור השתי קשתות היוצאות מקטב כל אחת מהן למקיפה, ותהיינה קשתי גׄהׄ הׄכׄ, יחד שוות לחצי קשת דׄכׄ. וקטב האחת סובב סביב קטב האחרת ונשוא על מקיפה והתנועה האחת כפל האחרת כמו שנתבאר.

B.1.II.2 ואם שמנו הנעתק על קשת כׄדׄ רׄלׄ נקודת כׄ קטב עגולה אמנם קטרה יקרה לו עקום. וזה כי כאשר יהיה
X.5 קטב הׄ על זׄ רׄלׄ חצי הׄזׄאׄ תהיה עגולת גׄחׄכׄ ממששת לקשת דׄכׄ

116a

על נקודת גׄ ויהיה הקטר הנזכר חותך קשת דׄכׄ על זויות נצבות על נקודת גׄ. וכאשר נשים זה הקטר מתנועע סביב [קטב][181] כׄ לצד תנועת קטב הׄ סביב קטב גׄ תנועה שוה לה, יהיה הקטר הנזכר, אבל הקשת אשר עליו, מדובק תמיד על קשת דׄכׄ עולה מכׄ לדׄ ויורד מדׄ אל כׄ.

B.1.II.2 ואם רצינו שתהיה זאת ההעתקה אשר לנקודת כׄ על קשת דׄכׄ בעגולות גדולות זאת ההעתקה
X.6 [אפשרית] [182] באחד משלש דרכים. ונחתוך קשת דׄכׄ לשני חצאין על גׄ ונרשום על קטב אׄ במרחק אׄגׄ

178. ו: אלמאכׄה.

179. ו: ודׄמׄאׄ.

180. א: מרכזי.

181. א: קטר.

182. א: איפשרית.

עגולת גׄחׄ הגדולה ועל קטב זׄ במרחק זׄכׄ עגולת כׄחׄ הגדולה ועל קטב אׄ במרחק אׄזׄ עגולת בׄהׄזׄ הקטנה. ונחשוב קטב עגולת כׄחׄ רׄלׄ זׄ במרחק סביב קטב אׄ על עגולת בׄהׄזׄ.

B.1.II.26/X.7 הראשון מהשלשה דרכים שתהיה נקודת כׄ מהגלגל אשר קטבו זׄ והמקבלת לה תקועות בגלגל שלישי בענין שלא תוכל לצאת מעגולה דׄגׄכׄ לצד מהצדדים. ויתחדש בזאת ההעתקה לכל נקודה שתורשם בגלגל אשר בו נקודת כׄ עיקום, זולתי נקודת כׄ והמקבלת לה, ואין בו נקודה תרשום [עגולה][183] שלימת הסבוב. וכאשר יגיע קטב זׄ אל בׄ תגיע נקודת כׄ אל דׄ וישתנה הנחת הגלגל ימין ושמאל לפי שהנחׄת קטביו משתנות בהעתקתם על עגולת מעברם, זולתי נקודת כׄ והמקבלת לה, לפי שהן תקועות כמו שאמרנו. ואין לה העתקה אלא על קשת דׄגׄכׄ.

B.1.II.26/X.8 והדרך השני שיהיו קטבי עגולת דׄגׄכׄ תקועים בגלגל עליון ממנו או מתחת לו בענין שיוכל להתנועע עליהם. ונקודת כׄ מהגלגל אשר בו דׄגׄכׄ והמקבלת לה תקועות בגלגל שלישי קטבו זׄ הסובב סביב קטב אׄ מגלגל רביעי. והגלגל אשר בו נקודת כׄ הוא האמצעי מן השלשה הראשונים, יתדבק באחד מהם בקטבי עגולת דׄגׄכׄ לא בזולתם, ובאחר בשתי נקודות כׄ והמקבלת לה לא בזולתם. וכאשר יתנועע קטב זׄ על עגולת בׄהׄזׄ סביב קטב אׄ, תעתק נקודת כׄ מהגלגל אשר קטבו זׄ על קשת דׄגׄכׄ לא תעבור ממנה, ושאר נקודות זה הגלגל ישתנה הנחתם כמו שנתבאר בדרך הראשון.

B.1.II.26/X.9 וכאשר נשים נקודה בגלגל האמצעי הנזכר אשר קטביו תקועים בנכח נקודת כׄ כאלו היא תקועה בה, תהיה מתנועעת על קשת בגלגלה דומה לקשת דׄגׄכׄ. ויתנועע הגלגל כלו זאת התנועה בעצמה כמו לו היתה זאת ההעתקה על קטבי עגולת דׄגׄכׄ לפי שנקודת כׄ אשר בגלגל אשר קטבו זׄ תשא אותה עמה. וישאר הנחת הגלגל כלו על עניינו לפי שקטביו תקועים בענין שלא ישתנה הנחתן ימין ושמאל.

B.1.II.26/X.10 והדרך השלישי והוא היותר שלם שתהיה זאת ההעתקה בתנועות סביביות מתדמות כמו שאר התנועות אשר אין בהן ספק. וזה יהיה כאשר נשים כי הגלגל אשר קטבו אׄ יעתיק עמו הגלגל זׄ בתנועתו. וירשום קטב זׄ בזאת התנועה עגולת בׄהׄזׄ הקטנה. ונשים תנועה אחרת לגלגל אשר קטבו זׄ הפנימי סביב קוטב זׄ לאחור, ותהיינה השתי תנועות שנית.

B.1.II.26/X.11 משל זה נשים קטב זׄ התנועע רביע זׄהׄ סביב קטב אׄ. ותהיה קשת טׄכׄחׄ היא קׄמׄגׄ ותהיה מקום נקודת כׄ על מׄ, ויהיה לה רוחב מנקודת טׄ שוה לרוחב אשר היה לה מנקודת גׄ. ויהיה חתוך חׄ על גׄ. ובתנועה השנית השוה לה אשר על קטב זׄ, תעתק נקודת כׄ רביע גׄכׄ לאחור, ותתנועע קשת מׄגׄ ותהיה על גׄ עצמה. וכאשר יתנועע הקטב רביע הׄבׄ והגלגל עצמו רביע אחר לאחור, גׄכׄ יגיע חתוך חׄ אל טׄ ותגיע הנקודה אל דׄ.

B.1.II.26/X.12 וברביע השלישי תהיה נקודת התכלית שמקבלת לנקודת

115b

כׄ מעגולת כׄחׄ על סׄ, מרחק סׄטׄ שוה לטׄמׄ, ותשוב הנקודה אל גׄ. וברביעי יהיה החתוך כמו שהוא עליו בצורה ותשוב הנקודה אל כׄ. אׄכׄ לא תשוב מרוץ ושוב על קשת דׄכׄ מן כׄ אל דׄ ומן דׄ אל כׄ, מבלתי שתעבור מזאת הקשת, וכל זה בתנועות סביביות שלימות הסבוב מתדמות. ודע כי בין שתי נקודות דׄגׄ ובין שתי נקודות גׄכׄ יקרה לנקודת כׄ הסרה קטנה ימין ושמאל מקו דׄכׄ יחייב אותה נטיית קטב זׄ מקטב אׄ. והנה יתבאר אחר זה איך יסתלק אלו ההסרות וזה מה שרצינו.

183. א: עגולת.

B.1.II.2 X.13 ומכאן נתבאר כי הנקודה אשר בעגולת טׄכׄ אשר נכח קטב עגולת דׄגׄכׄ רׄלׄ טׄ תרשום באלו השתי תנועות קשת מעגולה גדולה שוה לקשת דׄגׄכׄ, ולא תסור מרוץ ושוב עליה כמו שתרשום נקודת כׄ קשת דׄגׄכׄ, והיא קשת סׄטׄמׄ. וכבר ידעת כי לא תחתוך נקודת כׄ מקשת דׄגׄכׄ, באלו השתי תנועות המתדמות, קשתות שוות בזמנים שוים. כי מן המבואר שכל שתהיה הנקודה קרובה לקצוי הקשת רׄלׄ נקודׄתׄ דׄכׄ, תהיה התנועה הנמשכת לנקודה יותר מתונה. וכל מה שתתקרב אל הנקודה האמצעית, רצוני לומר גׄכׄ גׄ, תהיה התנועה יותר מהירה. והאריכות בביאור זה מותר.

B.1.II.2 X.14 ואם נשים המתנועע על עגולת כׄחׄ, רׄלׄ נקודת כׄ קטב עגולה, אמנם יקרה לקטרה עקום באלו השתי תנועות. וזה כי כאשר יתנועע קוטב זׄ רביע זׄהׄ, ותהיה נקודת כׄ מעגולת טׄכׄחׄ על נקודת מׄ מעגולת קׄמׄגׄ, ויתנועע גׄכׄ קטב עגולת לׄגׄמׄ רׄלׄ כׄ רביע מׄגׄ ויהיה על גׄ. אׄכׄ שתי נקודות חתוך עגולת לׄגׄסׄ עם עגולת טׄכׄחׄ, והם שתי נקודות לׄמׄ, אשר היה מן הראוי להיותן על שתי נקודות [עׄנׄ][184] מעגולת טׄגׄחׄ לולי העקום, היו על נקודות צׄסׄ מעגולת קׄמׄגׄ. ויהיה העקום קשת סׄנׄ ובכמו זה יתעקם קטר גׄכׄ, רׄלׄ יתר זאת הקשת, ויהיה כאלו הוא גׄכׄ קשת כׄפׄ שוה לסׄנׄ.

B.1.II.2 X.15 אׄכׄ כאשר יעתק קטב כׄ באלו השתי תנועות מכׄ לגׄ ברוחב, יתעקם הקטר הנזכר קשת [כׄפׄ].[185] וכאשר יתנועע קטב זׄ הרביע השני רׄלׄ הׄבׄ, יגיע קטב כׄ אל דׄ ויחזור העקום ויסתלק. וכאשר שב הקטב מדׄ לגׄ שב העקום בזה השעור בצד האחר. וכאשר שב הקטב אל כׄ יסתלק העקום ויסור.

B.1.II.2 X.16 וגם כן נרצה שיסתלק הרוחב אשר [קרה][186] לקטב כׄ באלו השתי תנועות מעגולת טׄכׄחׄ, והוא קשת כׄגׄדׄ, אבל יהיה נשאר במקומו מעגולת טׄכׄחׄ. ולא יסתלק העקום אבל שימצא כפל מה שהוא עליו קודם הסתלקות הרוחב. וזה יהיה בשנשים על אלו שני הגלגלים שני גלגלים אחרים, קטב האחד אׄ וקטב השני זׄ גׄכׄ, וסביב קטב אׄ עגולת [קׄטׄגׄ][187] וסביב זׄ עגולת טׄכׄחׄ

115a

הגדולׄתׄ. ונשים קטב עגולת טׄגׄחׄ העליונה והוא אׄ נעתק סביב קטב עגולת טׄכׄחׄ העליונה והוא זׄ, וקטב זׄ שוקע על עגולת אׄנׄ הקטנה. והגלגל נמשך עמו להפך תנועת קטב זׄ התחתון, סביב קטב אׄ אל נׄ. ויעתק גׄכׄ הגלגל אשר קטבו אׄ העליון סביב עצמו תנועה שוה לתנועת קטב אׄ על עגולת אׄנׄ.

B.1.II.2 X.17 אׄכׄ בזמן אשר התנועע קטב זׄ התחתון סביב קטב אׄ התחתון רביע זׄהׄ, ונתחדש לקטב כׄ רוחב כׄגׄ, ונתחדש לקטר גׄכׄ עקום קשת [כׄפׄ],[188] בזה הזמן נעתק קטב אׄ העליון סביב קטב זׄ העליון רביע אׄנׄ, ויתנועע הוא סביב עצמו כמו זה. ותהיה נקודת תכלית גׄ מעגולת טׄגׄחׄ מן העליון על נקודת תׄ רׄלׄ רביע, ונקודת חתוך טׄ העליונה על כׄ העליונה, ויעתק עמו השני גלגלים התחתונים בתנועתו. ויגיע קטב עגולת עׄדׄסׄ אשר היה על תׄ אל כׄ. ונקודת חתוך נׄ אשר היה ראוי שתהיה על מׄ לולא העקום השני, היתה על רׄ. ובכמו זה נתעקם קטר גׄכׄ מן יׄ לשׄ. אׄכׄ כבר שב קטב העגולה אל כׄ ונסתלק הרוחב שחייבו אותו השני גלגלים התחתיים מצד השני גלגלים העליונים. ונשאר העקום כפל מה שהיה ראשונה לפי שנתקבצו לקטר העקומים כלם.

184. אׄ: עׄכׄ.
185. אׄ: קׄפׄ.
186. אׄ: קרא.
187. אׄ: קׄטׄכׄ.
188. אׄ: קׄפׄ

B.1.II.26/X.18 וברביע השני תהיה ג̇ על כ̇ ויגיע [הקוטב][189] מצד התחתיים לד̇ ומצד העליונים לי̇ ונסתלק הרוחב ויסתלקו העקומים גכ̇. וברביע השלישי יסתלק הרוחב גכ̇ מצד כי המרחבים הנקנים לקטב כ̇ הפכיים ויתחדשו העקומים מן הצד האחר. וברביע הרביעי יסתלקו המרחבים והעקומים גכ̇ וזה מה שרצינו.

B.1.II.26/X.19 ואם נשים קטב ז̇ העליון נעתק להפך תנועת התחתון סביב קטב שלישי עליון נוטה מקטב א̇ התחתון כפל נטיית קטב ז̇ ממנו, ויהיה קטב ת̇, יהיה המתחייב מזה דבר אחד זולתי כי בעקר הראשון הארבעה גלגלים הם שלשה לפי שקטבי השנים אחד כמו שיתבאר אחריו בדומה לזה. וידוע כי כמו שנתחייב זה הענין לקטר גכ̇ כזה יתחייב לקטר העגולה הנרשמת על הנקודה אשר היתה נכח קטב עגולת דגכ̇, והיא ט̇, תהיה גדולה או קטנה. ר̇ל̇ שיקרה לו ענין המרחבים והעקומים לא שמרחבו יסתלק באלו התנועות כמו שנסתלק רוחב קטר גכ̇. ואמנם קטב עגולת דגכ̇ הוא שוקט וזה מבואר.

B.1.II.26/X.20 ואם נרצה הפך זה, רצוני לומר שישאר לקטב כ̇ ההסרה ברוחב אשר קנה אותה מצד השני גלגלים התחתיים, ויסור העקום אשר נתחדש לקטר, ר̇ל̇ שלא יהיה שם עקום כלל, נשים קטב א̇ העליון מתנועע גכ̇ סביב קטב ב̇ העליון על עגולה תורשם סביב קטב ב̇ במרחק בא̇ השוה לאז̇ להפך תנועת קטב ז̇ התחתון. המשל אם היתה תנועת קטב ז̇ התחתון מז̇ לצד נקודת ה̇, תהיה תנועת קטב א̇ העליון מן א̇ לצד המקביל לנקודת ה̇, ויעתיק עמו השני גלגלים התחתיים, ויעתק גכ̇ סביב עצמו לאחור כמו זה. ובזה יסתלק העקום כאלו אין לו עקום כלל וישאר ההסרה ברוחב כפל מה שהיתה. ואם תרצה איפשר לך לעשות זה בשלשה גלגלים בלבד, והם אשר קטבו ז̇ התחתון יתנועע סביב קטב א̇ מגלגל ממעל לו, וזה הקטב ר̇ל̇ א̇ יתנועע להפך הצד סביב קטב ב̇ ממעל לו חצי זה. והגלגל אשר קטבו ז̇ יתנועע סביב עצמו להפך תנועת קטב ז̇ גכ̇ בכמו זה, ר̇ל̇ חצי תנועת קטב ז̇ סביב קטב א̇ כמו שנתבאר במה שקדם. וכשתעיין תמצא כי עשות זה בארבעה גלגלים ישוב לזה בעצמו אשר בשלשה כי השנים מהארבעה קטבם אחד, והתנועה לצד אחד אכ̇ תהיה לתחתוני מהם כפלה.

B.1.II.26/X.21 וידוע כי לא יהיו שם ההסרות לימין ולשמאל הנזכרות מצד הקבלת ההסרות בכל נקודה

114b

ואם יתחדש עקום מה לקטר העגולה הנזכרת כפי הנזכר בלא רוחב. ונרצה שיתגדל העקום היותר גדול על מה שהיה שעור מוגבל, ונשים כי קטב ז̇ הפנימי יתרחק מקטב א̇ הפנימי שעור מוגבל על הקשת מעגולה גדולה העובר בהם. ויתהפך על זה הקשת וישוב אל המרחק הראשון בענין שיתחדש לגלגל רוחב בלבד בלי עקום. וזה במחברת אחת משלשה גלגלים כמו שנתבאר. ויתרחק גם כן בזה השעור קטב א̇ העליון מקטב ז̇ העליון וישוב אל המרחק הראשון שנית כמו שנתבאר. אם כל מה שיתרחקו הקטבים בדרך הנזכר יתגדל העקום לעגולה הנזכרת. וכאשר ישובו אל המרחק הראשון ישוב העקום למה שהיה ראשונה וזה בתכלית הבאור.

B.1.II.26/X.22 ואחר שקדם זה נרשום עגולת המזלות גטד̇ והעגולה הנושאת גהד̇ ועגולת המעבר לטנמ̇ קטבה ה̇ ממששת לעגולת המזלות על ט̇ והיא נקודת המהלך הגדול. המשל כשיתנועע קטב ה̇ רביע הד̇, יתנועע מרכז השמש רביע טנ̇. ונרשום העוברת בקטבים הטח̇, ונרשום על ח̇ ר̇ל̇ קטב הנשאת קשת אחב̇ תחתוך קשת הטח̇ על זויות נצבות, וקשת אח̇ שוה לחב̇ ונוציא קשת הזב̇ מעגולה גדולה. אומר כי איפשר לנו שנשים כי בזמן אשר בו יתנועע קטב ה̇ רביע הד̇, תתעקם נקודת המהלך הגדול מעגולת המעבר ר̇ל̇ ט̇ החלקים שנרצה ויהיו המשל מט̇ לז̇, ולא יקרה לקטב ה̇ הסרה כלל. וזה יתכן באחת משלשה צדדים.

189. א: הקט.

B.1.II.2 הראשון שנשים שני גלגלים [יעתיקו][190] קטב ה׳ מן ה׳ אל כ׳ ברוחב. ובזה הזמן תתעקם נקודת ט׳
X.23 מעגולת המעבר מ[ט][191] אל ו׳, ושני גלגלים על אלו [יעתיקו][192] הקטב בזה הזמן לצד [המקבל][193] ברוחב כמו זה ויהיה מן ה׳ לצ׳. ויעקמו נקודת ו׳ גם כן מן ו׳ אל ז׳ בשעור העקום הראשון כמו שנתבאר בהקדמה. ובזה הזמן נעתק קטב ה׳ בעגולתו הנושאת, ונשא עמו הארבעה גלגלים אשר תחתיו, מן ה׳ אל ד׳. ואחר שהסרות הקטב ברוחב הפכיות ר״ל הכ׳ הצ׳, ישאר קטב עגולת המעבר במקומו בלא הסרה. ותתעקם נקודת המהלך הגדול השני עקומים יחד. וברביע השני תהיה הסרת הקטב הראשונה מכ׳ לס׳, והשנית ההפכית לה מצ׳ לע׳, וישאר גם כן הקטב במקומו ויסתלקו העקומים. וברביע השלישי תהיה ההסרה הראשונה לקטב מס׳ לכ׳ והשנית מע׳ לצ׳ ויהיו העקומים לצד המקבלת. וברביעי ג״כ לפי שההסרות הפכיות, ישאר הקטב במקומו. ויסתלקו העקומים וההסרות הקטנות אשר לקטב ה׳ מקשת הס׳ ימין ושמאל כמו שקדם, לא יחייבו לקטב ה׳ הסרה ממקומו לפי שההסרות הקורות אותו בקשתות הס׳ הע׳ מקבילות.

B.1.II.2 והצד השני שנשים קטב עגולת המעבר ר״ל ה׳ יעתק על קשת הק׳ ויתהפך עליו כמו שקדם. וכבר בארנו
X.24 כי בזה יקרה לנקודה אשר היתה בכח קטב זאת הקשת ותהיה ח׳ על קשת אב׳, ושיתעקם קטר עגולת המעבר על המשל הנזכר. המשל בשני גלגלים התחתיים בזמן הרביע, תעתק נקודת ח׳ מח׳ אל ב׳ וקטב ה׳ מה׳ אל פ׳, ותתעקם נקודת ט׳ מט׳ לז׳. ובשני גלגלים העליונים תעתק נקודת ח׳ מח׳ לב׳ גם כן בתנועת הקטב להפך תנועת הקטב הפנימי ובעגולה אחרת כמו שקדם. ובכמו זאת הקשת תעתק נקודת ח׳ מן התחתיים אשר היתה על ב׳ לצד זה. ויעתק קטב ה׳ להפך מה׳ לש׳ כמו ההעתקה הראשונה, וקשת אב׳ העליונה נוטה על קשת אב׳ התחתונה, ומתחתכת עמה על נקודת ח׳. אם כן ההעתקות לקטב ה׳ הפכיות

114a

ושוות, וההסרות נמי הנזכרות הפכיות. א״כ ישאר במקומו עם השארות העקום בעניינו אבל יהיה כפלו ר״ל כפל קשת טז׳.

B.1.II.2 וברביע השני ישוב בשני הגלגלים הראשונים קטב ח׳ אל ב׳ ויעתק קטב ה׳ מפ׳ לק׳ קשת פק׳ שוה לפה׳
X.25 ויסור העקום. וכשנשים הגלגלים השניים יעתק קטב ח׳ מב׳ גם כן וקטבה ה׳ מש׳ לת׳ קשת שת׳ שוה להש׳. וישאר קטב ה׳ במקומו ויסור העקום.

B.1.II.2 וברביע השלישי יעתק קטב ח׳ בשני הגלגלים הראשונים מח׳ לא׳ ונתחדש העקום מן הצד האחר. ויעתק
X.26 קטב ה׳ מק׳ לפ׳ ובשניים תעתק נקודת ח׳ מח׳ לא׳. וכמו זה יעתק קטב ח׳ מן התחתיים, שהיה על א׳, לצד זה וקטב ה׳ מת׳ לש׳. ונעדר הרוחב וישאר הקטב במקומו תמיד מצד הפכיות ההעתקות. וישאר העקום בצד האחר, אבל כפל מה שהיה ראשונה. וברביע הרביעי ישוב קטב ח׳ בשניהם מא׳ לח׳ וישוב קטב ה׳ מפ׳ לה׳ ומש׳ לה׳ ויסור העקום ויסתלק. ובכל אלו השני צדדים, ואם עשינו זה בארבעה גלגלים, כבר נתבאר לך שאיפשר עשות זה בשלשה גלגלים.

B.1.II.2 והצד השלישי שנשים קשת מהעגולה הגדולה הנכוחית לעגולת המעבר רצה ושבה על קשת דומה
X.27 לכפל קשת טז׳ בזמנים הנזכרים, כמו שנתבאר בצורה הראשונה מן ההקדמה. ומבואר כי כמו זה

190. א: יעתקו.
191. א: ס.
192. א: יעתקו.
193. א: המקבלת.

יקרה לעגולת המעבר. וידוע גם כן כי השלשה הצדדין האלו שבות לעקר אחד והם לפי האמת צד אחד בלבד.

B.1.II.26/X.28 ואחר שכבר נתבאר זה, נשים כי זמן העתקת קטב עגולת המעבר רביע הד׳ שוה לזמן העתקת קטב ח׳ על הדרך האמור קשת חב׳, ר״ל חצי אב׳. ושוה גם כן לזמן העתקת קטב ה׳ קשת הכ׳ ר״ל חצי הס׳ בצד הראשון, ושיעור [הקשת][194] שווה לשעור החלוף הגדול. ויוסיף עליו מעט כשנקח בשעור החלוף הגדול ערכו אצל השלמתו לרביע. ונוסיף אותו עליו ושעור קשת הכ׳ בצד הראשון חצי זה.

B.1.II.26/X.29 אם כן כבר נשלם בזה כל מה שיראה מן החלוף בשמש כמו שהוא עליו. וכי הזמן אשר מן התנועה הגדולה עד האמצעית פחות מהזמן אשר מן האמצעית אל הקטנה, לפי שבזמן אשר בו יתנועע קטב ה׳ קשת פחותה מרביע, בכמו חלקי החלוף הגדול, ישלם החלוף הגדול בעקום הנזכר, ויהיה מעבר אמצעי. ובכל אחד מהצדדים הנזכרים, העדפת חלקי העקום תהיה יותר בחלקים היותר קרובים מנקודות המהלך הגדול או הקטן. וענין זה נצטרך אליו לסלק קירוב אם יפול בשמירת השמש עגולת המזלות. ואם הוא בלתי מורגש וכל זה מבואר והאריכות בבאורו מותר.

B.1.II.26/X.30 וההסרה הקטנה המתחדשת לשמש מעגולת המזלות ברוחב במקומות מועטים מצד הנחתינו העקום יותר [ב][195]חלקי החלוף הגדול אינה מורגשת כלל לפי שלא יגיע ההסרה

113b

ברוחב שלשה שניים. ולא יאות לזכור הסרה כזו כמו שלא יחוש בטלמיוס להסרה יותר גדולה מזאת במקומות רבים ואומר שהיא בלתי מורגשת. ועם זה כבר ידעת כי בעקרי בטלמיוס יתחדש מהסרה קטנה ברוחב הסרה גדולה באורך. אבל בעקרינו איפשר שלא יתחדש מהסרה ברוחב הסרה באורך כלל. ואמנם זכרנו כל זה כדי לרמוז אל העניינים האמִתיים בכל דבר וכאשר תרגיל עצמך בציור כל זה, תראה בו ענין מופלא אלקי והשבח לאל מגלה הרזים.

B.1.II.26/X.31 ואמנם בעקר השני אשר שמנו בו התנועות לצד אחד, כבר נמצאו שם שני עקומים עקום גדול ראשון ועקום קטן שני, האחד הפך האחר. ורצוני לומר בקטן העקום העומד במקום השתי עגולות הנזכרות, וראוי שיהיה שעור העקום הקטן השני או הקשת שהיא יתר הנקבץ מחצאי השני קטרים, השעור הנזכר במה שקדם.

B.1.II.26/X.32 ואחר שכבר נתבאר כל זה בהנחת תנועת האורך על קטבי הנשׂאת לקטבי עגולת המעבר בשני העקרים, ר״ל עקר השתי עגולות ועקר העקום, יראה ממה שנאמר כי זה בעצמו יתחייב כשנשים תנועת האורך על קטבי עגולת המזלות. וזה שנשים ראשונה כי תנועת האורך, ר״ל תנועת קטב עגולת המעבר עם השמש באורך, [היא][196] על קטבי עגולת המזלות והשמש סובב על עגולת מעברו נמשך לתנועת החלוף כמו שנתבאר, ובעגולות הנזכרות או בעקום הנזכר. ובזה ישלם כל מה שיראה בו מן החלוף, וכי הזמן אשר מן התנועה הגדולה עד האמצעית פחות מהזמן אשר מהאמצעית אל הקטנה, ושעור העקום או שעור שני חצאי השתי עגולות שוה לחלקי החלוף הגדול. וראה והבן איך ישמר מרכז השמש עגולת המזלות תמיד מבלתי שיפול בזה קרוב כלל.

194. א: קשת.
195. א: מ.
196. א: הי.

B.1.II.2 X.33 אמנם בעקר העקום יהיה זה בסבוב קטב הנושאת על העגולה המחייבת העקום. נרשום הצורה ויהיה [קטב הנושאת][197] בׄ סובב על עגולת בׄכׄסׄזׄ סביב קטב עגולת המזלות והוא אׄ בתנועה שוה לתנועת השמש על עגולת מעברו, ועגולת בׄכׄסׄזׄ שוה לעגולת המעבר. ונשים כי כשיהיה השמש על נקודת טׄ [והיא][198] נקודת המהלך הגדול מעגולת המעבר, אז השמש בעגולת המזלות כמו שהוא בצורה, וקטב הנושאת על בׄ. וכאשר יתנועע השמש רביע טׄנׄ ויתרחק ברוחב מעגולת המזלות בכמו קשת טׄהׄ, יתנועע קטב בׄ גם כן רביע בׄכׄ ויתנועע הגלגל אשר בו העגולה הנושאת בכמו זה לאחור. ויקנה כמו זה הרוחב להפך זה הצד וימשוך עמו העגולה הנושאת ועגולת המעבר, ויהיה קטב עגולת המעבר על עגולת המזלות. וכאשר יגיע השמש אל מׄ יגיע קטב בׄ לסׄ, ואז תהיה עגולת המעבר ממששת לעגולת המזלות בנקודת מׄ, וכזה בשאר חלקי העגולה. ולולא העקום יהיה השמש בזה בלבד שומר לעגולת המזלות תמיד.

B.1.II.2 X.34 ובעבור זה נשים קטב הנושאת סובב על קשת וׄכׄחׄ מעגולה גדולה ורץ ושב עליו בשני גלגלים, מבלתי שיתחייב לשמש תנועה באורך ולא עקום כמו שנתבאר. ושעור חצי זאת הקשת שוה לקשת היותר גדול מעגולה גדולה ייתר אותו העקום היותר גדול. והנקודה החותכת לקשת [וׄחׄ][199] אשר מעגולה גדולה לשני חצאין, היא דבקה תמיד על עגולת בׄכׄסׄזׄ ונעתקת סביב קטב אׄ, והגלגל אשר היא בו עמה, בכמו התנועה הנזכרת.

B.1.II.2 X.35 ובזמן אשר בו יתנועע השמש

113a

בתנועה האמצעית רביע טׄנׄ תתנועע נקודת כׄ רביע בׄכׄ ותעתיק עמה קטב הנושאת מבׄ לכׄ. ויתנועע הגלגל הנושא סביב קטביו וכמו זה להפך ויתחדש לזה העקום. ולפי שהשמש אז סר מנקודת נׄ לצד נקודת מׄ בשעור העקום היותר גדול, ובזה הזמן נעתק ברוחב קטב הנושאת מכׄ לוׄ בעגולה הגדולה בדרך הנזכר, ושב השמש לעגולת המזלות. וכאשר הגיע השמש אל מׄ, שב הקטב אל כׄ והגיע כׄ אל סׄ. וכאשר יגיע השמש בתנועה האמצעית אל לׄ תגיע נקודת בׄ אל זׄ. ולפי שהשמש סר מנקודת לׄ לצד נקודת מׄ, אז בשעור העקום היותר גדול בזה הזמן, נעתק ברוחב קטב הנושאת מכׄ לחׄ בעגולה הגדולה ונקודת חׄ תהיה אז בין שתי נקודות סׄזׄ. אׄכׄ ישוב השמש לעגולת המזלות. וכאשר יגיע השמש מלׄ לטׄ ישוב הקטב מחׄ לכׄ ונקודת כׄ תהיה אז על בׄ. ובזה נמצא השמש שומר לעגולת המזלות תמיד מבלתי קירוב כלל. לפי שבחלקים אשר יהלוך השמש בעגולת מעברו בענין שיתחדש רוחב קטן מעגולת המזלות, באלו העתים ילך קטב הנושאת בעגולת בׄכׄסׄזׄ בחלקים שוים להם במספר יתחייב מהם רוחב קטן שוה לו להפך זה הצד. וכאשר ילך השמש בחלקים יחייבו רוחב גדול גם הקטב.

B.1.II.2 X.36 וזה בעצמו יקרה בעקום לפי שכשיהיו חלקים מה מקשת וׄחׄ אשר מעגולה גדולה קרובים מנקודת בׄ אז יחייבו רוחב מעט כמו חלקי העקום כמותם כשיהיו קרובים מנקודת טׄ. וכאשר תהיה הקשת אצל נקודת כׄ אז יחייבו החלקים ההם בעצמם רוחב גדול כמו חלקים כמותם מהעיקום כשיהיה אצל המעבר האמצעי רׄלׄ נקודת נׄ או לׄ. ולפי שקשת דׄאׄ אשר מעגולה גדולה דבק בעגולת בׄכׄסׄזׄ על נקודת כׄ בלבד, ושאר חלקי הקשת נלוזים ממקיף עגולת בׄכׄסׄזׄ, ורוב זאת ההסרה היא אצל נקודות וׄחׄ. ויתחדש בזאת ההסרה עקום מה לעגולת המעבר נוסף על העקום הראשון, אבל כשתחשוב זה תמצא כי זה העקום המתחדש מזאת ההסרה בלתי מורגש כלל.

197. א: קטבה נושאת.
198. א: והי.
199. א: נׄחׄ.

B.1.II.26/X.37 ולפי שכבר נתבאר כי הגלגל אשר בו העגולה הנושאת יעתק סביב קטבו הנעתק על קשת ו׳ח׳ כמו התנועה האמצעית אשר לו סביב קטב המזלות, ר"ל להפך, ורצוני לומר באמרי תנועתו האמצעית תנועת נקודת כ׳ סביב קטב א׳ על עגולת ב׳כ׳ס׳ז׳. וכאשר נשים כי זה הקטב בעצמו ר"ל קטב הגלגל הנושא, נעתק על קשת ח׳ו׳ בענין שלא יתחייב לגלגל הנושא העתקה באורך ולא עקום כי אם העתקת הרוחב בלבד, יאות שנשים גם כן כי הגלגל הנושא יתנועע סביב קטבו כמו זה גם כן כמו שנתבאר בהקדמה.

B.1.II.26/X.38 וזאת התנועה בידינו לשום אותה לאיזה צד שנרצה כשנשים תנועת

112b

הגלגלים המחייבים זאת ההעתקה לאיזה צד שנרצה, כמו שנתבאר בהקדמה גם כן. וכאשר נשים זאת התנועה להפך צד התנועה הראשונה, תהיינה סביב קטב אחד מגלגל אחד, שתי תנועות שוות הפכיות. ויהיה שוקט בהכרח, ר"ל כי הגלגל הנושא לא יתנועע על קטביו. ויתחייב לשמש משתי ההעתקות אשר לו סביב קטב א׳ ועל קשת ו׳ח׳, מה שיתחייב לו אם נעתק סביב קטב א׳ לבדו. ונעתק גם כן סביב עצמו כמו זה להפך, ויתחייב לו גם כן תוספת הרוחב אשר תחייב אותו ההעתקה אשר על קשת ו׳ח׳. ויתנועע גם כן הגלגלים הנזכרים כלם סביב קטב גלגל המזלות, ר"ל א׳, לצד תנועת קטב הנושאת סביב קטב המזלות תנועה שוה לה. ויהיו לפי זה סביב קטב גלגל המזלות שתי תנועות שוות לצד אחד. והן לפי זה תנועה אחת שיעורה הנקבץ מן השתי תנועות.

B.1.II.26/X.39 ואמנם בעקר השתי עגולות, קטב הנושאת אינו סובב על עגולת ב׳כ׳ס׳ז׳, ואם לא כן יהיו בכאן שני עקומים. אבל הוא סובב על קשת ב׳א׳ס׳ בשני גלגלים כמו שקדם. ויקנה בזה הרוחב בעצמו אשר יחייב אותו העתקת אמצע השמש, ר"ל קטב העגולה האחת מהשתי עגולות על עגולת מעברו, לפי שכבר נתבאר כי ההעתקה אשר לקטב כל מה שיתקרב לקצות הקשת היא יותר מתונה. ויפחתו אז תוספות הרוחב כמו התוספות אשר לרוחב מצד העתקות השמש בקרוב לנקודות ט׳מ׳ מעגולת מעברו. ובעבור הרוחב ההווה משתי העגולות, נשים כי קטב א׳ ר"ל קטב המזלות אשר בגלגל הירח, נעתק על קשת א׳ע׳ בשני גלגלים על הדרך הקודם, שעורו שוה לקשת היותר גדול מעגולה גדולה, ייתר אותו קטר העגולה האחת משתי העגולות אשר עליהן [סבוב][200] מרכז השמש. ויעתיק עמו ברוחב הגלגלים אשר תחתיו כלם.

B.1.II.26/X.40 ובזמן אשר יתנועע אמצע השמש רביע ט׳נ׳, יעתק קטב הנושאת מב׳ לא׳ וקטב המזלות ברוחב קשת א׳ע׳. וכאשר יגיע השמש למ׳ ישוב קטב המזלות לא׳ ויגיע קטב הנושאת לס׳. וכאשר יגיע לל׳ ישוב לע׳. וכאשר יגיע השמש לט׳ ישוב לא׳, וקטב הנושאת לב׳ וישמור השמש בזה עגולת המזלות תמיד. ואם כן השמש כפי עקר השתי עגולות, גלגל המזלות נעתק בהעתקת קטבו על קשת א׳ע׳ וסביב קטביו, יהיה סביב קטב הנושאת על עגולת ב׳כ׳ס׳ז׳. וגלגל המזלות שוקט על כולם, קטבו א׳, וסִבוביו יהיה סיבוב הגלגלים כלם באורך.

B.1.II.26/X.41 וגם כן יתבאר זה הענין בירח ובשאר הכוכבים, ר"ל כשנשים תנועת האורך על קטבי עגולת הרוחב. וזה מבואר, ואין כוונתינו להאריך המאמר במה שהוא מבואר. ואחר שבארנו כי בזאת התכונה האחרונה יתחייב כל מה שיתחייב מעקרי בטלמיוס בשנשים שיעור העקום היותר גדול שוה לחלקי החלוף היותר

200. א: סביב.

גדול, אם כן בכל מקום במה שעתיד, נשים העקום היותר גדול שוה לחלקי החלוף היותר גדול. ולא נחוש כי בשני הדרכים הקודמים שמנו אותו יותר בשעור הנזכר.][201]

B.1.II.27 אׄכׄ כבר הצענו כל העקרים האמִתיים לתנועת השמש והחלופים הנראים בו, עם ההמשך לדעות בטלמיוס [].[202] ולא יתערב עם דבר מזה העדר האפשרות כלל לפי שאלו העקרים בנויים על פי התנועות כלם על כדורים שלמות הסבוב. ואין שם תנועה זולתי אשר היא שוה מתדמה בזמנים שוים מסבבת סביב קטבים מוגבלים, או נמשכת לתנועה סבובית מתרפקת עמה סביב קטביה גם כן

112a

מבלתי צורך בדבר מזה ליוצא מרכז ולא גלגל הקפה. וידוע גם כן ממה שאמרנו בשמש כי הלוחות אשר [חקק אותם][203] בטלמיוס לתנועות ולחלופים הן בעצמם הלוחות שאנו צריכים אליהם לחשבון באלו העקרים, זולתי כי אשר יקרא אצלו תנועה להפך יקרא אצלנו בעקר השני קצור. וגׄכׄ יתבאר זה לעתיד [בׄהׄ].[204]

B.1.II.28 ואמנם אלבטרוגי דעתו בזה החלוף דעת רחוק מהאמת, לפי שהוא סובר כי מרכז השמש יתנועע תמיד על עגולת נטייתו על שהוא נשוא עליה בתנועה. וכי סבת החלוף הנראה בו אמנם הוא מפני העתקת קטבי עגולת נטייתו על עגולת מעבר קטבי עגולת המזלות. וחשב כי כשנשים קטבי עגולת נטיית השמש יקצרו חתיכות שוות מעגולת מעבר קטבי המזלות המתדמות בזמנים מחולפים, כי כמו זה יקרה בשמש בעגולת המזלות שהוא יקצר בזמנים מחולפים קשתות שוות. וזה אמנם יהיה כפי מה שאמר אם לא היתה כוונתינו בחלוף זולתי שיקצר השמש חלקי המזלות בזמנים מחולפים באורך אפילו היה מרוחק ממנה ברוחב לפי שזה יתחייב מהנחתו שאפילו קצר השמש באורך הרובע אשר מהפוך הקור עד השואת הקיץ דרך משל בפחות מרביע ימי השנה, לא יחתוך מעגולת נטייתו רביע שלם בזה הזמן עד שיגיע לחתוך זאת העגולה עם [גלגל][205] המישור כמו שהגיע במזלות, לפי שאין שם דבר יחייב שיחתוך רובעי זאת העגולה בזמנים בלתי שווים.

B.1.II.29 ואם כן לא יהיה השמש תמיד על עגולת המזלות ואם לא יהיה בעת ההשואה על עגולת המזלות, רׄלׄ בעת היות השמש בראש טלה כמו שיתחייב מהנחתו, איך יודה בעיונים אשר הניח אותם בטלמיוס לעתי ההשואות אשר עיין אותם בטבעת הנחֹשת היושבת בשטח עגולת גלגל המישור, [וזה ידוע][206] כשתאיר הטבעת משני הצדדין בזאת העת ובנה על אלו העיונים גם כן, לפי שאי אפשר כפי הנחתו שימצא בזאת הכלי הזמנים אשר בין היות השמש בנקודות השוויים וההפוכים מחולף כלל, לפי שלא תאיר הטבעת משני צדדיה אלא בעת היות [מרכז][207] השמש בנקודות חתוך עגולת נטיית השמש עם עגולת המישור, ולא יגיע לאלו הנקודות זולתי בזמנים שוים.

B.1.II.30 וגם כן הוא ישים בסבת שהקטב מקצר עגולת מעברו בזמנים מחולפים שהוא נעתק על עגולת מוסרת הקטב. ולפי זה לא ימצא השמש תמיד על עגולת המזלות, והוא מודה כי השמש לא יצא מעגולת המזלות. והתמה איך שכח כל זה. ואני חושב כי כוונתו אמנם היא לתת עקרים לחלוף מה חדשו

201. ו: -.
202. ו: פי כל דלך.
203. ו: עמל עליהא.
204. ו: -.
205. ו: -.
206. ו: -.
207. ו: -.

[ובראׄ אותו][208] מלבו ואינו נמשך לדעות הקדמונים. ועם זה יבנה מופתיו על עיוני הקדמונים בו, ונפל מזה בנבוכה מבוארת הטעות והוא לא ירגיש. וכבר הוצאתי לאלו העקרים אשר הניח אותם זה האיש צדדים איפשר שיראה מהם הדרוש, אלא שלא ימלטו ממניעיות יותר חזקים ויותר זרים מן העקרים המוסדים עד עתה. ולזה הפלנו כל זה מזה הספר ונשוב למה שהיינו בו ונבאר שעור זה החלוף.

B.1.III.1 הפרק השלישי בבאור שעור החלוף היותר גדול הנראה בשמש ודרך חישוב החלופים החלקיים וביאור הנקודה מהמזלות אשר בה הקצור או המהלך היותר קטן. וזה יתכן

111b

כשנבאר ראשונה ערך חצי קטר גלגל השמש אצל חצי קטר עגולת מעבר המרכז. וכזה יאות לבאר הנקודות מהמזלות אשר בהן המהלך הגדול והקטן, רׄלׄ [הנקודה מהמזלות אשר יהיה מרכז השמש בה כשיהיה על נקודת המהלך הגדול][209] והנקודה מהמזלות אשר יהיה השמש בה כשיהיה על נקודת המהלך הקטן. וזה יתבאר בעיונים אשר ביאר בהם בטלמיוס שיעור הקו אשר בין שני המרכזים ומקום הרום מהמזלות. והוא ביאר זה כאשר מצא הזמן אשר מעת חנות השמש בנקודת השואת הקיץ עד [עת][210] חנותו בנקודת היפוך החום צׄדׄ יום וחצי. [ומהיפוך][211] החום עד [השואת][212] החורף צׄבׄ יום וחצי והשמש יחתוך בשוה בצׄדׄ יום וחצי צׄגׄ חלקים טׄ דקים ויחתוך בצׄבׄ יום וחצי צׄאׄ חלקים יׄאׄ דקים.

B.1.III.2 ותהיה עגולת מעבר המרכז אׄבׄגׄ והפרק המשותף לה ולשטח העגולה הנושאת אותה קו נׄלׄכׄ. ונשים השמש בעת השואת הקיץ על נקודת אׄ והתנועע עד עת הפוך החום קשת אׄבׄ שיעורו צׄגׄ טׄ [][213] ייתר מהמזלות גׄ טׄ לגרוע. והתנועע מהיפוך החום עד השואת החורף קשת בׄגׄ שיעורו צׄאׄ יׄאׄ ייתר מהמזלות אׄ יׄאׄ לגרוע גם כן. ונוציא מנקודת בׄ עמוד על הפרק המשותך יגיע למקיף עגולת המעבר מן הצד האחר והוא עמודת בׄלׄהׄ. ונוציא מנקודׄתׄ אׄגׄ על קו בׄהׄ עמודי אׄמׄ גׄזׄ ונדביק אׄהׄ גׄהׄ אׄגׄ, ונוציא מנקודת גׄ קו גׄטׄ [עמוד][214] על קו אׄהׄ, כשיצא על יושר, לפי [שזוית][215] אׄהׄגׄ נרוחת. ונוציא מנקודת דׄ רׄלׄ מרכז המזלות לנקודות אׄגׄ קוי דׄגׄ דׄאׄ ולנקודׄתׄ מׄזׄ קוי דׄמׄ דׄזׄ.

[וידוע כי השני קוים המקיפים בזוית חלוף, איזו קשת שתהיה מעגולת המעבר, הם השני קוים המדביקים בין מרכז גלגל המזלות ובין קצות העמוד היוצא מאחת קצות הנזכרת בשטח עגולת המעבר על הקו היוצא מן הקצה השנית, נצב על החתוך המשותף לעגולת המעבר ולעגולה הנושאת. המשל כי זוית חלוף קשת בׄאׄ יקיפו בה שני קוי דׄאׄ דׄמׄ, וידוע גם כן שקו אׄמׄ הדבק בין קצוי אלו השני קוים. אמנם כפי התכונה אשר שמנו בה תנועת האורך על קטבי המזלות, הוא חתוך משותף לעגולת המעבר ולעגולת המזלות כשיהיה השמש בנקודת אׄ. ואמנם אם שמנו אותה על קטבי הנושאת אינו החתוך המשותף הנזכר.

והזוית הנבדלת בחתוך המשותף הנזכר המדביק בין נקודת אׄ וקו בׄהׄ שהיא גדולה או קטנה מזוית אׄדׄמׄ, אינה זוית חלוף קשת אׄבׄ הראוי להוסיף אותה על מהלך האורך או לגרוע אותה ממנו לפי שמהלך האורך האמצעי יתגדל ויתקטן בסבת נטיית הנושאת מעגולת המזלות. והעיקום המתחדש

208. ו: -. א: ברה אותו.
209. ו: -.
210. ו: נקטה.
211. ו: ומן.
212. ו: -.
213. ו: דקׄ.
214. ו: עמוד.
215. א: שזויות. ו: זאויה.

לעגולת המעבר בסבת זאת הנטייה כשעור מה שתוסיף או תחסר זאת הזוית מזוית אבמ̇. אם כן, הזוית הצריך להוסיף אותה על מהלך האורך האמצעי או לגרוע אותה ממנו, בסבת קשת אב̇, היא זוית אדמ̇ עצמה והזויות הדומות לה בשאר הקשתות.][216]

B.1.III.3 [][217] ולפי שזויות אדמ̇, ר"ל אשר תגרע מהקצור כפי הדעת השני או אשר תגרע מהמהלך כפי הדעת הראשון, בסבת קשת בא̇ ידועה. אם כן ערך קו דא̇, ר"ל חצי קטר עגולת המזלות המדומה בגלגל השמש אצל העמוד היוצא מנקודת א̇ על קו דמ̇ ידוע. וזה העמוד אמנם כשיהיה קו בה̇ קטר לעגולת המעבר יהיה הוא בעצמו קו אמ̇. ואם לא יהיה קטר, לא יהיה קו אמ̇. אנחנו נשתמש בהוצאת הערך

111a

על שהוא קו אמ̇ ולא [נחוש][218] למותר אשר ביניהם לפי שההפרש ביניהם קטן. ואחר כן נחזור לבאר זה המותר ונתקן בו הערך.

B.1.III.4 ותהיה לפי זה ערך דא̇ אצל אמ̇ ידוע, ולפי שקשת בא̇ אבל זוית בהא̇ ידועה יהיה ערך אה̇ הקטר אצל אמ̇ ידועה, וערך דא̇ אצל אה̇ ידוע. וגם כן לפי שמקום נקודת ב̇ מהמזלות הוא נקודת ז̇ תהיה זוית גדז̇ ידועה. וערך גד̇ ר"ל דא̇ אצל העמוד ר"ל גז̇ על התנאי הנזכר ידוע. ולפי שקשת בג̇ אבל זוית בהג̇ ידועה, יהיה ערך גה̇ הקטר אצל גז̇ ידוע. אם כן ערך דא̇ אצל גה̇ ידוע. ולפי שזוית אהג̇ אבל גהט̇ ידוע יהיה ערך גה̇ הקטר אצל גט̇ אבל אצל הט̇ ידוע, ויהיה ערך גה̇ אצל טא̇ אבל גא̇ שהוא בכח שוה לגט̇ טא̇ ידוע. אכ̇ ערך דא̇ אצל [דג̇][219] ידוע וכבר היה ערך גא̇ אצל קטר עגולת המעבר ידוע לפי שקשת אבג̇ ידועה. אם כן ערך דא̇ ר"ל חצי קטר גלגל השמש אצל חצי קטר עגולת המעבר ידוע. ויצא בזה הערך אשר הוציא אותו בטלמיוס והוא כי חצי קטר עגולת המעבר חלק מכד̇ בקירוב מחצי קטר עגולת השמש. וזה יהיה אחר התקון אשר נזכיר אותו בקרוב.

B.1.III.5 ואמנם איך נדע החלק מהמזלות אשר יהיה בו השמש כשיהיה בנקודת הקיצור הקטן. זה יהיה בשנדע מרחק נקודת ב̇ מנקודת הקצור הקטן, וזה ידוע לפי שערך גה̇ אצל גא̇ ידוע, וקשת גה̇ ידועה אבל קשת בגה̇ אם כן חציה ר"ל בנ̇ ידועה. אם כן מרחק נקודת ב̇ מנקודת הקצור הקטן ר"ל נקודת ו̇ ידועה. וזה הוא מרחק נקודת הפוך החום מן הנקודה אשר יהיה השמש בה כשיהיה בנקודת הקצור הקטן לסדר המזלות, [אחר תקונו בחלוף אשר יחייב אותו],[220] ותהיה זאת הנקודה בה̇ חלקים בקרוב מתאומים כמו שיצא לבטלמיוס. ותהיה נקודת הקצור הגדול בחלק המקביל מהמזלות.

B.1.III.6 ואחר אשר בארנו זה נשוב לבאר שיעור המותרות אשר בין קוי אמ̇ גז̇ ובין העמודים היוצאים משתי נקודות אג̇ עלי שני קוים דמ̇ דז̇ כדי שנתקן בזה הערך. ונקדים ההקדמה אשר בארנו אותה בכלל הראשון והיא זאת. כשיפלו בעגולת שלש נקודות בכדי שתהיה כל אחת מהקשתות הנחתכות באלו הנקודות פחות מחצי עגולה ונדבק קו בין הנקודה הראשונה והשלישית, ר"ל יתר הקשת הנקבצת מה שתי קשתות, ואחר כן נדבק קו בין מרכז העגולה ובין הנקודה האמצעית מהשלשה יחתוך הקו הנזכר לשני חלקים, יהיה ערך בקע אחת מהקשתות הנחתכת אצל בקע האחרת כערך [החלק][221] מן היתר הנזכר הסמוך לזאת הקשת אצל חלקו השני.

.216 ו: -.
.217 ו: פלאן מוצע נקטה ב̇ מן אלברוג̇ הו נקטה מ̇.
.218 ו: נבאלי פי אלעאג̇ל.
.219 ו: אג̇.
.220 ו: -.
.221 ו: אלקוס.

B.1.III.7 ואחר שהקדמנו זה נחשוב כי קו אמֹ פרק משותף לעגולת המעבר ולעגולת המזלות. ונוציא אותו עד נקודת חֹ ויהיה קו אמֹחֹ יתר לקשת מעגולת המזלות אשר ייתר אותו קשת אבֹחֹ מעגולת

110b

המעבר. וכשיצא קו דמֹ על יושר יחתוך זאת הקשת מעגולת המזלות. אם כן כבר נפלו בזאת הקשת שלש נקודות השתים הם נקודות אחֹ והאמצעית הנקודה אשר יחתוך עליה קו דמֹ לזאת הקשת. והשתי קשתות הנחתכות באלו הנקודות ידועות לפי שהקשת אשר מנקודת אֹ עד הנקודה האמצעית היא בשיעור זוית [אכֹמֹ],[222] והקשת השנית היא בשיעור הזוית אשר תוסיף או תגרע בסבת קשת בחֹ. ותהיה ידועה לפי שקשת בוֹ ידועה, ובסבת הערך הקשת אשר מעגולת המזלות אשר ייתר אותו קשת בוֹ ידועה, וגם כן הקשת אשר תיתר אותו קשת וחֹ. תשאר הקשת אשר תייתר אותו בחֹ ידועה. ולפי ההקדמה ערך בקע הקשת אשר מנקודת אֹ עד הנקודה האמצעית והוא העמוד היוצא מנקודת אֹ על קו דמֹ והוא ידוע, אצל בקע הקשת השנית והוא העמוד היוצא מנקודת חֹ על זה הקו, והוא ידוע גם כן, כערך קו אמֹ אצל קו מחֹ. ואחר שכל קו אחֹ ידוע קו אמֹ ידוע. ובכמו זה יתבאר שיעור קו גזֹ וזה משֹ.

B.1.III.8 ואם נחזור להוציא הערך באלו השיעורים ויצא הערך אשר בין [][223] הקטרים, ונוציא שיעור המותרות פעם שנית. וחזרנו להוציא הערך פעם שנית אחר שנוסיף או נגרע על הקשתות שיעור החלוף המתחייב מאלו הקשתות מצד העתקת השמש סביב קטבי השתי עגולות הנזכרות [או מצד העקום הנזכר. והבן זה הענין ממני בכל מקום אזכור בו בזה הספר ענין השתי עגולות כי העיקום יעמוד במקומו, כדי שלא נחזור ענין אחר פעמים רבות][224] יצא הערך האמיתי. וידוע שאיפשר לנו לשום [עקר שורש מונח][225] לתנועות האמצעיות בעתי העיונים אשר זכר אותם לשווים או להפוכים להוציא בכל עת שנרצה המקומות האמצעיות לשמש בעגולת המזלות ובעגולת המעבר.

B.1.III.9 וכדי שיתבאר דרך חשוב החלופים החלקיים נשים עגולת מעבר המרכז אהֹגמֹ מרכזה זֹ הפרק המשותף לה ולשטח העגולה הנושאת כשתהיה ממששת לעגולת המזלות קו בזֹמֹ. ויהיה קו אזֹגֹ נצב עליו, ונקודות אגֹ אחת מהן נקודת הקצור הגדול והשנית נקודת הקצור הקטן. ותהיה נקודת הקצור הקטן נקודת גֹ ונשים נקודת המעבר האמצעי בֹ ונדביק בזֹ ודבֹ ודזֹ. ולפי שערך דבֹ אצל בזֹ ידוע וזוית זֹ נצבת, תהיה זוית בדזֹ והיא זוית החלוף הגדול ידועה. ונשים מרכז השמש על נקודת חֹ על שיהיה מרחקו [][226] גֹ, רלֹ נקודת הקצור הגדול, ידוע. ונוציא מנקודת חֹ עמוד חטֹ על קו אגֹ ונדביק דחֹ דטֹ. ולפי שקשת חגֹ אבל זוית חזגֹ רלֹ זוית התנועה האמצעית ידועה, יהיה ערך זחֹ אצל חטֹ ידוע [],[227] וזוית חטדֹ נצבת. אכֹ זוית חדטֹ רלֹ זוית החלוף המתחייבת מקשת חגֹ ידועה, וכזה יתבאר בשאר ההנחות. ובכמו זה יתבאר כי כשתהיה זוית החלוף ידועה תהיה זוית התנועה האמצעית ידועה. ותהיה מצד זה זוית התנועה האמיתית, רלֹ הנראת, ידועה בשנוסיף זוית החלוף על התנועה האמצעית כשיהיה השמש על קשת גהֹאֹ ונגרע

110a

אותה ממנה כשיהיה על קשת אבֹגֹ.

222. ו: אדֹמֹ.
223. ו: נצפי.
224. ו: -.
225. ו: אצל לתקייד.
226. ו: מן נקטה.
227. ו: פאדאֹ נסבה דחֹ אעני נצף אלקטר אלי חטֹ מעטאה.

B.1.III.10 ולפי שבהיות השמש על נקודת הקצור הגדול ר״ל נקודת ג׳, ונעתק עד נקודת ח׳ בשעור התנועה האמצעית, נעתק גם כן מרכז השמש מוסיף לזה הצד בסבת הענין הנזכר בהעתקתו סביב קטבי השתי עגולות הנזכרות. ושיעור זאת ההעתקה הוא בשעור קשת החלוף. בעבור זה נוסיף על קשת ח׳ג׳ [קשת][228] שוה לקשת החלוף אשר תחייב אותה ולחשוב שנית קשת החלוף אשר תחייב זאת הקשת הנקבצת והיא קשת החלוף האמיתי. ולא נסיר מלהוסיף זאת הקשת על התנועה האמצעית להוציא בה החלוף כל זמן שיתמיד השמש על הקשת המוסיף בקצור, ר״ל הקשת אשר בין המעברים הבינונים מצד המהלך הגדול. ולא נסור מלגרוע אותה בשאר העגולה ונחשוב החלוף אשר יחייב הנשאר והוא החלוף האמיתי. א״כ כבר הצענו כל מה ראוי לבנות עליו מענין תנועת השמש וחלופו וכל זה מדובק למה שיצא מעקרי בטלמיוס זולתי המעוט אשר הוא מסגולת אלו העקרים שלא יכנס הזק מצדו מבלתי שנצטרך בדבר ממנו לשום גלגל יוצא המרכז ולא גלגל הקפה. והוא ית׳ היודע.

[וכבר התבאר ממה שקדם כי מספר גלגלי השמש כפי עקר העקום חמישה גלגלים, שני גלגלים מהם ראשיים מחברת אחת משלשה גלגלים. הראשון, ר״ל התחתון מן הראשיים, הוא גלגל הנושא למרכז השמש והוא אשר בו עגולת המעבר. וממעל לו מחברת אחת משלשה גלגלים לעקום עגולת המעבר בעקום גלגלה כמו שנתבאר. ולמעלה ממנה הגלגל אשר בו העגולה הנושאת לקטבי עגולת המעבר המעתקת אותה באורך. ואמנם כפי עקר השתי עגולות העומדות במקום העקום הם ארבעה גלגלים, יותר תחתי מהם מתנועע בתנועתו המיוחדת לו ומתנועע ג״כ בכל תנועות הגלגלים הנזכרים אשר למעלה ממנו, וזה כשנשים תנועת האורך על קטבי הנושאת. ואמנם כשנשים אותה על קטבי המזלות, אז יהיו הגלגלים כפי עקר העקום חמשה גם כן. ואמנם כפי עקר השתי עגולות יהיו יותר מאלו, ולא יתעלם ממך מספרם וסדרם.][229]

נשלם המאמר הראשון [מן הכלל השני][230] מספר אור עולם והשבח לאל יתעלה.

B.2.0.1 המאמר השני בביאור היסודות אשר עליהם יאות לעשות בתנועות הירח והחלופים הנראים בו ובו חמשה פרקים.

B.2.1.1 [ ][231] הראשון בעקרים הנתונים לתנועות הירח הסבוביים. אמנם העתקת הירח על פי הדעת השני דומה להעתקת השמש אלא שקצורו יותר גדול מקצור השמש מצד שבירת הכח בסבת מרחקו מן המניע [ ].[232] ויראו בו גם כן חלופים רבים ויאות לנו שיהיו עקרינו הנתונים על זה בנויים על עיוני בטלמיוס כמו שעשינו בשמש. ונאמר כי בטלמיוס מצא הירח מתרחק לצד הצפון ולצד הדרום מעגולת המזלות חלקים שוים. אם כן הירח ראוי להיותו נעתק על עגולה נוטה על עגולת המזלות בשעור ריחוק הירח ברוחב מעגולת המזלות אשר מצאו קרוב מה׳ חלקים. נמצא גם כן כי חזרת הרוחב יותר מהירה מחזר האורך. וזה כי מצא בעיונו כי הירח משלים החזרה ברוחב בפחות

109b

מהזמן אשר ישלים בו החזרה באורך. וזה אי אפשר כפי עקריו זולתי בהעתקת נקודת החתוך [הנקרא ראש התלי][233] על חלקי גלגל המזלות לצד המערב, ר״ל להפך סדר המזלות. והביא ראייה

228. ו: -.
229. ו: -.
230. ו: -.
231. ו: אלפצל.
232. ו: אלאול.
233. ו: -.

גם כן על העתקת החתוך לפי שמצא הרוחב הגדול פעם לצד צפון ופעם לצד דרום בחלק [אשר][234] מהמזלות עם היות הירח בשניהם צפוני מעגולת המישור או דרומי. ופעם מצא שאין רוחב לו בזה החלק בעצמו.

B.2.1.2 והביא ראיה על זה גם כן לפי שמצא שהשמש והירח ילקו בכל אחד מחלקי המזלות עם שמירת הסדר, נמצא שיש בכאן שתי תנועות ברוחב. האחת תנועת הירח על קטבי גלגליו המיוחדים בו ר״ל על קטבי עגולת נטיית הירח. ובזאת התנועה יראה הירח פעם דרומי מעגולת המזלות ופעם צפוני, לא בחלק אחד מהמזלות עד כי כשיהיה הרוחב הגדול בדרום בחלק אחד יהיה הרוחב הגדול בצפון בחלק המקביל לו ומן הצד האחר מעגולת המישור. וזה אם היתה כל אחת מנקודות החתוך שוקטות שומרות חלק אחד מעגולת המזלות. ובזאת התנועה יהיה לירח רוחב בהקש לעגולת המזלות ובהקש לעגולת המישור יחד.

B.2.1.3 והתנועה השנית העתקת עגולת הרוחב עם נקודת החתוך על חלקי המזלות להפך סדרם שומרות רחבה מעגולת המזלות. ובזאת התנועה מעורבת בראשונה יהיה לירח רוחב. ויהיה הירח פעם בתכלית מרחקו בדרום ופעם בתכלית מרחקו בצפון [בחלק][235] אחד מהמזלות עם היות הירח [בשני העתים][236] בצד אחד מעגולת המישור. וזה לפי שהעתקת החתוך עם שמירת תכלית נטייתה מעגולת המזלות אינה זולתי העתקת קטבי גלגל הירח, ר״ל קטבי עגולת רוחב הירח על עגולת מעבר קטבה קטב גלגל המזלות, והקשת אשר מקטבה למקיפה שוה לתכלית נטיית עגולת הרוחב מעגולת המזלות. וכשיזדמן שיהיה קטב גלגל הירח בתכלית מרחקו מקטב הכל והיה הירח בתכלית הדרומי, דרך משל מעגולת נטייתו, יהיה הירח אז בתכלית הנטייה שאיפשר להיות לו מעגולת המזלות ומעגולת המישור. ותהיה אז נטיית הירח מעגולת המישור שתי הנטיות יחד לצד הדרום, ואז היה בתכלית הצפוני לצד הצפון על החלק המקביל לו.

B.2.1.4 ואמנם כשיזדמן להיות קטב גלגל הירח בתכלית קירובו מקטב הכל בעגולת מעברו והיה הירח בתכלית הצפוני, יהיה הירח אז בתכלית נטייתו לצד הצפון מעגולת המזלות בזה החלק בעצמו אשר נטה ממנו לצד הדרום ראשונה. ויהיה אז רוחב הירח מעגולת המישור לצד הדרום מותר הגדולה שבשתי הנטיות על הקטנה. ואם היה בתכלית הדרומי יהיה בהפך. ולזה קראתי זאת התנועה תנועת הרוחב אשר בהקש המזלות וקראתי הראשונה תנועת הרוחב הכללית. ובטלמיוס דייק אלו התנועות בעיוניו הנמשכים בלקיות הירח לפי שלא יכנס בהם שנוי מראה. ומצא תנועת האורך האמצעית ביום יג׳ י׳ לה׳ בקירוב ותנועת הרוחב יג׳ יג׳ מו׳ בקירוב, ר״ל [הקירוב][237] הכללית. ומצא תנועת החלוף, ר״ל תנועת הירח בגלגל ההקפה אשר הניחו יג׳ ג׳ נד׳ בקירוב ביום. ויהיה תוספת תנועת הרוחב על תנועת האורך ג׳ דקים ביום. ובזה השעור יהיה העתקת החתוך [ר״ל ראש התלי וזנבו][238] להפך כי התנועה אשר תהיה על עגולת הרוחב היא בעצמה תעשה תנועת האורך זולתי אלו השלשה דקים אשר יפחתו מהאורך לסבת העתקת החתוך [להפך].[239] וזה

234. ו: ואחד.
235. ו: פי.
236. ו: פי אלנקטתין.
237. ו: אלערץ׳.
238. ו: -.
239. א: להיפך.

109a

לפי שנטיית זאת העגולה מעגולת המזלות קטנה בכדי שאינה מחייבת תוספת ולא חסרון בתנועת האורך.

B.2.1.5 ונשים ראשונה העקר לתנועות האמצעיות באורך וברוחב. ותהיה עגולת המזלות המדומה בגלגל הכוכבים הקיימים עגולת גׄטׄדׄ קטבה בׄ שהוא מסבב על עגולת לׄבׄנׄ אשר קטבה קטב [גלגל][240] המישור והוא אׄ. ותהיה העגולה הנוטה אשר לירח המדומה בזה הגלגל גם כן עגולת גׄהׄדׄ קטבה מׄ וזה הקטב הוא מסבב תמיד על עגולת מעבר למׄנׄ קטבה קטב עגולת המזלות רלׄ בׄ. והעגולה העוברת בנקודות ההפוכים ובקטבים עגולת הׄטׄבׄאׄ, כי לפי הדעת הראשון אין צורך לדבר זולתי שנדמה אלו השתי עגולות בזה התואר בגלגל הירח על שיהיה נחלק כמו שאמרנו. ואמנם לפי הדעת השני נשים בגלגל הירח שתי עגולות אחרות במקום עגולות המזלות והעגולה הנוטה, כל אחת מהן בשטח חברתה מן [המדומות][241] בגלגל העליון, וקטב עגולת הרוחב סובב [גם כן][242] על עגולת מעבר קטבה קטב העגולה הנמשלת לעגולת המזלות דומה אל המדומה בגלגל העליון.

B.2.1.6 ונשים דרך משל כי הירח בתכליתו הדרומי ותהיה נקודת הׄ ומקומו מהמזלות נקודת הפוך הקור. ואז יהיו עגולות המזלות והרוחב המדומות בגלגל הירח בשטחי חביריהם המדומות בגלגל העליון. וגם כן קטביהן החבירים על בריח אחד ורוחב הירח מעגולת המישור שתי הנטיות יחד. ונשים כי ביום אחד המשל נעתק גלגל הירח בכלל נמשך לתנועה היומית וקיצר כל אחד מקטבי עגולת המזלות ועגולת הרוחב המדומה בו מקטב העליון קשת מעגולת מעבר קטב עגולת המזלות, שיעורו כפל הקצור האחרון. וחנה קטב עגולת המזלות על נקודת חׄ מעגולת מעבר הקטב וקטב עגולת הרוחב על נקודת כׄ. וכמו זה קצר נקודת היפוך הקור המדומה בגלגל הירח ונקודת התכלית הדרומי מעגולת הרוחב שתי קשתות מן העגולות הנכוחיות לעגולת המישור שיעור כל אחד מהם כפל הקיצור האחרון, רלׄ [קשת][243] שוה לקשת בׄחׄ. ותהיה המשל עגולת הרוחב שהיתה ראשונה מדומה בשטח עגולה גׄהׄדׄ היא עגולת סׄזׄ והתכלית הדרומי אשר היא על נקודת הׄ יהיה על נקודת סׄ.

B.2.1.7 ויאות לנו לבאר איך יחתוך מרכז הירח הרביע מעגולת רחבו אשר מנקודת סׄ עד חתוך עגולת הרוחב עם עגולת המזלות, רלׄ החתוך, בזמן אשר יחתוך רביע הׄדׄ כשמתחיל מנקודת הׄ להפך תנועת הכל על דעת בטלמיוס ועל דעתינו כפי הדעת הראשון. ועם זה לא יחתוך רביע שלם מעגולת המזלות המדומה בגלגל הירח בזה הזמן כמו שלא יחתוך רביע טׄדׄ מעגולת המזלות בזה הזמן על הדעת הראשון.

108b

ואמנם נתחייב זה על ההנחה הראשונה לפי שהושם בה כי [קצור][244] דׄ נעתק לצד תנועת הכל רלׄ לצד נקודת טׄ, אם כן הירח יחתוך רביע הׄדׄ בזמן אשר יחתוך בו פחות מרביע טׄדׄ.

B.2.1.8 וזה יהיה כפי זה העקר השני על זה התואר. כי מרכז הירח יתנועע על קוטב עגולת הרוחב, רלׄ קטב בׄ, ביום אחד קשת סׄזׄ שיעורו החלקים שאמר בטלמיוס שהם תנועת הרוחב והם יׄגׄ מׄוׄ בקירוב. וזאת התנועה אינה בזה השיעור בעגולת המזלות המדומה בגלגל הירח אבל היא פחות ממנה בגׄ דקים. וזה יהיה כשנשים נקודת סׄ, רלׄ התכלית הדרומי מעגולת הרוחב המדומה בגלגל הירח, מקצרת ביום אחד

240. ו: -.
241. ו: אללואתי.
242. ו: -.
243. ו: -.
244. ו: עקדה.

מהגיע למקומה יותר ממה שקצרה נקודת היפוך הקור המדומה בזה הגלגל ג׳ דקים, והתעות הנכחות אשר היא בין נקודת ס׳ ובין נקודת ההפוך. ונעשה מקום נקודת ס׳ על שלש דקים מראש גדי עם שמירת תכלית נטייתה. ר״ל כי נטיית נקודת ס׳ ברוחב מזאת הנקודה אשר היא עליה שוה לנטייה שהיתה לה מראש גדי ראשונה.

B.2.I.9 וכשקיצר קטב עגולת המזלות המדומה בגלגל הירח ר״ל ב׳ קשת ח״ב היה ראוי שלא יגיע קטב עגולת הרוחב ר״ל מ׳ לנקודת כ׳. ר״ל שיהיו השלשה קטבים על קשת אחד מעגולה גדולה כמו קשת אח״כס, לפי שהיה כן היה קצור התכלית הדרומי שוה לקצור ההיפוך. ויתחייב לזה שלא יעתק קטב מ׳ שומר להנחתו ממנו, אבל יהיה כאלו קטב עגולת המזלות יקדים אותו ויעזבהו לאחריו על מקיף עגולת למ״נ מאצל נקודת מ׳ לצד נקודת נ׳. וכשיגיע קטב ב׳ לנקודת ח׳ ודמינו עגולת למ״נ על קטב ח׳, ונקודת מ׳ על קטב כ׳, לא יגיע קטב עגולת הרוחב שהיה על נקודת מ׳ לנקודת כ׳ [כי אם][245] לנקודת ר׳ המשל. ולא יסור יעתק זה הקטב על זאת העגולה לזה הצד נמשך לתנועת קטבי עגולת המזלות המדומים בו, עד שיהיה תמיד קצור התכלית יותר מקצור ההיפוך בזה השיעור בכל יום.

B.2.I.10 ויתנועע מרכז הירח קשת ס״ז מעגולת הרוחב ויראה שהתנועע בעגולת המזלות קשת פחות ממנה בשעור הנזכר, ויגיע הירח לרוחב החלק מהמזלות אשר יגיע אליו כפי עקרי בטלמיוס. ויגיע אל החתוך גם כן בזמן אשר יגיע אליו ולרוחב זה החלק עצמו אשר עליו החתוך כפי עקריו. וישאר אחר זה קצור האורך האחרון השיעור הנזכר לחולשת הכח היורד עליו, ויהיה העתקת [ראש התלי וזנבו][246] עם שמירת תכלית הנטייה תמיד. וזה לפי שהעתקת קטבי עגולת הרוחב אמנם היא על עגולה קטבה קטב עגולת המזלות והיא עגולת למ״נ. ואין זאת ההעתקה [תנועה][247] להפך לפי שזאת ההעתקה אמנם היא נמשכת לתנועת קטבי עגולת המזלות לפי שכוונתה שיהיה קצורה פחות מקיצור עגולת הרוחב, ר״ל קצור הגלגל אשר היא בו, לפי שהתנועה אמנם היא לגלגל לא לעגולות הפשוטות. ויאות לך שתזכיר מה שאמרנו במאמר הראשון כי הגלגל מתחלק לשני גלגלים או שלשה או יותר למה שנצטרך אליו כשנשים לגלגל האחד שתי תנועות או שלשה על קטבים מחולפים.

B.2.II.1 הפרק השני בביאור החלוף הראשון האמיתי הנראה לירח. ואחר שזכרנו העקרים אשר בהם יראו התנועות האמצעיות באורך וברוחב על פי מה שהם [נשוב לבאר][248] הדרך [][249] אשר עליו ילך הענין בחלוף הראשון והוא לבדו אנו צריכין לדעתו בלקיות. וזה יהיה בדומה למה שעשינו בשמש על זה התואר.

B.2.II.2 תהיה העגולה

108a

הנוטה אשר לירח המדומה בגלגל העליון בעל הכוכבים, ר״ל עגולת הרוחב עגולת ג״ה״ד קטבה מ׳ סובב על עגולת קטבה קטב גלגל המזלות. ונשים בזה הגלגל עגולה נוטה עליה בשיעור חצי קטר עגולת המעבר אשר נשים אותה והיא עגולת ג״ט״ד קטבה סובב גם כן על עגולת מעבר קטבה קטב עגולת הרוחב. ויהיה קטב עגולת המזלות ב׳ סובב על עגולת ב״ח אשר קטבה קטב גלגל המישור ויהיה א׳.

245. ו: בל.
246. ו: אלעקדתין.
247. ו: -.
248. ו: פלנבין אלאן.
249. ו: אלאול.

ונדמה בגלגל הירח שתי עגולות במקום עגולת הרוחב והעגולה הנוטה עליה כל אחד מהן בשטח חבירתה [חׄ בירח][250] מן המדומות בגלגל הכוכבים [][251] החבירות על בריח אחד.

B.2.II.3 ונשים המשל כי נקודת הׄ התכלית הדרומי מעגולת הרוחב וגם כן נקודת טׄ מן העגולה הנושאת ושמקומותן מעגולת המזלות נקודת הפוך הקור. ונשים כי מרכז הירח סובב על עגולה [][252] קטבה נשוא על העגולה המדומה בגלגל הירח בשטח עגולת גטׄדׄ וסובב עליה. ותהיה זאת העגולה עגולת כׄלׄנׄעׄ וקטבה קׄ על שנשים קטב קׄ על נקודת טׄ, והקשת אשר מקטבה למקיפה שוה לקשת הדומה בקשת הׄטׄ, רׄלׄ תכלית נטיית העגולה הנושאת על עגולת הרוחב. וכאשר נשים קטב עגולת מעבר המרכז כאלו הוא על נקודת טׄ, ומרכז הירח בעגולת מעברו בתכלית הסמוך לעגולת המזלות רׄלׄ על נקודת כׄ, יהיה מרכז הירח בתכלית הדרומי מעגולת הרוחב. ויהיה מקומו מהמזלות נקודת הפוך הקור.

B.2.II.4 ונבאר ראשונה זה הענין על הדעת השני. ונשים המשל כי ביום אחד קיצר קטב עגולת המזלות המצויירת בגלגל הירח אשר [היה][253] על נקודת בׄ מהגיע למקומו קשת חׄבׄ מעגולת מעבר הקטב שיעורו כפל קיצור האורך האחרון. והעתיק עמו [שתי עגולות מעבר][254] קטבי עגולת הרוחב והעגולה הנושאת רׄלׄ עגולת מׄרׄ ועגולת שׄתׄ. ובזה הזמן יעתק קטב עגולת הרוחב להפך קשת מׄרׄ שיעורו השלש דקים. וחנה קטב עגולת הרוחב על נקודת רׄ וקטב העגולה הנושאת על נקודת שׄ. ותהיה לפי זה עגולת הרוחב שהיתה ראשונה בשטח

107b

עגולת גׄהׄדׄ היא עגולת צׄסׄ תכליתה הדרומי אשר היא ראשונה על נקודת הׄ חנה על נקודת סׄ. והתכלית הדרומי מהעגולה הנושאת אשר היה על נקודת טׄ חנה על נקודת זׄ וקטב עגולת מעבר המרכז אשר היה על נקודת טׄ חנה על נקודת זׄ גם כן. אם כבר קצרו שני תכליתי הדרום אשר לעגולת הרוחב ולעגולה הנושאת אשר על שתי הנקודות סׄזׄ, וקטב עגולת המרכז, מהגיע למקומותיהם יותר ממה שקיצר נקודת היפוך הקור המדומה בגלגל הירח שלשה דקים עם שהם שומרות מרחבם מעגולת המזלות. ונשים גם כן כי בזה היום נעתק קטב עגולת מעבר המרכז סביב קטב העגולה הנושאת אותו נמשך לתנועת הכללית מנקודת זׄ עד נקודת קׄ שיעור קשת זׄקׄ שוה לתנועת הרוחב ביום ואפילו ששתי נקודות לׄעׄ, רׄלׄ נקודות החתוך, שומרות תמיד למקיף עגולת צׄזׄ.

B.2.II.5 אמנם בסבת השני גלגלים הנושאים לשתי העגולות הנזכרות בשמש אשר כמותם צריך לשום בירח כמו שקדם יהיו [][255] נקודות החתוך שומרות למקיף עגולת גׄטׄדׄ בהקש למרכז הירח. וישאר לפי זה קיצור הירח השיעור אשר אמרנו שהוא קצור האורך האחרון. ונשים גם כן בזה היום נעתק מרכז הירח שהיה על נקודת כׄ על קטב עגולת המעבר לצד התנועה הכללית גם כן קשת כׄוׄ שעורו החלקים אשר אמרנו שהם לחלוף. אבל בהציענו העקר על זה התואר יתחייב שלא יגיע מרכז הירח למקיף עגולת הרוחב המדומה בגלגל הכוכבים. וזה לפי שחלקי הרוחב יותר מחלקי החלוף לפי שקשת זׄקׄ יׄגׄ יׄגׄ מׄוׄ וקשת כׄוׄ יׄגׄ גׄ נׄדׄ.

250. ו: -.
251. ו: ואקטאב.
252. ו: צגירה.
253. ו: כאן. א: היא.
254. ו: דאירה ממר.
255. ו: כאן.

B.2.II.6 והענין בירח אינו כן לפי שכבר יראה מצד הלקיות ושיעורם כי הירח לא יצא תמיד ממרחב עגולת הרוחב. ויראה מזה כי תנועת קטב עגולת מעבר המרכז על קטב העגולה הנושאת אותו שוה לתנועת החלוף, וכי קטב העגולה הנושאת ותכליתה הדרומי יעתק לזה הצד גם כן שומר מרחבו מעגולת הרוחב תמיד על קטב עגולת הרוחב בשעור מותר תנועת הרוחב על תנועת החלוף והוא י׳ דקים בקירוב. וזה יצוייר בזה התואר, נדמה קטב עגולת מעבר המרכז [והתכלית][256] הדרומי מן העגולה הנושאת על נקודת ז׳, ואז יהיה קטב העגולה הנושאת על נקודת ש׳ מעגולת מעברו. ובזה היום יעתק קטב העגולה הנושאת סביב קטב עגולת הרוחב ר׳ל׳ ר׳ לצד התנועה הכללית מנקודת ש׳ עד נקודת [כ׳][257] שיעור זה הקשת שוה למותר תנועת הרוחב על תנועת החלוף והוא עשרה דקים. ונעתק התכלית הדרומי אשר היה על נקודת ז׳ וחנה על נקודת פ׳ בכמו זה השיעור.

B.2.II.7 ובזה השיעור הוסיף הירח בתנועת האורך. ובזה היום נעתק קטב עגולת מעבר המרכז על קטב העגולה הנושאת לזה הצד בכמו תנועת החלוף ויהיה קשת פ׳ק׳. ובכמו אלו החלקים נעתק מרכז הירח לצד התנועה הכללית גם כן מנקודת כ׳ מעגולת מעברו עד נקודת ו׳ ר׳ל׳ בשיעור חלקי החלוף. ויהיה חסרון רוחב נקודת ו׳ מרוחב נקודת כ׳ שוה לחסרון רוחב נקודת ק׳ מן העגולה הנושאת מרוחב נקודת פ׳. ויגיע בזה מרכז הירח לרוחב החלק מעגולת הרוחב המצויירת בגלגל הכוכבים הדומה לחלק אשר הוא בו מעגולת הרוחב המדומה בגלגל הירח. ועם זה צריך לחבר בזה המקום הענין אשר זכרנו אותו בשמש מהשתי עגולות הדומה לזה הענין.

B.2.II.8 ונבאר זה הענין

107a

על הדעת הראשון. [][258] ונצייר כפי הדעת הראשון כי התכלית הדרומי מהעגולה הנושאת ר׳ל׳ נקודת ט׳ נעתקת להפך התנועה הכללית על חלקי עגולת הרוחב שומר מרחבה ממנה תמיד בשיעור מותר תנועת הרוחב על תנועת החילוף. ותהיה המשל זאת ההעתקה ביום אחד מט׳ לפ׳ בצורה השנית בהעתקת קטבה מש׳ עד ת׳, ויעתק קטב עגולת המעבר לזה הצד על העגולה הנושאת בשיעור תנועת החלוף. ויעתקו [ראשי התלי וזנבו][259] להפך גם כן השיעור הנזכר ממ׳ לר׳ בזאת הצורה הנזכרת. ויעתק הירח בעגולת מעברו ועל השתי עגולות הנזכרות בשיעור תנועת החלוף.

B.2.II.9 ואשר יתחייב מזאת ההנחה דומה למה שיתחייב מן ההנחה השני, לפי שאפילו נשים כי התכלית הדרומי אשר היה על נקודת ז׳ על עגולתו נעתק לצד התנועה הכללית השיעור הנזכרת שומרת מרחבה מעגולת הרוחב בהעתקת קטבה על שתי עגולות מעברם סביב קטבי עגולת הרוחב, ונעתק [קטב][260] עגולת מעבר המרכז על קטבי העגולה הנושאת בשיעור תנועת החלוף לזה הצד, ונעתק [ראש התלי וזנבו][261] בהעתקת הקטבים על עגולת מעבר מ׳ר׳ אשר קטבה קטב עגולת המזלות להפך השיעור הנזכר, קטב עגולת מעבר המרכז הגיע לנקודה [][262] בכל אחד משני העקרים. ולא יהיה ביניהם הפרש ברוחב כלל. וזהו המחייב [][263] הלקות והמחייב גם כן שיעורי הלקיות כפי המרחק מהחתוך.

256. ו: -.
257. ו: ת׳.
258. ו: ונקול אנא.
259. ו: אלעקדתין.
260. א: הקטב. ו: קטב.
261. ו: אלעקדתין.
262. ו: ואחדה.
263. ו: לחדות.

B.2.II.10 ואמנם ההפרש אשר ביניהם הוא מצד התוספת או החסרון באורך אשר מצד המצעדים כמו שבארנו במאמר הראשון. וההזק שאיפשר להכנס מזה הצד אמנם יפול על הרוב בקדימת עת הלקות או איחורו אשר כמו זה הלקות או יותר ממנו יפול פעמים רבות בהוצאת אלו העתים. ואמנם ההזק שאיפשר להכנס מזה הצד בשיעור הלקות אינו ממה שראוי לחוש עליו. אם כן לא ישאר בהם כי אם הברירה [][264] עם שכבר הודענוך איזה מהם יותר ראוי [לברור][265] בו, [והוא העקר אשר הושמו בו התנועות לצדדים מחולפים בעבור שמירת המרכז הרוחב.][266]

B.2.III.1 הפרק השלישי בשעור הגדול מזה החלוף. [ועתה][267] נבאר ערך חצי קטר גלגל הירח אצל חצי קטר עגולת המעבר כדי שנרמוז לדרך חישוב היותר גדול מזה החילוף. וזה יהיה בלקיות עצמם אשר

106b

[בהם][268] ביאר בטלמיוס ערך חצי קטר הנושא אצל חצי קטר גלגל ההקפה. וזה כי הוא לקח אלו הלקיות הירחיים והוציא החלק מהמזלות אשר היא בו אמיתית השמש באמצע הלקות הראשון ובאמצע השני ומצא המרחק בין שתי המקומות, והוא עצמו המרחק בין שני מקומות הירח להקבלתו לשמש. רל̇ התנועה האמיתית בזמן אשר היה בין שני הלקיות יותר מן התנועה האמצעית אשר לירח בזה הזמן הידוע אצלו. אם כן תנועת החלוף לירח אשר היתה ידועה מצד הידיעה בזמן תוסיף במהלך האורך [בכדי][269] תוספת המהלך האמתי על האמצעי בזה הזמן. ולקח גם כן החלק מהמזלות אשר היתה בו אמיתת השמש באמצע הלקות השלישי ומצא בו תנועת אמיתת השמש אבל אמיתת הירח על הקפות שלמות בזמן אשר בין השני והשלישי יגרע התנועה האמצעית לירח בזה הזמן חלקים ידועים. אם כן תנועת חלוף הירח הידועה בזה הזמן [היא המחייבת זה הגרעון].[270]

B.2.III.2 ואחר שהצענו זה תהיה עגולת מעבר המרכז בא̇ג̇ה̇, והפרק המשותף בינה ובין שטח העגולה הנושאת אותה כנ̇, ומקום הירח באמצע הלקות הראשון נקודת א̇. והתנועע עד הלקות השני על הקפות שלמות קשת א̇ג̇ב̇, מיתרת מהמזלות חלקים ידועים להוסיף, והוא כפי חשבון בטלמיוס ג̇ חלקים כד̇ דקים. וידוע כי קשת ב̇א̇ יחסר [חלקים][271] כמותם. והתנועע מן הלקות השני עד השלישי קשת ב̇א̇ג̇ על הקפות שלימות מיתרת מהמזלות לז̇ דקים לגרוע כפי חשבונו גם כן וישאר קשת א̇ג̇ יוסיף שני חלקים מז̇ דקים. ונוציא מנקודת הלקות השני עמוד על הפרק המשותף יגיע למקיף עגולת המעבר בצד המקביל והוא עמוד ב̇ל̇ה̇. ונוציא מנקודת א̇ג̇ [שני עמודים][272] על קו ב̇ה̇ והם א̇מ̇ ג̇ז̇ ונוציא א̇ה̇ ג̇ה̇ א̇ג̇. ונוציא מנקודת ג̇ [על קו][273] א̇ה̇ עמוד ג̇ט̇ ונוציא מנקודת ד̇ רל̇ מרכז המזלות עד נקודות הלקות הראשון והשלישי שני קוי ד̇א̇ ד̇ג̇ ועד נקודות נפילת עמודי א̇מ̇ ג̇ז̇ שני קוי ד̇מ̇ ד̇ז̇. ונקח עגולת הרוחב בזה המופת במקום עגולת המזלות לפי שאין ביניהם הפרש מורגש. ולפי שקו א̇מ̇ ידוע ומקום הלקות השני רל̇ נקודת ב̇ מהמזלות הוא בעצמו מקום נקודת מ̇ תהיה זוית א̇ד̇מ̇ רל̇ הגורעת מהקיצור בסבת קשת ב̇א̇ ידועה. וערך קו ד̇א̇ והוא חצי קטר גלגל המזלות המדומה בגלגל הירח אצל העמוד היוצא

.264 ו: אלא אכ̇תיאר אלמאכ̇ד.
.265 ו: אן יכ̇תה. א: לבאור.
.266 ו: -.
.267 ו: אן.
.268 ו: -.
.269 ו: מתל.
.270 ו: נקצהא.
.271 ו: -.
.272 ו: עמוד.
.273 ו: אלי.

מנקודת אׄ על קו דמׄ ידוע. וזה העמוד אמנם כשיהיה קו בהׄ קטר לעגולת המעבר יהיה הוא עצמו קו אמׄ. ואמנם כשיהיה זולת הקטר אינו קו אמׄ, ואנחנו נעשה בזה ובהוצאת המותר אשר בין העמוד ובין קו אמׄ [כמו][274] שעשינו בשמש. ויהיה לפי זה ערך דאׄ אצל אמׄ ידוע ולפי שקשת באׄ אבל זוית בהאׄ ידועה יהיה ערך אהׄ הקטר אצל אמׄ ידוע, אם כן ערך דאׄ אצל אהׄ ידוע. ולפי שמקום נקודת בׄ הוא נקודת זׄ תהיה זוית גדזׄ

106a

ידועה, אם כן ערך דגׄ רלׄ דאׄ אצל העמוד רלׄ גזׄ ידוע. ולפי שקשת באגׄ [אבל][275] זוית בהגׄ ידועה יהיה ערך גהׄ הקטר אצל גזׄ ידוע אם כן ערך דאׄ אצל גהׄ ידוע. ולפי שקשת אגׄ אבל זוית אהגׄ ידועה יהיה ערך הקטר אצל גטׄ אבל אצל הטׄ תשלומו ידוע. וישאר ערך גהׄ אצל טאׄ אבל אצל גאׄ אשר בכח שוה לגטׄ טאׄ ידוע וערך דאׄ אצל אגׄ ידוע, וכבר היה ערך גאׄ אצל קטר עגולת המעבר ידוע לפי שקשת אגׄ ידוע. אכׄ ערך דאׄ רלׄ חצי קטר [עגולת][276] המזלות אצל חצי קטר עגולת המעבר ידוע.

B.2.III.3 ויצא בזה הדרך הערך אשר יצא לבטלמיוס אחר התקון אשר יתבאר בקרוב והוא ערך סׄ אצל חמשה ורביע בקירוב זולתי התוספת או החסרון אשר מצד חישוב הזמנים בחלוף הימים בלילותיהם, כי לא יהיה לפי העקר השני כמו שהוא כפי עקרי בטלמיוס לפי שכבר נתבאר כי מסגולת זה העקר שיהיו שתי המזלות הסמוכות לנקודת השווי מוסיפין במצעדים. ולפי זה לא יהיה שעור זה החלוף השעור אשר זכר. ואם זה החלוף הוא כמו שאמר [][277] נצטרך לומר כי הטעות הנופל מזה הצד כפי זה העקר לא יזיק למיעוטו עם שאיפשר להיות המתחייב מזה העקר אמת. וזה הטעות הקטן ישיג העקר האחר. ואמנם לפי הדעת הראשון אין בזה הפרש כלל. והאל יתעלה הוא היודע.

B.2.III.4 ואמנם איך נדע מרחק נקודת כל לקות מנקודת הקצור הקטן הוא על זה הדרך. לפי שערך גהׄ אצל גאׄ ידוע יהיה קשת גהׄ ידוע אבל כל קשת באגהׄ ויצא פחות מחצי עגולה, וחציו רלׄ בנׄ ידוע. אם כן מרחק נקודת בׄ מנקודת המהלך הקטן או הקצור הקטן והיא נקודת וׄ ידוע. וכאשר נעשה בשלשה לקיות האחרונים שהביא אותם בטלמיוס כמו זה המעשה אז נוכל לתקן חזרת החלוף בזה הדרך עצמו אשר תקנו הוא.

B.2.III.5 ואחר זה נשוב לבאר המותרות [הנזכרות][278] בכמו זה המעשה אשר עשינו בשמש. ונוציא הערך פעם שנית ונוציא המותר פעם שנית בזה הערך ונחזור להוציא הערך אחר שנוסיף או נגרע על הקשתות החלוף המתחייב מהן. כמו שנזכר בשמש וכל מה שנחזור [בזה המעשה][279] יהיה יותר קרוב אל האמת [].[280] וידוע גם כן כי דרך חישוב חלוף הירח הוא בעצמו הדרך בשמש ונעתיק אותו לזה המקום כדי שלא נחזור לזכור הענין האחד פעמים רבות.

B.2.IV.1 הפרק הרביעי [יבאר][281] כי חלוף שעורי קטרי הירח אצל הראות לא יתחייב ממנו בהכרח חלוף מרחקי הירח מהארץ. ואחר זה נקח חלוף מה למראה הירח במרחק אחד מנקודת נוכח הראש ונוציא מצד זה מרחק הירח בהקש אל חצי קטר הארץ כמו שעשה זה בטלמיוס. ואחר זה נוציא שאר חלופי

274. ו: עלי קיאס.
275. א: אצל. ו: בל.
276. ו: פלך.
277. ו: אעני עלי אלאצל אלאול.
278. ו: -.
279. ו: -.
280. ו: ודמׄאׄ.
281. ו: -.

המראה החלקיים מצד הידיעה במרחק הירח. ונבאר מצד זה מרחק השמש ומרחק נקודת קצה הצל, וערכי קטרי השמש והירח והארץ וגם כן ערכי הגופים קצתם.

B.2.IV.2 ונוציא מצד זה גבולי הלקיות וגם כן עת כל לקות ושעורו ושעור שהות מי שיש לו שהות מהם בדרך אשר עשה זה בטלמיוס זולתי שאנחנו נוציא מה שנוציא משנוי המראה על שמרכז הירח במרחק אחד ממרכז הארץ תמיד כפי המתחייב מאלו העקרים. ואנחנו נזכיר זה [בה][282] בדרך היותר

105b

קצר שנוכל. וגם כן נעשה על כי קטר הירח ייתר זוית אחד אצל הראות בכל מקום לפי שתוספת הזוית הזאת בקצת המקומות ובקצת העתים איפשר ליחס אותו לסבות אחרות כמו שנוי האויר או לחותו יחייב אותו מבטי הירח בעתים ידועים. לפי שזה אי אפשר כפי העיון הטבעי ליחס אותו לריחוק הירח פעם וקירובו פעם למרכז הארץ, לפי שהקרוב והריחוק אי אפשר שידומה זולתי בהנחת גלגל יוצא המרכז או בהנחת גלגל הקפה וכבר נתבאר מניעות אלו העקרים.

B.2.IV.3 והראיה על אמיתת מה שאמרנו כי זה החלוף אשר לשיעורי הקטרים לא ימצא הולך תמיד על סדר אחד אצל הכל, [כי][283] בעלי זאת החכמה אשר שמו אלו העקרים חולקים במה שמצאו משעורי אלו הקטרים בראות עד כי אברכס יאמר שהוא [מצא][284] קטר הירח ייתר זוית שוה לאשר ייתר אותה קטר השמש אצל הראות בהיותו במעבר הבינוני מגלגל הקפתו. ובטלמיוס יאמר כי הוא מצא אותו ייתר כמו זאת הזוית בהיותו במרחק הרוחק ובנה על מה שמצא מזה. ואיך ייחס זה [הטעות][285] לכמו אברכס כי בזה העיון אין צורך למדידת חלקי השרביט, ובטלמיוס עצמו יודה זה. ואיך יפול בזה טעות כלל. והיה ראוי עליו שיאמר כי אברכס השיג האמת [][286] וכי המחייב לזה החלוף הוא מה שאמרנו. אבל שומו אלו העקרים הכריחו אותו לשום קטר הירח מיתר [זוית מחולפות][287] בהיותו במקומות מחולפים. ואחר כן זה מצא חלוף בשיעור זאת הזוית למעיינים וייחס אותו לאלו העקרים. ויהיה זה השעור מאמרינו מספיק בחלופי הירח האמתיים הצריך ידיעתם בלקיות.

B.2.V.1 הפרק החמישי בשני החלופים הנשארים לירח ר״ל החלוף העורך אצל בטלמיוס לשמש והוא אשר ישתמש בו ביוצא המרכז והחלוף העורך אצלו לעקום קטר גלגל ההקפה. אמנם העקרים אשר עליהם [נבנה][288] החלוף הראשון והוא לבדו הצריך ידיעתו בלקיות כבר דברנו בהם מה שיספיק. ואמנם העיון בשאר החלופים הנראים בו בזולת אלו העתים מן הראוי היה לשני דברים, האחד מהם כי אלו החלופים יעדרו בעתי הדבוקים אבל בעתים אשר יעדר בהם הראשון. אם כן לא יכנס הזק מזה הצד בלקיות. והשני כי הדברים אשר בם יתבארו אלו החלופים הם העיונים בכלים וכבר יכזבו פעמים רבות אלו העיונים מצד שנוי המראה ומצד המעיינים גם כן.

B.2.V.2 כי מצאנו אבן אלהיתם ביאר במאמר הרביעי מספרו בחכמה המראות כי השגת הראות לכוכבים אינו ביושר אמנם ישיג אותם דרך עקלתון. ואמר כי המעיין הכוכבים הקיימים בכלי הנקרא בעל הטבעות ימצא מרחק הכוכב האחד מקטב העולם אצל זריחתו פחות ממרחקו ממנו כשהוא באמצע השמים,

282. ו: פימא בעד.
283. ו: פאנא נגֿד.
284. א: מצד. ו: וגֿד.
285. ו: אל.
286. ו: פי מא וגֿדה.
287. ו: זאויה מכֿתלפה.
288. ו: יגֿרי אמר אלקמר פי.

וזה אי אפשר להיות בכוכבים הקיימים אם הראות משיג אותם על יושר. ואמר גם כן כי הירח ימצא בעיון בכלי כשהוא זורח, מרחקו מנֹכח הראש פחות מהשעור [ ][289] למרחק בעת הזאת על פי החשבון. ויראה מזה כי אור הירח לא יגיע לנקבי הכלי על יושר. ואחר שלא ישיג הראות הכוכבים על יושר ולא בהיפוך הניצוץ לפי שאין בשמים [ובארץ][290] ולא באויר גשם עב בהיר יתהפכו ממנו הצורות ישאר שישיג בעקום, לפי [שהראות][291] לא ישיג דבר מן הדברים הנראים כי אם באחד מאלו

105a

השלשה דרכים. ואמר גם כן כי אין בשמים מקום ישיג הראות הכוכב אשר יהיה בו ביושר זולתי כשיהיה הכוכב על נקודת נֹכח הראש או קרוב מאד ממנה. ואחר שזה כן ידוע כי מה שאמר בטלמיוס וביאר אותו מענין אלו החלופים אינו כדי לסמוך עליו לפי שאלו העיונים אמנם הם על פי כי השגת הראות הכוכבים הוא על יושר. עם כל זה זכרנו העקרים אשר בהם יראו אלו החלופים מבלתי יוצא המרכז ומבלי גלגל הקפה מצד מעלת אלו העקרים אשר לנו בזה וגודל עניינם ואני מתחיל מעתה בביאור זה.

B.2.V.3 כי בטלמיוס מצא אצל קבוץ הירח עם השמש או [הקבלתו חלוף][292] אחד לירח בלבד פעם יוסיף רל֞ כי המהלך האמיתי יקדים המהלך האמצעי ופעם יגרע. ומצא אותו שומר סדרו תמיד כפי החלק אשר הוא בו מעגולת חלופו. וכבר זכרנו בעקרים אשר עליהם ילך הענין בזה החלוף. ואמנם במרחקים אשר ביניהם ובפרט אצל רביעו לשמש אם היה החלוף הראשון מחייב תוספת מצא אותו מוסיף יותר ואם היה מחייב מגרעת מצא גורע יותר בשיעור המגרעת והתוספת תמיד. ומצאו רוב זה התוספת או המגרעת מהחלוף הראשון שני חלקים וחצי בקירוב.

B.2.V.4 ונשים עגולת הרוחב אד֞ג קטבה מ֞ והעגולה הנושאת אב֞ג קטבה ש֞ ועגולת מעבר המרכז חה֞טס֞ קטבה ב֞. ונשים בסבת זה החלוף כי מרכז הירח אינו סובב על זאת העגולה עצמה. אבל הוא סובב על עגולה קטנה ממששת לה והקשת אשר מקטבה למקיפה שוה לחצי רוב זה החלוף השני. ותהיה עגולת הו֞ד קטבה ז֞ סובב תמיד על עגולת נז֞מ֞ אשר קטבה ב֞ גם כן ותהיה נטיית עגולת אב֞ג על עגולת אד֞ג מוסיף על מה שאמרנו בשעור קטר זאת העגולה. ונאמר אמנם בקבוץ וההקבלה מרכז הירח יהיה על הנקודה מעגולת מעברו הממששת לעגולת חה֞טס֞ והיא נקודת [כ֞][293] תמיד עד שיהיה החלוף אשר יקרה אצל הקבוץ [והקבלה][294] כפי מה שתחייב בעגולת חה֞טס֞ [לא יותר].[295] ואמנם כשיהיה הירח בריבוע השמש והוא המקום אשר בו זה החלוף השני היותר גדול שיוכל להיות אז יהיה הירח על נקודת ד֞ מעגולת מעברו רל֞ על מקיף עגולת [לד֞ם][296] אשר קטבה ב֞ גם כן.

B.2.V.5 וכשיהיה הירח אז בנקודת הקצור הגדול רל֞ ה֞ או בנקודת הקצור הקטן רל֞ ס֞ בכדי שיעדר החלוף הראשון אז לא יהיה שם חלוף כלל כמו שיתחייב מעקרי בטלמיוס לפי שמקום נקודת ד֞ מהמזלות הוא בעצמו מקום נקודת ה֞. ואמנם כשיהיה הירח על נקודת ט֞, והיא הנקודה אשר בה היותר גדול שיוכל

289. ו: אלדי יחצל.
290. ו: -.
291. א: שהאות. ו: אלבצר.
292. א: הקבלת וחלוף. ו: אסתקבאלה אלאכתלאף.
293. ו: ה֞.
294. א: 2.
295. א: לאיובה. ו: לא גיר.
296. ו: לד֞כ֞.

להיות מן החלוף הראשון להוסיף במהלך, אז תהיה נקודת ד׳ אשר בה הירח על נקודת כ׳. ויגדל בזה החלוף בשיעור קשת ט״מ״כ ואז יהיה זה התוספת

104b

היותר גדול שיוכל להיות. וכאשר יהיה בנקודת ח׳ אשר בה החלוף היותר גדול לגרוע יהיה הירח על נקודת ל׳ ויגדל החלוף ג״כ לגרוע בזה השיעור. כאשר יהיה בין שתי נקודות ה״ט או ה״ח יהיה תוספות החלופים על שיעור החלוף הראשון. וכאשר יהיה הירח בין הקבוץ והריבוע או בין ההקבלה והרִבוע ר״ל בין שתי נקודות ד״ה מעגולת הו״ד אז יהיו התוספות בזויות החלוף יותר פחותים. ויחתוך [הירח][297] עגולת מעברו ר״ל עגולת הו״ד שתי פעמים בזמן החדש האמצעי, ר״ל כי הגלגל אשר זאת העגולה מדומה ישלים סבובו על קטב ז׳ והירח תקוע בו שתי פעמים בזמן החדש האמצעי. וכבר ידעת כי כל עגולה נזכור אותה בכאן היא בגלגל מיוחד והתנועה היא לגלגלים לא לעגולות הפשוטות, האחד מהם מקיף בחבירו. כל מה שיראה החלוף השני בזה העקר הוא מוסכם למה שיראה ממנו מעקרי בטלמיוס לא יעדר ממנו דבר עם המרחק אשר בין שני אלו העקרים באמיתות [][298] תכלית המרחק כמו שהוא גלוי וידוע מהם.

B.2.V.6 וצריכים שנשים כי קטב ז׳ נעתק על שתי העגולות המתוארות במאמר הראשון בדומה לזה הענין בשמש בסבת היות מרכז הירח על עגולת הרוחב תמיד. [וכבר קדם איך לא ישתנה הנחת קטר עגולת הו״דת באלו ההעתקות או נשים במקום זה העקום הנזכר. וכבר יתבאר גם כן איך תהיה הנחת עגולת הו״דת בעקר השתי עגולות כהנחתה בעקר העקום בכל המקומות, ר״ל כאלו היא נעתקת סביב קטב ב׳ תמיד. וכאשר יהיה הירח ברִבועים, ר״ל בנקודת ד׳, צריך שיהיה העקום יותר גדול ממה שהוא עליו בדבוקים, ר״ל בנקודת ה׳. ושיתגדל העקום כל מה שסר הירח מנקודת ה׳ לצד נקודת ד׳ כדי שישמור הירח עגולת הרוחב תמיד. וכבר נתבאר במה שקדם איך יתגדל העקום ויתקטן איזה שיעור שנרצה וגם כן שעור השתי עגולות][299] כי מה שיתחייב בקטב ז׳ יתחייב במרכז הירח באיזה מקום שיהיה מעגולת הו״דת. זהו העקר המושם ממנו לזה החלוף השני.

B.2.V.7 ואמנם החלוף השלישי כבר יתבאר מציאותו בזה העקר בעצמו המונח ואין צורך לתוספת דבר ר״ל תוספת תנועה ולא עיקום קטר ולא דבר מאלו הנמנעות המוצעות מבטלמיוס לשני החלופים האלה, כי [הסברות][300] הזרות המשיגות לעקריו בשני החלופים האלה אינם ממין הנמנעות המשיגות לשאר עקריו המונחות ממנו. אבל ישיג אותם מהעדר האיפשרות מה שאין הדעת סובלו בכדי שהיה לו יותר טוב השתיקה על שאנחנו כבר זכרנו סברתו במה שקדם. וכפי מה שהיתה כונתו אין ראוי להאשימו על זה. ונאמר כי הראיה המבוארת על אמיתת זה העקר היות זה החלוף השלישי נמשך מהשני בשהוא הכרחי ממנו. ר״ל שאי אפשר מציאותו בלתי במקומות אשר בהם יראה בלבד. ואמנם איך יתחייב זה הוא על זה הצד.

B.2.V.8 כבר אמרנו כי לא [יקרה][301] בעת הקבוץ וההקבלה זולתי החלוף הראשון. ואמנם בשאר התמונות ובפרט בעת הרִבוע אמנם יקרה בו נוסף על הראשון יותר מה שיוכל להיות מהחלוף השני כשיהיה

297. א: השמש. ו: אלשמס.
298. ו: ופי סהולה אלאחתמאל ללעקל.
299. ו: -.
300. ו: -.
301. א: יקרא. ו: יערץ̇.

הראשון נמצא כמו שנזכר. ובזאת העת גם כן יעדר החלוף השלישי. ואמנם בשאר התמונות [ובפרט בעת הרִבוע][302] כבר יקרה בו חלוף שלישי והוא אשר העריכו אותו לעקום קטר גלגל ההקפה והוא על זה הדרך.

B.2.V.9 כשיהיה הירח בעגולת חלופו במקום התוספת במהלך או התוספת

104a

בקצור, והיה הירח במה שבין הקבוץ והרבוע [הראשון][303] או במה שבין ההקבלה והריבוע [השני],[304] יגדל החלוף. רצוני לומר כי הקצור באורך לאמיתת הירח יהיה יותר ממה שיחייבו אותו החלופים הנשארים. וכאשר יהיה הירח במה שבין הריבוע הראשון וההקבלה או במה שבין הריבוע השני והקבוץ אז יהיה הענין בהפך. ואמנם כשיהיה הירח בעגולת חלופו על הקשת הגורע מן המהלך ר״ל הגורע מן הקצור אז יהיה הענין בכל זה בהפך. ר״ל כי יגרע המהלך ממה שיחייבו אותו החלופים הראשונים בהנחות הראשונות ויגדל בשניות. ואומר כי זה יתחייב מצד העתקת הירח על העגולה הפועלת החלוף השני ר״ל עגולת הו״ד. וזה כי כשיהיה הירח בקשת [][305] המהלך ר״ל קשת ס״ה״ט ויהיה המשל על נקודת ה׳ כאשר בצורה ואז יהיה עת קבוץ אמצעי או הקבלה. כל זמן שתתמיד נקודת ה׳ להיות נעתקת לצד נקודת ט׳ נמשכת לחלוף הראשון והיה הירח נעתק על קטב ז׳ מנקודת ה׳ עד נקודת ו׳ ועד ד׳, ר״ל מן הקבוץ ועד הרִבוע הראשון או מן ההקבלה עד השני, יהיה החלוף הולך וגדל לפי שבזאת ההעתקה ימהר הירח להתקרב לצד נקודת ט׳. ורוב זה החלוף יהיה כשיהיה הירח באמצע הדִבוק והרבוע ר״ל בהיותו על נקודת ו׳. וכאשר יהיה בשני ההנחות השניות ר״ל בהיות הירח [על][306] קשת ד״ת״ה יהיה בהפך. ואמנם כשיהיה הירח בקשת הגורע ר״ל על קשת ט״ס״ה אז יתחייב שיהיה כל זה בהפך ומ״ש.

B.2.V.10 ויראה מזה כי לא ימלט החלוף השני באלו ההנחות מחלוף שלישי [][307] יחייב אותו העתקת הירח על זאת העגולה. וכאשר נשים ההעתקה כפי מה שזכרנו יראה זה החלוף השלישי כפי מה שהוא עליו. ויראה ג״כ כי שיעור זה החלוף על הדרך הזאת לא יהיה בינו ובין שיעור החלוף אשר שם בטלמיוס בסבת העקום מוחש כלל. ויתבאר גם כן שלא [יקרה][308] בדבוקים האמִתיים מחייבו הלקיות חלוף מורגש מצד זה העקר המונח לשני החלופים האלה, וזה יתבאר בכמו המופת אשר בו ביאר בטלמיוס. וזה הוא מה שיראה לנו בחמלת האל יתעלה עלינו מאמִתת העקרים הראויים להציעם לאלו החלופים מבלי שום צורך לדבר מן הנמנעות המוצעות מבטלמיוס וזולתו מזה. וזה דבר לא יצא כי אם מאתו ית׳ אשר לו השבח וההודאה.

B.2.V.11 והנה נשאר עלינו לבאר איך ישמור הירח תמיד מרחב עגולת הרוחב למה שהושם מן ההסרה ברוחב בסבת עגולת הו״ד המונחת לחלוף השני והשלישי. ונחזור הצורה שעברה ונשים עגולת מעבר קטב עגולת הרוחב, ר״ל הגלגל אשר היה בו עגולת ל״מ קטבה [ס׳],[309] והוא קטב גלגל המזלות, ועגולת מעבר קטב הגלגל אשר בו העגולה הנושאת עגולת מ״ש קטבה ס׳. ונשים קטב ס׳ סובב על עגולת ס״מ

302. ו: סוא הדה אלתלאתה.
303. ו: אלתאני.
304. ו: אלאול.
305. ו: אלזיאדה או פי.
306. א: כל. ו: עלי.
307. ו: כיף מא כאן.
308. א: יקרא. ו: יכון.
309. ו: ח׳.

הממששת לעגולת לֹמֹ ותהיה שוה לעגולת הֹוֹדֹ. [והעתקת [קטבי][310] סֹטֹ סביב קטב עגולת הרוחב רֹלֹ מֹ אחת].[311] ויהיה העתקת קטב סֹ על עגולת סֹמֹ שוה לתנועת הירח על עגולת הֹוֹדֹ בכדי שבהיות מרכז הירח על נקודת [][312] רֹלֹ על מקיף עגולת הֹטֹ, יהיה קטב סֹ על נקודת מֹ רֹלֹ עגולת הרוחב. יהיה אז הירח על עגולת הרוחב. וכאשר יהיה על נקודת דֹ רֹלֹ בריבוע יהיה קטב סֹ על נקודת סֹ מעגולת מֹסֹ. [וכל עוד שהירח נעתק על חצי הֹוֹדֹ, יעתק קטב סֹ על חצי סֹטֹ המקביל לו].[313] ובזה הדרך יהיה הירח על עגולת הרוחב תמיד ויחתוך קטב סֹ לעגולת מֹסֹ שתי פעמים בזמן החדש האמצעי.

B.2.V.12 וצריך שתדע כי

103b

זאת התנועה אשר הצענו אותה לקטב סֹ על עגולת מֹסֹ היא במקום שאינה מחייבת תנועה לכוכב באורך זולתי ברוחב לבד. וזה איפשר רֹלֹ שיעתק קטב סֹ על עגולת מֹסֹ נמשכת לתנועת הגלגל אשר בו תדומה עגולת מֹסֹ על קוטב נֹ בכדי שלא יעתק הגלגל אשר קטבו סֹ [][314] בזאת התנועה כלל. [ואם יעתק באורך כבר, איך ישאר על הנחתו מבלתי שיהיה לו הסרה באורך כלל. ואם קרה לו עיקום קטן אינו מורגש כלל, ואם תרצה לסלק הקירוב אשר בשיעור זה החלוף השלישי, כי זה החלוף אמנם כפי מה שימצא בעיוני בטלמיוס, וזולתו לא ימצאו לא פחות ולא יותר ממה שהוא עליו בעקרינו אלה. ואמנם כפי מה שיתחייב מעקריו הנמנעות, יתחייב שיהיה יותר ממה שהוא עליו בעקרינו חמשה דקים. ובידך לתקן זה בשתשים עקום לעגולת הֹוֹדֹ כמו שנתבאר, ויהיה העקום לקטר כֹהֹ כלו על קטב הֹ לצד מקום הירח מעגולת הֹוֹדֹ בשעור שנצטרך אליו. ובזה יהיה גם כן החלוף היותר גדול קרוב מהשלושים רֹלֹ במרחק מאה ועשרים מנקודת הֹ וזה מבואר. ואמנם כפי מה שיתחייב מהעיונים, בטלמיוס עצמו יודה שלא ימצאו עיונים יהיה זה החלוף יותר גדול ממה שהוא עליו בריבועים. אבל הוא שם עקום קטר גלגל ההקפה עקר לזה החלוף, ונמשך מזה כי החלוף היותר גדול בשלושים. בנה עליו מבלתי עיון ולא מופת אחר וזה טעות גמור.][315] ואמנם דרך ביאור חישוב [][316] החלופים ביחד זה יהיה בדרך אשר בארנו בו החלוף [][317] וביאורו קרוב. וצריך שתהיה נזכר למה שזכרנו פעמים רבות מחלוק הגלגל למספר גלגלים בסבת הענין הנזכר, כי בזה יסתלק כל הספקות שאיפשר שיסופק בהם בזה המקום. [וידוע ממה שקדם כי גלגלי הירח כפי העקום הם חמשה גלגלים, ראשיים למהלכים ולחלופים, וארבע מחברות לעקומים והמרחבים כמו שנתבאר. הראשון מן הראשיים רֹלֹ התחתיי הוא הגלגל הנושא למרכז הירח, והוא אשר בו העגולה אשר יתחייב ממנה החלוף השני והשלישי. והשני והוא על הראשון הוא הנושא לקטב זה הגלגל ומעתיק אותם בתנועתו, והוא הגלגל אשר בו עגולת המעבר. ועל השני שלש מחברות לעקום עגולת המעבר כמו שנתבאר. ועל החמישי הגלגל אשר בו העגולה הנושאת לקוטבי עגולת המעבר המעתקת אותה באורך. ועל הששי מחברת משני גלגלים להסיר הרוחב המתחדש לירח מצד הגלגל המושם לחלוף

310. א: קטב.
311. ו: -.
312. ו: הֹ.
313. ו: -.
314. ו: פי אלטול.
315. ו: -.
316. ו: מקדאר.
317. ו: אלאול.

103a

השני והשלישי עם השאר הגלגלים הפנימיים ממנה על הנחתם בלא עקום מורגש כלל.

אבל אם נצטרך לעקום כמו שרמזנו אליו אז נצטרך למחברת רביעית על הראשון ועל אלו הגלגל המעתיק לתכליתיות הנושאת, אבל מה שבתוכה בשיעור מותר תנועת הרוחב על תנועת החלוף. ועליו הגלגל המעתיק לתכליתיות העגולה הנושאת ועגולת הרוחב בשעור מותר תנועת הרוחב על תנועת האורך והיא השלש דקים בכל יום. ואמנם הגלגל אשר בו תצוייר עגולת הרוחב ועגולת המזלות בירח, יספיק שנצייר אותה בגלגל המזלות עצמו מבלתי שנשים גלגל בזה התואר בירח. ואלו הגלגלים הם כולם סביב מרכז אחד, וכל גלגל תחתיי מהם יתנועע תנועתו המיוחדת לו ויתנועע גם כן בכל תנועות הגלגלים הנזכרים אשר ממעל לו.

ואמנם כפי עקר השתי עגולות העומדות במקום העקום, אמנם יחסרו מזה המספר גלגל אחד וזה מבואר, וזה כשתהיה תנועת האורך על קטבי עגולת הנושאת לקטבי עגולת המעבר. אבל אם שמנו אותה על קטבי עגולת [המזלות][318] כדרך שבארנו בשמש בתכונה האחרונה, יהיו גלגלי הירח כפי עקר העקום בזה המספר גם כן, אבל אינם הנזכרים. וזה כי השלשה מאלו הנזכרים אין אנו צריכין אליהם אז, והם הגלגל האחד מן המחברת הראשונה והששית והמעתיק לתכליתיות הנושאת. ובמקום אלו השלשה יש בכאן בהכרח שלשה גלגלים אחרים, והם שני גלגלים לתקון רוחב העקום וגלגל אחר בו עגולת הרוחב והוא המחייב תנועת האורך. ומזה יתבאר מספר הגלגלים בעקר השתי עגולות.][319]

B.2.V.13 והשבח לאל בה

B.3.0.1 המאמר השלישי בהעתקת הכוכבים הקיימים ובו שני פרקים.

B.3.1.1 הפרק הראשון בדעת אלבטרוגי בזאת ההעתקה ובטולו. יאות לנו לדבק בזה המאמר בהעתקת הכוכבים הקיימים כפי הסדר אשר סדרו הקדמונים ובטלמיוס. וכפי מה שבארנו מהפרק ההפכות בתנועה הסבובית, כי עקריו בזה אמִתיים לפי שאין צורך בזה ליצא המרכז ולא לגלגל הקפה. ואמנם [בארנו][320] זה כפי הדרך השני למה שקדם ממאמרינו. [ודרך][321] העתקת אלו הכוכבים אצל אלבטרוגי הוא זה, כי הכוכבים הקיימים כולם מלתחמה [כולם][322] בכדור אחד. וזה הגלגל יתנועע לצד תנועת הכל נמשך לעליון המתנועע התנועה היומית ועל קטבין ויקצר בתנועתה זאת [מהגיע][323] למקום העליון מחמת הסבה הנזכרת קצת הקצור. ולפי שכוונתו ההדמות בעליון ובקשת זה השלימות לפי שהוא תכליתו, יתנועע על קטבין המיוחדים תנועה אחרת לזה הצד. ועם זה ישאר לו קצת הקצור באורך על דעת בטלמיוס שהוא יעתק להפך שיעור מורגש. או ישלים בתנועתו תנועת העליון ולא ישאר לו קיצור באורך כלל על דעת האחרונים. ואמנם יהיה לו העתקה ברוחב יחייב אותה העתקת זה הגלגל על קטבין בסבת ההשלמה.

B.3.1.2 ואמר גם כן כי תנועת האיחור והקדימה אשר שמו אותה האחרונים כפי עיונם אינה תנועה לפי האמת. אבל התנועה היא שוה תמיד לצד אחד זולתי כי כל מה שיתמיד הכוכב [נעתק][324] על חלקי הרביע

318. א: -.
319. ו: כמלת אלמקאלה אלתאניה מן אלג'מלה אלתאניה מן כתאב נור אלעאלם.
320. ו: סלכנא.
321. ו: ואלוג'ה אלדי עליה אלאמר פי.
322. ו: -.
323. א: מרגיע. ו: אלוצול.
324. א: מעתה. ו: מנתקלא.

אשר באמצעותו נקודת השווי מה שיעלה עם זה הרביעי [מעגולת][325] המישור הוא פחות. ולזה יראה לכוכב איחור להפך וקראו זה הקצור איחור. וכל מה שיתמיד נעתק

102b

ברביע הנמשך לו מה שיעלה עמו יותר ממנו ולזה יראה לכוכב [][326] תוספת לצד תנועת הכל וקראו אותו קדימה. וזה יראה בין שיהיה לו קצור על דעת בטלמיוס או לא. ועל זה הדעת כשיהיה הכוכב על הנקודה הדומה להיפוך החום יראה נעתק לנקודת השואת הקיץ וממנה להיפוך הקור וממנה להשואת החורף בהפך תשומת הקדמונים. זה הוא דעת זה האיש בזאת ההעתקה.

B.3.1.3 ואם נודה כי זה הגלגל יקצר מהעליון כפי מה שסובר בטלמיוס שהוא נעתק להפך יהיה הקצור שוה להעתקה להפך אשר ישים אותה או פחות ממנה או לא ישאר לו קצור באורך כלל, עם שהוא בלתי ראוי בחק העקר אשר יבנה עליו, הוא כבר יודה העיונים אשר בהם נמצאת זאת ההעתקה ר״ל העתקת הרוחב. אבל אי אפשר לו להכחישו וכבר נמצא לעיונים הנמשכים שהכוכבים נעתקים בהפך מה שהניח לפי שכבר נמצא בעיון כי הסמאך אלאעזל שהוא קרוב מנקודת השואת החורף ריחוקו מנקודת ההשואה לצד הצפון חלקים מה נמצאו זה המרחק יתמעט בהמשך הזמן.

B.3.1.4 וגם כן הכוכבים אשר הם [קרובים][327] לנקודת השואת הקיץ נמצאו מתרחקים לפי אורך הזמן מנקודת זה השווי לצד צפון. ודרך כלל כי בטלמיוס כשהקיש עיוניו לעיוני אברכס וטימוכאריס [וארסטלס][328] ומילאוש המהנדס ובהקש קצת אלו העיונים אצל קצת ואצל עיונים אחרים מן המעיינים המפורסמים בזה הזמן הקודם כי הכוכבים אשר בחצי הכדור אשר בו נקודת השואת הקיץ מתרחקים לצד הצפון תמיד בהמשך הזמן מהמקומות אשר היו בהם ראשונה, ואשר הם בחצי השני בהפך. ויתחייב מזה שיראה הכוכב נעתק כפי מה שיתחייב שיעתק כפי עקריו. ודרך כלל כי כל מה שיראה מהעתקת כל אחד מאלו הכוכבים בכל אחד מהעקרים יהיה על [הפך][329] מה שיראה באחר בין שיהיו הכוכבים אשר על חגורת המזלות או חוץ ממנה זולתי בשני כוכבים בלבד, והם אשר היו על שתי ההיפוכים בתחלת ההעתקה. כי אלו השני כוכבים יראו תמיד על מקום אחד בכל אחד משני העקרים. ונשים לזה משל.

B.3.1.5 ותהיה עגולת משוה היום גטד קטבה הצפוני א ועגולת המזלות גהד קטבה הצפוני ב [ונחשוב][330] זאת העגולה [][331] העובר על אמצע המזלות הנעתקת עגולת נזו קטבה ש ונדמה אותה שהיא היתה ראשונה בשטח עגולת גהד ונקודת ז על נקודת ה אשר היא נקודת היפוך הקור.

B.3.1.6 ונשים כי הגלגל אשר היא בו נעתק בזמן מה נמשך לתנועת הגלגל העליון וקיצר מהגיע למקומו קשת זה מהעגולה הנכחית לעגולת המישור. ונעתק זה הגלגל על קטביו המיוחדים בו בבקשו להשיג העליון קשת זכ עד שהכוכב אשר היה על נקודת ז, ר״ל הנקודה הדומה להיפוך הקור היה על נקודת כ. והכוכב אשר היה על נקודת ל מזאת העגולה ר״ל עגולת נזו והיה מרחקו מנקודת נ ר״ל הנקודה הדומה להשואת החורף קשת לנ היה על נקודת מ, ר״ל מתרחק ממנו לצד הצפון. והכוכב אשר היה על נקודת ס מזאת העגולה והיה מרחקו מנקודת ו, ר״ל הדומה להשואת הקיץ, קשת וס היה על נקודת ע, ר״ל [מנקודת

325. ו: מן.
326. ו: תטפיף.
327. א: כוכבים. ו: אלתי באלקרב.
328. א: אכסטלס.
329. ו: -.
330. ו: ונתוהם.
331. ו: גיר מנתקלה אבדא. ולתכן אלדאירה.

השווי][332] לצד הדרום. אכֿ הכוכב אשר על נקודת ההיפוך, כל מה שיראה בו כפי כל אחד משני העקרים [יראה בו כפי העקר השני לפי שבין][333] שיעתק הכוכב אשר על נקודת הֿ להפך לנקודת כֿ או יעתק לזה הצד מנקודת זֿ כמו זה הקשת, בכל אחר

102a

משני העקרים יראה הכוכב על נקודת כֿ בעת אחד. וכזה יתחייב בהפוך האחר. ואמנם בשאר הכוכבים אשר על זאת העגולה זה יהיה בהפך לפי שהכוכב אשר היה על נקודת לֿ מעגולת גֿהֿדֿ ונשים המשל שהוא הסמאך אלאעזאל, והוא כאשר בו היה העיון אחר זמן נמצא על נקודת מֿ, רלֿ מתקרב לנקודת גֿ אשר היה נקודת השואת החורף. ומן הראוי היה זה על דעת בטלמיוס נמשך להעתקת זה הגלגל להפך, רלֿ מנקודת הֿ לצד נקודת כֿ, וכבר ביארנו כפי עקרי אלבטרוגי כי מן הראוי היה שיתרחק מזאת הנקודה לצד הצפון רלֿ מנקודת לֿ עד נקודת מֿ מעגולת נֿזֿוֿ.

B.3.I.7 וגכֿ הכוכב אשר הוא על נקודת סֿ מעגולת גֿהֿדֿ והיה מרחקו לצד הצפון מנקודת [דֿ][334] רלֿ השואת הקיץ, קשת דֿסֿ, ונמצא אחר זמן על נקודת עֿ רלֿ מתרחק ממנה לזה הצד יותר מזה והוא המתחייב לסברתו וכבר נתבאר כפי עקרי אלבטרוגי שיהיה ראוי להיות בהפך. וגם כן יתבאר זה החלוף או דומה לא באיזה מקום הונח הכוכב מגלגלו בין היה על עגולת חגורתו או חוץ ממנה. ואחר שהדבר כן מה הועיל לנו זה האיש בשומו זה העיקר אחר שאין שם דבר ממנה שיראה לחוש דבר למה שיראה מעקרו. ומזה יראה [מה שאמרנו פעמים אחרות][335] מכוונות זה האיש. ואני מתחיל מעתה לזכור העיקר האמיתי לזאת ההעתקה על הדעת השני רלֿ שיהיה מה שיתחייב ממנו מסכים עם למה שיראה בעיונים הקדמונים לא יעדר ממנו דבר.

B.3.II.1 הפרק השני בדרך העתקת אלו הכוכבים. ונאמר כי הדבר אשר בסבתו הסכימו הקדמונים בשאלו הכוכבים הנקראים קיימים כולם בכידור אחד אמנם הוא לפי שהם מצאו אותם שומרים מצבם קצתם אצל קצת תמיד [וכי][336] ההעתקה אחת ומתדמה בכולם וגזרו שהיא לכדור אחד. ואין בכאן דבר ימנע שנאמר כי לכל כוכב מהם גלגל מיוחד או לכל שנים מהם ויהיה לכלם העתקה אחת ושוה. אבל מצד הסכמתם בעקר [][337] זאת ההעתקה היה לפי זה המאמר אשר הם כלם בכידור אחד מספיק, עם שישאר על הטבעי החקירה אחר הסבה אשר בעבורה היה זה הריבוי מהכוכבים בכדור אחד ולשבעת הכוכבים הנשארים שבעה כדורים. ולא זה בלבד אבל שיהיה כל כידור מהם נחלק לשנים או יותר כמו שנתבאר. וכבר חקר ארסטו בזה ונתן בו סבות [בלתי אמִתיות].[338] וכבר זכרנו מה שיקרה מן הנמנע והכזב העיונים הרבים המדוייקים מצד הנחת הכוכבים הקיימים כלם בכידור אחד עם היות כל התנועות לצד אחד. ואם שמנו לכל כוכב או שני כוכבים מהם כידור אחד

332. ו: אעני מתקארבה. גם כן א בשוליים: כאן היתה העגולה בטופס ולא היה לי מקום פה לעשותו והפוך הדף ותמצאנה.

333. ו: שי ואחד לאנה סוא.

334. ו: -.

335. ו: מא קלנאה גיר מא מרה.

336. ו: כאן.

337. ו: אלדי עליה יגֿרי אלאמר פי.

338. ו: ואלבעד ען אליקין טֿאהר פיהא.

101b

שיהיה זה מחוייב בהם ושלא יהיה מה שיראה לחוש מהעתקתם על מה שהוא עליו בזולת זה הדרך כפי עקרים האמִתיים הנזכרים، הרוחנו ענין גדול בסילוק זה המקום מן החקירה וזה יהיה במה שאתאר.

B.3.II.2 ונזכר ראשונה הדרך [הנכון לתנועת][339] הכוכבים הקיימים אשר בשטח העגולה העוברת באמצע המזלות ר׳ל העגולה אשר ירשום אותה השמש בקצורו האחרון. ויאות שיובן זה הענין ממנו בכל מקום אזכור בו עגולת המזלות ומזה יתבאר לאשר אינם על זאת העגולה. ונאמר כי השני כוכבים אשר היו על שתי נקודות ההיפוכים בהתחלה אחת מן ההעתקה כבר נתבאר מעניינם כי מה שיתבאר בהם בזה העיקר מסכים עם מה שיראה בעיון، אם כן אלו השנים בלבד יאות שיהיה בכדור אחד. ונאמר כי הכוכב אשר על זאת העגולה אשר היה מרחקו בתחלה ה׳ חלקים מנקודת היפוך הקור לצד נקודת השואת הקיץ המשל כי הוא בכידור אחר נטייתו כמו נטיית הראשון והוא בעגולה האמצעית ממנו، ומרחקו מן הנקודה הדומה להיפוך הקור ה׳ חלקים גם כן בצד המקביל. ר׳ל כי כשנשים הנקודה הדומה לנקודת [היפוך][340] הקור מזאת העגולה על נקודת היפוך הקור עצמו יהיה מרחק הכוכב מנקודת היפוך הקור לצד נקודת החורף ה׳ חלקים. והכוכב המקביל לו כשיושם בזה הכידור עצמו، יתחייב בו מה שנתחייב בראשון.

B.3.II.3 ולזה יתחייב שיהיו אלו השנים [בלבד][341] בכדור אחד על התואר הנזכר. על זה המשל ילך הענין בשאר בין שיהיו אשר על עגולת המזלות או חוץ ממנה בשנשים כל אחד מהחצונים ממנה בעגולה קטנה נכוחית לעגולה הגדולה אשר בזה הכדור הדומה לעגולת המזלות، מרחקה מזאת הגדולה שוה למרחקה ממנה אם היו הכוכבים כולם בכדור אחד. ועל שיהיה [כל אחד מהם][342] בכדור אחד ונטיות הכידורים כלם אחד، ר׳ כי קטביהם כלם סובבים על שתי עגולות קטביהם קטב [גלגל][343] המישור. ותהיה העתקת אלו הכידורים כלם [ר׳ל אשר הם בעגולת המזלות][344] שוה אצל החוש. ונשים לזה משל בעגולות ואותיות.

B.3.II.4 ותהיה עגולת המזלות ד׳ט׳ז׳ קטבה הצפוני ג׳ על עגולת ג׳ב׳ אשר קטבה קטב [גלגל][345] המישור והוא א׳. ותהיה נקודת ד׳ נקודת היפוך הקור ונקודת [כ׳ היפוך][346] החום ונקודת השואת הקיץ ש׳ והחורף ק׳. ונשים כוכב על נקודת ט׳ מרחקו מנקודת ד׳ לצד נקודת ש׳ ר׳ל נקודת השואת הקיץ ה׳ חלקים. וכבר יראה זה הכוכב בהמשך הזמן כפי עקרי בטלמיוס על נקודת ז׳ ר׳ל קרוב לנקודת השואת הקיץ שיהיה צפוני ממה שהיה ראשונה. ונשים עגולת ה׳ח׳ז׳ בגלגל אחר קטבו ב׳] על זאת הנטייה והנקודה הדומה לנקודה היפוך הקור ה׳ ולהיפוך החום ל׳. ואם שמנו זה הכוכב כפי עקרינו על נקודת ח׳، מרחקו מנקודת

339. ו: אלדי עליה יג׳רי אלאמר פי.
340. א: ההיפוך.
341. ו: -.
342. ו: כל אתנין מתקאטרין.
343. ו: -.
344. ו: -.
345. ו: -.
346. ו: -.

ה̇ ה̇ חלקים לצד הנקודה המקבלת, ר̇ל̇ השואת החורף [][347]. כשתהיה נקודת ח̇ על נקודת ט̇, ר̇ל̇ [נקודת][348] הכוכב, בכל אחד משני העקרים

101a
יראה הכוכב בעת אחד על נקודת ז̇ לפי שבזמן אשר יעתק הכוכב להפך קשת ט̇ז̇ כפי עיקרי בטלמיוס ויקצר זה הגלגל כפי עקרינו קשת ח̇ט̇ הנכוחית. ויעתק הכוכב על קטביו קשת ח̇ז̇ וזה בעצמו יתחייב בכוכב המקביל לו. ויהיה המשל על נקודת נ̇ ובעת היות נקודת ח̇ על נקודת ט̇ תהיה נקודת מ̇ על נקודת נ̇, וכפי כל אחד משני העקרים יהיו בעת אחד על נקודת ס̇. וכזה יתחייב בשאר, וקטבי אלו הגלגלים כלם הם תמיד על עגולת ב̇ג̇, ולא יהיה בזה הפרש זולתי [החלוף][349] אשר מצד המצעדים וכבר נתבאר זה במאמר הראשון ומש̇ל̇. [ואמנם הכוכבים שהם חוץ מעגולת המזלות, קצורם ותנועתם האחרונה אינה בשיעור אחד בכולם. אבל הקיצור בכל אחד מהם וגם כן תנועתו האחרונה הוא בשיעור שיגיע הכוכב למקומו מהמזלות כפי עקרי בטלמיוס, וזה השיעור מחולף בהם. ולא יכנס בזה הזיק גדול מאד שזה הענין לא יתכן בקצת הכוכבים ריחוקם מקטב המזלות פחות מתכלית הנטייה כמו בזולתם. ובזה ישמרו הכוכבים החצונים מעגולת המזלות מצבם קצתם על קצת עד גבול מוגבל. ויספיק לנו שיראו כל הכוכבים שומרים מצבם כפי כל אחד מהעקרים בענין אחד למספר אלו שנים רב, ואם לא ישמרו אותו לעולם. ומי יודע איך יפול הדבר עולמים אשר הם לפנינו כי האחרונים הכחישו זאת התנועה ואמרו כי הכוכבים יש להם הליכה וחזרה. ומדעתי כי הענין כמו שהם סוברים, אבל העיקר לזאת התנועה אינו העקר הנמנע היותו אשר שמו לה האחרונים ואזרקיל עמהם. אבל הוא העקר האמיתי אשר הצענו אותו במה שקדם בעקרי החילופים, לפי שתנועת ההליכה והחזרה אשר שמו אותה האחרונים כבר יתבאר לך ממה שקדם בעקרי החלופים, איך תהיה בתנועות סבוביות דומות שלימות הסבוב. וזה במחברת מג̇ גלגלים או ד̇ מרכזם מרכז העולם, כי מה שיתחייב בכוכב אחד שיעתק על קשת מה ויחזור עליה ולא יעבור ממנו. בזה יתחייב בשאר הכוכבים, לפי שכלם בכדור אחד, ואם תרצה להמשך לדעת בטלמיוס ותכחיש זאת התנועה.][350]

B.3.II.5 [][351] איפשר לנו לומר כי זה אמנם הוא בדמיון וכי הוא מצד תוספת מצעדי קצת הכוכבים על קצת כפי קירובם מנקודת ההיפוך או השווי. וגם כן לא ימנע שיהיו קצת אלו הכוכבים גדולי הקיצור מקצת מצד חילוף מרחקי הכדורים אשר הם עליהם מהעליון ולא הושג לחוש זה החלוף עד עתה לקטנותו, כי הקצור עצמו לא נשלם ענינו עד עתה [][352] חילוף הקיצור. ואיפשר כי זה [החילוף הטעה][353] האחרונים עד ששמו תנועת האיחור והקדימה. ומה שאמרנו יספיק בהשלמת העיקר האמיתי לזאת ההעתקה. והשבח הגדול לאל אשר הגיענו לזאת התכלית.

347. ו: אדא כאנת נקטה ה̇ מטאבקה עלי נקטה ד̇.
348. ו: מוצ̇ע.
349. ו: -.
350. ו: -.
351. ו: ואמא חרכה אלאקבאל ואלאדבאר אלמוצ̇ועה ען אלמתאכ̇רין.
352. ו: פצלא̇ ען. א: ?
353. ו: אלתקציר אכ̇תלאף והם.

4

Translation of the Significant Insertions in the Hebrew Recension of *The Light of the World*

From 124a–123b

§A.II.5/X

(See figure 4.1.) The two arcs *ZK KL* from small circle *ZKL* are similar to the two arcs *AE ED* from great circle *AED*, each to its counterpart. I say that the ratio of the sine of arc *AE* to the sine of arc *ED* equals the ratio of the sine of arc *ZK* to the sine of arc *KL*, and that the ratio of the sine of arc *AE* to the sine of arc *AD*, all of it, equals the ratio of the sine of arc *ZK* to the sine of arc *ZKL*. We propose *M* as the center of the two circles, and we extend lines *MZA MKE MLD*, and we extend the sines *AB DG ZH LT AZ ZK*. The ratio of *AB* to *ZH* is equal to the ratio of *AM* to *MZ*, that is to say *DM* to *LM*, rather *DG* to *LT*. And, vice versa, *AB* to *DG* is equal to *ZH* to *LT*, and also the ratio of *AB* to *ZH* is equal to the ratio of *AM* to *ZM*, [123b] rather *AW* to *ZN*. And, vice versa, *MK* to *MW* is equal to *ZH* to *ZN*, and that is what we wanted.

From 123b–a

A.II.7/X

Hence, it has been shown, through the ratio that has been compounded, that if *BZE* were a small circle arc with its pole on circle *ADB*, upon which the perpendiculars fall, then this will follow. That is to say that the ratio will be compounded as mentioned above, as if arc *BZE* were a great circle arc. We will have need of this matter in the future. Also, when the ratio is divided, it does not follow necessarily that all of these [123a] four arcs are great circle arcs because this is precisely what would follow when we propose that one of these four arcs is a small circle arc if

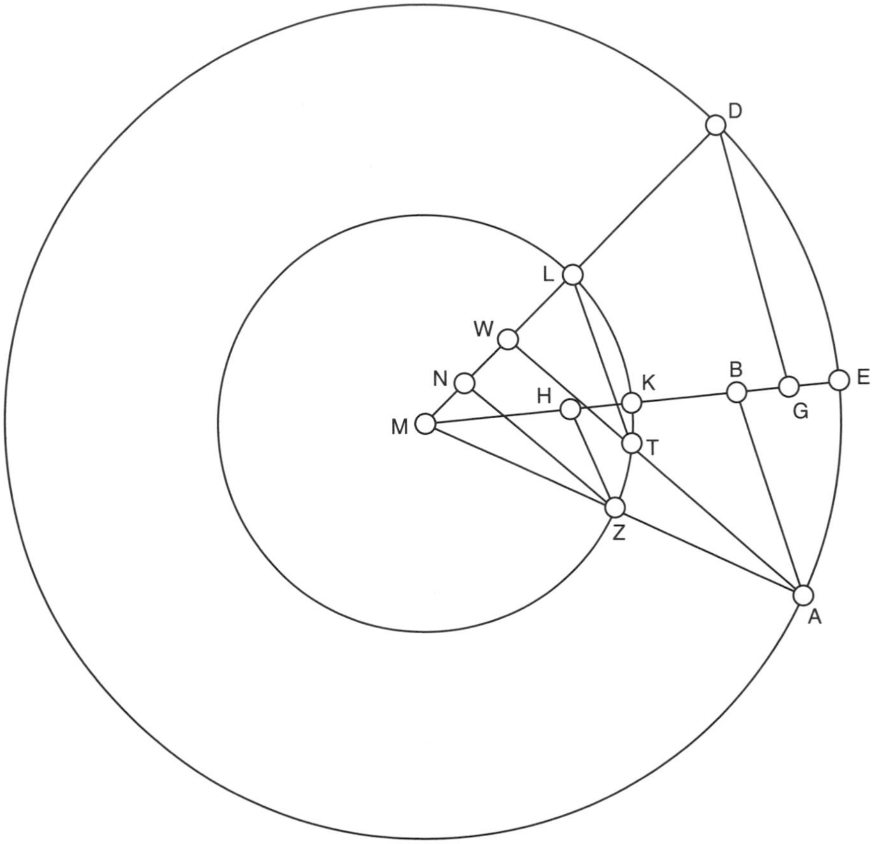

FIGURE 4.1

this arc is not from the circle upon which the perpendiculars fall. Rather, its pole must be upon the arc from the circle upon which the perpendiculars fall.

From 121b
§A.IV.2/X

(See figure 4.2.) It is clear from what has been explained in the preliminaries that when we propose that point *A* from the equinoctial in the first figure is a pole, and we trace, on the distance of arc *AB*, a quadrant of a small circle, namely arc *BS*, then we can take arc *SW* from it to be known since the inclination of point

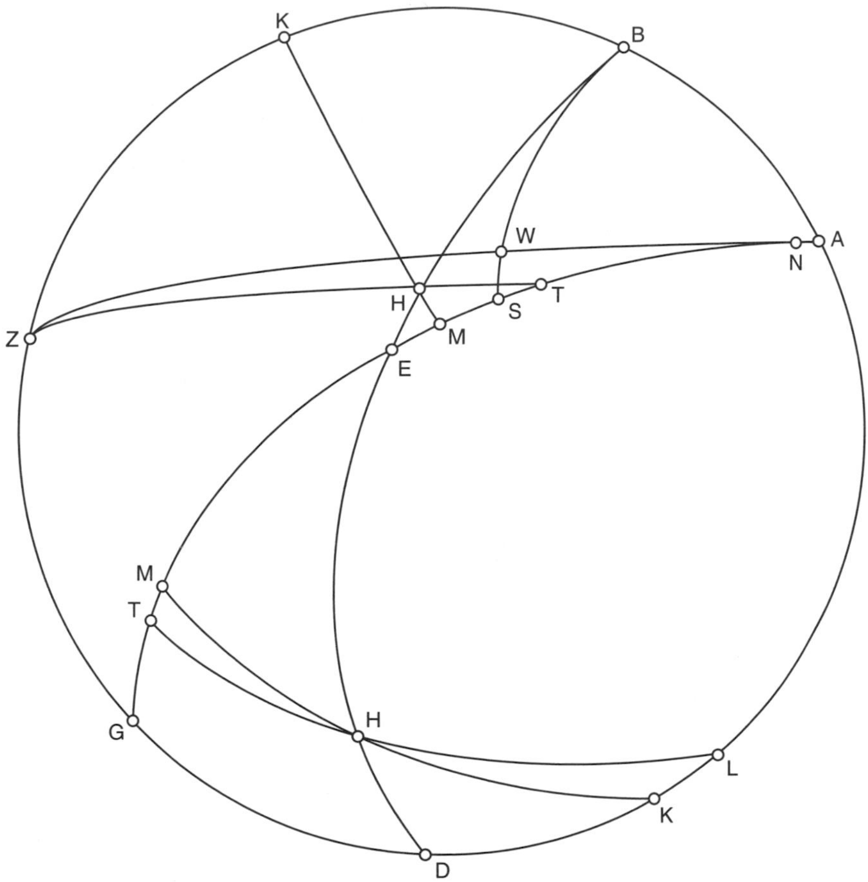

FIGURE 4.2

W from the equinoctial, namely arc *WN*, will also be known precisely through this way. And when we propose that [the ratio of the sine of] arc *SW*, similar to arc *EH*, to the sine of arc *AB* is compounded from the ratio of the sine of arc *HE* to the sine of arc *EB* and from the ratio of the sine of arc *ZT* to the sine of arc *TH* that is unknown, and if it is also compounded from the ratio of the sine of arc *SW* to the sine of arc *SB*, that is to say the ratio of arc *EH* to the sine of arc *EB* and from the ratio of the sine of arc *WT*, rather *ZT*, to the sine of arc *NW* the unknown, then the sine of arc *TH* remains equal to the sine of arc *WN*, and that is what we desired.

From 120b
§B.0.3/X
The fourth treatise on the motions of the five planets—Saturn, Jupiter, Mars, Venus, and Mercury—and their anomalies, and in it are 10 chapters. The first chapter is about the treatise's preliminaries. The second chapter is about the explanation of the true hypothesis that is our thesis for the first anomaly, which is related to the degrees of the ecliptic, found for the three upper planets. The third chapter is about the second anomaly found for these three planets, which is found to be according to the distance of the planet from the Sun, and [about] proposing the true hypotheses for it and for the particular matters that are observed with it that necessitate its largest [value]. The fourth chapter is about explaining the first anomaly related to the ecliptic and showing the point from the ecliptic where there is the least speed for each of the three upper planets. The fifth chapter is about showing the measure of the circle that produces the anomaly that is related to the Sun for each one of the three upper planets. The sixth chapter is about how we derive from the motion in revolution the true motions. The seventh chapter is about the model for the planet Venus. The eighth chapter is about showing the point on the ecliptic in which there is the least speed and showing the ratio of the radius of the orb of Venus to the radius of the circle that produces the anomaly related to the ecliptic, and showing the measure of the circle that produces the anomaly that is related to the Sun. The ninth chapter is about the model for Mercury, called ʿUṭārid. The 10th chapter is about the latitude of the five planets.

From 117b
§B.1.II.20/X
We have explained this matter here, and we transfer it to the lunar model so that we do not repeat the single matter many times. It falls to you to divide the orb of the Sun into a number of orbs, in the above-mentioned way, for the reason that each of these two circles needs to be proposed in a separate orb with one surrounding the other. It appears from this also that the period of time which is from the greatest motion to the mean motion is less than the period of time that is from the mean motion to the smallest motion, by the measure that Ptolemy showed. This is shown when we propose that the measure of the two half arcs would be more than what was proposed for the measure shown with the hypothesis of the slant.

From 116b
§B.1.II.26/X
The Sun is also found, through this second hypothesis, preserving the ecliptic in §B.1.II
a way that can be taken as an example only when the two points of the greatest X.1
and least speed are at the two points of the solstices. So when they are in the rest

of the degrees of the ecliptic, in that our thesis is that when the mobile is in the plane in which the pole of the circle of the path is fixed, at that time the mobile is at the greatest extreme with respect to the equinoctial, but not to the ecliptic. From there the motion begins, with the motion of the two circles beginning when the inclination of the mobile from the ecliptic and from the point of intersection between the circle of the path and the ecliptic, in which the Sun is at that time, is the least possible. It is known that the measure of the two circles changes according to the position of the greatest speed on the ecliptic.

Indeed, when the point of the greatest speed changes its positions in the degrees of the ecliptic, as is the case with the Moon in order to preserve the latitude, we need to add other things that are difficult to picture, and it is not our intention to enlarge the contents of the book with this. Thus, we have chosen the first hypothesis since there is no difference whether or not these points are the points of the solstices, or if they differ through the degrees of the ecliptic because the aforementioned demonstration applies to whatever is proposed, and that is what we wanted. §B.1.II.26/X.2

All of this will be found through another more concise and more complete way without the thesis of these two small circles, and so that the center of the Sun is carried on the circle of the path itself. And for this we present this lemma: §B.1.II.26/X.3

(See figure 4.3.) We propose some point or pole or center of a planet on the surface of an orb, and it is point *K*. I say that it is possible that this point moves on great circle arc *DK*, whose measure is whatever degrees we desire, and that it reaches *K* from *D* on great circle *DGK* through motions in revolution. It returns from *K* to *D* without completing a circle, neither circle *DGK* nor another, and without ever departing from so doing. And if we want this motion to be upon small circles, then this has already been explained through the hypothesis of the anomaly when our thesis was two equal small circles on arc *DK*, with each one of them being in its own orb, with the measure of the two arcs *GE* and *EK* going out from the poles of each of them to its circumference together being equal to half of arc *DK*. The pole of one revolves around the pole of the other and is carried on its circumference, with one movement being twice the other as has been shown. §B.1.II.26/X.4

And if we propose that the mobile is on arc *KD*, that is to say that point *K* is the pole of a circle, then the diameter slants. This is because when pole *E* is on *Z*, that is to say half[-way] on *EZA*, circle *GHK* is tangent to arc *DK* [116a] at point *G*, and the above-mentioned diameter [*GEK*] intersects arc *DK* at right angles on point *G*. When our thesis is that this diameter is moving about pole *K*, in the direction of the motion of pole *E* about pole *G*, through a motion equal to it, then the above-mentioned diameter, rather the arc upon which it is, will always be superimposed upon arc *DK*, ascending from *K* to *D* and descending from *D* to *K*. §B.1.II.26/X.5

And if we want this movement of point *K* upon arc *DK* to be through great circles, then this motion is possible through one of three ways. We bisect arc *DK* §B.1.II.26/X.6

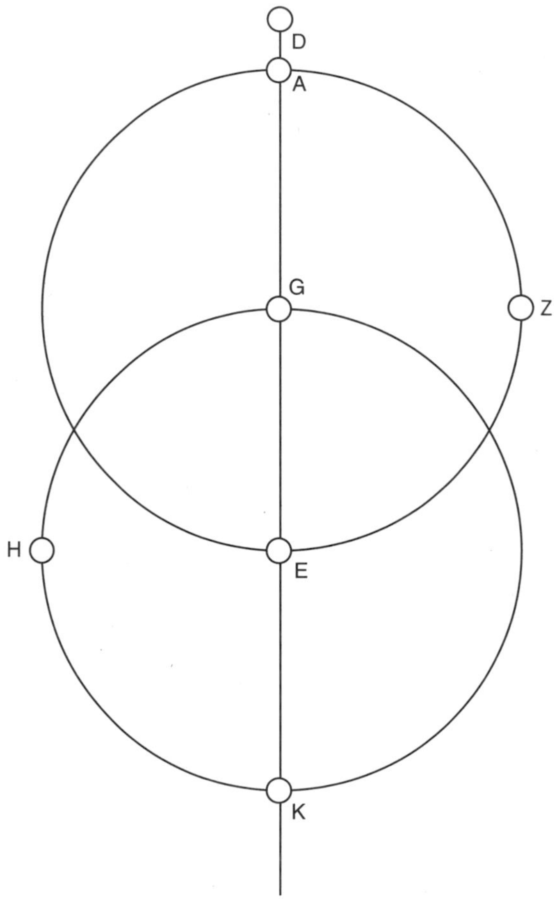

FIGURE 4.3

at *G*, and we trace about pole *A* at distance *AG* the great circle *GH*, and upon pole *Z* at distance *ZK* the great circle *KH*, and upon pole *A* at distance *AZ* the small circle *BEZ*. We intend the pole of circle *KH*, that is to say *Z*, to be, at a distance, about pole *A* upon circle *BEZ*.

The first of the three ways is that point *K* from the orb whose pole is *Z* and §B.1.II.2
the point opposite it are fixed in a third orb so that it cannot depart from circle X.7
DGK in any direction. A slant is created through this motion for each point that is traced on the orb in which there is point *K*, apart from point *K* and the point opposite it, and there is no point on it [the orb] that traces a circle of full revolution. When pole *Z* reaches *B*, point *K* reaches *D*, and the position of the orb to the right and left varies according to how the positions of its poles vary through their

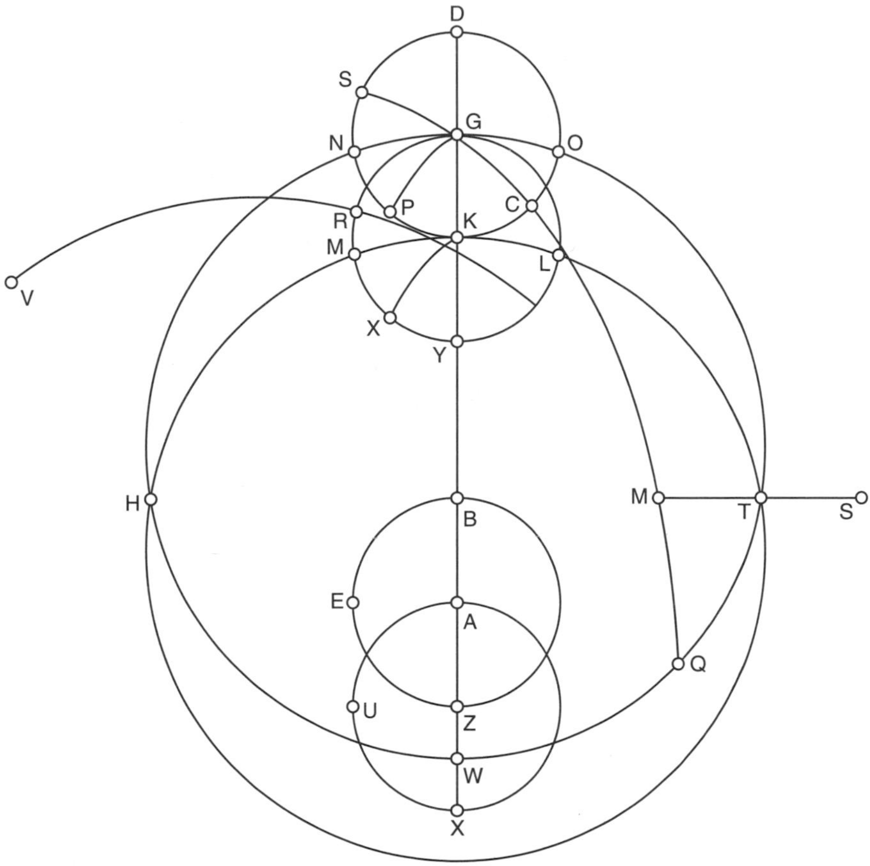

FIGURE 4.4

movement on the circle of their path, excluding point *K* and the point opposite it, since they are fixed as we said. It [*K*] has no motion except on arc *DGK*.

(See figure 4.4.) The second way is that the poles of circle *DGK* are fixed in an orb above or below it so that the orb can move about them. Point *K*, and the point opposite it from the orb which has in it *DGK*, are fixed in a third orb whose pole is *Z*, which revolves about pole *A* from a fourth orb. The orb which has point *K* in it is the middle of the first three, connected to one of them by nothing but the poles of circle *DGK*, and to the other by nothing but point *K* and the point opposite it. When pole *Z* moves on circle *BEZ* about pole *A*, point *K* from the orb whose pole is *Z* moves on arc *DGK* without passing from it, and the positions of the rest of the points of this orb vary as was shown in the first way. §B.1.II.26/X.8

And when we propose a point in the above-mentioned middle orb, whose §B.1.II.2
poles are fixed facing point *K* as if it is fixed in it, then it [the point] will be mov- X.9
ing in its orb on an arc similar to arc *DGK*. The entire orb moves with this motion itself as if this motion were about the poles of circle *DGK*, according to how point *K*, which is in the orb whose pole is *Z*, carries it with it. The position of the orb remains as it is, as its poles are fixed in such a way so that their position does not vary to the right or the left.

The third way, it being the most complete, is that this movement be through §B.1.II.2
uniform revolutions like the rest of the motions about which there is no doubt. X.10
This will be when our thesis is that the orb whose pole is *A* moves with it the orb [whose pole is] *Z* in its motion. Pole *Z* traces through this motion small circle *BEZ*. Our thesis is that there is another motion for the orb whose inner pole is *Z* about pole *Z* in reverse, with the two motions being a second [motion].

An example of this is that we propose pole *Z* to have moved through quadrant §B.1.II.
ZE about pole *A*. Arc *TKH* becomes *QMG*, and the location of point *K* is upon *M*, X.11
with it having a latitude from point *T* equal to the latitude that it had from point *G*. The point of intersection *H* is upon *G*. And through the second motion that is equal to it, which is about pole *Z*, point *K* is conveyed through quadrant *GK* in reverse, and it moves through arc *MG* and comes to be upon *G* itself. When the pole moves through quadrant *EB*, and the orb itself [moves] through another quadrant in reverse, then the point of intersection *H* also reaches *T*, and the point reaches *D*.

In the third quadrant, the point of the extremity that is opposite point [115b] *K* §B.1.II.
from circle *KH* is upon *S*, at a distance of *ST* equal to *TM*, and the point returns X.12
to *G*. In the fourth [quadrant], the point of intersection is as in the figure, and the point returns to *K*. Therefore, it [the point] does not cease going back and forth on arc *DK* from *K* to *D* and from *D* to *K*, never passing from this arc, with all of this being through uniform, complete revolutions. Know that between the two points *DG* and between the two points *GK*, there occurs for point *K* a slight deviation to the right and to the left from line *DK*, which is necessitated by the inclination of pole *Z* from pole *A*. Moreover, it shall be explained after this how these departures [can] disappear, and that is what we wanted.

Hence, it has been shown that the point that is on circle *TK* facing the pole of §B.1.II.
circle *DGK*, that is to say *T*, traces through these two motions a great circle arc X.13
equal to arc *DGK*, and it is arc *STM*, without ceasing to go back and forth upon it just as point *K* traces arc *DGK*. You already have known that point *K* from arc *DGK*, through these two uniform motions, does not traverse equal arcs in equal periods of time. For it is clear that whenever the point is near the edges of the arc, that is to say points *D* and *K*, the motion that belongs to the point is slower. And whenever it approaches the middle point, that is to say also *G*, the motion is faster. This lengthy explanation is permissible.

If we propose that the mobile is upon circle *KH*, that is to say that point *K* is the pole of a circle, then its diameter slants through these two motions. This is because if pole *Z* moves through quadrant *ZE*, and point *K* from circle *TKH* is upon point *M* from circle *QMG*, then the pole of circle *LGM*, that is to say *K*, also moves through quadrant *MG* and is upon *G*. The two points of the intersection of circle *LGM* with circle *TKH*, points *LM*, which would properly be on points *ON* from circle *TGH* were it not for the slant, come to be upon points *CS* from circle *QMG*. The slant is arc *SN*, and the diameter also slants through an arc of that amount, that is to say the chord of this arc, as if it were also arc *KP* equal to *SN*. §B.1.II.26/X.14

Therefore, when pole *K* moves, through these two motions, from *K* to *G* in latitude, the aforementioned diameter slants through arc *KP*. When pole *Z* moves through the second quadrant, that is to say *EB*, then pole *K* reaches *D*, and the slant reverts and disappears. And when the pole returns from *D* to *G*, the slant returns by this amount on the other side. When the pole returns to *K*, the slant disappears and departs. §B.1.II.26/X.15

Also, we want the latitude that occurs for pole *K*, namely arc *KGD*, from circle *TKH* from these two motions to disappear, rather for it [pole *K*] instead to remain in its position on circle *TKH*. The slant will not disappear but rather will be found to be double what it was before the latitude disappeared. This is through us proposing two more orbs upon these two orbs, with the pole of one being *A* and the pole of the second also *Z*; about pole *A* is great circle *QTG*, and about *Z* is great circle *TKH*. [115a] Our thesis is that the pole of upper circle *TGH*, namely *A*, is moving about the pole of upper circle *TKH*, namely *Z*, with pole *Z* being fixed for small circle *AU*. The orb is pulled with it, in the direction opposite the motion of lower pole *Z*, about pole *A* to *U*. The orb whose upper pole is *A* moves about itself with a motion equal to the motion of pole *A* upon circle *AU*. §B.1.II.26/X.16

Therefore, in the time period in which lower pole *Z* moved about lower pole *A* through quadrant *ZE*, and in which latitude *KG* was created for pole *K*, and in which diameter *GK* slanted by arc *KP*, in this time period upper pole *A* moved about upper pole *Z* through quadrant *AU*, with it [upper pole *A*] moving about itself an equal amount. The point of the extremity *G* from circle *TGH* from the upper [orb] comes to be on point *V*, that is to say at a quadrant, and the upper point of intersection *T* comes to be on upper [point] *K*, with the two lower orbs moving with it in its motion. The pole of circle *ODS*, which had been on *V*, reaches *K*. The point of intersection *N*, which properly would have been on *M* were it not for the second slant, is on *R*. Through the likes of this, diameter *GK* slants from *Y* to *X*. Therefore, the pole of the circle has returned to *K*, and the latitude that the two lower orbs necessitated for it, with respect to the two upper orbs, disappeared. The slant remains double what it had been at first on account of how all of the slants of the diameter have accumulated. §B.1.II.26/X.17

In the second quadrant, *G* is upon *K*, and the pole, due to the lower [orbs], §B.1.II.26/X.18

reaches *D* and, due to the upper [orbs], reaches *Y*, so the latitude disappears as do the slants. In the third quadrant, the latitude also disappears on one hand, because the latitudes acquired by pole *K* are opposites, but, on the other hand, the slants are renewed. In the fourth quadrant, the slants and, as well, the latitudes disappear, and this is what we wanted.

If our thesis is that upper pole *Z* is moving opposite the motion of the lower §B.1.II.2
[orb] about a third upper pole, pole *V*, inclined from lower pole *A* by double the X.19
inclination of pole *Z* from it, then one thing follows, except that with the first hypothesis, the four orbs are three according to how the poles of the two orbs are one as will likewise be shown afterward. It is known that just as this matter follows for the diameter, it therefore follows for the circle, be it large or small, traced about the point that was facing the pole of circle *DGK*, namely *T*. That is to say that latitudes and slants occur for it, not that its latitude disappears through these motions as the latitude of diameter *GK* disappears. Indeed, the pole of circle *DGK* is fixed, as is clear.

If we want the opposite of this, that is to say that the deviation in latitude that §B.1.II.2
pole *K* acquired from the two lower orbs remains, but that the slant created for X.20
the diameter departs, that is to say that there would be no slant at all, then our thesis is that upper pole *A* is moving opposite the motion of lower pole *Z* about upper pole *B* upon a circle traced about pole *B* at a distance *BA* equal to *AZ*. For example, if the motion of lower pole *Z* were from *Z* toward point *E*, then the motion of upper pole *A* would be from *A* in the opposite direction to point *E*, and it would move the two lower orbs with it, and it would also move about itself in reverse. Through this the slant disappears as if it [the diameter] had never had a slant at all, and the deviation in latitude remains twice what it had been. If you wish, it is possible for you to do this with only three orbs, they being the one whose lower pole is *Z* moving about pole *A* from the orb above it, with this pole, that is to say *A*, moving in the opposite direction about pole *B*, which is above it, with half of this [motion]. The orb whose pole is *Z* moves about itself with the likes of this, opposite the motion of pole *Z*, that is to say with half of the motion of pole *Z* about pole *A*, as was explained previously. When you look carefully, you find that doing this with four orbs reverts to [doing] precisely this with three orbs, because two of the four have the same pole, and the motion in one direction therefore will be double for the lower [orb] of them.

It is known that there will not be the aforementioned departures to the right §B.1.II.2
and the left because of their opposition at every point, [114b] even if some slant X.21
were created for the diameter of the aforementioned circle without a latitude, as has been mentioned. We want the greatest slant to increase over what it was by a certain amount, so we propose that inner pole *Z* is at a certain distance from inner pole *A* upon the great circle arc that passes through them. Then it [inner pole *Z*] reverses upon this arc and returns to the first distance so that only a

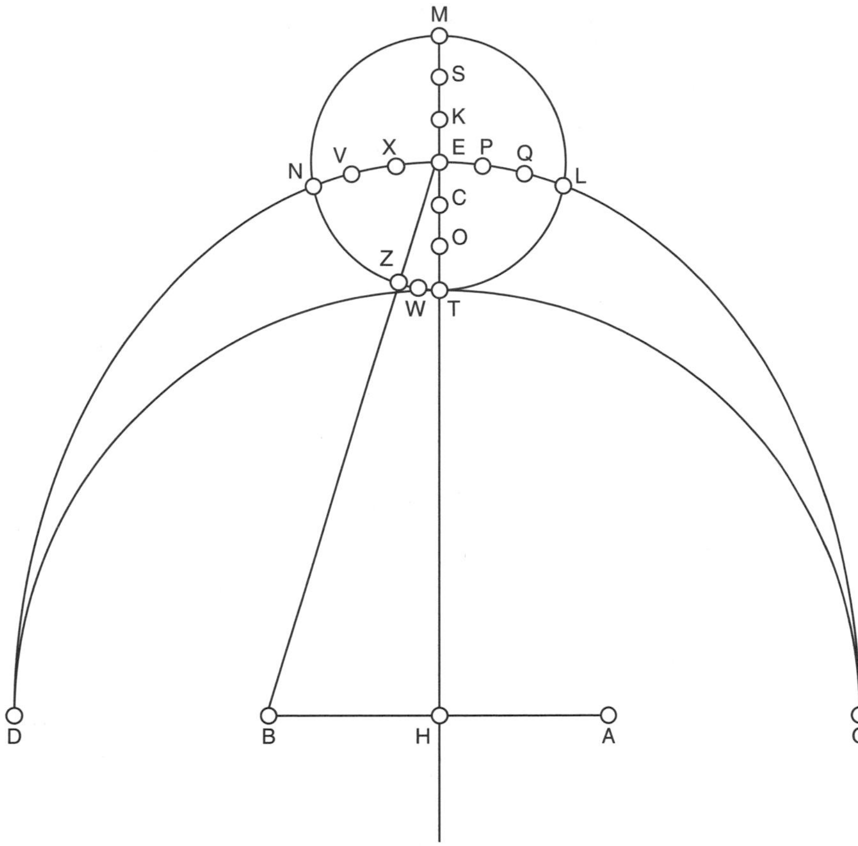

FIGURE 4.5

latitude without a slant is created for the orb. This is through a single compound of three orbs as has been shown. Upper pole *A* is distant by this amount from upper pole *Z* and reverts to the first distance a second time as has been explained. Whenever the poles become distant in the aforementioned way, the slant grows greater for the aforementioned circle. When they revert to the first distance, the slant returns to what it was at first. This has been completely explained.

(See figure 4.5.) After this has preceded, we trace the circle of the ecliptic *GTD*, the deferent circle *GED*, and the circle of the path *LTNM*, whose pole is *E*, tangent to the ecliptic at *T*, which is the point of the greatest speed. When pole *E* moves through quadrant *ED*, for example, the center of the Sun moves through quadrant *TN*. We trace, passing through the poles, [the arc] *ETH*, and we trace about *H*, that is to say the pole of the deferent, arc *AHB* intersecting arc *ETH* at §B.1.II.26/ X.22

right angles, with arc *AH* being equal to *HB*, and we extend great circle arc *EZB*. I say that it is possible for us to propose that in the period of time in which pole *E* moves through quadrant *ED*, the point of greatest speed on the circle of the path, that is to say *T*, slants by the degrees that we desire, for example from *T* to *Z*, without any deviation at all occurring for pole *E*. This is possible through one of three ways.

The first is our thesis that there are two orbs moving pole *E* from *E* to *K* in §B.1.II.2
latitude. In this period of time, point *T* from the circle of the path slants from *T* X.23
to *W*, with two orbs upon these likewise moving the pole in this period of time in the opposite direction in latitude, from *E* to *C*. Also, they impart a slant to point *W* from *W* to *Z* on the measure of the first slant, as was shown in the lemma. And in this time period pole *E* moves on its deferent circle from *E* to *D*, carrying with it the four orbs beneath it. As a result of how the departures of the pole in latitude, that is to say *EK EC*, are opposite, the pole of the circle of the path remains in its place without a deviation. The point of greatest speed slants with the two slants together. In the second quadrant, the first departure of the pole is from *N* to *S*, and the second is the [departure in latitude] opposite it from *C* to *O*; thus, the pole also remains in its position, and the slants disappear. In the third quadrant the first departure for the pole is from *S* to *K* and the second is from *O* to *C*, so the slants are to the analogous side. And in the fourth [quadrant] also, according to how the departures are opposite, the pole remains in its position. The slants and the small departures of pole *E*, to the right and left of arc *ES*, disappear as has preceded, necessitating for pole *E* no deviation from its position, because of how the departures causing it in arcs *EC EO* are opposite.

The second way is our thesis that the pole of the circle of the path, that is to say §B.1.II.2
E, is moving on arc *EQ*, and reversing upon it, as has preceded. We have already X.24
shown that through this [a slant] occurs for any point that was potentially a pole of this arc, namely *H* on arc *AB*, so the diameter of the circle of the path slants according to the aforementioned example. For example, through the two lower orbs in the time period of the quadrant, point *H* moves from *H* to *B*, and pole *E* from *E* to *P*, and point *T* slants from *T* to *Z*. Point *H* moves from *H* to *B* through the two upper orbs, also through the movement of the pole opposite the motion of the inner pole, and through another circle as had preceded. By an equal arc, point *H* of the lower [orbs], that was upon *B*, moves in this direction. Pole *E* moves in reverse from *E* to *X* equal to the first motion, with upper arc *AB* being inclined to lower arc *AB*, intersecting it upon point *H*. If the motions for pole *E* are thus opposite [114a] and equal, then the departures, also the above-mentioned ones, are opposite. Therefore, it remains in its position with the slant remaining as it was but double, that is to say twice arc *TZ*.

In the second quadrant, pole *H* returns, through the first two orbs, to *B*, and §B.1.II.2
pole *E* moves from *P* to *Q* through arc *PQ* equal to *PE*, and the slant departs. X.25

When we propose the second two orbs, pole *H* moves from *B* as well and its [the circle of the path's] pole *E* moves from *X* to *V* through arc *XV* equal to *EX*. Pole *E* remains in its position, and the slant departs.

In the third quadrant, pole *H* moves through the first two orbs from *H* to *A*, and the slant is created on the other side. Pole *E* moves from *Q* to *P*, and through the second two [orbs], point *H* moves from *H* to *A*. Likewise, pole *H* from the lower [orbs], which had been on *A*, moves in this direction, and pole *E* from *V* to *X*. The latitude vanishes, and the pole always remains in its position due to the opposition of the motions. The slant remains on the other side but twice what it was at first. In the fourth quadrant, pole *H* returns through both [orbs] from *A* to *H*, and pole *E* returns from *P* to *E* and from *X* to *E*, and the slant departs and disappears. With each of these two ways, even if we were to do it with four orbs, it has already been explained to you that it is possible to do it with three orbs. §B.1.II.26/X.26

The third way is our thesis that a great circle arc facing the circle of the path is going and returning in the aforementioned time periods upon an arc twice arc *TZ*, as was shown in the first figure from the lemma; that something like this occurs for the circle of the path has been explained. It is known as well that these three ways come back to a single hypothesis, so they are, truly, only a single way. §B.1.II.26/X.27

Now that this has been shown, our thesis is that the time period of the motion of the pole of the circle of the path through quadrant *ED* is equal to the time period of the motion of pole *H* in the said way through arc *HB*, that is to say half of *AB*. And it is also equal to the time period of the motion of pole *E* through arc *EK*, that is to say half of *ES* in the first way [of the three], with the measure of the arc being equal to the greatest anomaly. It increases over it by a little when we take with the measure of the greatest anomaly its ratio to its complement. We add it to it, and the measure of arc *EK* in the first way is half of this [the greatest anomaly plus its ratio to its complement]. §B.1.II.26/X.28

If this is so, all that is proper to the solar anomaly, such as it is, has been completed. The time period from the greatest motion to the mean motion is less than the time period that is from the mean motion to the least motion, since in the time period in which pole *E* moves through an arc that is less than a quadrant, equal to the degrees of the greatest anomaly, it completes the greatest anomaly with the above-mentioned slant at the mean passage. In each of the aforementioned ways, the difference [imparted by] the degrees of the slant will be more at the degrees [of longitude] closest to the points of greatest and least speed. We will need this matter to eliminate any approximation for the Sun's preserving the ecliptic. If it [the approximation] is not sensed, then all of this is explained, and lengthy explanation is permissible. §B.1.II.26/X.29

The small deviation created for the Sun from the ecliptic in latitude in a few places, due to our thesis that the slant is greater at the degrees of the greatest anomaly, is not to be sensed at all, as the deviation does not reach [113b] 3" of §B.1.II.26/X.30

latitude. It is not appropriate to mention a deviation like this, as Ptolemy did not sense a deviation greater than this in many places, and I say that it is not sensed. Despite this, you already know that with Ptolemy's hypotheses, a large deviation in longitude is created from a small deviation in latitude. But with our hypotheses, it is possible that no deviation in longitude is created from the deviation in latitude. Indeed, we have mentioned all of this in order to allude to the truth in everything, and when you accustom yourself to picturing all of this, you will see in it a wondrous, divine matter. Praise to God the revealer of mysteries.

Through, however, the second hypothesis, in which our thesis is that the §B.1.II.26
motions are in a single direction, there exist two slants, a first large slant and a X.31
second small slant, one being the opposite of the other. My intent is to say that the smallness of the slant that takes the place of the two above-mentioned circles, which is properly the measure of the second small slant or the arc which is the chord of the sum of the radii of the two, is the previously aforementioned amount.

Since all this has already been shown with the thesis that the motion in longi- §B.1.II.26
tude is about the poles of the [orb] carrying the poles of the circle of the path with X.32
both hypotheses, that is to say the hypothesis of the two circles and the hypothesis of the slant, then it appears from what has been said that this itself follows when our thesis is that the motion in longitude is about the poles of the ecliptic. Our thesis is, first, that the motion in longitude, that is to say the motion of the pole of the circle of the path with the Sun in longitude, will be about the poles of the ecliptic with the Sun revolving on the circle of its path following the motion in anomaly as has been shown, either through the aforementioned circles or the aforementioned slant. Through this, the entire visible anomaly is brought about, because the time period that is from the greatest motion to the mean motion is less than the time period that is from the mean motion to the smallest motion, with the measure of the slant or the measure of the radii of the two circles being equal to the greatest anomaly. See and understand how the center of the Sun always preserves the ecliptic without any approximation befalling it.

(See figure 4.6.) With the hypothesis of the slant, however, this is through the §B.1.II.2
revolution of the pole of the deferent upon the circle that necessitates the slant. X.33
We draw the figure, and let the pole of the deferent *B* be revolving upon circle *BKSZ* about *A*, the pole of the ecliptic, with a motion equal to that of the Sun on the circle of its path, with circle *BKSZ* being equal to the circle of the path. Our thesis is that when the Sun is on point *T*, the point of the greatest speed on the circle of the path, then the Sun is in the ecliptic as it is in the picture, with the pole of the deferent being on *B*. When the Sun moves through quadrant *TN* and becomes distant in latitude from the ecliptic by an arc equal to *TE*, pole *B* moves also through quadrant *BK*, with the orb of the deferent circle moving backward by an equal amount. It acquires an equal latitude in the opposite direction and

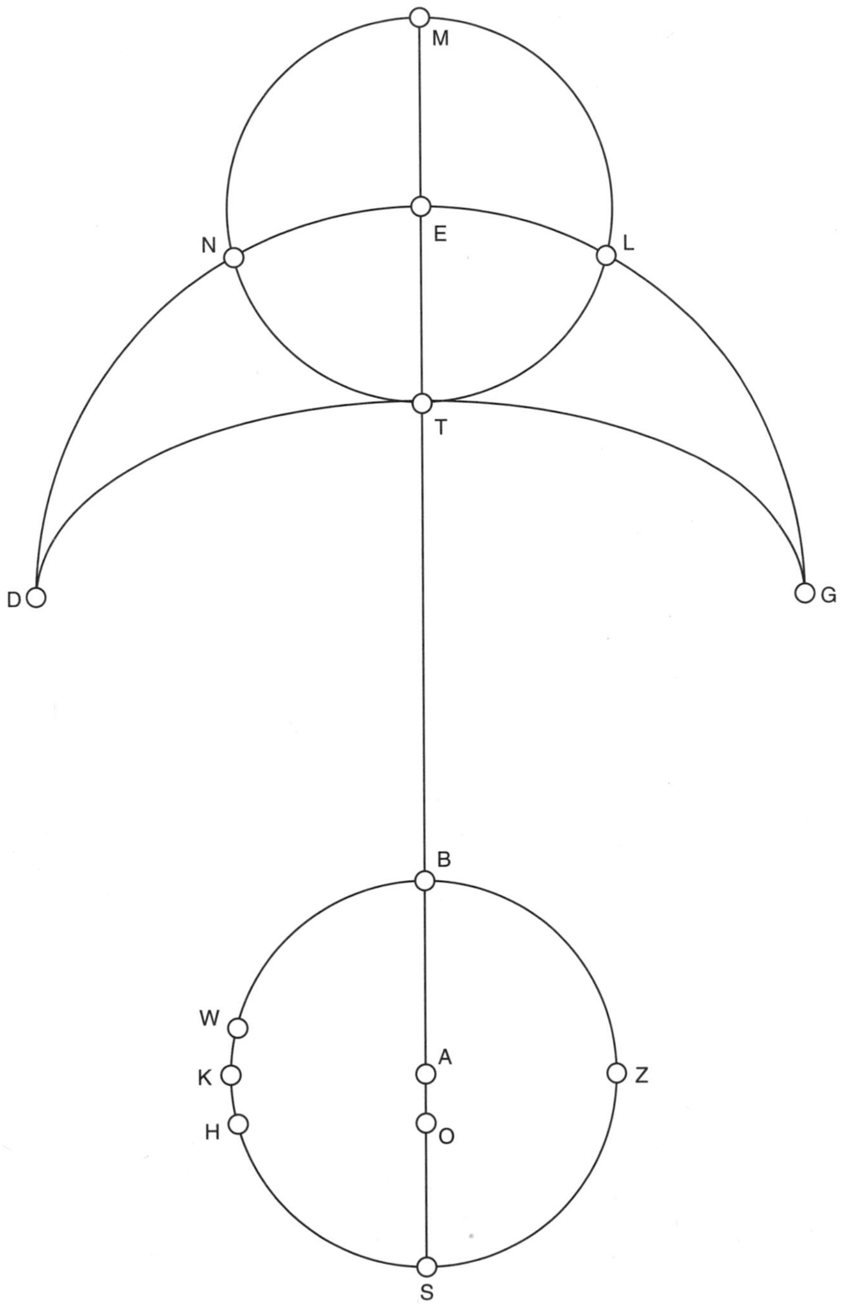

FIGURE 4.6

pulls with it the deferent circle and the circle of the path, with the pole of the circle of the path being on the ecliptic. When the Sun reaches *M*, pole *B* reaches *S*, so the circle of the path is tangent to the ecliptic at point *M*, and likewise with the rest of the degrees of the circle. Were it not for the slant, the Sun through this alone would always preserve the ecliptic.

Because of this, our thesis is that the pole of the deferent revolves upon great circle arc *WKH*,[1] going back and forth upon it through two orbs, without a motion in longitude or a slant being necessitated for the Sun, as has been shown. The measure of half of this arc is equal to the largest great circle arc subtended by the chord of the largest slant. The point bisecting great circle arc *WH*, along with the orb in which it is, is always attached to circle *BKSZ*, and moves about pole *A* through the above-mentioned motions. §B.1.II.2 X.34

In the period of time in which the Sun moves [113a] with the mean motion through quadrant *TN*, point *K* moves through quadrant *BK*, moving with it the pole of the deferent from *B* to *K*. The deferent orb moves about its poles an equal amount in the opposite direction, thereby creating the slant. According to how the Sun traveled at that time from point *N* in the direction of point *M* by the amount of the greatest slant, the pole of the deferent moved in latitude in this time period from *K* to *W* on a great circle in the aforementioned way, and the Sun returns to the ecliptic. And when the Sun reaches *M*, the pole returns to *K*, and *K* reaches *S*. When the Sun reaches *L* through the mean motion, point *B* reaches *Z*. And according to how the Sun departs from point *L* in the direction of point *M*, then by the measure of the greatest slant in that period of time, the pole of the deferent moves in latitude from *K* to *H* on the great circle, with point *H* being at that time between points *S* and *Z*. Therefore, the Sun returns to the ecliptic. And when the Sun reaches *T* from *L*, the pole returns from *H* to *K*, with point *K* being at that time on *B*. That way, the Sun is found always preserving the ecliptic without any approximation. According to the degrees through which the Sun moves on the circle of its path when a small latitude from the ecliptic will be created, at these times the pole of the deferent moves in circle *BKSZ* through an equal number of degrees, necessitating a small latitude equal to it in the opposite direction. And when the Sun goes through degrees that necessitate a great latitude, so does the pole. §B.1.II.2 X.35

This itself occurs through a slant according to how whenever a certain number of degrees from great circle arc *WH* are near point *B*, at that time they necessitate a small latitude of the number of degrees equal to the slant when they are near point *T*. When the arc is at point *K*, at that time those degrees themselves necessitate a great latitude equal to the slant at the mean passage, that is to say point *N* or *L*. Considering how great circle arc *DA* is superimposed upon circle *BKSZ* at point *K* alone, with the remainder of the degrees of the arc deviating from the §B.1.II.2 X.36

1. Great circle arc WKH is not shown in the MS figure.

circumference of circle *BKSZ*, then the maximum of this departure is at points *W* and *H*. Through this departure, a slant for the circle of the path is created that is added to the first slant, but when you calculate it, you find that this slant created from this deviation is not sensible at all.

As it has been shown already that the orb with the deferent circle moves about its pole, which moves upon arc *WH* equal to the mean motion that it has about the pole of the ecliptic, that is to say in reverse, then it is my intent to say that its mean motion is the motion of point *K* about pole *A* upon circle *BKSZ*. When our thesis is that this pole itself, that is to say the pole of the deferent orb, moves upon arc *HW* such that there is no motion in longitude nor a slant necessitated for the deferent orb, for it is a movement in latitude alone, then it is best that our thesis be also that the deferent orb moves about its pole like this, as was explained in the lemma. §B.1.II.26/X.37

It falls to us to propose this motion in whichever direction we wish, when we propose the motion [112b] of the orbs necessitating this motion in whichever direction we wish, as was also explained in the lemma. If our thesis is that this motion is in the opposite direction of the first motion, then there will be, about a single pole from a single orb, two equal opposite motions. It [the pole] will be necessarily fixed, that is to say that the deferent orb does not move about its poles. There follows for the Sun, from the two motions it has about pole *A* and upon arc *WH*, that which would follow were it moved about pole *A* alone. It would also move about itself like this in the opposite [direction], with an increase in latitude following like the increase in latitude necessitated by the motion on arc *WH*. The aforementioned orbs also move about the pole of the ecliptic, that is to say *A*, in the direction of the motion of the pole of the deferent about the pole of the ecliptic with a motion equal to it. Accordingly, there will be two equal motions in a single direction about the pole of the ecliptic. Accordingly, they are a single motion whose measure is the sum of the two motions. §B.1.II.26/X.38

Indeed, through the hypothesis of the two circles, the pole of the deferent does not revolve on circle *BKSZ*, and if it were not like this there would be two slants. Rather it revolves on arc *BAS* through two orbs as has preceded. It acquires through this the very latitude necessitated by the movement of the mean Sun, that is to say the pole of one of the two circles upon the circle of its path, according to how it has already been shown that the motion for the pole is slower whenever it comes near to the edges of the arc. Then the increases in latitude decrease like the increases in latitude that are due to the motions of the Sun when it approaches points *T* and *M* on the circle of its path. Because of the latitude arising from the two circles, our thesis is that pole *A*, that is to say the pole of the ecliptic that is in the orb of the Moon, moves upon arc *AO* by two orbs in the preceding way, with each [orb's] measure being equal to the larger great circle arc, the chord of which is the diameter of the one of the two circles upon which is the revolution of the center of the Sun. It moves with it in latitude all of the orbs that are beneath it. §B.1.II.26/X.39

In the time period in which the mean Sun moves through quadrant *TN*, the §B.1.II.
pole of the deferent moves from *B* to *A*, and the pole of the ecliptic [moves] in X.40
latitude through arc *AO*. When the Sun reaches *M*, the pole of the ecliptic returns to *A*, and the pole of the deferent reaches *S*. When [the Sun] reaches *L*, it [the pole of the ecliptic] returns to *O*. And when the Sun reaches *T*, it [the pole of the ecliptic] returns to *A*, with the pole of the deferent [returning] to *B*, and through this the Sun always preserves the ecliptic. If the Sun is like this, according to the hypothesis of the two circles, then the orb of the ecliptic is conveyed by the motion of its pole upon arc *AO* and about its poles, which will be about the pole of the deferent upon arc *BKSZ*. The orb of the ecliptic, whose pole is *A*, is fixed upon all of them, and its revolutions are the revolution of all of the orbs in longitude.

This matter is clear also for the Moon and the rest of the planets, that is to say §B.1.II.
when our thesis is that the motion in longitude is about the poles of the circle of X.41
the latitude. This having been explained, our intention is not to elongate things with that which has been explained. After we have explained that, through this final model there follows all that follows from the hypotheses of Ptolemy, in that our thesis is that the measure of the greatest slant is equal to the degrees of the greatest anomaly, then if it is so in each future position, our thesis is that the greatest slant is equal to the degrees of the greatest anomaly. We do not sense that in the two previous ways we have proposed it to be more by the aforementioned measure.

From 111b
§B.1.III.2/X

It is known that the two lines that enclose an angle of anomaly, whichever arc it may be from the circle of the path, are the lines that connect the center of the orb of the ecliptic and the ends of the perpendicular that extends from the abovementioned one of the ends in the plane of the circle of the path upon the line that extends from the second end, perpendicular to the common intersection of the circle of the path and the deferent circle. For example, an angle of anomaly is arc *BA* enclosed by the two lines *DA DM*, and it is known, also, that line *AM* connects the edges of these two lines. Indeed, according to the model in which our thesis was that the motion in longitude to be about the poles of the ecliptic, it is the common intersection of the circle of the path and the circle of the ecliptic when the Sun is at point *A*. And if our thesis was that it was about the poles of the deferent, then it is not the aforementioned common intersection.

The angle that is separated off by the aforementioned common intersection that connects *A* and line *BE*, whether it is larger or smaller than angle *ADM*, is not arc *AB*, the angle of the anomaly that is supposed to be added to or subtracted from the speed in longitude so that the mean speed in longitude increases or decreases because of the inclination of the deferent from the ecliptic. The slant that is created for the circle of the path because of this inclination is equal to the amount by which this angle exceeds or falls short of angle *ABM*. If this is so, then the angle that needs to be added to or subtracted from the mean speed in longitude, because of arc *AB*, is angle *ADM* itself and the angles that resemble it from among the rest of the arcs.

From 110b
§B.1.III.8/X
Or due to the aforementioned slant. Understand this from me, so that we do not have to repeat a single matter many times, that wherever I mention in this book the matter of the two circles, the slant takes its place.

From 110a
§B.1.III.10/X
It has already become clear from what has preceded that the number of orbs of the Sun, according to the hypothesis of the slant, is five orbs, two of them being principal [orbs] with a compound of three orbs. The first, that is the lowest of the principal [orbs], is the deferent orb for the center of the Sun, and it is that in which there is the circle of the path. Above it is a compound of three orbs for the slant of the circle of the path, through the slant of its orb, as has been shown. Above it is the orb in which there is the circle that carries the poles of the circle of the path that moves it in longitude. According, however, to the hypothesis of the two circles that takes the place of the slant, there are four orbs, with the lowest of them moving with the motions particular to it and also with all of the motions of the aforementioned orbs above it, when our thesis is that the motion in longitude is about the poles of the deferent. When our thesis, however, is that it [the motion in longitude] is about the poles of the ecliptic, then the orbs are, as according to the hypothesis of the slant, five as well. And according to the hypothesis of the two circles, they would be more than these, and their order and number is not hidden from you.

From 107a
§B.2.II.10/X
It being the hypothesis in which the movements were proposed in different directions in order for the center [of the Sun] to preserve the latitude.

From 104b
§B.2.V.6/X
(See figure 2.23.) It has already preceded how the position of the diameter of circle *EWDV* will not change through these motions, or the aforementioned slant would be our thesis in this place. It will be shown, as well, how the position

of circle *EWDV* with the hypothesis of the two circles is everywhere equal to its position with the hypothesis of the slant, that is to say as if it always moved about pole *B*. When the Moon is at quadratures, that is to say at point *D*, then the slant needs to be greater than it was at the conjunctions, that is to say that at point *E*. The slant needs to get bigger as the Moon travels from point *E* in the direction of point *D* so that the Moon always stays on the circle of latitude. It has already been demonstrated, previously, how the slant can get bigger or smaller by whichever measure we wish, and likewise, the measure of the two circles.

From 104a
§B.2.V.11/X
As long as the Moon moves on half-[circle] *EWD*, pole *S* should move on the half-[circle] *ST* that corresponds to it.

From 103b
§B.2.V.12/X
If it already moves in latitude, then how can it remain in its position without it having any deviation in longitude? And if a small slant occurred, then it would not be sensed at all, even if you want to remove the approximation that is on the measure of this third anomaly, for this anomaly is according to what was found with Ptolemy's observations, with others finding nothing less and nothing more than what it is through these hypotheses of ours. According to what follows from his impossible hypotheses, it is necessary that it be more than it is in our hypotheses by 5'. It is up to you to correct that by proposing a slant for circle *EWD*, as has been shown, and the slant is for all of diameter *KE* upon pole *E* in the direction of the position of the Moon on circle *EWD* by the necessary measure. Through this, as well, the largest anomaly will be nearer to the trines, that is at a distance of 120° from point *E*, and this is clear. As for what follows from the observations, Ptolemy himself admits that there do not exist observations in which this anomaly is greater than what it is at quadratures. Rather, he proposed, as a hypothesis for this anomaly, a slant of the diameter of the orb of the epicycle, and it follows from this that the greatest anomaly will be at trines. He depended on it without either an observation or another demonstration, and this is a total error.

From 103b–a
§B.2.V.12/Xb
It is known from what has preceded that the orbs of the Moon, according to the [hypothesis of the] slant, are five principal orbs for mean movements and anomalies, and four compounds for the slants and the latitudes as has been explained. The first of the principal orbs, that is to say the lowermost, is the orb that carries the center of the Moon, and it is that in which there is the circle from which

follows the second anomaly and the third. The second [principal orb], and it is upon the first, carries the pole[s] of this orb and moves them through its motion, it being the orb in which there is the circle of the path. Above the second orb is a compound of three orbs for the slant of the circle of the path as has become clear. Above the fifth [principal orb] is the orb in which is the circle that carries the poles of the circle of the path, moving them in longitude. Above the sixth [principal orb] is a compound of two orbs to remove the latitude created for the Moon by the orb proposed for the second [103a] anomaly and the third, with the interior orbs remaining in their position without any sensible slant at all.

But if we need a slant as we alluded, then we need a fourth compound upon the first, and upon these is the orb that is the mover of the extremes of the deferent, but with what is inside it being on the measure of the surplus of the motion in latitude over the motion in anomaly. Above it is the orb that is the mover for the extremes of the deferent circle and the circle of latitude on the measure of the surplus of the motion in latitude over the motion in longitude, and it is the 3' per day. As for the orb in which the circle of latitude and the parecliptic for the Moon are imagined, it is sufficient that we imagine it in the ecliptic itself without proposing an orb of that description for the Moon. All of these orbs are about a single center, and each orb that is underneath them moves with the motion that is particular to it and also moves with all the motions of the above-mentioned orbs that are above it.

According to the hypothesis of the two circles that take the place of the curve, they subtract from this number a single orb, and this is clear, and is so that the motion in longitude is about the poles of the circle carrying the poles of the circle of the path. But if our thesis is that it is about the poles of the circle of the ecliptic like the way that we demonstrated with the Sun in the last model, then the orbs of the Moon would also be of this number according to the hypothesis of the slant, but they are not the aforementioned. This is because we do not need the aforementioned three, they being the single orb from the first compound orb and the sixth [compound orb] and the mover for the extremes of the deferent orb. In place of these three orbs, there are, out of necessity, three other orbs, and they are two orbs to correct the latitude from the slant and another orb with the circle of the latitude in it, it being that which necessitates the motion in longitude. From this the number of orbs for the hypothesis of the two circles is explained.

From 101a
§B.3.II.4/X

As for the stars that are outside the ecliptic, their lag and final motion are not all of a single measure. Rather the lag and also the final motion of each of them are on the measure by which the star reaches its position with respect to the ecliptic according to the hypotheses of Ptolemy, and this measure varies. No very great harm enters from this matter not being possible with some of those stars, whose distance from the pole of the ecliptic is less than the maximum inclination, as it is with the rest of them. Through this, the stars that are outside the ecliptic preserve their positions with respect to each other to a certain point. It is enough for us that all of the stars are observed preserving their positions according to each one of the hypotheses in a single manner for a great number of years, even if they do not preserve it forever. And who knows how things will fall out in the times ahead of us, for the recent astronomers denied this motion and said that the stars had a motion in trepidation. From my knowledge, the matter is as they explain, but the true hypothesis for the motion is not the hypothesis whose existence is impossible that was the thesis of the recent astronomers, Ibn al-Zarqāl among them. Rather it is the true hypothesis that was our thesis previously with the hypotheses for the anomalies, according to which the motion in trepidation that was the thesis of the recent astronomers might be explained to you through what has preceded of the hypotheses for the anomalies, how they are through uniform perfect revolutions. This is through a compound of three or four orbs whose center is the center of the Earth, for it is obligatory for a single star that it move and return on some arc without passing from it. Thus, it follows for the rest of the stars, so that all of them are in a single orb, even if you want to follow Ptolemy's doctrine and deny this motion.

5

Commentary on the Judeo-Arabic Text

50v

The opening stanza of poetry

The poem on 50v of Vatican MS Ebr. 392, before the beginning of the text itself on 51r, contains an acrostic of Ibn Naḥmias's name (יוסף ן נחמיש), indicating that he was the likely author.[1]

This poem, not found in the Bodleian MS containing the Hebrew recension, is as follows:

ידידי קום קרא ספר לאורו   מעון קדם יכונן את אשורו
ותאר הגבול לזבול ועלה   עד שת עם צבא מרום מדורו
סמכתיהו במוסדי המדידה   והשיעור להראות את יקרו
פעולתו לפניו היא בעין כל   חכם לבב והן אתר שכרו
ובונים פחדו לבא בלי שוט   בשרביטו חצרו וחדרו
נצור משפט גבולי הנטיה   לכל גלגל גלילי ועד חגורו
חדל יוצא במרכזו להמציא   נשים תחתיו עגולת מעברו
מסלות כוכבי לכת ושבת   סדורות בו לאיש די מחסורו
יכונן מעגלות כוכבי מבוכה   וגם שמש יישר את הדורו
שמור דרכי והתבונן אשר לא   לחנם הלכו גוים לאורו

1. The observation about the acrostic is found in Benjamin Richler (ed.), paleographical and codicological descriptions by Malachi Beit-Arié (with Nurit Pasternak): *Hebrew Manuscripts in the Vatican Library: Catalogue* (Vatican City: Biblioteca Apostolica Vaticano, 2008): p. 339.

Translation:

Rise up, my friend, read a book in light of itself
A dwelling place of the eternal, establishing its validity
It described the boundary of the heavens and went up
Until it placed his dwelling with the hosts of high
I strengthened it with the foundations of measurement
And the measure to indicate its glory
His deed before him is in the eye of everyone
Wise of heart, and here is the site of his reward
We understand his fear [of God] to come without a whip
Through his scepter is his court and his wisdom
Preserves the judgment of the boundaries of the inclination
For each orb of his orbs up to its belt
Desist from creating eccentrics
In its stead we will place the circle of its path
The paths of the planets and fixed stars
Sufficiently ordered in it for a man
The circles of the planets will be rightly founded
And also the Sun straightening its elegance
Follow in my path and reflect
For it is not for nothing that nations walked in its light

51r

practitioners of mathematics . . . physics 0.1

The Judeo-Arabic term for the practitioners of mathematics was *ahl al-ʿilm al-riyāḍī* and for the practitioners of physics was *ahl al-ʿilm al-ṭabīʿī*. The debate between a mathematical approach to astronomy, understood by its adherents to be focused on prediction and retrodiction of observations, and a physical approach to astronomy, which was ostensibly Aristotelian, as well as their integration, had a history.[2] To partisans of the mathematical approach, the predictive and retrodictive accuracy of the eccentric and epicycle had not yet been matched by models of homocentric orbs. That shortcoming of homocentric models meant that the approach of the practitioners of mathematics, to its partisans, was truer. To partisans of the physical approach, mathematical astronomers relied on hypotheses, namely the eccentric and epicycle, the existence of which had not been demonstrated conclusively either observationally, or, more important for

2. See Gerhard Endress: "Mathematics and Philosophy in Medieval Islam," in Jan P. Hogendijk and Abdelhamid I. Sabra (eds.): *The Enterprise of Science in Islam: New Perspectives* (Cambridge: MIT Press, 2003): pp. 121–76.

Ibn Naḥmias, metaphysically.[3] To Ibn Bājja, Biṭrūjī,[4] and probably Ibn Rushd,[5] the approach and results of mathematical astronomy were not on a par with the truth of Aristotelian philosophy.[6]

At least for Ibn Naḥmias, the distinction between the approaches of the practitioners of mathematics and the practitioners of physics was not as absolute as the practitioners of physics portrayed it to be in several respects, nor was the need to choose a side as pressing. First, Ibn Naḥmias's argument against the existence of the epicycle cited Ibn Rushd's and Aristotle's argument from observation (and not from metaphysics) that the Moon, and the other planets by extension, did not roll.[7] Second, Biṭrūjī had acknowledged a connection between

3. Aristotle, in *Metaphysics* Book 12, Chapter Eight, posited a cosmos of homocentric orbs.

4. On Biṭrūjī, see Julio Samsó: "Biṭrūjī: Nūr al-Dīn Abū Isḥāq [Abū Jaʿfar] Ibrāhīm ibn Yūsuf al-Biṭrūjī," in Thomas Hockey et al. (eds.): *The Biographical Encyclopedia of Astronomers* (New York: Springer, 2007): pp. 133–4, and the bibliography.

5. Ibn Rushd's intellectual biography was complicated; he did not always hold epicycles and eccentrics in contempt. On that, see Juliane Lay: "*L'Abrégé de l'Almageste:* Un Inédit d'Averroès en Version Hébraïque," in *Arabic Sciences and Philosophy* VI (1996): pp. 23–61, at p. 25. On Averroes's *Physics,* see, now, Ruth Glasner: *Averroes' "Physics": A Turning Point in Medieval Natural Philosophy* (Oxford: Oxford University Press, 2009). Also, the first text in Vatican MS Ebr 392, the codex with *The Light of the World,* is an excerpt from a summary of the *Almagest,* but on the basis of a comparison of its contents with the table of contents (Lay, "*L'Abrégé,*" pp. 35–6) of Ibn Rushd's *Compendium,* one can conclude that the summary of the *Almagest* in Vatican MS Ebr. 392 is not a version of Ibn Rushd's *Compendium of the Almagest* in a variety of Arabic in Hebrew characters.

6. Endress, "Mathematics and Philosophy," pp. 151–7. On p. 151, Endress referred to Ibn Bājja's *Kalām fī al-hay'a* and its critique of mathematical astronomy: "In trying to build a universal proof of their *hay'a,* they set up a syllogism in which the inferences of such findings will form the middle term of a syllogism; hence, their results must be at variance with the principles of physical science." Astronomers whom Ibn Naḥmias would have classed as proponents of a mathematical approach developed their own arguments for why hypotheses such as the epicycle and eccentric (and others) were correct. The most extensive such arguments were put forth by Mu'ayyad al-Dīn al-ʿUrḍī (d. 1266). See Y. Tzvi Langermann: "Ibn Kammūna and the New Wisdom of the Thirteenth Century," in *Arabic Sciences and Philosophy* XV (2005): pp. 277–327, at pp. 293–5.

Biṭrūjī had entitled his own book *Kitāb al-hay'a* (translated as *On the Principles of Astronomy*). On Biṭrūjī's use of the word *hay'a,* see Julio Samsó: "On al-Bitrujī and the Hay'a Tradition in al-Andalus," article XII in *Islamic Astronomy and Medieval Spain* [Aldershot: Ashgate (Variorum), 1992]. On the history of the discipline of *ʿilm al-hay'a,* see also George Saliba: *Islamic Science and the Making of the European Renaissance* (Cambridge and London: MIT Press, 2007): 18–9, and passim. One of the earliest original compositions on astronomy from Andalusia had *hay'a* in the title. On that, see Josep Casulleras: "The Contents of Qāsim ibn Muṭarrif al-Qaṭṭān's *Kitāb al-Hay'a,*" in Maribel Fierro and Julio Samsó (eds.): *The Formation of al-Andalus, Part 2: Language, Religion, Culture and the Sciences* (Aldershot, UK: Ashgate, 1998): pp. 339–58.

7. For Ibn Naḥmias's argument, see *The Light of the World,* §B.4.I.2: "Aristotle shows in the second book of *De Caelo* that the planets do not have a rolling motion. What he said, according to what Ibn Rushd summarized, is as follows . . ." Ibn Naḥmias was likely referring to *De Caelo,* 290a, where Aristotle explained that the stars did not revolve in place on their axes.

theory and observation and alleged that Ptolemy did not always rely just on observations (perhaps implying that Ptolemy's theoretical shortcomings were due to neglect of observations in certain cases): "I say that they (claimed to) base their statements concerning these (motions) on observation and nothing else. But though they made it a principle (*aṣl*) in some cases, (they neglected it in others) . . ."[8] Conversely, there were observations for which Ptolemy's mathematical astronomy could not yet account. Though Ptolemy's models were the standard of predictive accuracy that both Ibn Naḥmias and Biṭrūjī sought, Ptolemy's lunar model predicted variations in the Moon's observed size that could not be observed. Biṭrūjī himself had acknowledged both Ptolemy's predictive accuracy and the fact that the *Almagest* predicted observations that were not observed.[9] Ibn Naḥmias, too, pointed to a disagreement between Hipparchus and Ptolemy over the Moon's visible diameter.[10]

Third, in Book 2, Chapter 24 of *The Guide of the Perplexed*, Maimonides described a perplexity that arose from the competing truth claims of the two approaches.[11] Maimonides refuted the existence of the epicycle and eccentric while acknowledging their predictive accuracy.[12] Maimonides saw that the position of the mathematical astronomers made the same claim to reality as the position of the philosophers, since the mathematicians' use of the epicycle and eccentric saved the phenomena: "Furthermore, how can one conceive the retrogradation of a star, together with its other motions, without assuming the existence of an epicycle? On the other hand, how can one imagine a rolling motion in the heavens or a motion around a center that is not immobile? This is the true perplexity."[13] Tzvi Langermann has pointed out that it is possible to

8. Al-Biṭrūjī (trans. and comm. Bernard R. Goldstein): *Al-Biṭrūjī: On the Principles of Astronomy* (New Haven and London: Yale University Press, 1971), vol. 1: p. 72. The translation of the word *aṣl* is complex (see my comments on "two hypotheses" in §0.1), but here Goldstein's translation uses "principle."

9. Biṭrūjī (trans. and comm. Goldstein), *On the Principles*, pp. 59–61, esp. p. 61: "nor do they produce agreement with what he observed and measured."

10. Ibn Naḥmias, *The Light of the World*, §B.2.IV.3.

11. Here Maimonides famously noted Ptolemy's problems in computing planetary distances, meaning that Ptolemy did not fully succeed in their own endeavor. See Maimonides (trans. Shlomo Pines, intro. Leo Strauss): *The Guide of the Perplexed* (Chicago and London: University of Chicago Press, 1963), vol. 2: p. 324. The important problem of planetary distances was also acknowledged by an astronomer concerned with predictive accuracy. See B. R. Goldstein and Noel Swerdlow: "Planetary Distances and Sizes in an Anonymous Arabic Treatise Preserved in Bodleian Ms. Marsh 621," in *Centaurus* XV (1971): pp. 135–50. George Saliba ["The First Non-Ptolemaic Astronomy at the Marāgha School," in *Isis* LXX (1979): pp. 571–6, at pp. 571–2] identified the author of this text as 'Urḍī.

12. Maimonides, *Guide*, vol. 2, pp. 322–7, esp. p. 324. For the Judeo-Arabic, see Maimonides (ed. S. Munk): *Dalālat al-ḥā'irīn* (Jerusalem, 1929): pp. 225–9, esp. p. 226.

13. Maimonides, *Guide*, vol. 2, p. 326.

read *Guide* II, 24 as a rejection only of epicycles that move on an eccentric.[14] This commentary will show that *The Light of the World* was, above all, an attempt to resolve this "true perplexity," inasmuch as Ibn Naḥmias saw predictive accuracy as a truth worth pursuing and that Ibn Naḥmias understood there to be a legitimate competition between the two approaches. Fourth, other scholars besides Maimonides, Ibn Bājja, Ibn Ṭufayl, Ibn Rushd, and Biṭrūjī, such as Levi ben Gerson (d. 1344) as well as those connected to the intellectual tradition of the Marāgha Observatory in Iran valued physical consistency.[15] A concern with harmonizing mathematical and physical approaches to astronomy extended beyond Andalusia and did not have to be an exclusive choice.

Ibn Naḥmias marveled that Ptolemy, and those who followed him, achieved a high degree of predictive and retrodictive accuracy with flawed hypotheses.[16]

14. Y. Tzvi Langermann: "The 'True Perplexity': The *Guide of the Perplexed*, Part II, Chapter 24," in Joel Kraemer (ed.): *Perspectives on Maimonides: Philosophical and Historical Studies* (Oxford and New York: Oxford University Press, 1991): pp. 159–74, at p. 168. Also: "There then follow two objections aimed specifically at the epicycles, but these, I emphasize, are brought in the name of Ibn Bājja." Langermann has now developed his thinking on the matter in "My Truest Perplexities," in *Aleph* VIII (2008): pp. 301–17. Gad Freudenthal has described ("Maimonides on the Knowability of the Heavens and Their Mover [*Guide* 2:24]," in *Aleph* VIII [2008]: pp. 151–7) the entire question as one of the knowability of God from the study of the heavens. Sālim Yafūt ("Ibn Bājja wa-'ilm al-falak al-Baṭlamyūsī," in Sālim Yafūt [ed.]: *Dirāsāt fī tārīkh al-'ulūm wa-'l-ibistīmūlūjiyyā* [Rabat, 1996]: pp. 65–73, at p. 72) also saw Ibn Bājja as a critic of Ptolemaic astronomy but acknowledged texts (p. 67) by Ibn Bājja that accepted the eccentric and epicycle in an apparent attempt to save the phenomena rather than describe reality.

15. There is a substantial amount of literature on the crucial achievements of the astronomers associated with the Marāgha Observatory. The most important books on the subject are F. J. Ragep (ed., trans., and comm.): *Naṣīr al-Dīn al-Ṭūsī's "Memoir on Astronomy" (al-Tadhkira fī 'ilm al-hay'a)* (New York and Berlin: Springer-Verlag, 1993); George Saliba: *Islamic Science and the Making of the European Renaissance* (Cambridge and London: MIT Press, 2007); and George Saliba: *A History of Arabic Astronomy* (New York: NYU Press, 1994). Bernard Goldstein has produced the most comprehensive studies of Levi ben Gerson's astronomy. To begin with, see Bernard R. Goldstein: *The Astronomy of Levi ben Gerson (1288–1344)* (New York, Berlin, Heidelberg, Tokyo: Springer-Verlag, 1985).

16. For Ibn Naḥmias's own statement, see *The Light of the World*, §0.7: "all this with the hypotheses upon which this is based being imagined so far from the truth." Maimonides had noted this same point ([trans. Pines] *Guide*, vol. 2, pp. 325–6). See also, Langermann, "The 'True Perplexity,'" p. 159. According to Langermann (p. 165), "Maimonides does regard the true configuration of the heavens as something humanly attainable." But although Ibn Naḥmias's proposal of new theories would suggest that humans are capable of finding at least truer representations of the heavens, Ibn Naḥmias did not cite the end of *Guide* II.24, where Maimonides noted such a possibility.

In the Islamic East, al-Sayyid al-Sharīf al-Jurjānī (d. 1413) and 'Alī Qushjī (d. 1474) made the related point that the agreement of the hypotheses with observations was both a source of wonder and an argument for the hypotheses' veracity in order to defend theoretical astronomy against the (plausible) accusation that astronomy's theories could not be demonstrated deductively. On Jurjānī and Qushjī, see F. J. Ragep: "Freeing Astronomy from Philosophy: An Aspect of Islamic Influence on Science," in *Osiris*, 2nd series, XVI (2001): pp. 49–71, p. 57 (for Jurjānī) and p. 55ff. (for Qushjī).

Ibn Naḥmias's goal was to show that Ptolemy's predictive and retrodictive accuracy was attainable without epicycles and eccentrics. Gad Freudenthal has written that "what Maimonides apparently intended was that a future cosmology, based on (dialectical) demonstrations, will be found that would be in contradiction neither with the principles of (Aristotelian) science, nor with those of mathematical astronomy. . . . Still, it would not provide *dioti* explanations of *all* celestial phenomena."[17] *The Light of the World* can be understood as a step toward Maimonides's goal, as Ibn Naḥmias subtly acknowledged, at times, the imperfections of his system, and his model for the motion of the fixed stars was a case where the true model was likely beyond the human ken.

some of them founded the matter upon two hypotheses with regard to which some of the practitioners of physics completely disagreed
The Judeo-Arabic original includes the partitive *minhum* (some of them) here; that expression is absent in the Hebrew recension. The inclusion of the partitive communicates that some, but not all of the practitioners of mathematical astronomy employed the hypotheses of the eccentric and epicycle. Ibn Naḥmias, conceivably, could have been referring to Levi ben Gerson, who denied the epicycle but accepted the eccentric. Further on, there is another partitive, *baʿḍ* (*some* of the practitioners of physics), that could connote some, several, or even one practitioner of physics. More interestingly, these distinctions between the mathematical and physical approaches were blurred further when some of the mathematicians used observations to make a physical statement that celestial motions could be in opposite directions, contra Aristotle's position in *De Caelo* 270a. For example, Ptolemy wrote in *Almagest* I.8 that the two primary motions of the heavens were clearly in opposite directions.[18] See also the comments on §0.2.

two hypotheses
Aṣl (pl. *uṣūl*), which I have translated as "hypothesis," is a word that is difficult to translate. A. I. Sabra has explained that an *aṣl* is a low-level proposition that makes up a larger model;[19] that is the sense in which Ibn Naḥmias has just used the word because, in the fourth treatise on the planets, Ibn Naḥmias began to use the word *hay'a* (configuration) for the complete models. Notably, in the

17. Gad Freudenthal: "'Instrumentalism' and 'Realism' as Categories in the History of Astronomy: Duhem vs. Popper, Maimonides vs. Gersonides," in *Centaurus* XLV (2003): pp. 227–48, at p. 241.

18. Gerald Toomer: *Ptolemy's Almagest* (London: Duckworth, 1984): p. 45. See also *The Light of the World*, §0.11: "As for the matter of opposite motions, it would have been necessary to leave them as they are according to Ptolemy due to how he showed the lack of opposition in the motions."

19. A. I. Sabra: "Configuring the Universe: Aporetic, Problem Solving, and Kinematic Modeling as Themes of Arabic Astronomy," in *Perspectives in Science* VI (1998): pp. 288–330, at pp. 293 and 313, n. 33.

Judeo-Arabic original and in the recension, Ibn Naḥmias also broke down his own models into their component *uṣūl.*[20] For instance, he presented his solar model sequentially (§B.0.1), beginning with the hypothesis for mean motion, then moving to the hypothesis for the anomaly. There is a precedent for different astronomy texts using *aṣl* differently; Mu'ayyad al-Dīn al-'Urḍī (d. 1266) used *aṣl* to mean postulate or axiom, whereas Quṭb al-Dīn Shīrāzī used *aṣl* solely in the sense of a building block of a larger model.[21]

As for the question of why *aṣl,* in the sense of a low-level proposition or building block, is best rendered in English as "hypothesis," Gutas and Endress's research has found that the word *aṣl* was used to translate the Greek word ὑπόθεσις.[22] Toomer, in his translation of the *Almagest* rendered that Greek word as "hypothesis."[23] A dimension of a hypothesis is that it can be disproved, as the Ptolemaic hypotheses were.[24] With the same hesitation Toomer conceded, I am going to translate *aṣl* in that sense as "hypothesis," but I recognize that "principle" would be a valid alternative. Indeed, the Arabic translations of Ptolemy's *Planetary Hypotheses* make the picture more complex. Ptolemy's *Planetary Hypotheses* was rendered in Arabic as *Kitāb al-Manshūrāt* (The book of sawed-off orbs) or *Kitāb al-Iqtiṣāṣ* (The book of the account) and not *Kitāb al-uṣūl.*[25] Still, in the *Planetary Hypotheses,* the epicycle and eccentric were presented as solid

20. Ragep has, in comparison, translated aṣl as "model." See F. Jamil Ragep: "Ibn al-Haytham and Eudoxus: The Revival of Homocentric Modeling in Islam," in Charles Burnett, Jan P. Hogendijk, Kim Plofker, and Michio Yano (eds.): Studies in Honour of David Pingree (Leiden: E. J. Brill, 2004): pp. 786–809, at p. 791. There Ragep calls "a system" the orbs that account for the motions of a given planet, whereas a model (*aṣl*) is a component (e.g., an eccentric or epicyclic orb) of a system.

21. Robert Morrison: "Quṭb al-Dīn al-Shīrāzī's Hypotheses for Celestial Motions," in *Journal for the History of Arabic Science* XIII (2005): pp. 21–140, at pp. 23–4. See also G. Saliba, *The Astronomical Work of Mu'ayyad al-Dīn al-'Urḍī* (Kitāb al-Hay'a): *A Thirteenth-Century Reform of Ptolemaic Astronomy* (Beirut, 1990): pp. 115–6.

22. Gerhard Endress and Dimitri Gutas (eds.): *A Greek and Arabic Lexicon: Materials for a Dictionary of the Mediaeval Translations from Greek into Arabic* (Leiden: E. J. Brill, 1992-), vol. 1, fascicle 2, pp. 224–31. There were other possible Arabic translations of the word ὑπόθεσις; see note 23.

23. See Toomer, Ptolemy's "Almagest,", pp. 23–4 and 141ff. (where the word *hypothesis* is applied to the epicycle and eccentric). Pedersen (*Survey of the Almagest,* p. 134) called the epicycle and eccentric "hypotheses" but more frequently (e.g., pp. 137–9), "models."

24. There is precedent for describing, and then rejecting, *uṣūl.* In *Nihāyat al-idrāk,* Shīrāzī included, in his list of hypotheses (*uṣūl*), hypotheses the existence of which he rejected. See Quṭb al-Dīn Shīrāzī, *Nihāyat al-idrāk fī dirāyat al-aflāk,* Istanbul Pertev Paşa MS 381, pp. 34–37.

25. Fuat Sezgin: *Geschichte des arabischen Schrifttums* (Leiden: E. J. Brill, 1978), vol. 6: pp. 94–5. A Hebrew translation of the *Planetary Hypotheses* by Qalonymos b. Qalonymos survives in a single MS, Paris BNF Hébreu 1028, with the title *Sippur 'inyanei ha-kokabim ha-nebukim.* See Moritz Steinschneider: *Die Hebräischen Übersetzungen des Mittelalters und die Juden als Dolmetscher* (Berlin, 1893; repr. Graz, 1956): pp. 538–539. Steinschneider explained that the Hebrew *sippur* (report, description) rendered the Arabic *iqtiṣāṣ.*

orbs on the one hand, and Ptolemy acknowledged the possibility of substituting sawed-off orbs (*manshūrāt*) on the other hand, implying that the orbs of the epicycle and eccentric were hypothetical.[26]

In *The Light of the World* itself, the best argument for translating *aṣl* as "principle" would be §0.2, where Ibn Naḥmias described the *uṣūl* that he was about to present as "certain"; and in many cases, starting in §0.9, he referred to the *uṣūl* of *The Light of the World* as "true." But these statements of confidence contrast with the more contingent or hypothetical attitude found in the presentation of the *uṣūl* themselves, where Ibn Naḥmias presented alternative *uṣūl* and rarely said that a single *aṣl* was exclusively true. In addition to the possibility that Ibn Naḥmias was not consistent, he may have been asserting that his *uṣūl* were true according to the requirements of philosophy and physics but not always those of mathematics.

There are three more specific ways in which Ibn Naḥmias's *uṣūl,* however probable, were, in the end, hypothetical. Evidence for the first way is found in the Judeo-Arabic and Hebrew versions. For much of *The Light of the World,* Ibn Naḥmias described as *uṣūl* the opinions about whether models should entail apparently opposite motions. While Ibn Naḥmias demonstrated the *aṣl* of allowing apparently opposite motions, Ibn Naḥmias also frequently presented his models according to the *aṣl* of lag. Ibn Naḥmias did not go so far as to prove that lag was false; rather, lag remained hypothetical. When Ibn Naḥmias discussed the *uṣūl* of opposite motions, there was no absolutely correct answer: "You know from what we said previously that the first hypothesis is worthier due to what we demonstrated regarding the lack of opposition in circular motion."[27] The second way pertains only to the Hebrew recension in which *aṣl* was translated by *'iqqar,* as well as by *shoresh* and *y^{e}sod,* which suggest, in addition, possible translations of "basis" or "principle." Like the Judeo-Arabic original, the Hebrew recension did use *'iqqar* to describe the components of the models (e.g., the *'iqqar* of the slant, the *'iqqar* of the two circles). When concluding the discussion of the solar model, the Hebrew recension explained that the orbs of the Sun, according to the *'iqqar* of the slant numbered five, whereas their number was different according to the *'iqqar* of the double circles.[28] The Hebrew recension did not come to a conclusion about which *'iqqar* was better; each remained hypothetical. Insertions at the end of the Hebrew recension's discussion of the lunar model, in §B.2.V.12/Xb, provide related evidence for how there remained a dimension in which the

26. Bernard Goldstein: "The Arabic Version of Ptolemy's *Planetary Hypotheses,*" in *Transactions of the American Philosophical Society,* new series, LVII (1967), no. 4: pp. 3–55: for example, p. 24 (for the epicycle and eccentric) and p. 50 (for the *manshūrāt*).

27. *The Light of the World,* §B.1.I.5.

28. *Or ha-'olam,* MS Bodleian Canon Misc. 334, 110a §B.1.III.10/X.

models were hypothetical, albeit probable. In this version of the lunar model, the first, lowest, orb was the orb that moved the center of the Moon, which I take to be the mean Moon. According to the recension, to account for the second and third lunar anomalies, the Moon also had to move on a small circle found on that first orb. At most, the recension has shown that it was hypothetically possible to account for the Moon's motions without varying the Earth–Moon distance. The recension *never* fully explained the construction of the model. A third dimension in which Ibn Naḥmias understood his models to be conjectural emerges in the comments on §B.1.I.2.

In order to emphasize the literal meaning of "hypothesis" as that upon which something is set forth, I frequently translate the Judeo-Arabic *waḍʿ* (otherwise: to put forward, to set forth) as thesis. I often translate the associated verb *waḍaʿ* as "to propose" or "to make () as a thesis."

explain the impossibility of these two hypotheses for the celestial bodies with only physical evidence

The Judeo-Arabic word translated by "physical evidence" is *dalāʾil ṭabīʿiyya,* that is, evidence from physics. Ibn Naḥmias distinguished between *dalāʾil ṭabīʿiyya* and *kalām burhānī* (demonstrative discourse),[29] meaning that this physical evidence did not provide a demonstration.

One type of physical evidence against the existence of a rolling motion is that the same side of the Moon is always visible, an observation that an epicycle does not entail.[30] Aristotle, who favored an astronomy similar to that of Eudoxus's concentric orbs, was the source for this argument.[31] This evidence was important to Gersonides,[32] an astronomer very much concerned with predictive accuracy, but also with harmonizing the mathematical and physical approaches,[33] as observations became a way to make an argument that impinged on metaphysics.

Ibn Rushd gave an additional, metaphysical argument against the existence of the eccentric and epicycle when he wrote, in his long commentary on Aristotle's *Metaphysics* (expounding upon Book 12, Chapter Eight): "And the doctrine of the eccentric sphere or the epicyclic orb is unnatural. As for the epicyclic orb, it is essentially impossible and that is because a body that moves in circular

29. On *kalām burhānī,* see *The Light of the World,* §0.2.

30. Ibn Naḥmias modified this very argument in *The Light of the World,* §B.4.I.2.

31. Aristotle, *De Caelo,* 290a. See also Paul Lettinck: *Aristotle's "Meteorology" and Its Reception in the Arab World* (Leiden, Boston, Köln: Brill, 1999), p. 576. On Aristotle's theory of homocentric orbs, see Thomas Heath: *Aristarchus of Samos: The Ancient Copernicus* (Oxford: Clarendon Press, 1913; repr., Mineola, NY: Dover Publications, 1981 and 2004), pp. 190–248.

32. Goldstein: *The Astronomy of Levi Ben Gerson,* p. 117. See also Goldstein: "Theory and Observation in Medieval Astronomy," in *Isis* LXIII (1972): pp. 39–47.

33. Goldstein, *The Astronomy of Levi Ben Gerson,* p. 7.

motion indeed revolves about the center of the universe, not external to it, since it is the thing revolving that makes the center. So if there were a circular motion external to this center, then there would have to be another center external to this center. . . . And it resembles the case of the eccentric sphere which Ptolemy imposes, and that is if there were many centers then there would have to be heavy bodies outside the place of the Earth."[34]

What was said about the impossibility of opposite motions in the heavens is invalid 0.2
In §0.1, Ibn Naḥmias mentioned mathematicians who allowed the existence of opposite motions in the heavens. Here, Ibn Naḥmias took up the philosophical question of whether the apparently opposite motions that the mathematicians allowed as a compromise were actually opposite. A major impetus for the composition of *The Light of the World* was that, in Ibn Naḥmias's view, Biṭrūjī's *On the Principles* had imposed an unnecessary restriction upon itself by holding that apparently opposite motions in the heavens were impossible and consequently constructing his models so that they would not produce apparently opposite motions.[35] The passage begins with a reference (*what was said*) to practitioners of physics, including Biṭrūjī. Biṭrūjī's ostensible justification was that since the mover of the heavens is one, then only a single motion could originate from that

34. Ibn Rushd (ed. Maurice Bouyges): *Tafsīr mā ba'd al-ṭabī'a* (Beirut, 1948): pp. 1661–2 (my translation). Ibn Rushd also rejected the eccentric in his *De Caelo*. See Ibn Rushd: "Kitāb al-Samā' wa-'l-'ālam," in *Rasā'il Ibn Rushd* (Hyderabad: Dā'irat al-ma'ārif, 1947): pp. 1–79, at p. 7 (each of the six treatises in the book, Ibn Rushd's short or middle commentaries, is paginated separately). Ibn Rushd explained that if a body whose natural motion is revolution moves toward or away from the center, then there must be a mover to compel this motion. See also ibid., p. 42, where Ibn Rushd said that an eternally moving mobile must move about something stationary: the Earth. Based on what Ibn Rushd has said, it would appear (ibid., p. 46) that it would be impossible for the celestial bodies to vary in their motion.

But Samuel Ibn Tibbon's Hebrew version of Aristotle's *Meteorology* ([ed. and intro. Resianne Fontaine]: *Otot ha-Shamayim: Samuel Ibn Tibbon's Hebrew Version of Aristotle's "Meteorology"* [Leiden, New York, and Köln: Brill, 1995]: p. 49) for which Samuel Ibn Tibbon had consulted Ibn Rushd's epitome of Aristotle's *Meteorology*, implied an eccentric orb for the Sun: "For the Sun approaches the Earth and then recedes and therefore the Sun is the cause of generation and corruption." Of course, Ibn Tibbon (ibid., p. 52) used Ibn Rushd's *Compendium of the "Almagest"* (which allowed eccentrics and epicycles) to resolve complicated passages, so one should not expect agreement on the existence of the eccentric. Still, Ibn Tibbon's *Meteorology* was not an isolated case. Ibn Rushd's *Kitāb al-Kawn wa-'l-fasād* (in *Rasā'il Ibn Rushd*, p. 29) has a similar passage in which Ibn Rushd related the Sun's distance from the Earth to the occurrence of certain weather phenomena. Similarly, in *Kitāb al-Āthār al-'ulwiyya* (in *Rasā'il Ibn Rushd*, p. 8), Ibn Rushd remarked that eccentricity was no folly (*fa-inna khurūj al-markaz lā yakūn 'abath^{an}*). Ibn Rushd came to homocentricity later in life.

35. Biṭrūjī ([trans. and comm. Goldstein], *On the Principles*, p. 5) berated Ptolemy for treating the daily motion as two separate motions. Since according to Biṭrūjī there was only one mover, there should be only one motion.

mover.[36] Throughout *The Light of the World,* Ibn Naḥmias took pains to argue (a) that apparently opposite celestial motions were not truly so and should therefore not be excluded, and (b) that even if one did exclude them, most of Ibn Naḥmias's models could be presented according to the hypothesis of lag.

With respect to the question of the possibility of opposite motions in the heavens, Ibn Naḥmias could have been participating in a broader, long-running conversation about interpreting Aristotle that continued after Biṭrūjī's death.[37] Aristotle had hinted, but not stated explicitly, in *De Caelo* 270a that the categories of "up" and "down" were not applicable to the motion of the celestial orbs.[38] Definitions of "up" and "down" depended on motions in a straight line, motions that did not exist in the heavens. Ibn Bājja, a philosopher cited by Maimonides in the *Guide* (but not by Ibn Naḥmias in *The Light of the World*), had said that there could not be true opposition in circular motion.[39] Ibn Rushd, in his Middle Commentary (*talkhīṣ*) on the *Categories* (Kitāb al-Maqūlāt), wrote that the opposite of a motion upward was a motion downward and that the opposite of a motion in place was either rest in place or a motion in the direction of something other than that place.[40] Moving toward a clarification, Ibn Rushd wrote in his *De Caelo* that "there is no opposition between the place of the sphere and up and down."[41] Then, a contemporary of Gersonides, Yedaiah Ha-Penini (d. ca. 1340),

36. Biṭrūjī (trans. and comm. Goldstein), *On the Principles,* p. 73 and pp. 76–9.

37. Even Biṭrūjī's earliest readers disagreed with some of his positions. On early responses to Biṭrūjī in Hebrew, see James T. Robinson: "The First References in Hebrew to al-Biṭrūjī's *On the Principles of Astronomy,*" in *Aleph* III (2003): pp. 145–63.

38. I described this debate in "Andalusian Responses to Ptolemy in Hebrew," in Jonathan P. Decter and Michael Rand (eds.): *Studies in Arabic and Hebrew Letters in Honor of Raymond P. Scheindlin* (Piscataway, NJ: Gorgias Press, 2007): p. 75.

39. Ibn Bājja: "Fī al-Faḥṣ 'an al-quwwa al-nuzū'iyya," in 'Abd al-Raḥmān Badawī (ed.): *Rasā'il Falsafiyya li-'l-Kindī wa-'l-Fārābī wa-Ibn Bājja wa-Ibn 'Adī* (Beirut: Dār al-Andalus, 1983): pp. 147–56, at p. 153. Ibn Bājja may also have been more of a follower of Ptolemy than Maimonides acknowledged. See the references in Miquel Forcada: "Ibn Bājja and the Classification of the Sciences," in *Arabic Sciences and Philosophy* XVI (2006): pp. 287–307, at p. 305.

40. Ibn Rushd (Maḥmūd Qāsim ed., Charles Butterworth and Aḥmad 'Abd al-Majīd Harīdī comm.): *Talkhīṣ Kitāb al-Maqūlāt* (Cairo: General Egyptian Book Organization, 1980): pp. 151–2. In his *De Substantia Orbis* (On the substance of the celestial sphere), Ibn Rushd ([Arthur Hyman ed.]: *Averroes' De Substantia Orbis* [Cambridge, MA, and Jerusalem: The Medieval Academy of America and the Israel Academy of Sciences and Humanities, 1986]: p. 75ff.) wrote that the celestial body cannot move up or down.

41. Ibn Rushd, "Kitāb al-Samā' wa-'l-'ālam," pp. 10–1. Also Ibn Rushd: "*[L]aw kānat al-ḥaraka al-mustadīra tuḍādd al-ḥaraka al-mustadīra la-kānat al-ṭabī'a qad fu'ilat bāṭil*[an]": [were a circular motion to oppose a circular motion, then nature would have been made for naught]. In his (Jamāl 'Alawī (ed.) *Talkhīṣ al-Samā' wa-'l-'ālam* [Fes: Kulliyat al-ādāb, 1983]: p. 93), Ibn Rushd said that not every dissimilarity (*khilāf*) is an opposite (*ḍidd*), and he added that opposites were the ultimate in

held that "up" and "down" were not applicable to the orbs.[42] Ḥasday Crescas (d. 1410/11) took it to be a settled point that the rotational motion of the orbs in place could never, appearances aside, be in opposite directions.[43] The fact that debates about Ibn Rushd, including his ideas about opposite motion in the heavens, moved to southern France, in the milieu of Gersonides,[44] means that there was a broader philosophical context and consensus for Ibn Naḥmias's position on the admissibility of apparently opposite motions in the heavens.[45]

Thus, Ibn Naḥmias has addressed, implicitly, Ptolemy's statement, in *Almagest* I.8, that the two primary motions of the heavens were clearly in opposite directions. In §0.4, §0.5, and §0.6 Ibn Naḥmias pointed to the futility of determining whether motions on great circle paths that intersected at a 90° angle were in the same or opposite directions. For Ibn Naḥmias, Ptolemy's findings were correct only with regard to appearances since *The Light of the World* held that the apparent opposite motions in the heavens were not, in fact, opposite.[46]

*Aristotle had shown that by inference (*bi-dalāʾil*) only*
This could be a reference to *De Caelo* 270b: "for throughout all past time, according to the records handed down from generation to generation, we find no trace

dissimilarity, and were there truly opposition in circular motion, then circular motions would not be of the same species (*nawʿ*).

42. Ruth Glasner: *A Fourteenth-Century Scientific-Philosophic Controversy: Yedaiah Ha-Penini's Treatise on Opposite Motions and Book of Confutation* [in Hebrew] (Jerusalem: World Union of Jewish Studies, 1998): p. 92. According to Penini, the definitions of *up* and *down* depended on motions in a straight line, motions that did not occur in the celestial realm. The scholars who participated in the debate that Glasner described were in Provence, a region whose scholars were connected intellectually to the Iberian Peninsula (see Ilan, "Rᵉdipat ha-emet," p. 12). For survey articles on science in the Provençal Jewish community, see Bernard R. Goldstein: "Scientific Traditions in Late Medieval Jewish Communities," in G. Dahan (ed.), *Les Juifs au regard de l'histoire: Mélanges en l'honneur de M. Bernhard Blumenkranz* (Paris: Picard, 1985), pp. 235–47; and Bernard R. Goldstein: "The Role of Science in the Jewish Community in Fourteenth-Century France," in *Annals of the New York Academy of Sciences* CCCXIV (1978): pp. 39–49.

43. Hasdai Crescas (E. Shweid intro.): *Seip̱er Or Ha-shem* (Jerusalem: Makor, 1970; facsimile edition of the Ferrara, 1555 edition): third page of the second section of the first discourse. See also Hasdaï Crescas (trans, notes, and preface Éric Smilevitch): *Lumière de l'éternel (Or Hachem)* (Paris: Éditions Hermann, 2010): p. 377.

44. Glasner, *A Fourteenth-Century*, p. 15 and 22–4. Here Glasner argued that Penini understood his commentary on Ibn Rushd's *Short Commentary on the "Physics"* to be in conversation with Gersonides. See also Glasner: "The Evolution of the Genre of Philosophical-Scientific Commentary: *Hebrew Supercommentaries on Aristotles's 'Physic s,'*" in Freudenthal, *Science in Medieval Jewish Cultures*, pp. 182–206, at p. 189.

45. See Ruth Glasner: "Levi Ben Gershom and the Study of Ibn Rushd in the Fourteenth Century," in *Jewish Quarterly Review*, new series, LVIII (1995), pp. 51–90.

46. *The Light of the World*, §0.11.

of change either in the whole of the outermost heaven or in any one of its proper parts."[47] Aristotle's remark might account for the aforementioned debate about interpreting Aristotle in that if celestial motions were sometimes in opposite directions and sometimes not, the change of direction would be a source of further changes that would themselves have been observed.

51v

the impossibilities and the violations of truth that follow from these circumstances of which Rabbi Moses, peace be upon him, reminded us in The Guide of the Perplexed 0.3

Maimonides mentioned the impossibilities of the epicycle and eccentric in Book II, Chapter 24 of the *Guide*, noting that the predictive accuracy of the mathematical astronomers was, in his time, attainable only by admitting the physically impossible hypotheses of the epicycle and eccentric.[48] For instance, the rolling of the epicycle meant that it would change its place completely, which was antithetical to the unchanging essence of a celestial orb. Also, the motion of epicycles was not about the center of the world, nor about any other fixed point. In terms of observations' implications for the physical structure of the heavens, Maimonides found that the eccentricities computed by Ptolemy meant that the center of Mars's eccentric was wholly outside the orb of Mercury. Hence, the observational data according to which Ptolemy constructed his system led to the apparent contradiction of orbs that did not nest. And if these orbs did not share a center but were somehow still nested, then there could be no way for the outer orb to move without moving the inner orb. See also the comments on §0.15.

If we were to admit that there is opposition 0.4

I translate both the Judeo-Arabic (*in*) and Hebrew (*im*) with the unlikely condition because the result of the discussion is that there cannot be true opposition in circular motions. Ibn Naḥmias granted for the purposes of argument that there may be opposition in circular motions and proposed the case of two motions inclined by 90° to each other. By fixing the poles of Venus in the ecliptic, the revolution of the orb of the ecliptic from east to west would carry the poles of the orb of Venus with it. Then, because the path of Venus would be perpendicular to the ecliptic, it would be impossible to state meaningfully whether one orb was moving in a direction opposite to the other.

Langermann has observed that by taking the position that there were not, in

47. See also *Physics* 226b; and Ruth Glasner, *Averroes' "Physics": A Turning Point in Medieval Natural Philosophy* (Oxford, UK: Oxford University Press, 2009): p. 68, on how the motions of the heavens are uniform and eternal.

48. Maimonides (trans. Pines), *Guide*, vol. 2, pp. 322–7.

fact, opposite motions in the heavens, Ibn Naḥmias could improve consonance with observations without dependence on the epicycle and eccentric.[49] That is, Langermann has interpreted the admission of apparently opposite motions as a philosophical concession to the exigencies of mathematical precision. Although eccentrics would obviously be inadmissible in a system of homocentric orbs, Ibn Naḥmias did propose small circles on the surface of the orb that, he wrote, took the place of the epicycle (§B.1.II.3). The motion of a pole on these small circles would sometimes appear to be in the opposite direction of the motion of the orb. Ibn Rushd, too, took more than one position on the existence of the epicycle and the eccentric, and Juliane Lay has written that Ibn Naḥmias perhaps knew of Ibn Rushd's *Compendium of the Almagest* in which Ibn Rushd accepted the eccentric and epicycle.[50]

52v

things are themselves in that manner 0.7

Ibn Naḥmias has criticized Ptolemy for presuming the truth of his hypotheses just because they account for the observations: "Because of the possibility of these things appearing to the senses according to how he structured them, these hypotheses became accepted by all." Miquel Forcada has shown that other scholars in Andalusia paid attention to the nuances of methods of demonstration as a way to assess competing truth claims.[51] A similar point, about how the predictive accuracy of Ptolemy's models could seduce one into presuming the physical reality of those models, appeared in Isaac ben Samuel Abū al-Khayr's 1497 commentary on Farghānī's *Elements of Astronomy*.[52] Ibn Naḥmias's motivation was also to show, as he believed Maimonides had intimated, that there are no correct, accurate observations that cannot, in principle, be explained through homocentric models. More specifically, Ibn Naḥmias acknowledged room for improving the conformity of his models with observations of the motions in latitude and longitude but rejected any observations of planets' varying distances, for such variations undercut the foundations of homocentric astronomy.

Abū Bakr Ibn al-Ṭufayl 0.9

Biṭrūjī mentioned Ibn Ṭufayl (d. 1185–6) in his *On the Principles of Astronomy*, saying that Ibn Ṭufayl had said that he had devised an astronomy that did not

49. Langermann, "'True Perplexity,'" p. 173.

50. Lay, "*L'Abrégé de l'Almageste*," p. 31, n. 24.

51. Forcada, "Ibn Bājja and the Classification of the Sciences," p. 305.

52. Isaac ben Samuel Abū al-Khayr, *Peirush al-Fargani*, Bodleian Neubauer 2015 (IMHM 19300), fol. 2a.

rely on epicycles and eccentrics and had promised to write a book on the subject.[53] Astronomy was the subject of a portion of Ibn Ṭufayl's *Ḥayy ibn Yaqẓān*;[54] in chapter seven, in a passing remark, Ibn Ṭufayl said that the heavenly bodies always move about the center.[55] At one point, however, Ibn Ṭufayl alluded to the possibility of heavenly bodies moving about variable centers.[56] Although there is no evidence of a treatise on astronomy by Ibn Ṭufayl, earlier scholarship in the history of astronomy had attributed much of Biṭrūjī's astronomy to Ibn Ṭufayl.[57] But Ibn Naḥmias used *'athara 'alā* (*pagash* in the Hebrew: to come across or to discover) to describe Ibn Ṭufayl's activities, meaning that Ibn Naḥmias might have thought that Ibn Ṭufayl had not invented his own models and, thus, that homocentric models had a history that stretched back before Ibn Ṭufayl.

nothing that follows from those proposed hypotheses corresponds to that which appears to the senses, until now 0.10

Goldstein's analysis of Biṭrūjī's *On the Principles of Astronomy* found widespread discrepancies between available observations and the positions predicted by Biṭrūjī's models.[58] But Biṭrūjī himself scolded Ptolemy for making assumptions that were not consistent with his (Ptolemy's) own observations.[59] Thus Biṭrūjī, in order to argue that his models were truer than those of Ptolemy, pressed in

53. Biṭrūjī (trans. and comm. Goldstein), *On the Principles*, p. 61. "He promised to write a book about it. . . . I continued to think about these matters after I had heard this from him, searching the sayings of the ancients, and finding nothing in them but a few hints." Miquel Forcada ["Ibn Ṭufayl: Abū Bakr Muḥammad ibn 'Abd al-Malik ibn Muḥammad ibn Muḥammad ibn Ṭufayl al-Qaysī," in Hockey et al. (eds.): *Biographical Encyclopedia of Astronomers*, p. 572] concurred that there is no evidence that Ibn Ṭufayl ever produced his promised treatise on astronomy, meaning that Biṭrūjī invented his own models.

54. Ibn Ṭufayl (ed. Fārūq Sa'd): *Ḥayy ibn Yaqẓān*, 5th ed. (Beirut: Dar al-āfāq al-jadīda): pp. 166–77. Cf. the Gauthier edition (Beirut: Imprimerie Catholique, 1936), pp. 75–82. Miquel Forcada (."Ibn Ṭufayl," in Hockey et al. (eds.):. *The Biographical Encyclopedia of Astronomers*, p. 572) wrote that Ibn Ṭufayl admitted the possibility of orbs moving about more than one center. But he also said (p. 79 Gauthier edition, p. 130 Goodman translation) that the sizes of the planets never vary, meaning that their distance from the observer never changes.

55. Ibn Ṭufayl, *Ḥayy ibn Yaqẓān*, p. 188. Cf. p. 79 in the Gauthier edition.

56. Ibn Tufayl (trans., intro., and notes Lenn Evan Goodman): *Ibn Tufayl's Hayy Ibn Yaqzān* (Chicago and London: University of Chicago Press, 2009): pp. 146–7. "In addition, Hayy prescribed himself circular motion of various kinds. Sometimes he would circle the island, skirting along the beach and roving in the inlets. Sometimes he would march around his house or certain large rocks a set number of times, either walking or at a trot. Or at times he would spin around in circles until he got dizzy." (Cf. p. 116 of the Gauthier edition.)

57. J. L. E. Dreyer: *A History of Astronomy from Thales to Kepler* (Cambridge: Cambridge University Press,1906; repr., New York: Dover Publications, 1953): p. 264. "Ibn Tofeil was therefore probably the real author of the fairly elaborate system, which his pupil worked out."

58. See, for example, Biṭrūjī (trans. and comm. Goldstein), *On the Principles*, vol. 1, pp. 8–9.

59. Biṭrūjī (trans. and comm. Goldstein), *On the Principles*, vol. 1, p. 61.

certain cases for additional, better observations that would confirm the superior (!) predictive accuracy of Biṭrūjī's models. In this passage, Ibn Naḥmias has observed that Biṭrūjī's call for renewing observations contradicted Biṭrūjī's stated desire to present only a qualitative account that accepted the *Almagest* as the standard of mathematical astronomy.[60] Despite what Ibn Naḥmias has said here, he would, in the case of the lunar model (see, e.g., §B.2.IV.3), reject certain observations, such as those of the Moon's changing size, for which homocentric models could not possibly account.

53r

So, what makes it out of the question to say that each one of the orbs that is beneath the uppermost moves in the direction of the uppermost orb following its motion and about its poles 0.12

Ibn Naḥmias has presented, and accepted as a conceivable explanation, Biṭrūjī's theory that all orbs beneath the uppermost, responsible for the daily motion, incur a lag (*taqṣīr*) of a given amount behind the motion of the uppermost as the lower orbs are moved through an attraction to the uppermost orb: "since its goal and intention is toward the uppermost and to resemble it, since it is its goal."[61] The idea that a lower orb would lag behind the motion of an upper orb, the orb that moved it, was found also in the *Epistles of the Brethren of Purity,* a text known to medieval Jewish philosophers.[62] Abū al-Barakāt al-Baghdādī's (d. 1164) *Kitāb al-mu'tabar* (The book of that which has been established by personal reflection) was another possible source for the concept of lag.[63] There are references in the

60. See Biṭrūjī (trans. and comm. Goldstein), *On the Principles,* vol. 1, p. 76. "It is not our intention to present the amounts of their motions, nor to indicate any other accidental property of theirs, nor to bother with their particular phenomena (*'umūr*) or the complete computation for their motions. To do so would necessitate great length and thorough study and a new presentation of observations." See also ibid., p. 72, where Biṭrūjī criticized earlier astronomers for relying on observations in some cases and, then, ignoring inconvenient observations in other cases.

61. *The Light of the World,* §0.13.

62. The *Rasā'il* of the Brethren of Purity related increased lag to an orb's distance from the uppermost (Ikhwān al-Ṣafā': *Rasā'il* [Beirut: Dār Ṣādir, n.d.], vol. 2 of 4: p. 35). On the connection that Ibn Naḥmias drew between lag or direct motion and distance from the uppermost, see *The Light of the World,* §0.14. For Jews' knowledge of the *Rasā'il,* see Berman, Lawrence: "Brethren of Sincerity, Epistles of," Michael Berenbaum and Fred Skolnik (eds.): *Encyclopaedia Judaica*, 2nd ed. (Detroit: Macmillan Reference, 2007), pp. 170–1.

63. See Shlomo Pines: "Notes on Abu'l Barakāt's Celestial Physics," in *Studies in Abū 'l-Barakāt al-Baghdādī: Physics and Metaphysics* (Jerusalem: Magnes Press; Leiden: E. J. Brill, 1979): pp. 175–80, at p. 177: "Each sphere has a desire (*yashtāq*) for the sphere that is above it, that is similar to the desire of the Earth, the water and the other (elements) for their natural places." See also Shlomo Pines: "Études sur Awḥad al-Zamān Abu'l-Barakāt al-Baghdādī," in *Studies in Abu'l-Barakāt,* pp. 1–95, at p. 78. Biṭrūjī compared the force that the uppermost sphere communicated to the lower spheres to the force imparted through projectile motion.

Kitāb al-īḍāḥ li-Ariṣṭūṭālīs fī al-khayr al-maḥḍ, attributed to Proclus and based on his *Elements of Theology*, to the idea that a desire for the uppermost orb could move the lower orbs.[64] Ibn Rushd, in his *Kitāb al-Samā' wa-'l-'ālam* (*De Caelo*) also mentioned the possibility of motion through desire for the mover.[65] And in his *Talkhīṣ al-Samā' wa-'l-'ālam*, Ibn Rushd said that the differences in the planets' motions (i.e., their speeds) was due to the planets' proximity or distance from the highest orb.[66] Ibn Naḥmias agreed with Biṭrūjī that the concept of lag did help establish the order of the planets.[67]

53v

It moves also about its own two poles 0.13

The distinction between Ibn Naḥmias's models and those of Biṭrūjī was that, in Ibn Naḥmias's models, through the desire for the uppermost, each orb might also move on its own *in the direction* of the uppermost orb about its own two poles (§0.13): "It moves also about its own two poles in that direction as if its intention was that it make up for its lag behind the uppermost through the first motion, since its goal and intention is toward the uppermost and to resemble it, since it is its goal." Ibn Naḥmias has taken Biṭrūjī's explanation for why all of the celestial motions must be in the same direction and has modified it to explain how two different, and apparently opposite, motions could somehow arise from the same mover. There is a partial theoretical parallel for the orb's two distinct

For more information on theories of motion contemporary with or just earlier than *The Light of the World*, see also Paul Lettinck: *Aristotle's "Physics" and its Reception in the Arabic World: With an Edition of the Unpublished Parts of Ibn Bājja's Commentary on the "Physics"* (Leiden, New York: E. J. Brill, 1994): pp. 665–7. Ibn Bājja, in his commentary on the *Physics*, followed Aristotle's view of projectile motion. See also Julio Samsó: "On al-Biṭrūjī and the hay'a Tradition in al-Andulus," in Julio Samsó: *Islamic Astronomy and Medieval Spain* (Aldershot: Ashgate Publishing, Variorum, 1994): article XII, pp. 1-13, at pp. 9–13.

64. Proclus: "*Kitāb al-īḍāḥ li-Ariṣṭūṭālīs fī al-khayr al-maḥḍ*," in 'Abd al-Raḥmān Badawī (ed.): *al-Aflāṭūniyya al-muḥdatha 'ind al-'Arab* (Kuwait: Wikālat al-maṭbū'āt, 1977): pp. 1–33, at p. 23. Specifically, Proclus argued that while not all things desire (*tashtāq ilā*) the intellect (*al-'aql*), they do desire the best (*al-khayr*). Gad Freudenthal has written on how the celestial bodies, according to Ibn Rushd, could engage in behavior appropriate to animate bodies. See Gad Freudenthal: "The Medieval Astrologization of Aristotle's Biology: Averroes on the Role of the Celestial Bodies in the Generation of Animate Beings," in *Arabic Sciences and Philosophy* XII (2002): pp. 111–37.

65. Ibn Rushd: "Kitāb al-Samā' wa-'l-'ālam," p. 41. On p. 49, he referred to inclination (*mayl*) as one of the two forces (*quwwatān*) that moved the orbs beneath the uppermost; the form (*ṣūra*) was the other force.

66. Ibn Rushd (ed. and intro. Jamāl al-Dīn al-'Alawī): *Talkhīṣ al-Samā' wa-'l-'ālam* (Fes: Kulliyyat al-ādāb, 1984): p. 242. Concupiscence (*shahwa*) played a role: "Wa-innamā hunāka shahwa wa-ikhtiyār, fa-kullamā kānat shahwatuh wa-ikhtiyāruh li-'l-iqtidā' bi-'l-muḥarrik al-awwal akthar, kānat ḥarakatuh li-'l-mutaḥarrik bih abṭa'."

67. Biṭrūjī (trans. and comm. Goldstein), *On the Principles*, p. 63.

motions with Shīrāzī's proposal in *Faʿalta fa-lā talum* that the fixed stars could be embedded in and move with Saturn's parecliptic, while the whole cosmos moves with the daily motion.[68] Ibn Naḥmias noted that an orb, even with the additional motion about its own poles, would nevertheless still fall short of the motion of the uppermost.

Therefore, the closer its position to the uppermost [orb], the less its final lag 0.14
Although the starred (*mukawkab*) orb is below the uppermost, there was disagreement over whether or not it has any lag at all. But the model that Ibn Naḥmias proposed for the fixed stars' motion (see §B.3.II.3) suggests that there was not necessarily a single starred orb and that all of the orbs carrying the fixed stars lagged (or had a direct motion).

Also, our greatest goal is to explain that a heavenly orb must necessarily have its concave surface be wholly tangent to the convex surface of that beneath it 0.15
This is another ramification of the exclusion of epicycles and eccentrics. Ibn Naḥmias has pointed out that if all of the orbs are homocentric, there would be no mechanical way for an outer orb to move an inner; hence the importance of the orbs' desire for the uppermost. Investigating how an outer orb might move an inner orb is reminiscent of Muḥammad ibn Mūsā's acknowledgment of and inability to solve the problem of how the outer of two concentric orbs could move the inner.[69] But even with *The Light of the World*'s focus on all questions of celestial mechanics, including lag, Ibn Naḥmias concluded the paragraph by explaining that *The Light of the World* will begin with the Sun's motions, and not with those of the uppermost orb, in order to facilitate a comparison with texts in the tradition of the *Almagest* that accepted epicycles and eccentrics. Biṭrūjī's *On the Principles* had been organized differently, beginning with the upper orbs and moving down.

Ibn Naḥmias might also have been responding to Maimonides's conclusion, among his doubts about the state of astronomy, that "necessity obliges the belief that between every two spheres there are bodies other than those of the spheres. Now if this be so, how many obscure points remain?"[70]

68. Quṭb al-Dīn Shīrāzī, *Faʿalta fa-lā talum,* Fatih MS 3175/2, 168r. See also Ragep, *Naṣīr al-Dīn al-Ṭūsī's "Memoir on Astronomy,"* p. 390.

69. George Saliba: "Early Arabic Critique of Ptolemaic Cosmology: A Ninth-Century Text on the Motion of the Celestial Spheres," in *Journal for the History of Astronomy* XXV (1994): pp. 115–41, at p. 126. Muḥammad ibn Mūsā eschewed recourse to the motion of the orbs occurring through souls.

70. Maimonides (trans. Pines), *Guide,* p. 324. Those intervening bodies would serve to unroll

54v

The fourth chapter is about the partial declinations and the rising times at sphaera recta 0.18

Some of the information promised in this chapter, namely the particular things known via a knowledge of rising times, is delivered only in §B.4.III.1.[71] Because §B.4.III.1 starts at the top of a page in the Vatican MS, it is possible that this chapter is a result of an earlier nonextant MS or the Vatican MS itself becoming jumbled.

the assumed premises whose presentation is necessary for this science A.I.1

A, C, D, E, F, G, H, I, J, K, L, M come from *Almagest* I.3.[72] N, O, P, Q come from *Almagest* I.4.[73] R, T, U, V come from *Almagest* I.5.[74] W comes from *Almagest* I.6.[75] X comes from *Almagest* I.7.[76] S is defined in *Almagest* I.8.[77] B (the near-equivalence of the local latitude with the North Celestial Pole) is defined in *Almagest* II.1.[78] As Ibn Naḥmias said in §A.I.2, B and M are preliminaries that demonstrate that the heavens are spherical, the goal of *Almagest* I.3 (in both the original Greek and Anatoli's Hebrew translation).[79] Jābir Ibn Aflaḥ also discussed physical evidence for the Earth being in the center without motion.[80] In a few spots, the Hebrew recension places the premises' letters in different places.

the effects of the complex of orbs of an upper planet on a lower planet. Aristotle had made a similar proposal in *Metaphysics,* Book 12, Chapter Eight.

71. For more on rising times, see E. S, Kennedy: "A Survey of Islamic Astronomical Tables," in *Transactions of the American Philosophical Society,* new series, XLVI (1956), part two: pp. 123–77, at p. 140.

72. Toomer, *Ptolemy's "Almagest,"* pp. 38–40. On *E,* Ptolemy wrote, "at the very moment of their disappearance, at which time they are gradually obstructed and cut off, as it were, by the Earth's surface." For *F,* Ptolemy considered the consequences of the distances from the Earth to the fixed stars varying (but the distances do not vary). For *G,* Ptolemy wrote (p. 40) that the aether, "because of the likeness of its parts moves in a circular and uniform fashion."

73. Toomer, *Ptolemy's "Almagest,"* pp. 40–1.

74. Toomer, *Ptolemy's "Almagest,"* pp. 41–2.

75. Toomer, *Ptolemy's "Almagest,"* p. 43.

76. Toomer, *Ptolemy's "Almagest,"* pp. 43–5. On page 45, Ptolemy raised the possibility that if the Earth did move, nothing would ever be seen to be moving in the opposite direction of the Earth's motion, since the Earth's velocity would have to be so great.

77. Toomer, *Ptolemy's "Almagest,"* p. 47.

78. Toomer, *Ptolemy's "Almagest,"* p. 75.

79. In Anatoli's translation, *Almagest* I.3 begins at Paris BNF MS Hébreu 1018, 2a.

80. Jābir Ibn Aflaḥ (trans. Jacob Ben Maḵir): *Qiṣṣur al-Majisṭi* (the Hebrew translation of *Iṣlāḥ al-Majisṭī*), Paris BNF MS Hébreu 1025, 23a–29b.

55v

From some of them what is sought is shown through a regular syllogism and with some of them through a reductio ad absurdum[81] A.I.2

The premises now play the role of premises in syllogisms. Ibn Naḥmias did not spell out how a reductio ad absurdum would lead to the conclusions that the heavens and the heavenly motions are spherical, but Ptolemy had at times,[82] as had Ibn Rushd in his *Talkhīṣ Kitāb al-Qiyās,* a commentary on the *Prior Analytics.*[83]

And it is shown through premise X that the Earth does not move about its center

Other scholars in Islamic societies had been interested in the question of whether such observational evidence existed. For instance, al-Bīrūnī, in his *al-Qānūn al-Masʿūdī,* considered the Earth's rotation in place and concluded that the question could be determined through observation.[84] Jamil Ragep has traced arguments for and against the Earth's rotation through the work of later astronomers in Islamic societies.[85] There were, as well, alternative cosmologies. The Ikhwān al-Ṣafāʾ held that the Sun was the center of the cosmos inasmuch as there were equal numbers of orbs above and below the Sun.[86] The Ikhwān al-Ṣafāʾ also held that the Earth could move about the center of the cosmos (which was not the Sun), though this motion was not felt.[87] Ptolemy himself had referred to ancients, perhaps Heraclides of Pontos (late fourth century BCE), who asserted that the Earth rotated on its axis, and/or Aristarchos of Samos, who proposed a heliocentric cosmos.[88] This reference to the lack of observational evidence for the

81. On the terminology, see F. W. Zimmermann: *Al-Fārābī's Commentary and Short Treatise on Aristotle's "De Interpretatione"* (New York and Oxford: Oxford University Press for The British Academy, 1991): p. 198, n. 6. See also trans. and notes Asad Q. Ahmed: *Avicenna's "Deliverance": Logic* (Karachi and New York: Oxford University Press, 2011): pp. 79–80.

82. Toomer, *Ptolemy's "Almagest,"* for example, p. 39.

83. Ibn Rushd (ed. Maḥmūd Qāsim; edition completed by Charles Butterworth and Aḥmad ʿAbd al-Majīd Harīdī): *Talkhīṣ Kitāb al-Qiyās* (Cairo: General Egyptian Book Organization, 1983): p. 314. This paragraph in Ibn Rushd's *Talkhīṣ* corresponds to Aristotle's *Prior Analytics* 62b:29–38.

84. F. Jamil Ragep: "Ṭūsī and Copernicus: The Earth's Motion in Context," in *Science in Context* XIV (2001): 145–63, at pp. 151–2.

85. Ragep: "Ṭūsī and Copernicus," pp. 145–63.

86. Ikhwān al-Ṣafā,' *Rasā'il,* vol. 2, p. 30.

87. A. Bausani: "Die Bewegungen der Erde im *Kitāb Ikhwān aṣ-ṣafāʾ*: Ein Vor-Philolaisch-Pythagoräisches System?" in *Zeitschrift für Geschichte der arabisch-islamischen Wissenschaften* I (1984): pp. 88–99, at pp. 97–8. Bausani posited that this cross-like (*kreuzförmigen*) motion had pre-Pythagorean origins. Bausani cited Susanne Diwald's partial 1975 translation of the *Epistles.* For the citation to an Arabic edition different from that which Diwald used, see Ikhwān al-Ṣafāʾ: *Rasā'il,* vol. 3 of 4, p. 326.

88. Toomer, Ptolemy's "Almagest," p. 44 (see also p. 44, n. 41). On Aristarchos's heliocentric system, see B. L. van der Waerden: "The Heliocentric System in Greek, Persian, and Hindu Astronomy," in David A. King and George Saliba (eds): *From Deferent to Equant* (New York: Annals of the New York Academy of Sciences, 1987): 525–45, at pp. 525–9. See also Heath, *Aristarchos.*

Earth's daily rotation is evidence for Ibn Naḥmias's greater attention to the form and contents of the *Almagest.*

56r

Let us first present the method for knowing the measures of the chords A.II.1

By following the physical preliminaries with a chapter on chords, Ibn Naḥmias continued to hew a path much closer to that of the *Almagest* than did Biṭrūjī's *On the Principles,* which did not discuss chords. Ibn Naḥmias's presentation of his own models, though, almost never relied on the chord function, even when he did engage in quantitative analysis, as the replacement of the chord function by the sine occurred in the Arabic *Almagest* translations.[89] The only time Ibn Naḥmias discussed a chord in the presentation of his models was §B.1.III.6, where he determined the measure of the greatest solar anomaly by attempting to transfer Ptolemy's calculations to a homocentric system. Thus, the production of the chord table is more evidence for Ibn Naḥmias's concern for the form of the *Almagest.*

if we wanted to divide a known number into mean and extreme ratio

Book II, proposition 11 of Euclid's *Elements* (and Book VI, proposition 30) demonstrated that dividing a line into mean and extreme ratio meant that the ratio of the lesser part to the greater equaled the ratio of the greater part to the whole.[90] This was the golden proportion.

The demonstration in *The Light of the World,* which more closely resembles the one in Book VI of the *Elements,* is as follows:

T.P. $GE > BE$ and $BG.BE = GE^2$

$BG(BE + GE) + GA^2 = BA^2 = (BD + DA)^2 = BD^2 + DA^2 + 2BD.DA$

$BG.BE + BG.GE + GA^2 = BD^2 + DA^2 + 2BD.DA$

(because $BD = GA$) $BG.BE + BG.GE = DA^2 + 2BD.DA$

(because $BD = BG/2$ and because $DA = GE$) $BG.BE = GE^2$

Therefore, $GE > BE$ and $GE:BG = BE:GE$

The last page of Profiat Duran's response to the Hebrew recension criticized Ibn Naḥmias for confusing lines and numbers in this demonstration, in that what holds for lines would *not* hold for (irrational) numbers.[91] Duran's accu-

89. George Saliba: *Islamic Science and the Making of the European Renaissance* (Cambridge: MIT Press, 2007): pp. 87–8. The Hebrew translation by Anatoli that exists in Paris BNF MS 1018 had not replaced the chords with sines to the same extent.

90. Euclid (intro., trans., and comm. Thomas L. Heath): *The Thirteen Books of the Elements* (Cambridge: Cambridge University Press, 1908; repr., Mineola, NY: Dover Publications, 1956): pp. 267–8. Heath noted (p. 268) that a closely related proposition had been demonstrated in Euclid's *Data.*

91. Duran, *Response,* Bodleian MS Canon. 334, 100a. See also p. 398 of this book.

sation was correct in that Euclid's demonstration did not address irrational numbers, though by Ibn Naḥmias's lifetime, scholars had addressed the problem algebraically.[92]

56v

as was shown in the 15th part of Euclid A.II.2

This paragraph presents Ptolemy's lemma that was the foundation for the construction of the chord table;[93] the lemma itself resembled Euclid's *Elements*, Book VI, proposition D of proposition 16.[94] Ptolemy wrote that the sides of the inscribed hexagon and decagon represent the extreme and mean ratios of the same straight line.[95] Ibn Naḥmias's wording is unclear, but by the "greatest part" he must have meant the side of the hexagon, as the side of the hexagon was the radius to which he referred. A possible referent for the phrase "the 15th part" is proposition 15 of Book IV of the Hebrew translation of the *Elements*, where Euclid found the measure of the arc that the side of an inscribed hexagon subtended in a circle.[96] Euclid demonstrated elsewhere in the *Elements* that the square of the side of the pentagon (see *Elements*, XIII.10) is the sum of the square of the decagon and the square of the hexagon.[97] The paragraph concludes by finding the measure of a chord based on the lemma.[98]

Ibn Naḥmias has reproduced Ptolemy's demonstration except in the following respects:

A. The proof begins by making angle *ABE* equal to *DAG*. Though the Arabic and Hebrew MSS of *The Light of the World* give *DAG*, the *Almagest* (I.10; p. 50) has *DBG*.[99] Ibn Naḥmias's error mattered because one obtains *ABD* = *EBG* by adding *EBD* to both *ABE* and *DBG*. The translation reflects my emendation of the MSS, an emendation justified by a rationale, subtending the same arc, given in the text.

92. For the relevant history, see Roger Herz-Fischler: *A Mathematical History of the Golden Number* (Mineola, NY: Dover Publications, 1998): pp. 121–33.

93. See also Toomer, Ptolemy's "Almagest," pp. 50–1.

94. Euclid (intro., trans., and comm. Heath), *Elements*, vol. 2 of 3: p. 225.

95. Toomer, Ptolemy's "Almagest," p. 49. Cf. Euclid (intro., trans., and comm. Heath), *Elements*, XIII.9.

96. Euclid (trans. Moses ibn Tibbon): *Seiper ha-yesodot*, Vienna Nationalbibliothek 193 (IMHM F 1456), 18a–19a. This is the Moses Ibn Tibbon translation of the *Elements*. See also Euclid (intro., trans., and comm. Heath), *Elements*, vol. 2: pp. 107–9.

97. Euclid (trans. Moses ibn Tibbon): *Seiper ha-yesodot*, Vienna Nationalbibliothek 193 (IMHM F 1456), 72b-73a. See also Neugebauer, *History of Ancient Mathematical Astronomy*, p. 22.

98. Toomer, *Ptolemy's "Almagest,"* p. 51. This is *Almagest*, I.10.

99. See also Ptolemy (trans. Jacob Anatoli), *Almagesṭi*, Paris BNF Hébreu 1018, 6b. Anatoli's Hebrew translation of the *Almagest* follows Ptolemy on this point.

B. The translation reads, "*BDA* equals *BGE*." Both the Arabic and Hebrew MSS of *The Light of the World* give *GDE* for *BGE*. The *Almagest* explains that the equality of *BDA* and *GDE* results from these angles subtending the same arc.

Euclid had demonstrated that A.II.4

In this demonstration, *Z* is the center of the circle. The demonstration is more concise than that found in the original *Almagest*.[100] The reference to Euclid may be to proposition D, a lemma introduced by Ptolemy, based on Book VI, proposition 16 of the *Elements*.[101]

57r

Likewise we know the rest by decreasing [increments of] half of a degree

Summarizing *Almagest* I.9–I.11, Ibn Naḥmias has explained how to produce a table of chords.[102]

And we also present as a lemma A.II.5

Except for the substitution of sines for chords, Ibn Naḥmias has followed *Almagest*, I.13.[103]

57v

let great circle arcs GD *and* BE A.II.7

This passage presents the Menelaos Configuration in terms of sines rather than chords (as in *Almagest* I.13).[104] The perpendiculars to the surface (arcs *GL*, *AH*, and *DT*) are half the chords of twice the arcs; that is to say they are visual representations of the sines. Ibn Naḥmias never actually used the Menelaos theorem in *The Light of the World*, so like the production of the chord table, Ibn Naḥmias's goal seems to have been a formal similarity to the *Almagest*. The paragraph concludes with a reference to Ibn Muʿādh's (fl. c. 1079) work on spherical trigonometry (*Kitāb Majhūlāt qisī al-kura*), Ibn Naḥmias's probable source for this information about the ratios between the sines of arcs.[105] Jābir ibn Aflaḥ's (fl.

100. Toomer, Ptolemy's "Almagest," pp. 52–3.

101. Euclid (intro., trans., and comm. Heath), *Elements* vol. 2, pp. 225–8.

102. Toomer, *Ptolemy's "Almagest,"* pp. 47–60. See the 10th chapter of Book One of Anatoli's Hebrew *Almagest* translation (Paris BNF MS Heb. 1018, pp. 12–17) for how to determine the chords of a circle, and the eleventh chapter (p. 17) for how to construct a chord table.

103. Toomer, *Ptolemy's "Almagest,"* pp. 65–6.

104. Anatoli's *Almagest* translation (Paris BNF 1018, pp. 9–10) used chords (I.12). See also Otto Neugebauer: *History of Ancient Mathematical Astronomy* (Berlin, New York: Springer-Verlag, 1975), 3 vols., vol. 1: pp. 26–30; and Pedersen, *Survey of the Almagest*, pp. 72–6. The figure in *The Light of the World* is not precisely the same as the one in Toomer, *Ptolemy's "Almagest,"* pp. 64–9.

105. Emilia Calvo: "Abū ʿAbd Allāh Muḥammad ibn Muʿādh al-Jayyānī," in Hockey: *The*

12th century) *Iṣlāḥ al-Majisṭī* (Correction of the *Almagest;* trans. into Hebrew as *Qiṣṣur al-Majisṭi*) was a possible additional conduit of information about Ibn Muʿādh to Ibn Naḥmias, although Jābir did not cite Ibn Muʿādh by name in the *Islāḥ*.[106] I have translated both *tarkīb* and *ta'līf* as "compounding" because, in both cases, Ibn Naḥmias intended the multiplication of ratios.[107]

58r

We say that for every four quantities A.II.10

This is a reference to, and summary of, the ratios between the sines of four different arcs on the surface of a sphere, also known as the law of sines.[108]

58v

And this is how to draw up the table A.II.11

For the table itself, see table 5.1. The basis of the table is that the product of the first (quantity) and the fourth combined with the sixth is equal to the product of the second with the third, then with the fifth, or ADW = BGE. Thus, A/B = GE/DW. So, A/B = G/D * E/W= G/W * E/D. Those ratios account for the first two rows. Then, for row three, divide both sides first by G and then multiply by B, producing: A/G = B/D E/W = B/W E/D. The ratios that cannot be compounded (AD, AW, BG, BE, GE, and DW), and which are indicated within an 'x' in the table, are those involving only the quantities which fall on the same side of the equation.

The third chapter about the instruments needed for the observations A.III.1

This section of *The Light of the World*, on instruments, had no parallel in Biṭrūjī's *On the Principles of Astronomy*. It is a further reflection of Ibn Naḥmias's greater

---

Biographical Encyclopedia of Astronomers, pp. 652–3. Ibn Muʿādh lived in the eleventh century, for he observed a solar eclipse in 1079. Ibn Muʿādh referenced *Kitāb Majhūlāt* in his *Tabulae Jaen*. See M. V. Villuendas: *La trigonometria europea en el siglo XI: Estudio de la obra de Ibn Mu'ad: 'el Kitab mayhulat'* (Barcelona: Memorias de la Real Academia de Buenas Letras de Barcelona, 1979).

106. Emilia Calvo: "Jābir ibn Aflaḥ," in Hockey: The Biographical Encyclopedia of Astronomers, pp. 581–2. For the demonstration in the Hebrew translation of Jābir's Recension of the Almagest, see *Qiṣṣur al-Majisṭī*, Paris BNF 1024, 11a–14a. The Latin translation of Jābir's *Iṣlāḥ al-Majisṭī* reads: "per quatuor numeros proportionales, no per sex numeros sicut perparantur in figura sectore." For this Latin, see Lorch, "The Astronomy of Jābir ibn Aflaḥ," in Centaurus XIX (1975): pp. 85–107, at p. 95. There also exist Hebrew translations of Jābir's commentaries on Thābit ibn Qurra's treatise on the Menelaos Theorem. See Lorch: "The Astronomy," pp. 93–4.

107. Cf. Euclid (intro., trans., and comm. Heath), *Elements*, Book V, note on Definitions 9 and 10 (vol. 2, p. 132).

108. See al- Bīrūnī (ed. and trans. Marie-Thérèse Debarnot): *Kitāb Maqālīd 'ilm al-hay'a* (Damascus: Institut Français de Damas, 1985): p. 3. Bīrūnī's introduction to the *Maqālīd* recounts the development of the law of sines from the Menelaos Configuration.

TABLE 5.1

The Number	The compounded ratio		The ratios that are compounded from it			
1	A	B	G	D	E	W
2	A	B	G	W	E	D
3	A	G	B	D	E	W
4	A	G	B	W	E	D
x	A	D	x	x	x	x
5	A	E	B	W	G	D
6	A	E	B	D	G	W
x	A	W	x	x	x	x
x	B	G	x	x	x	x
7	B	D	A	G	W	E
8	B	D	A	E	W	G
x	B	E	x	x	x	x
9	B	W	A	G	D	E
10	B	W	A	E	G	B
11	G	D	A	B	W	E
12	G	D	A	E	W	B
x	G	E	x	x	x	x
13	G	W	A	B	D	E
14	G	W	A	E	D	B
15	D	E	B	A	G	W
16	D	E	B	W	G	A
x	D	W	x	x	x	x
17	E	W	A	B	D	G
18	E	W	A	G	D	B

concern with predictive accuracy and with following the form of the *Almagest*. The idea that instruments complete and perfect the science of astronomy had parallels in the astronomy of Islamic societies.[109] The first instrument is the meridional armillary found in *Almagest* I.12.[110] The second instrument Ibn Naḥmias called *dhāt al-ḥalaq* (lit. ringed instrument) but is the armillary sphere that

109. Robert Morrison: *Islam and Science: The Intellectual Career of Niẓām al-Dīn al-Nīsābūrī* (Oxon: Routledge, 2007): p. 22.

110. Toomer, *Ptolemy's "Almagest,"* p. 61. For more on Ptolemy's instruments, see Derek J. Price:

Ptolemy called an astrolabe in *Almagest* V.1.[111] The third is the parallactic rods (*Almagest* V.12),[112] and the fourth is the dioptra (described in *Almagest* V.14).[113] §A.III.3 follows the last two paragraphs of *Almagest* V.1 quite closely. Elsewhere, *The Light of World* seems to be abridging the text of the *Almagest*.

59v

We take the arc that is upon it A.III.4

This process is spelled out in detail in *Almagest* V.12.[114]

60r

We know the arc that is between the solstices, through the instrument which was made for that, to be 47°43′ A.IV.1

This yields, almost, the Ptolemaic parameter of 23;51,20° for the obliquity.[115] Biṭrūjī had stated only that the obliquity was "about 24°."[116]

60v

we extract the rising times of each degree of the circle in [each] latitude A.IV.3

(See figures 2.10 and 2.11.) The rising time is the arc of the equinoctial that rises with a given arc of the ecliptic.[117] Rising times were significant for Ibn Naḥmias's discussion of the motion of the fixed stars and for the distinction between attributing all motions to lag, or allowing apparently opposite motions in the heavens. A choice between lag and allowing apparently opposite motions would affect the direction of the motion of the ecliptic and therefore whether the ecliptic was moving in the direction of increasing or decreasing rising times. Rising times were important for timekeeping, and increasing or decreasing rising times would correlate with whether the days were getting longer or shorter. But while the great circles of the ecliptic and equinoctial would be moving in different directions, presuming one allows apparently opposite motions in the heavens, when one computes the rising time of a given ecliptic arc at a given moment, the instanta-

---

"Precision Instruments: To 1500," in Charles Joseph Singer et al (eds): *A History of Technology* (Oxford: The Clarendon Press, 1954–1978), vol. 3 of 5: pp. 582–619, at pp. 587–92.

111. Toomer, *Ptolemy's "Almagest,"* pp. 217–9.

112. Toomer, *Ptolemy's "Almagest,"* pp. 244–7.

113. Toomer, *Ptolemy's "Almagest,"* p. 252.

114. Toomer, *Ptolemy's "Almagest,"* p. 246.

115. Toomer, *Ptolemy's "Almagest,"* p. 63. In fact, the text of the *Almagest* reads "the arc between the solstitial points, is always greater than 47 2/3° and less than 47 3/4°." The arc 47; 43° lies between those extremes. Anatoli's translation lists 47;42,40° (Paris BNF MS Hébreu 1018, p. 22).

116. Biṭrūjī (trans. and comm. Goldstein), *On the Principles*, p. 153.

117. See Neugebauer, *History of Ancient Mathematical Astronomy*, pp. 30–4. Cf. Toomer, *Ptolemy's "Almagest,"* pp. 76–80.

neous rising time would be the same whether one was attributing all motions to lag or allowing apparently opposite motions in the heavens.

In this way it is possible to draw up tables A.IV.4
This paragraph follows *Almagest* II.9. §B.4.III.7–11 continues Ibn Naḥmias's discussion of rising times.

61r

the tables of the rising times that Ptolemy drew up A.IV.5
Ptolemy's table of rising times at *sphaera recta* (right ascensions) was in *Almagest* I.15.[118]

with each of those years being 365 days A.V.1
This information summarizes that found in *Almagest* III.1.[119] An Egyptian year was a 365-day year, and after 300 years, the solstices would fall on the 74th day, not on the 75th day as would be otherwise predicted. Thus, the accurate length of the year would exceed 365 days by less than one-quarter day. The 74 or 75 days were the margin by which the time between solstices exceeds 300 years of precisely 365 days.[120]

61v

Ibn Sīnā said that this is not its meaning. A.V.2
Ibn Sīnā's *Talkhīṣ al-Majisṭī* is Ibn Naḥmias's source for the discussion of the problem of a nighttime equinox and the subsequent solution, using two successive meridian transits.[121]

62r

The third treatise is about the movement of the fixed stars B.O.3
This is the final treatise to which Ibn Naḥmias referred in the introduction to the second treatise of *The Light of the World*. Still, on 83v of the Judeo-Arabic MS, a fourth treatise about the five planets commenced. The Hebrew recension mentions that fourth treatise, on the five planets, at this point (§B.o.3) in the

118. See Toomer, *Ptolemy's "Almagest,"* p. 72. In Anatoli's translation (Paris BNF Hébreu MS 1018), *Almagest* I.15 and I.16 are folded into I.14.

119. See Toomer, Ptolemy's "Almagest," pp. 137–8. See also Anatoli's translation (Almagesṭi, Paris BNF Hébreu MS 1018), pp. 56–7.

120. Ptolemy (Toomer, *Ptolemy's "Almagest,"* p. 140) determined the length of a year to be 365; 14, 48°.

121. Ibn Sīnā (ed. Ibrāhīm Madkūr, Muḥammad Madwar, and Imām Ibrāhīm Aḥmad): *Al-Shifā': 'Ilm al-Hay'a* (Cairo: General Egyptian Book Organization, 1980): pp. 153–4.

Hebrew recension, although the Hebrew recension terminates after the third treatise.[122]

That is imagined according to what I describe. B.1.1.2

Ibn Naḥmias first defined "pole" in §0.5; there, the poles were simply the two endpoints of the axis of an orb's rotation. Here, he expands the definition of "pole." A pole is a point 90° away from a point on a certain great circle of an orb. The motion of the pole entails the motion of the related great circle. If there is more than one great circle imagined on the surface of an orb, then there can be more than one pole imagined on that orb. In some cases, the great circles were abstractions for the equators of orbs.

The term "imagine" (Ar. *yutawahham;* Heb. *yᵉṣuyyar*) should be understood as closer to the sense of "conjectured" and not as imaginary in the modern sense. Freudenthal has noted that scientists' own philosophical positions on the ontological status of the models they proposed, when those positions are discernable, can be relevant.[123] Ibn Naḥmias's use of the word *yutawahham* indicates that he was not taking an absolute position on the reality of his models. Instead, he was arguing that one could aspire to retain the truth of mathematical astronomy without sacrificing philosophic truth, and that his models met the standards for philosophic truth. Ibn Naḥmias was conjecturing how the motions could occur on the surface of an orb without proposing, at least in the Judeo-Arabic original, physical movers. The fact that Ibn Naḥmias acknowledged his models' predictive imprecisions reflected his acknowledgment that predictive accuracy was a dimension of reality. As a point of comparison, the famous Ashʿarī *mutakallim* ʿAḍud al-Dīn al-Ījī (d. 1355) described mathematical concepts such as the great circles on orbs and their poles as imaginary (*umūr mawhūma*) and as "'neither an object of belief nor subject to affirmation or negation.'"[124] Ibn Naḥmias's position was distinct from Ījī's in that Ibn Naḥmias has laid out criteria, primarily homocentricity and predictive accuracy, according to which the truth of models could be evaluated.

122. *Or ha-ʿolam*, MS Bodleian Canon Misc. 334, §B.0.3/X.

123. See also Freudenthal. "'Instrumentalism,'" p. 243: The "relevant ideals are those that have come to be called 'realist' and 'instrumentalist' (or: fictionalist). Their continued use seems to me warranted. Provided we state precisely what we mean by them: they must reflect meta-theoretical beliefs or ideals held by historical actors, not by the historian."

124. Sabra, "Science and Philosophy," pp. 37–8. Jurjānī wrote in response (p. 39) that the astronomers' assumptions were not mere imaginings, but "notions" by which "'the conditions of (celestial movements are regulated in regard to speed and direction. . . . Discovery is made of the characteristics of the celestial orbs and the Earth, and of what they reveal of subtle wisdom and wondrous creation—things that overcome whoever apprehends them with awe, and, facing him with the glory of their creator.'"

62v

following [62v] *the motion of the universe from the east to the west*
(See figure 2.12.) The axis of the orb running through pole *B* (the pole of the ecliptic) is inclined to the axis running through point *A* (the pole of the equinoctial), just as the axis of the Sun's motion in longitude is inclined to the axis of the Sun's daily motion, a motion occasioned by the daily rotation of the heavens. The resulting lag, arc *TH* equal to arc *KB*, is the first component of the west-east mean motion of the Sun in longitude and is twice the final lag.

The position of the circle of the Sun at that time B.1.I:3
The circle of the Sun is a great circle that is 90° away from pole *B*. When pole *B* was at its starting point, the circle of the Sun coincided with the ecliptic. After pole *B* lagged through arc *KB*, and moved to point *K*, the position of the circle of the Sun coincides with circle *ZH*. Then, the motion from point *H* to point *Z*, in the direction of the motion of the universe, is out of a desire to compensate for a loss of power descending from the uppermost orb. Following Goldstein's translation of Biṭrūjī's *On the Principles*, I have translated the Judeo-Arabic *quwwa* (Hebrew: *ko'aḥ*) as "power" so as to avoid the modern connotations of "force."[125] The motion between point *H* to point *Z* is called the motion in latitude, Ibn Naḥmias's description for any motion that moved a celestial body to the north or south of the ecliptic. This motion in latitude (point *H* to point *Z*), which occurred on the circle of the Sun, is equal to the final lag in longitude, 59' per day, and served theoretically to keep the Sun in the ecliptic. Though my analysis, as well as the text itself, describes these motions as occurring sequentially, it is most accurate to conceive of them all occurring simultaneously. Once the Sun reaches point *Z*, it has completed its daily motion in longitude.

(See figure 5.1.) Although Ibn Naḥmias wrote that the final lag in longitude was equal to the motion in latitude, Ibn Naḥmias did not investigate whether his model for the Sun's mean motion in fact worked as he claimed it would. The following analysis will show that the motion in latitude would have to vary from day to day if the mean Sun was to remain in the ecliptic.[126] I have posited in figure 5.1 a lag of 90° for the pole on the circle of its path, which should yield 45° of mean motion, or a final lag of 45°, for the Sun. Once the pole of the Sun at *B* moves 90° to point *K*, and the Sun from point *T* to point *H*, the new position of the circle of the Sun, represented by arc *ZH*, intersected the ecliptic. Ibn Naḥmias would then have the Sun move about its own pole at *K* from point *H* to point *Z*, through an angle equal to the final lag (45°).

125. Biṭrūjī (trans. and comm. Goldstein), *On the Principles*, p. 63.

126. Robert Morrison: "The Solar Model in Joseph Ibn Joseph Ibn Nahmias' *The Light of the World*," in *Arabic Sciences and Philosophy* XV (2005): pp. 57–108, at pp. 80–2.

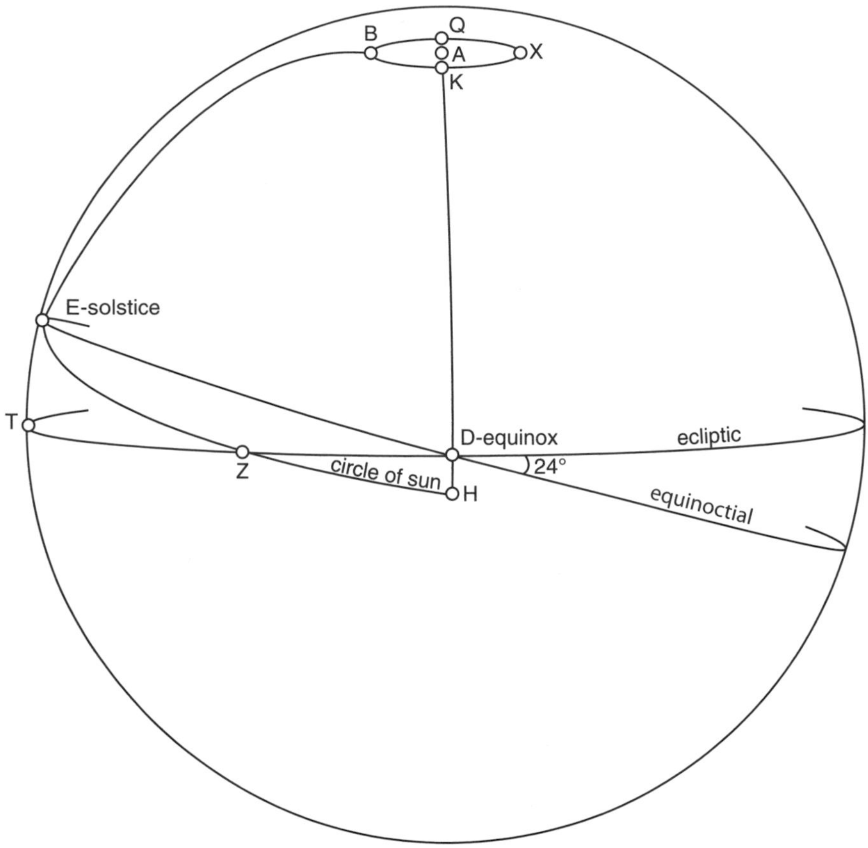

FIGURE 5.1

We consider spherical triangle *DZE*. Angle *D* is 24°, the obliquity of the ecliptic, and *z* (side *ED*), the distance on the equator from a solstice to an equinox, is 90°. Because arc *ZT* is 45° and because there is 90° between a solstice and an equinox, *e* (side *DZ*) is 45°. We can compute *d* (side *EZ*) with the cosine rule for sides:

$\cos d = \cos z \cos e + \sin z \sin e \cos D$
$\cos d = \sin 45° \cos 24°$
meaning $d = 49;45,40°$

Arc *EH* is 90° because the Sun has moved 90° on the circle of its path from point *E* to point *H*. So, arc $HZ = 90° -$ arc $EZ = 40;14,20°$. The mean motion in longitude (45°) does not equal the motion in latitude as Ibn Naḥmias alleged.[127] Additionally, the ratio of the motion in latitude to the motion in longitude would vary depending on the motion in longitude. Inspection shows that with a mean motion of 90°, 180°, and 270°, the sum of the mean motion plus the motion in latitude is the lag of the pole on the circle of the path of the pole. Thus, the model functions as described for the solstices and equinoxes.

63r

There is no difference between the two [opinions] except with the matter of rising times B.1.I.4

When Ibn Naḥmias mentioned the two opinions, he meant whether the motions were understood to be always in one direction or could appear to be in opposite directions. The difference between the two cases is whether the Sun is understood to be moving forward or backward, that is, toward increasing or decreasing rising times. For more on rising times, see the comments on §A.IV.3.

Ibn Naḥmias has also explained that if one is willing to follow the first opinion (or hypothesis) and accept the existence of heavenly motions that might appear to be in opposite directions, although they are not truly opposite, then the model for mean motion can be simplified. The great circle of the ecliptic can revolve about its own pole, in the direction apparently opposite to the daily motion, by the measure of the Sun's mean motion in longitude.

and it is that the center of the Sun is not carried in [63v] its particular motion upon the inclined circle B.1.II.2

This is a restatement of the challenge that Ibn Naḥmias faced: the Sun did not move with uniform motion through the ecliptic, but it was never observed departing from the ecliptic. He had to explain how the Sun could move at varying speeds on a great circle while remaining equidistant from the observer. Varying the distance of the observer from the Sun through the eccentric hypothesis, or through the mathematically equivalent epicyclic hypothesis, were not options. Biṭrūjī had tried to account for these observed variations in the Sun's motion by displacing the pole of the circle of the pole of the Sun from the pole of the equinoctial.[128] Ibn Naḥmias rejected that component of Biṭrūjī's solar model because the distance between the pole of the Sun's circle and the pole of the Sun meant that the Sun should be observed straying from the ecliptic.[129] Put differently, Ibn Naḥmias's criticism of Biṭrūjī's solar model was that Biṭrūjī's explanation for the

127. *The Light of the World*, §B.1.I.4.

128. Biṭrūjī (trans. and comm. Goldstein), *On the Principles*, vol. 1, p. 13 ("[H]e simply places Ptolemy's eccentric model for the Sun about the north pole of the equator, where the north pole of the equator plays the role of the observer in the Ptolemaic model") and pp. 136–9.

129. *The Light of the World*, §B.1.II.28–9.

Sun's variable motion in the ecliptic necessitated that the Sun leave the ecliptic. Also, Biṭrūjī's solution would produce apparent variations in motion only if the observer were at the pole of the equator, not at the orb's center. Otherwise, the Sun's motions would vary only with respect to their being measured with respect to the ecliptic, not with respect to the actual path of the Sun. To account for variations in the Sun's motion, Ibn Naḥmias introduced a small circle, that is, any circle on the surface of an orb other than a great circle, at the orb's equator, called the circle of the path of the center of the Sun.

63v

Let that small circle be called the circle of the path of the center B.1.II.3

Here, the center of the Sun moves on a small circle called the circle of the path of the center (*dā'irat mamarr al-markaz*), which is at the orb's equator. Biṭrūjī had employed a small circle called *dā'irat al-mamarr* (literally: the circle of the path) near the pole of the orb,[130] which Goldstein translated as "polar epicycle."[131] *The Light of the World*'s model for the solar anomaly took the model for the mean motion as a given, probably because of the simplicity of explaining the mean motion according to the first opinion (or hypothesis) in which there is no barrier to hypothesizing apparently opposite motions in the heavens. Ibn Naḥmias wrote (§B.1.I.5) that the first opinion (or hypothesis) was preferable.

I have chosen not to call these small circles epicycles because that would imply that they would move by being embedded in the wall of a larger orb (which cannot be the case in a homocentric model). Ibn Naḥmias explained that these circles *took the place* (emphasis mine) of the epicycle. In addition, in the Judeo-Arabic version of *The Light of the World*, there is no mover for these small circles.[132] Ibn Naḥmias did assert, briefly, at the very end of the section on the solar model, that additional orbs caused these small circles to move,[133] and there was certainly technical precedent for a mechanism in which a pole moving on a small circle moves a point at the equator in a small circle.[134] But a pole moving in a small circle moves a point at the equator in a noncircular, lenticular shape.[135] Thus, the circle

130. Biṭrūjī (trans. and comm. Goldstein), *On the Principles*, vol. 2, p. 263.

131. Biṭrūjī (trans. and comm. Goldstein), *On the Principles*, vol. 1, p. 111.

132. See, however, *The Light of the World*, §B.4.I.3. Physical mechanisms were implicit, though. Here Ibn Naḥmias clarified that the planets could be carried by combinations of orbs with the outer one moving the poles of the inner orb. In §B.1.II.10, he wrote that the small circle had two poles, indicating at least some attention to the physical questions.

133. *The Light of the World*, §B.1.II.27.

134. Bernard R. Goldstein: "On the Theory of Trepidation according to Thābit ibn Qurra and al-Zarqāllu and Its Implications for Homocentric Planetary Theory," in *Centaurus* X (1964): 232–47, at p. 233.

135. Ragep, *Tadhkira*, p. 402. The motion of a pole in a small circle about the pole of another orb

of the path of the center could not have been produced by a motion at the pole. Quṭb al-Dīn Shīrāzī recognized this fact in his model for trepidation, stating that the motion of the equinoxes would trace an ovoid (*ihlījī*) figure.[136] The Hebrew recension did propose physical mechanisms for the motions of some of the small circles that Ibn Naḥmias posited here, though not for the circle of the path of the center.[137] At this point in the text, though, Ibn Naḥmias wrote only that the small circle rotated about its own poles, meaning that the poles of the circle of the path of the center would be embedded at diametrically opposite points in the inclined circle carrying the circle of the path of the center.

64r

And because the matter of the Sun is the opposite of that B.1.II.7

The measure of the motion of the small circle, intended to account both for the anomaly in longitude and to keep the Sun in the ecliptic, is the same as the mean measure of the motion of the lag, 59' per day. Thus, although Ibn Naḥmias did refer to "lag" (*taqṣīr*), the fact that the poles of the circle of the path of the center are 90° from the pole of the circle of the Sun made the distinction between lag and direct motion semantic at best. Indeed, Ibn Naḥmias added that "this motion can be proposed in any direction we wish."[138]

Thus, there occurs for the Sun, on account of this movement, an increase in its final lag B.1.II.8

The motion of the Sun on the circle of the path of the center must serve, ideally, both to keep the Sun in the ecliptic *and* provide for a variation in the longitude.

64v

And it might have been possible that this anomaly B.1.II.9–10

(See figure 5.2.) Here, Ibn Naḥmias explores a dimension of the model for the solar anomaly that the Hebrew recension pursued further. This connection is evidence for how the Hebrew recension, if not the work of Ibn Naḥmias himself, was in the spirit of the Judeo-Arabic original. The key is that when one explains the motions according to the second opinion (or hypothesis), great circle *SZ* is slanted (or inclined) to the circle carrying the circle of the path of the center of the Sun by as much as 2;30°. Thus, it would be possible to think of points *T* and *N* as having acquired the inclination of great circle *SZ*, rather than remaining

produces an oval shape at the equator. "It is clearly impossible to move a sphere in such a way that every point on it describes a small circle."

136. Quṭb al-Dīn al-Shīrāzī, *al-Tuḥfa al-shāhiyya*, Paris BNF MS Arabe 2516, 15v.

137. *Or ha-ʿolam*, MS Bodleian Canon Misc. 334, §B.1.II.26/X.1ff.

138. *The Light of the World*, §B.1.II.7.

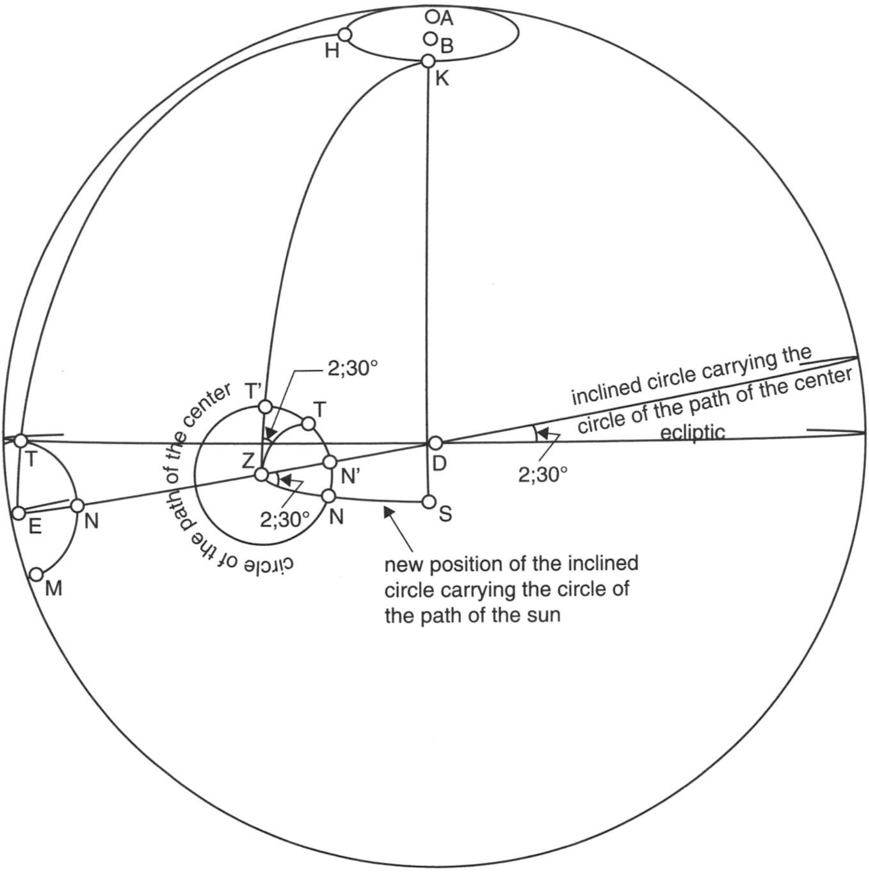

FIGURE 5.2

superimposed on great circle *DZE*. In the figure, points *T* and *N* retain their inclinations with respect to great circle *SZ*, whereas points *T'* and *N'* are aligned with great circle *DZE*. Ibn Naḥmias claimed that the maximum effect of points *T* and *N* being inclined, or slanted, with great circle *SZ* would be akin to an additional revolution of 5° on the circle of the path of the center. In fact, as the commentary on the additions in the Hebrew recension (§B.1.II.26/X.14) will show, the maximum slant would be 2;30°, after the pole rotates 90° on small circle *HK*.

Because Ibn Naḥmias preferred to explain things according to the first opinion (or hypothesis), which admitted of apparently opposite motions in the heavens, there was actually no need for great circle *SZ* and the slant that ensued. Indeed, a few lines down, he clarified that points *L* and *N* would never leave great circle

DZE: "our thesis is that points LN never depart from circle *DZE*, not even by a little bit."[139]

65r

If our thesis is that the hypothesis for this anomaly is the first hypothesis B.1.II.11–12

That is, apparently opposite motions are not, in fact, opposite. Just as Ibn Naḥmias remarked in the chapter on the hypotheses for the mean solar motion (cf. comments on §B.1.I.4) that it would be possible to account for the Sun's mean motion with a single orb if apparently opposite motions were allowed in the heavens, allowing apparently opposite motions in the heavens had an effect on the motion of the circle of the path of the center of the Sun. Specifically, following the first opinion (or hypothesis) meant that there would be no slant. In §B.1.II.12, he also identified the slant as a distinct source of variable motion but explained that it resulted only from the second opinion (or hypothesis).

The demonstration of that B.1.II.13

(See figure 5.3.) This comment first assesses Ibn Naḥmias's own explanation for why the model for the solar anomaly did not keep the Sun in the ecliptic, which Ibn Naḥmias also called the circle of the Sun's declination, and then provides a more accurate and systematic assessment.[140] Ibn Naḥmias argued that as the center of the circle of the path of the Sun moved from point *E* to point *Z* through the model for mean motion, the center, now at point *Z*, has gotten closer to the ecliptic (great circle *DBA*), by the measure of arc *TB*. Then, after point *T* revolved, its position relative to the ecliptic (equivalent to its Φ coordinate in a system of spherical coordinates)[141] would be the same as point *B*'s. That is, as the center of the Sun, at point *T*, rotates on the circle of its path, it would approach great circle arc *DBA* also by the measure of arc *TB*, so point *B*'s position relative to the ecliptic could not be the same as point *G*'s. In Ibn Naḥmias's reckoning, the motion of the Sun on the small circle would be insufficient to bring the center of the Sun to point *G*, from point *T*, because point *G* was closer to great circle *DZE* than point *B*. Ibn Naḥmias's general conclusion, that his hypothesis for the anomaly did not fully account for observations, was correct, but his reasoning was imprecise. Still Ibn Naḥmias's argument communicated how the maximum displacement of the Sun from the ecliptic will be at 90° and 270° of mean motion. There, point *Z* will

139. *The Light of the World*, §B.1.II.12.

140. See also Morrison: "The Solar Model," pp. 90–2. Ibn Naḥmias was also correct to say (§B.1.II.14) that the Sun came closer to the latitude of its mean position than it did to the latitude of its true position.

141. The commentary uses the upper case Φ and Θ to denote the spherical coordinates and the lower case θ to denote the mean motion.

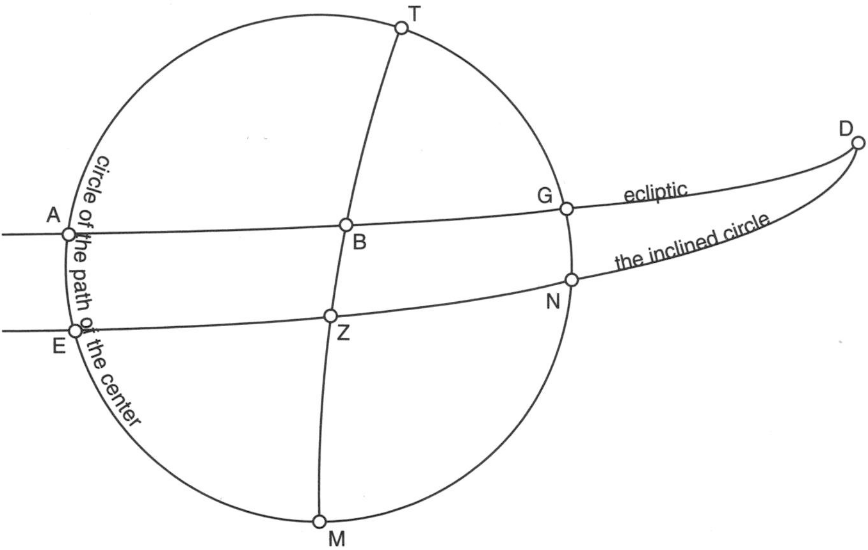

FIGURE 5.3

rest on point *D* and the Sun will rest on the inclined circle carrying the path of the center.

(See figure 5.4.) For a more precise analysis, we calculate the displacement of point *T*, the position of the Sun after the center of the circle of the path of the center (point *Z*) has moved θ° toward point *D*, and after the Sun has rotated the same amount on the circle of the path of the center. Arc *TS* is a segment of a great circle, perpendicular to the inclined circle carrying the circle of the path of the center, and intersects the ecliptic at point *Q*. Note that the circle of the path of the center is not a great circle and is therefore not an equator of an orb. For Φ(*T*), we draw spherical triangle *PTZ*, with *P* being the pole. Side *t* (side *PZ*) is 90°, angle *Z* is θ, and *p* (side *ZT*) is r (in this case 2;30°). We apply the cosine rule for sides:

$$\cos z = \cos t \cos p + \sin t \sin p \cos Z$$
$$\text{since } t = 90°,\ \cos z = \sin p \cos Z, \text{ thus}$$
$$z = \Phi(T) = \arccos(\sin r \cos\theta) \qquad (1)$$

for Θ(T), we solve for *t* (side *ZS*) in triangle *ZTS*. Side *s* (side *TZ*) is *r*, *z* (side *TS*) is *z* = 90–Φ(T), *Z* is 90°–θ, and *S* is 90°. Again applying the cosine rule for sides:

$$\cos s = \cos z \cos t + \sin z \sin t \cos S$$
$$\text{Since } s = r \text{ and } S = 90°,\ \cos r = \cos(90-\Phi(T)) \cos t$$
$$\text{Thus, } \cos r = \sin(\Phi(T)]\cos t \text{ and } t = \arccos[\cos r/\sin(\Phi(T))]$$
$$\text{So } \Theta(T) = \theta + t = \theta + \arccos[\cos r/\sin(\Phi(T))] \qquad (2)$$

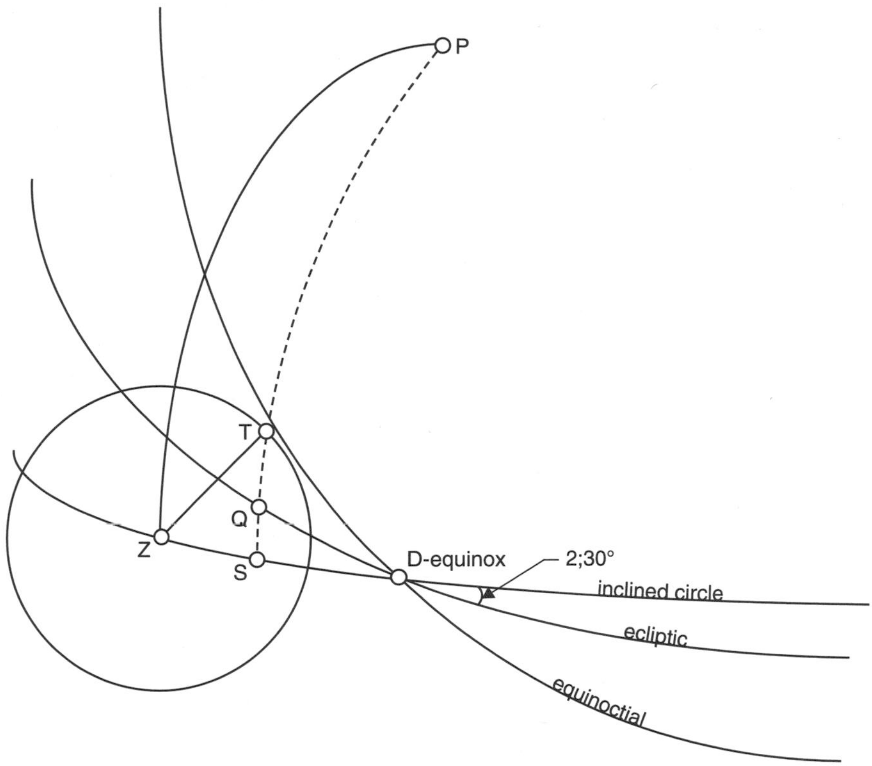

FIGURE 5.4

Then we compute the Φ coordinate of point *Q* which is on the ecliptic and which has a Θ coordinate of Θ(*T*). We consider triangle *DSQ*, where *D* is 2;30°, *S* is 90°, and *q* (side *DS*) is 90°–Θ(*T*). We apply the rule of four parts:[142]

$\cos q \cos S = \sin q \cot d - \sin S \cot D$

since $S = 90°$, $0 = \sin(90° - \Theta(T))\cot d - \cot D = \cos(\Theta(T))\cot d - \cot D$

when $\cot d = \cot D / \cos(\Theta(T))$ or $\tan d = \cos(\Theta(T)) \tan D$ and $d = \arctan[\cos(\Theta(T)) \tan D]$

then $\Phi(Q) = 90° - \arctan[\tan D \cos(\Theta(T))]$ (3)

142. W. M. Smart: *Text-Book on Spherical Astronomy*, 6th ed. (Cambridge: Cambridge University Press, 1960): p. 12. The four parts formula is derivable from the law of cosines. On the origin of the law of cosines, see Glen Van Brummelen: *Heavenly Mathematics: The Forgotten Art of Spherical Trigonometry* (Princeton, NJ, and Oxford, UK: Princeton University Press, 2013), pp. 94–8.

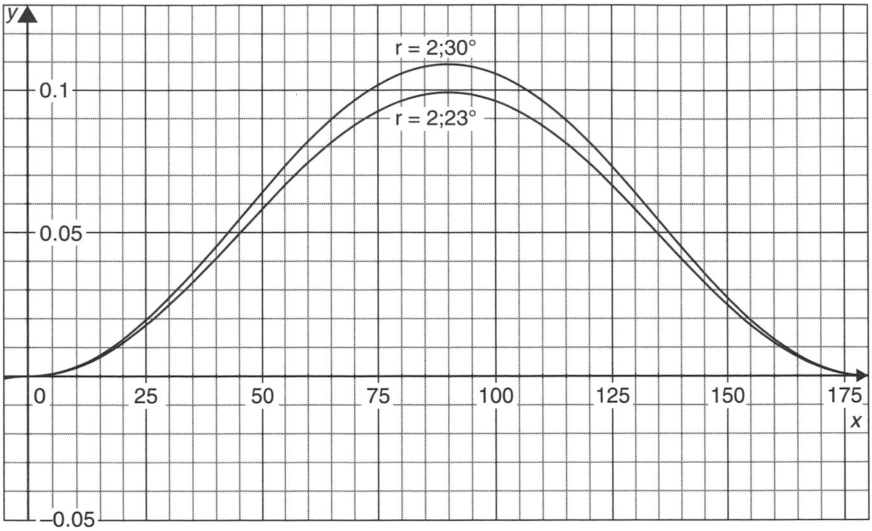

GRAPH 5.1

(See graph 5.1). Thus, T and Q will have different Φ coordinates. At $\theta = 45°$ the deviation will be 0;3,16°; at $\theta = 90°$ of mean motion, the deviation will be at its maximum and will be 0;6,33°. This is not a large error, and it does improve on a displacement from the ecliptic of over a degree at times in Biṭrūjī's model, presuming one accepts Ibn Naḥmias's hypothesis for the solar mean motion. Given values for r and D, we can use equations (1) to (3) to find $\Phi(Q)$ and $\Phi(T)$ as functions of θ. Graph 5.1 shows the difference $\Phi(Q)$ and $\Phi(T)$ for $r = D = 2;30°$ (Ibn Naḥmias's parameter, equaling the Ptolemaic eccentricity) and for $r = D = 2;23°$ (the parameter of the maximum solar equation in the *Almagest*).

Ibn Naḥmias never mentioned an amount of error that he would find acceptable. Another problem that Ibn Naḥmias did not raise at this point in the text was that the hypothesis of the circle of the path of the center cannot reproduce the asymmetries of the Ptolemaic solar model in that the sum of the days of spring and summer in the northern hemisphere is greater than the sum of the days of fall and winter.

65v

Since that has become clear, it is incumbent on us to show how the center of the Sun retains forever the circumferences of these two circles B.1.II.1

In an attempt to compensate for the remaining displacement of the Sun from the ecliptic, Ibn Naḥmias introduced into his solar model double circles, reminiscent of the Ṭūsī couple of Naṣīr al-Dīn al-Ṭūsī (1201–74).[143] This first version, the detailed account of which begins in §B.1.II.17, with arc *KBS* always parallel to *HAC*, is different from the Ṭūsī couple, either in the *Tadhkira* or in Ṭūsī's earlier *Taḥrīr al-Majisṭī* (Recension of "The Almagest").[144] But Ibn Naḥmias presented more than one version of his double-circle hypothesis, and the second version turns out to be more similar to the double circles in Ṭūsī's *Tadhkira* and virtually identical to the double circles used by Giovanni Battista Amico (d. 1538).[145] Full analysis of this first version of the double-circle hypothesis is found in the comments on §B.1.II.18.

66r

arc KBS *is forever parallel to arc* HAC *through this movement* B.1.II.18

(See figure 5.5.) The stipulation that arcs *KBS* and *HAC* remain parallel distinguishes this first version of the double-circle hypothesis. The figure shows the solar model, modified with the first version of the double-circle hypothesis, after 45° of mean motion. While I will describe the motions sequentially, they actually occur simultaneously. In figure 5.5, *Z*, the center of the circle of the path of the Sun, revolves 45° clockwise in the direction of the equinox with the mean motion. In addition, the revolution of the circle of the path of the center has carried pole *A* of one of the double circles, coincident with the Sun at point *K*, 45° clockwise. Then, small circle *HBC* revolves 45° counterclockwise about pole

143. Ibn Naḥmias did not mention Ṭūsī's name, so one cannot pronounce unequivocally about whether or not Ibn Naḥmias devised his double-circle hypothesis independently. There is some evidence of connections between Jews in Provence and Andalusia and developments in the Islamic East. Gersonides was certainly aware of the Pseudo-Ṭūsī commentary on Euclid's *Elements,* and he may have been aware of the authentic Ṭūsī commentary, too. See Tony Lévy: "Gersonide, Commentateur d'Euclide," in Gad Freudenthal (ed.): *Studies on Gersonides* (Leiden, New York, Köln: E. J. Brill, 1992): pp. 90–1. There was also the possibility of exchanges between other scholars in Andalusia and the Islamic East. See M. Comes: "The Possible Scientific Exchange between the Courts of Hulaghu of Maragha and Alphonse 10th of Castille," in *Science, Techniques et Instruments dans le Monde Iranien* (Tehran, 2003): pp. 29–50. Finally, some of R. Israel Israeli's (of Toledo) synagogue poetry may have been written in the East (see Ilan, "Rᵉdipat ha-emet," p. 51). If Israel Israeli did spend time in the Islamic East, a path for Ṭūsī's ideas is clear.

144. In *The Light of the World,* the two circles revolved on the surface of a sphere, whereas in *Taḥrīr al-Majisṭī,* the circles moved in a plane. The Ṭūsī couple has its own history. Recently, Ragep has dated the *Ḥall-i Muʿīniyya,* with the origin of the fully-formed Ṭūsī couple, to 1245, before the 1247 *Taḥrīr al-Majisṭī.* See F. J. Ragep: "The Origins of the Ṭūsī Couple Revisited," in *Proceedings of the "Scientific and Philosophical Heritage of Naṣīr al-Dīn al-Ṭūsī" Conference,* February 23–24, 2011 (Tehran: Miras-e Maktoob, in press).

145. On Amico, see Mario di Bono: "Copernicus, Amico, Fracastoro, and Ṭūsī's Device: Observations on the Use and Transmission of a Model," in *Journal for the History of Astronomy* XXVI (1995): pp. 133–54. The double-circle device in *Taḥrīr al-Majisṭī* proposed motions in a plane, whereas Ibn Naḥmias's and Amico's double circles proposed motions on the surface of an orb.

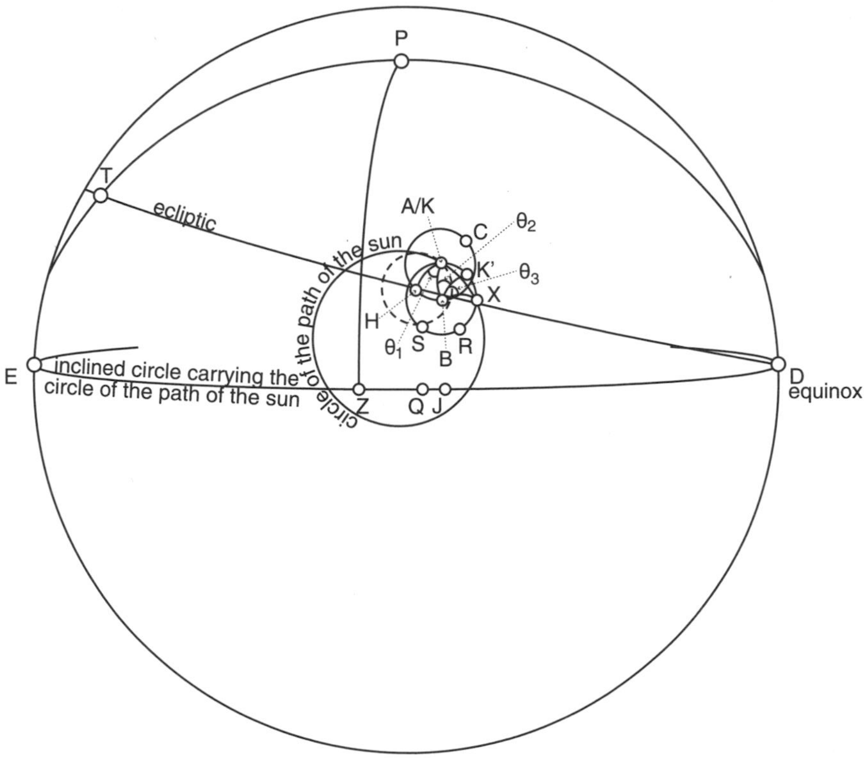

FIGURE 5.5

A, carrying the pole of circle *SRK* from *H* to point *B*. If arcs *KBS* and *HAC* are supposed to be parallel, then the first motion, indicated with θ_1, of pole *H* to point *B* causes point *K*, the location of the Sun, to move from *A* to *K'*.[146] This motion is indicated with θ_2. That motion of point *K* from point *A* to point *K'* keeps arcs *HAC* and *KBS* parallel. The stipulation that these two arcs are parallel would mean, additionally, that point *K*'s separation from point *A* would be tantamount to an additional revolution on the measure of the first. Finally, point *K* would revolve 45° (indicated with θ_3) about pole *B* to arrive at point *X*, supposedly the intersection of circle of the path of the Sun with the ecliptic. In this first version of the double-circle hypothesis, all of the revolutions of the small circles are on the same measure (θ), equal to the mean motion and the motion in anomaly. Thus the principle of attraction to the motion of the uppermost orb could plausibly be the cause of all of the motions.

146. Morrison, "The Solar Model," p. 93.

The ability of this version of the double-circle hypothesis to keep the Sun in the ecliptic depended on the size of the two small circles. In the text, Ibn Naḥmias wrote that the radius of each of the small circles was half of the maximum solar anomaly, which he defined as 2°.[147] The radius of each of the small circles would then be 1°. On the basis of that statement, one could conclude that Ibn Naḥmias actually wanted the double-circle hypothesis to take the place of the circle of the path of the center. The Hebrew recension explored at greater length the possibility of eliminating the circle of the path of the Sun and having the variations in the solar longitude result only from the double circles. But the figures in the MSS of *The Light of the World* (as well as some comments he made later in the text) suggest that Ibn Naḥmias's initial intention was to add the double-circle hypothesis to a solar model that included the circle of the path of the Sun.

To calculate the Sun's final position, we presume that the radii of the small circles are great circle arcs and that θ_1, θ_2, and θ_3 are all 45°. The radius of the circle of the path of the center is 2;30° and the radii of the double circles are 1°. First, we use equations (1) and (2) from §B.1.II.13 to determine the spherical coordinates of point *A* on the circle of the path of the center (where the Sun begins) after 45° of mean motion. *A*'s coordinates are $\Theta = 46;46°$ and $\Phi = 88;14°$. To determine point *H*'s coordinates, we make point *P* the pole and draw spherical triangle *PZH* with $Z = 45°$, p (side *ZH*) = 1;30°, and h (side *PZ*) = 90°. We apply the cosine rule for sides:

$$\cos z = \cos h \cos p + \sin h \sin p \cos Z:$$
$$\cos z = \cos 90° \cos 1;30° + \sin 90° \sin 1;30° \cos 45°$$
$$z = 88;56° = \Phi(H)$$

To determine $\Theta(H)$, we draw spherical triangle *ZHQ*, where point *Q* is a point on the inclined circle with $\Theta(H)$. Side q (side *ZH*) =1;30°, z (side *HQ*) = $90°-\Phi(H)$ = 1;04°, q (side *HB*) = 1;30°, and $Q = 90°$. According to the cosine rule for sides:

$$\cos q = \cos h \cos z + \sin h \sin z \cos Q$$
$$\cos 1;30° = \cos 1;04° \cos h$$
$$h = 1;04°, \text{ and since } \Theta(Z) = 45°, \Theta(H) = 45° + 1;04° = 46;04°$$

This determination of the Θ and Φ coordinates of point *H* has presumed that arcs *HAC* and *KBS* are both parallel. That this cannot be the case becomes apparent when we consider spherical triangle *ZAJ*, where point *J* is the intersection of arc *AB*, were it extended to great circle arc *EZD*, with great circle arc *EZD*. Since angle *HZD* must be 45°, and angle *ZAJ* must be 45°, then angle *ZDA* cannot be

147. *The Light of the World*, §B.1.II.17. The following calculations presume a 2;30° radius for the circle of the path of the Sun to facilitate comparisons with the unmodified model for the Sun's motion in anomaly.

exactly 90°. Therefore points *B* and *A* do not share the same Θ coordinate, meaning that point *B* is not directly below point *A*. Put differently, if the diameters of the double circles were great circle arcs and both angles *HZD* and *ZAJ* were 45°, then the diameters of the double circles would not be perfectly parallel. Or if the diameters of the double circles were parallel, then they would not be great circle arcs. Given how small the radii are, however, the deviation, once computed, would be negligible.

But if we do assume that arcs *KBS* and *HAC* remain, somehow, parallel, then once point *K* leaves point *A*, point *K* would have moved on circle *KRS*, to its new position *K*'; arc *K*'B will make a 45° angle with arc *AB*. Two more questions emerge. The first is that Ibn Naḥmias has not explained what forces point *K* to separate from point *A*; point *K* could just as easily have remained superimposed on point *A* with pole *H* moving to point *B*. Second, Ibn Naḥmias did not explain what kept the two arcs parallel in the first place. I speculate that he introduced the requirement of being parallel in order for the second rotation to be, in effect, twice the first, without having to propose a physical mover that moves one orb twice the measure of the revolution from *H* to point *B* about pole *A*.

To determine the Φ coordinate of point *B*, we accept Ibn Naḥmias's stipulation of the parallelism of arcs *KBS* and *HAC*, and we first solve triangle *PAH*. Side *h* (side *PA*) = 88;14°, *a* (side *PH*) is 88;56°, *p* (side *AH*) is 1°. Angle *P* is 0;42° (the difference in Θ coordinates between poles *H* and *A*). Then, we apply the cosine rule for sides:

$\cos a = \cos h \cos p + \sin h \sin p \cos A$
$\cos 88;56° = \cos 88;14° \cos 1° + \sin 88;14° \sin 1° \cos A$
Angle *A* (i.e., angle *HAB*) = 134;31°
thus angle *PAB* = 45° + angle *HAB* = 179;31°.

Again, we apply the cosine rule for sides:

$\cos a = \cos b \cos p + \sin b \sin p \cos A$
$\cos a = \cos 88;14° \cos 1° + \sin 88;14° \sin 1° \cos 179;33°$
$a = \Phi(B) = 89;14°$

Because *AZD* = 45° and *HAB* = 45°, Θ(*B*) is therefore very close to Θ(*A*). After point *K* moves from point *K*' through θ_3, it arrives at point *X* with coordinates of Θ(X) = Θ(*B*) +1° = 47;46° and Φ(*X*) = Φ(*B*) = 89;14°. Using the parametric equations (1) and (2) from §B.1.II.13, we can compute the Φ coordinate of the point on the ecliptic with Θ = 47;46°; it is Φ = 88;19°. Thus, the Sun is displaced from the ecliptic by 0;5°.

(See figure 5.6.) While Ibn Naḥmias did not say so, one could markedly improve the accuracy of this model by decreasing the radii of the two small circles. We can compute the dimensions of a double-circle hypothesis necessary to compensate for a maximum displacement of 0;7°, the maximum displacement

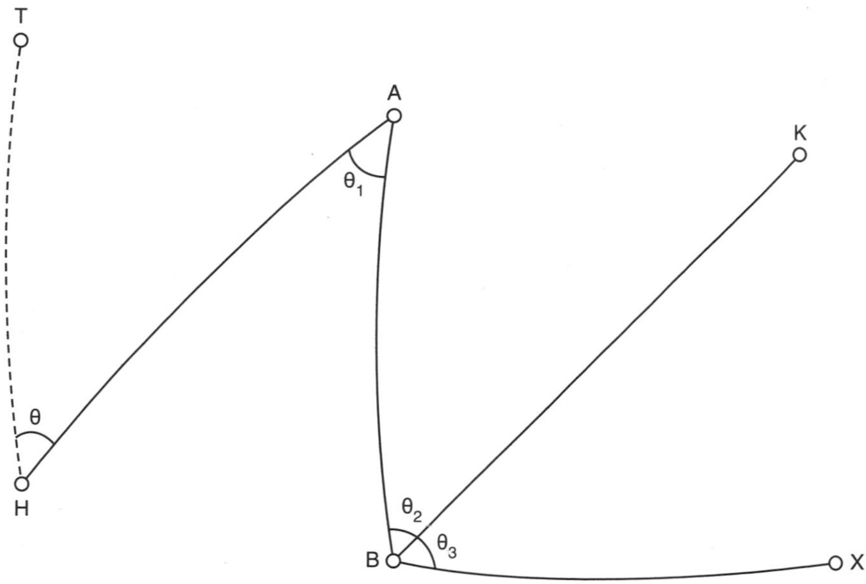

FIGURE 5.6

(at θ = 90°) found with the original solar model. Each line in the figure represents the length of a radius of a circle of the double-circle hypothesis. The revolution at point *B* is given as 2θ, accepting Ibn Naḥmias's implication that the separation of point *K* from point *A* is equivalent to an additional revolution on the measure of θ°. At θ = 90°, arc *AH* will be perpendicular to the dashed line, indicating that the circle of the path of the center has revolved 90°. Also, arcs *AB* and *BX* will form one long arc, arc *AX*, which needs to compensate for the 0;7° displacement. Each radius, that is, arc *AB* and arc *BX*, should therefore be 0;3,5°.

One can compute how well double circles with radii of 0;3,5° will compensate for the Sun's displacement from the ecliptic at θ = 45°. If the radii of the double circles in figure 5.5 are 0;3,5°, then Θ(*A*) = 46;46° and Φ(*A*) = 88;14°. Θ(*B*) = 46;46° and Φ(*B*) = 88;17°. Θ(*X*) = 46;49° and Φ(*X*) = 88;14°. By solving a right spherical triangle, one finds that the corresponding point on the ecliptic with a Θ of 46;49° has a Φ of 88;17°. There is a discrepancy of 0;3° in the Φ coordinate at θ = 45°. Radii of 0;3,5° are an improvement on radii of 1° at all positions.

If we made each of these two motions not equal to the movement of pole Z but more than it B.1.II.19

This is an adjustment of this first version of the double-circle hypothesis in which the measure of the two circles' motion is no longer linked precisely to the motion of the uppermost orb. Still, in this version of the double-circle hypothesis, the two circles do still each revolve by the same measure. Ibn Naḥmias proposed this change in order to account for the Sun's outstanding deviation from the both the circle of the path of the center and the circle of the path. This proposal indicated that Ibn Naḥmias was aware of the imperfections of the version of the double-circle hypothesis that he had proposed so far, and also how deviations resulting from those imperfections were inverted in the other quadrants.

66v

So the motion of the center of the Sun about point B *in this time period needs to be twice the aforementioned amount* B.1.II.2

(See figure 5.7.) This is Ibn Naḥmias's preferred version of the double-circle hypothesis. *Z*, the center of the circle of the path of the center of the Sun has moved with the mean motion, and *A* has revolved with the motion of the circle of the path of the center of the Sun. In this version, one circle revolves with twice the angular velocity of the other, and in the opposite direction. The upper small circle revolves θ about *A*, moving the center of the lower small circle from point *H* to point *B*. Then the lower small circle revolves 2θ about its pole, now at point *B*. Point *X* marks the Sun's final position after the revolution of 2θ from point *A*. Arc *AN* is the diameter of the circle *HBC*, and in the figure arc *AN* is perpendicular to arc *HC*. The Sun is supposed to oscillate on arc *AN*.

To assess whether *X* is on *AN*, we solve spherical triangle *ABX* when angle *ABX* (i.e., 2θ) is 90°. Arcs *BA* and *BX* are equal because they are both radii of the small circles. Since angle *HAB* is 45°, because it is the measure of the motion (θ) of one of the double circles, and angle *NAC* is 90°, because arc *AN* is perpendicular to arc *HC*, then angles *BXA* and *BAX* would have to be 45° for point *X* to be on arc *AN*. If angle *BAX* is 45°, then spherical triangle *ABX* would have to contain 180°, which is impossible. Therefore the angles at *A* and *X* will be close to 45° but will not be exactly 45°, meaning that *X* will not be found precisely on arc *AN*.

At first, this version of the double-circle hypothesis seems most similar to the rudimentary Ṭūsī couple found in Ṭūsī's *Recension of "The Almagest"* (Taḥrīr al-Majisṭī) in that there is no explanation given for how the points move on the small circles.[148] Only the Hebrew recension raised the question of physical movers for the double circles. Later versions of the Ṭūsī couple, particularly those that appeared in Ṭūsī's *Tadhkira*, were different in that they addressed the question of physical movers, proposing two orbs with one twice the size of the other. But, in a different and important respect, Ibn Naḥmias's double-circle hypothesis was

148. See George Saliba: "The Role of the *Almagest* Commentaries in Medieval Arabic Astronomy: A Preliminary Survey of Ṭūsī's Redaction of *Almagest*," in *Archives Internationales d'Histoire des Sciences* XXXVII (1987): pp. 3–20.

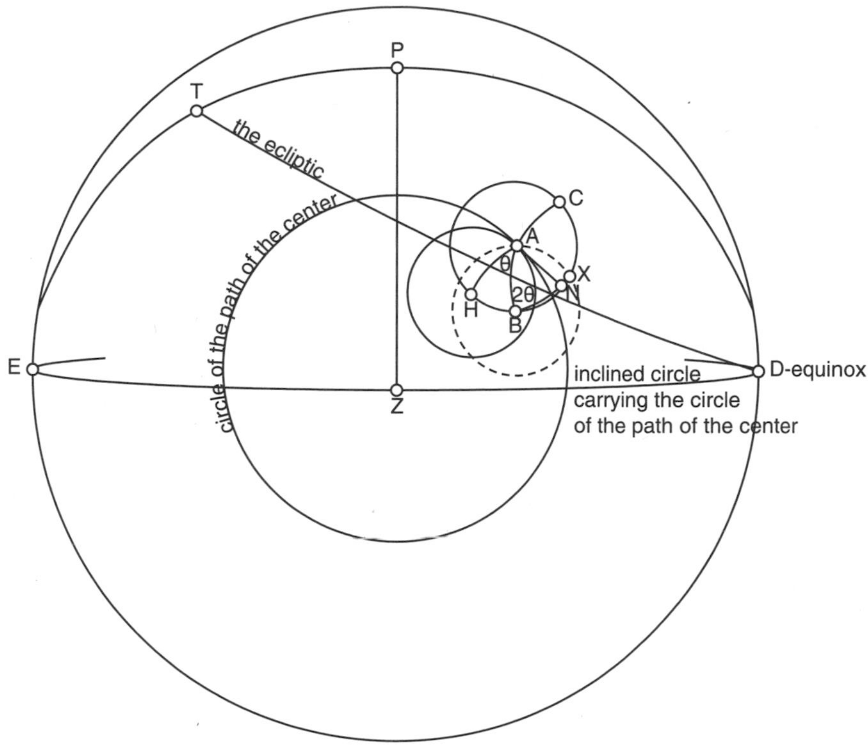

FIGURE 5.7

more similar to the spherical Ṭūsī couple in the *Tadhkira* than to the rudimentary Ṭūsī couple in *Taḥrīr al-Majisṭī* in that the point was moving on the surface of an orb, not in the plane.[149] We move to formulating parametric equations that describe the path of the oscillating point *X:*

First, to find side *AX* in triangle *AXB,* we use the rule of cosines for sides:[150]

$\cos b = \cos x \cos a + \sin x \sin a \cos B$

Because side a = side $x = r$, and because angle B is 2θ, then

$\cos^{-1}(\cos^2 r + \sin^2 r \cos 2\theta) = \text{side } AX = b$ (1)

149. Physically, however, there were some important differences, above all that the spherical Ṭūsī couple involved a third orb to eliminate displacements caused by the motions of the two orbs that made up the couple.

150. The origins of the law of cosines are complex. See Van Brummelen: *Heavenly Mathematics*, pp. 94–8.

We can then use the law of sines to determine angles $AXB = XAB$:

$$\sin AX/\sin 2\theta = \sin r/\sin XAB$$
$$\sin XAB = \sin r \sin 2\theta/\sin AX$$
$$XAB = \sin^{-1}(\sin r \sin 2\theta/\sin AX)$$
$$\text{Angle } XAN = XAB-(90°-\theta) \text{ or } AXB-(90°-\theta) \qquad (2)$$

In order to write equations that describe the x and y coordinates of point X, we solve spherical triangle AXN, with angle ANX=90°. Side n is b in equation (1). Angle A is angle XAN in equation (2). Side a (side XN) is the y-coordinate:

$$\sin a/\sin A = \sin n/\sin 90°$$
$$\sin a = \sin n \sin A$$
$$\text{side } a = \sin^{-1}(\sin n \sin A) = y \qquad (3)$$

To solve for the x coordinate, side AN, we use the cosine rule for sides:

$$\cos n = \cos a \cos x + \sin a \sin x \cos N$$
$$\cos x = \cos n/\cos a = \cos AN \qquad (4)$$

We also apply the law of cosines:

$$\cos A = -\cos X \cos N + \sin X \sin N \cos a$$
$$\sin X = \cos A/\cos a = \cos XAN/\cos XN \qquad (5)$$

And, according to the law of sines, $\sin x/\sin X = \sin n/\sin 90° = \sin n$ (6)

Combining equations (5) and (6):

$$\sin x/(\cos A/\cos a) = \sin n$$
$$\sin x = \sin n \cos A/\cos a$$
and, with equation (4), $\cos x=\cos n/\cos a$
thus, $\tan x = \sin n \cos A/\cos n = \tan n \cos A$
so, $x = \tan^{-1}(\tan n \cos A)$ = the x coordinate (7)

These equations are equivalent to the parametric equations derived by Saliba and Kennedy in their 1991 analysis of the spherical Ṭūsī couple. In that publication, PC was side AX (side n), and angle δ was XAN (side A).[151] Ibn Naḥmias's double circles, therefore, superficially resembled the rudimentary Ṭūsī couple found in *Taḥrīr al-Majisṭī* in its lack of physical movers and basis in the motion of two small circles, but were mathematically equivalent to the spherical Ṭūsī couple found in the *Tadhkira*. Saliba and Kennedy showed that the displacement of X from arc AN is minuscule.

151. George Saliba and E. S. Kennedy: "The Spherical Case of the Ṭūsī Couple," in *Arabic Sciences and Philosophy* I (1991): pp. 285–91, at p. 289. Their equations were $x=\tan^{-1}(\tan PC \cos\delta)$ and $y=\sin^{-1}(\sin PC \sin\delta)$.

(See graph 5.2.) The comments on §B.1.II.18 have shown that the double-circle hypothesis can solve the displacement of the Sun from the ecliptic that remains after the Sun's motion on the circle of the path of the center. Ibn Naḥmias would, eventually, consider dispensing with the circle of the path of the center and using the double-circle hypothesis to account for the entire solar anomaly. Graph 5.2 computes the x-coordinate for a solar model without a circle of the path of the center, but with double circles of a radius of 1;11,30°. As there is no circle of the path of the center, the x-coordinate would yield the Ptolemaic maximum solar equation of 2;23° at 90° of mean motion. While the double-circle hypothesis does account well for the longitudes, it cannot reproduce the asymmetries of the Ptolemaic model. The maximum deviation from the Ptolemaic position occurred after 135° of mean motion and is 0;3,12°. There is, again, no evidence that Ibn Naḥmias carried out the level of analysis found in the commentary, but the analysis does support his claim that he improved upon Biṭrūjī's models. These results also show why later scholars might have been interested in *The Light of the World.*

Second, let us assume according to the second opinion B.1.II.21
The second opinion is the one in which apparently opposite motions are unacceptable, meaning that it is difficult to say that one motion is twice the measure of the other *and in the opposite direction*. The issue that Ibn Naḥmias addressed was how, in this version of the solar model, the diameter of the circle of the path of the center of the Sun maintains the inclination to arc *ED* that it acquired through the model for the Sun's mean motion. This proposal that arc *LN* retains its inclination to arc *ED* is analogous to the hypothesis of the slant that would emerge in the Hebrew recension (see comments on §B.1.II.26/X.14).

67r

with it [arc GL*] equaling twice the inclination of point* E *from the equinoctial* B.1.II.22
(See figure 5.8.) This is Ibn Naḥmias's attempt to modify the double-circle hypothesis to follow the second opinion (or hypothesis), in which apparently opposite motions are unacceptable. This proved to be a difficult goal to achieve. The measure of arc *GL* is twice the inclination of point *E* from twice the inclination of the inclined circle carrying the pole of the circle of the path to the equinoctial, or about 53° [2*(24° + 2;30°)].[152] A parameter of 53° would have a major impact on the final position of the Sun on the circle of the path of the center because when one adds the motion in anomaly on the circle of the path of the center, the Sun will have gone well to the south of the ecliptic. To counteract that displacement, Ibn Naḥmias wrote, "the center of the Sun was conveyed by the two spheres

152. The Hebrew recension, with the hypothesis of the slant, modified things by making the maximum value for this motion the inclination of *E* to the *ecliptic*, making arc *GL* much smaller.

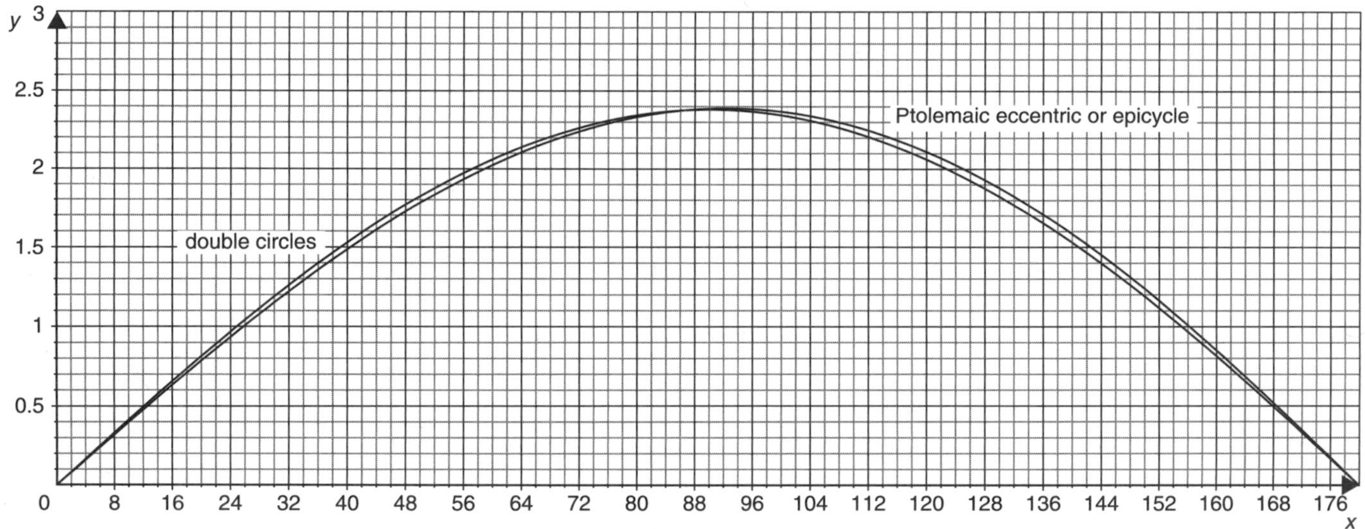
y
3
2.5
2
1.5
1
0.5
0
8
16
24
32
40
48
56
64
72
80
88
96
104
112
120
128
136
144
152
160
168
176
x
double circles
Ptolemaic eccentric or epicycle

GRAPH 5.2

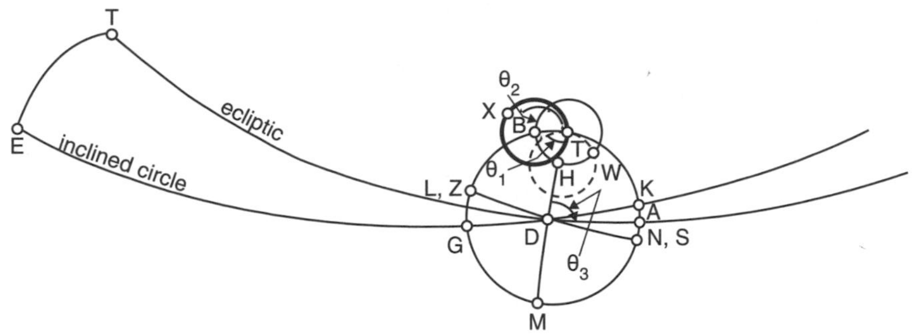

FIGURE 5.8

described above through the difference of *TA* from a quarter circle toward point *Z* in this period of time."

Let it be, according to this description B.1.II.23

As the pole of the path of the center of the Sun moves with the mean motion of θ (90° in the figure) to point *D*, arc *LN* retains its inclination to the inclined circle and has revolved 26;30°, which is the inclination of the inclined circle, with respect to the equinoctial (see comments on B.1.II.22). Before the revolution of small circle *LTNM* through an angle equal to the mean motion in longitude, the center of the Sun would arrive at point *T*, which bisects arc *LN*. Although all of the motions in this model should occur simultaneously, Ibn Naḥmias's explanation implied that he conceived of the two circles moving, in order to cancel out the effect of the slant of the diameter of the circle of the path of the Sun, before the double circles' motion with the circle of the path of the center.[153] The problems with this version of the double-circle hypothesis mean that this is a case where Ibn Naḥmias was unable to explain something through the hypothesis of lag.

The Sun must move from point *T* to point *X* by the difference of arc *TA* from θ (in this case 90°) in preparation for the 90° rotation of small circle *LTNM*. The pole of small circle *TW* revolves about point *T* through a 90° angle, from point *H* to point *B*.[154] The dashed line in the figure indicates the old position of small circle *HBW*, and the dark, solid line indicates the new position of small circle *HBW*. Then, the center of the Sun revolves about point *B* from point *T* to point *X*, through a 180° angle. So, the motions of the two small circles have brought the

153. That is why Ibn Naḥmias wrote (*The Light of the World*, §B.1.II.24), "Also in this period of time, it [the center of the Sun] moved about the pole of the circle of the path through a quarter of a circle and came to point A because arc *XTA* is a quadrant."

154. *The Light of the World*, §B.1.II.24.

point of the Sun's greatest lag (the Sun's position) to point *X*. The circle of the path of the center then would need only to revolve, with the Sun's mean motion (90° in this case), about its pole at point *D*, in order to move the point of the Sun's greatest lag to point *A* on the ecliptic.

The location of point *Z*, beyond the fact that it is a point on small circle *LTNM*, is difficult to pin down from the text. The text says that points *L* and *N* retained their positions on great circle *SZ*, which would suggest that point *L* rests on point *Z*. But if small circle arc *GL* is twice the inclination of point *E* from the equator, and point *T* bisects arc *LN*, then small circle arc *ZT* cannot be exactly equal to small circle arc *GL* because small circle arc GT was a quadrant of small circle *LTNM*. Point *Z* would have to be several degrees closer to point *G* than to point *L* because twice the inclination of point *E* from the equator is greater than 45°. While the precise location of point *Z* is not essential to understanding the operation of this version of the model, the difficulty in determining point *Z*'s location might be a reason why Ibn Naḥmias rejected this version of the double circles. In addition, any version of the double-circle hypothesis would seem to contradict the second opinion (cf. §B.1.II.21) in which apparently opposite motions were forbidden.

The measure of the arc from the circle of the path

This determination of the radius of the small circles for this final version (the presentation of which began in §B.1.II.21) of the double-circle hypothesis was confusing due to the fact that the double circles would have to be rather large. Ibn Naḥmias first explained that the radii of the small circles should be equal to θ less small circle arc *TA*, expressed in terms of degrees of revolution about point *D*. He then wrote that the radius of each of the small circles needs to be half of twice the distance from point *E* or point *D* to the ecliptic less the maximum deviation, 0.5*(2*TD*– *AK*)). By maximum deviation, Ibn Naḥmias must have meant the maximum deviation from the ecliptic in his original model for motion in anomaly, which occurred at the equinoxes. Using arcs *TD* and *AK* in the same calculation was confusing, though, because while *TD* was part of a great circle arc, *AK* was an arc on *LTNM*, a small circle. This error could be rectified by changing small circle arc *AK* to angle *KDA*. More important, though, since *LTNM* was not a great circle, *LTNM* could not both bisect one of the double circles and pass through the center of the other, which would be necessary if the double circles moved on the circumference of small circle *LTNM* with the motions Ibn Naḥmias prescribed.

67v

I think the first hypothesis is more appropriate B.1.II.26

Ibn Naḥmias's preference for the first hypothesis means that he has preferred using the two circles that might appear to move in opposite directions in order to convey the center of the Sun in the same direction of the circle of the path of the center, and without points *L* and *N* retaining their positions on great circle *SZ*. He has also remarked that were it not for how "the time period from the greatest lag to its mean is less than the time period from its mean to its smallest" it would have been possible to account for the solar anomaly with simply the double-circle hypothesis moving on the ecliptic. In graph 5.2, by comparing the true positions for 90° and 93° of mean motion, we can see that the Ptolemaic model and the solar model with just the double-circle hypothesis predict, as Ibn Naḥmias has said, slightly different locations for the point of maximum anomaly. In the graph, the sum of the radii of the double circles was 2;23°, the maximum Ptolemaic solar equation.

(See figure 5.6.) Ibn Naḥmias has now said that the solar model with a double-circle hypothesis *and* a circle of the path of the center will do more to reproduce the asymmetry of the Ptolemaic model. We will now assess Ibn Naḥmias's claim.

The effect of the double-circle hypothesis on longitude is equal to angle *APX* because *X* is the final location of the Sun with a double-circle hypothesis, and *A* would be the location of the Sun on the circle of the path of the center of the Sun without the double-circle hypothesis. In triangle *PAZ*, *a* (side *PZ*) is 90°, *p* (side *AZ*) is 2;30° (the radius of the circle of the path of the center, and *z* (side *PA*) is Φ(A). Angle *Z* is θ. In the following equations, *r* is the radius of each of the double circles. With the law of sines:

$\sin A = \sin\theta/\sin(\Phi(A))$

$A = \arcsin(\sin\theta/\sin(\Phi(A)))$

Since arc *NA* is defined as orthogonal to arc *CAH*, angle *PAN* = $90° + A = 270°-A$

$PAN = 270°-\arcsin(\sin\theta/\sin(\Phi(A)))$ (1)

Angle $PAX = PAN-NAX = 270°-\arcsin(\sin\theta/\sin(\Phi(A)))-NAX$

As *NAX* is computed in equation (2) in §B.1.II.20, $PAX = 270°-\arcsin(\sin\theta/\sin(\Phi(A))) -\sin^{-1}(\sin r \sin 2\theta/\sin AX)+(90°-\theta)$ (2)

In triangle *PAX*, *x* (side *PA*) is Φ(*A*) and *p* (side *AX*) is equation (1) in §B.1.II.20. Angle *A* = angle *PAX* (which is computed in equation [2] above). The radius of each of the small circles is 0;3,5°. By applying the cosine rule for sides, we have:

$\cos a = \cos x \cos p + \sin p \sin x \cos A$

$\cos a = (\cos^2 r + \sin^2 r \cos 2\theta) \cos(\Phi(A)) + \sin p \sin(\Phi(A)) \cos[270°- \arcsin(\sin\theta/\sin(\Phi(A)))-\sin^{-1}(\sin r \sin 2\theta/\sin(\cos^{-1}(\cos^2 r + \sin^2 r \cos 2\theta)))+(90°-\theta)]$

$a = \arccos((\cos^2 r + \sin^2 r \cos 2\theta) \cos(\Phi(A)) + \sin p \sin(\Phi(A)) \cos[270°-\arcsin(\sin\theta/\sin(\Phi(A)))-\sin^{-1}(\sin r \sin 2\theta/\sin(\cos^{-1}(\cos^2 r + \sin^2 r \cos 2\theta)))+(90°- \theta)])$

To determine angle *P*, we apply the law of sines:

$\sin p / \sin P = \sin a / \sin A$

The results of the calculations showing the effect of the double-circle hypothesis (with radii of 0;3,5°) on the anomaly are in graph 5.3. The computations in the graph show that angle *APX* is greatest at 45° and 135° of θ, meaning that while the double-circle hypothesis and the circle of the path of the center introduce distinct variations into the longitude, the variations are symmetrical. Thus, the addition of the double-circle hypothesis to the circle of the path of the center of the Sun does not help mirror the asymmetries of the Ptolemaic solar model. Moreover, there remains the structural problem that the period of the variations imparted by the double circles must be calibrated with the displacement from the ecliptic occasioned by the circle of the path of the center. The intersection between the ecliptic and the inclined circle carrying the circle of the path of the center could be allowed to vary. See also the comments on §B.1.III.2.

By no means should those hypotheses be sullied by the impossible B.1.II.27
Because Ibn Naḥmias's hypotheses do a better job of saving the phenomena than Biṭrūjī's, while also remaining in accord with Aristotelian philosophy, Ibn Naḥmias saw himself as having responded to the challenge of Maimonides's *Guide* II, 24.[155] More important, he claimed that all of the motions are established upon concentric spherical orbs. The tendency of the Judeo-Arabic version not to specify the physical movers begins to address the question of whether Ibn Naḥmias was interested in rediscovering Eudoxus's lost models, or whether his more proximate source was Ibn al-Zarqāl's models for trepidation which featured a pole rotating about the orb's pole that moved a point at the orb's equator.

Certainly, the Judeo-Arabic original of *The Light of the World* did mention physical movers at times; here is a quotation from the section on the five planets: "If we need our thesis for one planet to be many orbs, then each of them is according to the aforementioned description, with one being inside the other moving through its motion, and moving also with a motion particular to it about its poles."[156] But if Ibn Naḥmias's initial motivation had been primarily Eudoxan, the complexes of nesting orbs might have been laid out in the Judeo-Arabic original. Also, Ibn Naḥmias's models are different from most reconstructions of Eudoxus's models in that Ibn Naḥmias's models depend heavily on having poles separated by around 90°, whereas the separation between the poles in most

155. See Langermann, "My Truest Perplexities," p. 313.
156. *The Light of the World*, §B.4.I.3.

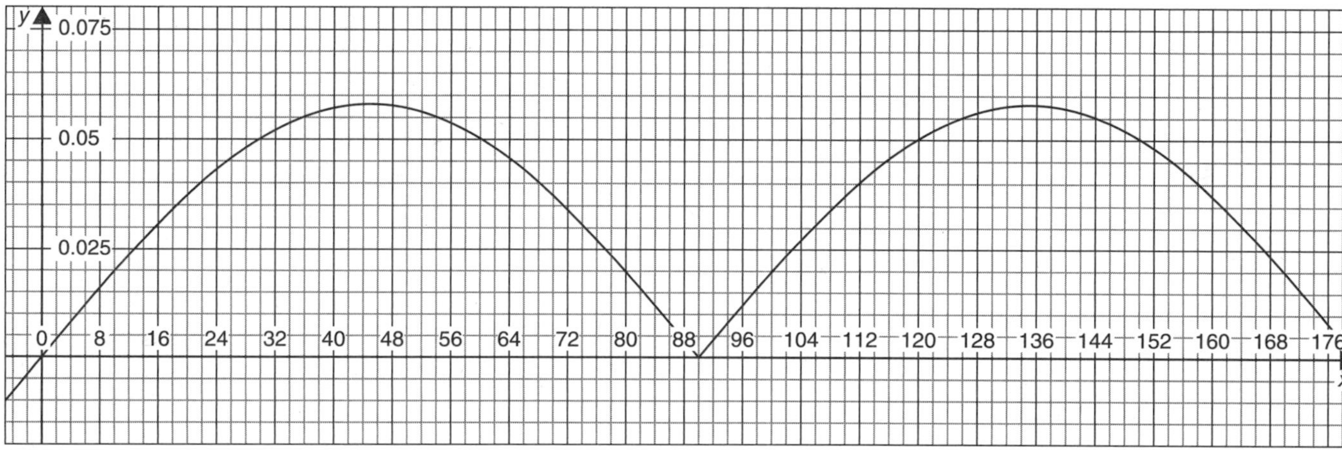
y
0.075
0.05
0.025
0
8
16
24
32
40
48
56
64
72
80
88
96
104
112
120
128
136
144
152
160
168
176
x

GRAPH 5.3

reconstructions of Eudoxus's models is generally less than 45°.[157] The introduction of Eudoxan elements in the Hebrew recension could be understood as stemming from a desire to propose physical movers for components of models derived from Ibn al-Zarqāl. But *The Light of the World* was a novel attempt to transpose aspects of Ptolemaic and post-Ptolemaic astronomy, for example, the epicycle and the Ṭūsī couple, onto the surface of an orb. In contrast to Ibn Rushd's position that there was a lost Aristotelian astronomy that needed to be recovered, Ibn Naḥmias followed Maimonides's position that it was possible to devise a homocentric astronomy endowed with predictive accuracy.

68r

As for al-Biṭrūjī B.1.II.28

(See figure 5.9.) Goldstein's study of Biṭrūjī showed that Biṭrūjī's solar model failed to keep the Sun in the ecliptic except at the solstices and equinoxes.[158] In this passage, we find some of Ibn Naḥmias's criticisms of Biṭrūjī's solar model. First, Ibn Naḥmias has noted that in Biṭrūjī's model the poles of the ecliptic are not the same as the poles of the circle of the Sun's declination, meaning that the circle of the Sun's declination was not coincident with the ecliptic. Thus, the Sun could not always be observed in the ecliptic. Second, Biṭrūjī's models would cause the Sun to traverse unequal arcs of the ecliptic in equal times only from the perspective of an observer at the north pole of the orb.[159] While the Sun might lag through the quadrants of the ecliptic in unequal times, there was nothing that made the Sun traverse equal portions from its inclined circle in unequal times. The only way the Sun might traverse unequal portions of the ecliptic in equal times would be by having the Sun depart from the ecliptic. Having the observer be slightly eccentric to the north pole of the orb, as Biṭrūjī theorized, would seem to be a significant step toward an eccentric, as the distances from the observer to all of the Sun's positions would not be equal. But Biṭrūjī has proposed that the pole of the circle of the path of the pole of the circle of the Sun is not coincident with the pole of the equator. As the radius of the path of the pole of the circle of the Sun must be 24° (the obliquity of the ecliptic), there would be times when the inclination of the circle of the Sun to the equator would not equal the obliquity of the ecliptic.

157. See Ido Yavetz: "On the Homocentric Spheres of Eudoxus," in *Archive for History of Exact Sciences* LII (1998): pp. 221–78; and Goldstein, "On the Theory of Trepidation."

158. Biṭrūjī (trans. and comm. Goldstein), *On the Principles*, vol. 1, pp. 10–14. The deviation could be over 1° at times.

159. Morrison, "The Solar Model," p. 105. See also Biṭrūjī (trans. and comm. Goldstein), *On the Principles*, vol. 1, p. 13: "[T]he north pole of the equator plays the role of the observer in the Ptolemaic model."

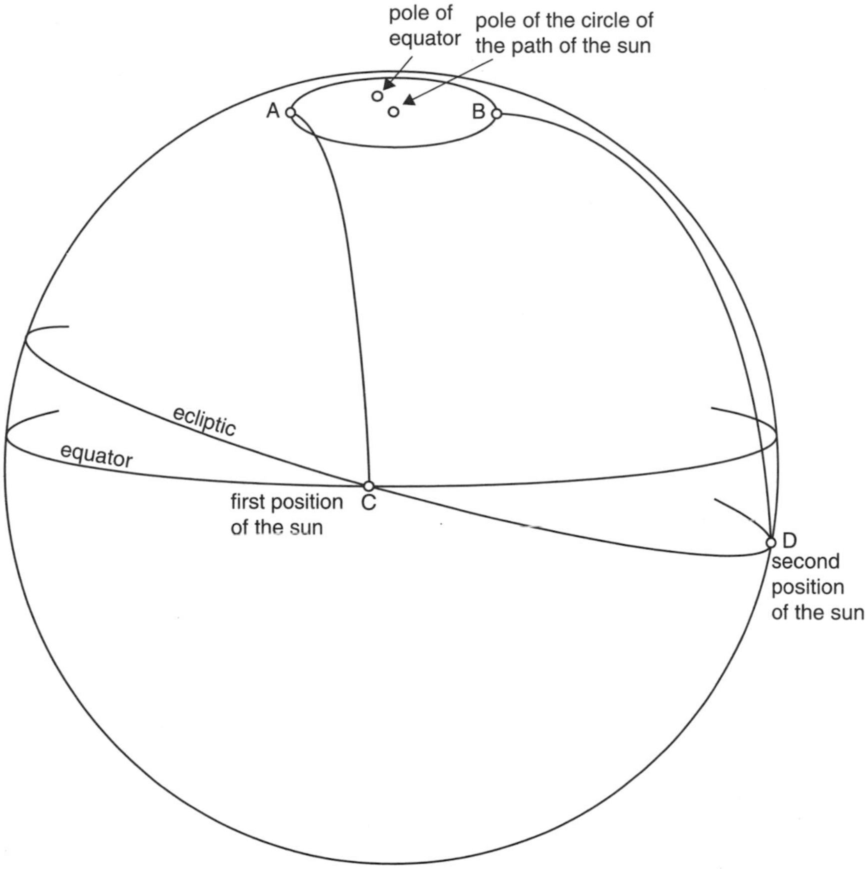

FIGURE 5.9

the times of the equinoxes which he observed with the brass ring B.1.II.29
This is a reference to the meridional armillary described in *Almagest* I.12.[160]

68v

The third chapter, about showing the measure of the greatest visible solar anomaly B.1.III.1
(See figure 5.10.) The number of days per season and the degrees traversed mirror the figures given in *Almagest* III.4, which presumed the eccentric hypothesis.[161]

160. Toomer, *Ptolemy's "Almagest,"* pp. 61–2.
161. See Toomer, *Ptolemy's "Almagest,"* pp. 154–5. See the summary of the *Almagest* in Vatican

69r

and it moves to the summer solstice through arc AB B.1.III.2

Here Ibn Naḥmias has defined the lag as the extent to which the Sun exceeds its mean motion, not the Sun's total motion per day. Though not identified as such, line *AG* must represent the ecliptic because the ecliptic and the Sun intersect at and only at the solstices. Point *D* is the center of the ecliptic.

Biṭrūjī's method of calculating the maximum solar anomaly had differed from Ptolemy's.[162] Ibn Naḥmias's comment that arc *NLK* was the portion common to the circle of the path of the center and the plane of the deferent circle indicated that he was trying to transpose Ptolemy's method to the surface of an orb, and not just repeating Ptolemy's calculations that yielded the conclusion that the solar eccentricity was 2;30°. In light of Ibn Naḥmias's interest in harmonizing *The Light of the World* with the *Almagest* as much as possible, integrating this section of the *Almagest* into a homocentric system must have bedeviled Ibn Naḥmias, as Ptolemy's own determination of the Sun's maximum anomaly was connected to the determination of the solar eccentricity, something the existence of which Ibn Naḥmias and Biṭrūjī had, after all, rejected. In *The Light of the World,* the measure of the Sun's anomaly, and therefore the location of the point of the greatest anomaly on the ecliptic, seemed constrained by the role of the circle of the path of the center in keeping the Sun in the ecliptic. In order for Naḥmias's model to locate the points of the Sun's greatest motion where those points were observed to be, the intersection of the inclined circle carrying the circle of the path of the center with the ecliptic would have to be somewhere other than the equinoxes. Only the Hebrew recension broached that point.

In his analysis, Ibn Naḥmias tried to use the measures of arcs on the circle of the path of the Sun, as seen from point *D,* to determine the measure of the radius of the circle of the path of the Sun. All of the observations that Ptolemy mentioned were, after all, in a single plane, but Ibn Naḥmias's circle of the path of the center was perpendicular to the plane of the circle carrying the circle of the center of the Sun in a homocentric model. Thus, contradictions emerged. For example, Ibn Naḥmias stated, "Therefore, the ratio of *DA* to *AM* is given," with point *M* being the location of point *B* on the ecliptic. Since Ibn Naḥmias had to have been discussing a small circle on the surface of the orb, then the ratio of *DA* to *AM*

MS Ebr., fol. 3r–3v. There one finds the same values as in Toomer, *Ptolemy's "Almagest."* See also Anatoli's translation (*Almagesti*, Paris BNF MS Hébreu 1018), pp. 66–7.

162. Biṭrūjī (trans. and comm. Goldstein), *On the Principles,* vol. 1, pp. 135–40. For Biṭrūjī, the Sun's eccentricity was produced by the location of the pole of the circle of the path of the pole of the Sun relative to the pole of the celestial equator. Thus, Biṭrūjī transferred the Ptolemaic eccentricity to the vicinity of the orb's pole.

would always be the same. And if segments such as arc *AH* were actually chords of the small circle, then point *M* could not be the point that corresponds to the position of point *B* on the ecliptic. Otherwise, for such statements to have been truly meaningful, Ibn Naḥmias would have had to imply that the circle of the path of the center had become tantamount to a Ptolemaic epicycle, causing the distance of the Sun from the observer to vary. In that case, Ibn Naḥmias's determination of the maximum solar anomaly would be confused in a different respect.

So because the position of point B on the ecliptic is point M B.1.III.3
This statement, if it corresponds with the figures in the MSS, implies that the path of the center of the circle of the path of the Sun and the ecliptic intersect at the vernal equinox but not at the autumnal equinox.

In addition, there are a number of places in this passage where Ibn Naḥmias, in his attempt to follow Ptolemy's computations from the *Almagest*, has confused small circles and great circles. For example (see figure 5.10), he stated that angles *BEA* and *BEG* are given. While arc *AM* can be known, as arc *AMH* is potentially a great circle arc, angles *BEA* and *BEG* are angles at the circumference of small circle *ABG*, meaning that their arcs are incommensurable with great circle arcs. Another problem was how both arcs *AE* and *BE* could be the diameter of the small circle. The source of this confusion may be that, in III.4 of the *Almagest*, *AE*, *BE*, and *GE* are all radii of the concentric.

69v

That the radius of the circle of the path B.1.III.4
If the radius of the circle of the Sun is 60, then 1⁄24 of its radius, the radius of the circle of the path, is 2;30°.[163] Biṭrūjī had computed the equivalent arc to be 2;22,8° which was the maximum solar equation in the Ptolemaic system.[164]

As for how we know the degree of the ecliptic in which the Sun is B.1.III.5
(See figure 5.10.) This is the point where the Sun's motion is the least, and that point was, according to Ptolemy, 5;30° of Gemini. Ibn Naḥmias simply stated, with no further explanation, that this point (*W*) was 5° from the head of Gemini.[165] Because the apogee, in the Ptolemaic system, was also the point of the Sun's greatest distance from the observer, the definition of the apogee would have to be (but

163. This is the Ptolemaic parameter for the solar eccentricity and for the radius of the solar epicycle. See Toomer, *Ptolemy's "Almagest,"* pp. 156–7.

164. Biṭrūjī (trans. and comm. Goldstein), *On the Principles*, vol. 1, p. 138. The Anatoli translation of the *Almagest* (Paris BNF Hébreu 1018) computed the maximum solar equation to be 2;23°.

165. For Ptolemy's calculations and conclusion, see Toomer, *Ptolemy's "Almagest,"* III.7.

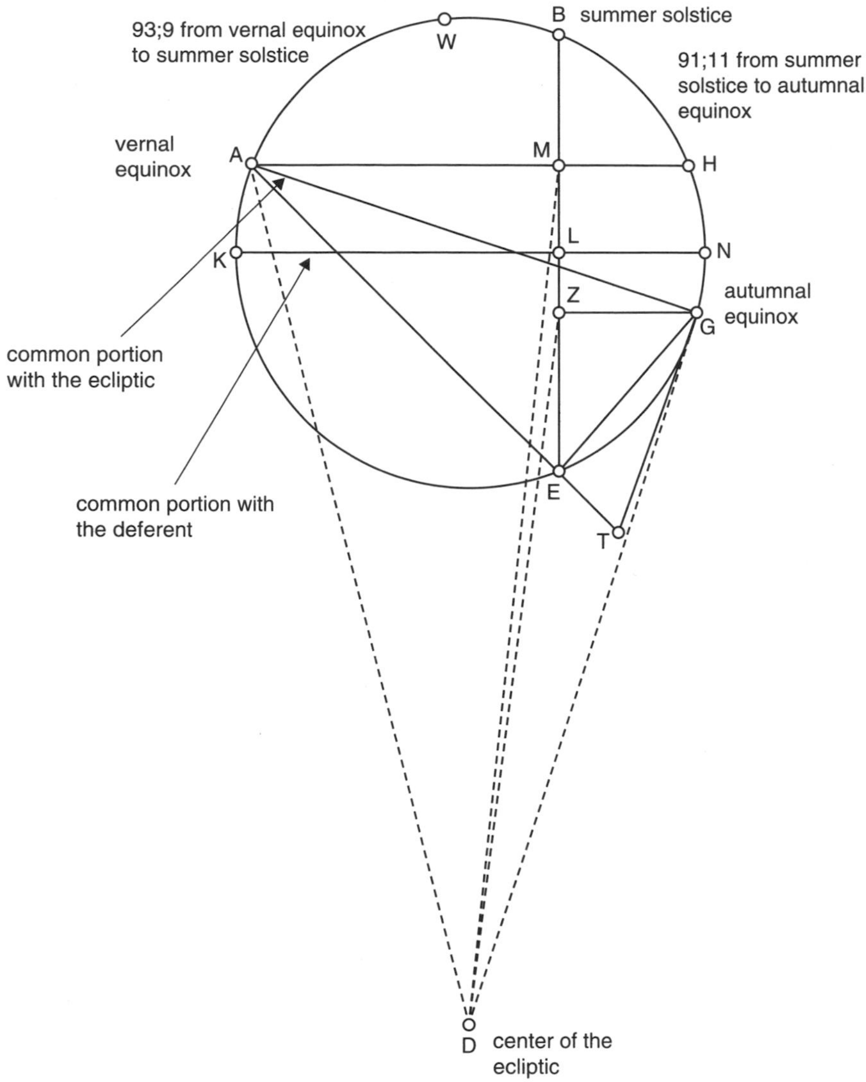

FIGURE 5.10

was not) reconsidered in a homocentric system. Moreover, in Ibn Naḥmias's solar model, for Gemini 5;30° to be the point of the least motion, Ibn Naḥmias would have to allow the intersection of the inclined circle carrying the circle of the path of the center with the ecliptic not to be at the equinoxes, a suggestion that only the Hebrew recension made.

We present for that the lemma that we proved in the first treatise B.1.III.6
Ibn Naḥmias had set forth this lemma originally in §A.II.5, and it was a restatement of a lemma found in the *Almagest,* though in terms of chords, not sines.[166] The problem was that Ibn Naḥmias applied the result of the lemma, a ratio between arcs on a plane circle, to a combination of small circle and great circle arcs. Measurements of small circle arcs cannot be used to compute radius *GZ*, part of a great circle arc.

70r

If we repeat the derivation of the ratio B.1.III.8
Ibn Naḥmias never specified the ratio he actually extracted, suggesting that he never actually repeated this calculation.

In order to explain a way to calculate the individual anomalies B.1.III.9
(See figure 5.11.) The individual anomalies are the anomalies calculated for any given point in the Sun's mean motion. Ibn Naḥmias has stated that the deferent circle of the path of the center of the Sun is tangent to the ecliptic. Since the deferent circle and the ecliptic are great circles that do not intersect at a 90° angle, Ibn Naḥmias must have meant to say that the circle of the path of the center was tangent to the ecliptic.

Although Ibn Naḥmias did not identify point *D* in this figure, it would have to be a point from which the Sun's motion was observed, which in a homocentric system would be the center of the cosmos. Angle *HZG* occurs at the center of the circle of the path of the center of the Sun, a circle responsible for the Sun's motion in anomaly. Therefore, by angle *HZG,* Ibn Naḥmias must have meant the uniform motion about the circle of the path of the center of the Sun. But, again, Ibn Naḥmias has confused small circles and great circles when he said that arc *HG,* a small circle arc, was given. There was no way to equate small circle arcs with great circle arcs. In addition, to determine the true motion, one must know arc *HT,* but *HT* is a small circle arc.

Ibn Naḥmias's exposition has also mixed up spherical and plane geometry. He wrote: "And let us propose that the center of the Sun is on point *H*, provided that its distance from point *G*, I mean the point of greatest lag, is given. We extend from point *H* perpendicular *HT* upon line *AG*, and we connect *DH* and *DT*. Because arc *HG*, rather angle *HZG*, I mean the angle of the mean motion is given, the ratio of *ZH* to *HT* is given."[167] Ibn Naḥmias did not acknowledge that triangles *BZD* and *HZD* are identical. Both triangles share *DZ*, and since points *B* and *H* are both points on an orb with its center at point *D*, arc *BD* = arc *HD*.

166. Toomer, *Ptolemy's "Almagest,"* pp. 65–6.
167. *The Light of the World*, §B.1.III.9.

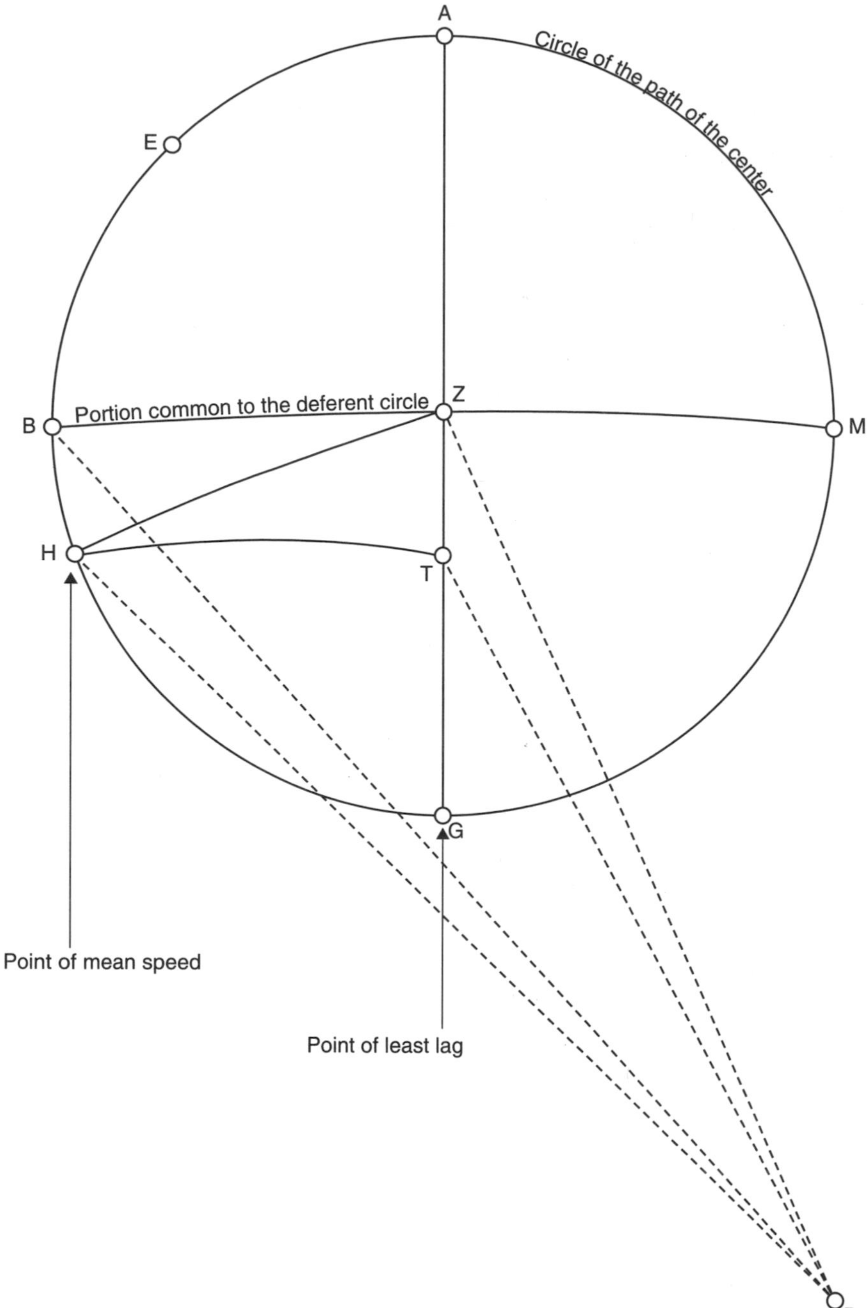
A
Circle of the path of the center
E
Z
Portion common to the deferent circle
B
M
H
T
G
Point of mean speed
Point of least lag
D

FIGURE 5.11

Finally, arcs *HZ* and *BZ* are both radii of the (small) circle of the path of the center of the Sun.

These attempts at calculations, though unsuccessful, constitute more formal evidence for Ibn Naḥmias's desire to reproduce the predictive accuracy of the *Almagest* with homocentric astronomy.

71r

this movement resembles the movement of the Sun B.2.I.1

Here Ibn Naḥmias has said, following Biṭrūjī, that the difference in the lags (or direct motion) does vary with the distance from the prime mover.[168] But once a complex of orbs became necessary to account for a planet's motions, how and why each orb of that complex was or was not affected differently by a desire to attain the uppermost orb was a question that Ibn Naḥmias did not attempt to answer.

it was found to complete the return in latitude in a period of time less than that

Ibn Naḥmias has explained how Ptolemy calculated that the Moon's motion in longitude differed slightly from the Moon's motion in latitude. Biṭrūjī, despite including a quote from the *Almagest* that distinguished the period of return in longitude and that in latitude, necessitating the motion of the nodes, wrote shortly thereafter, "[B]ecause of this, [the ancients] allowed its return in latitude to be somewhat different from its return in longitude, though in truth they are one and the same."[169] Goldstein noted that, while Biṭrūjī evinced some awareness of Ptolemy's lunar model, he disregarded much of it.[170] Ibn Naḥmias's lunar model took note of these different parameters (13;10° in longitude versus 13;13° in latitude) and attempted to incorporate (see §B.2.I.7) the 3' daily motion of the nodes of the lunar path that accounts for the difference between two parameters. Ibn Naḥmias did not explain how or why the orb or orbs composing the lunar model could move with these slightly different in parameters.

(See figures 5.12 and 5.13 [a three-dimensional rendering of 5.12].) [171] In addition, there was a problem with Biṭrūjī's lunar model that Ibn Naḥmias did not mention. In both figures, point *S* is the center of the path of the pole of the Sun, and point *C* is the center of the uppermost orb and the pole of the celestial equa-

168. Biṭrūjī (trans. and comm. Goldstein), *On the Principles*, vol. 1, p. 63. "The motion of the sphere closest to (the sphere of daily motion) is most similar to its motion in swiftness and power, whereas that which is furthest is slowest and least powerful."

169. Biṭrūjī (trans. and comm. Goldstein), *On the Principles*, vol. 1, pp. 143–4.

170. Biṭrūjī (trans. and comm. Goldstein), *On the Principles*, vol. 1, p. 37.

171. Figure 6.13 is an adaptation of figure 6.12, the figure found in Biṭrūjī (trans. and comm. Goldstein), *On the Principles*, vol. 1, p. 147.

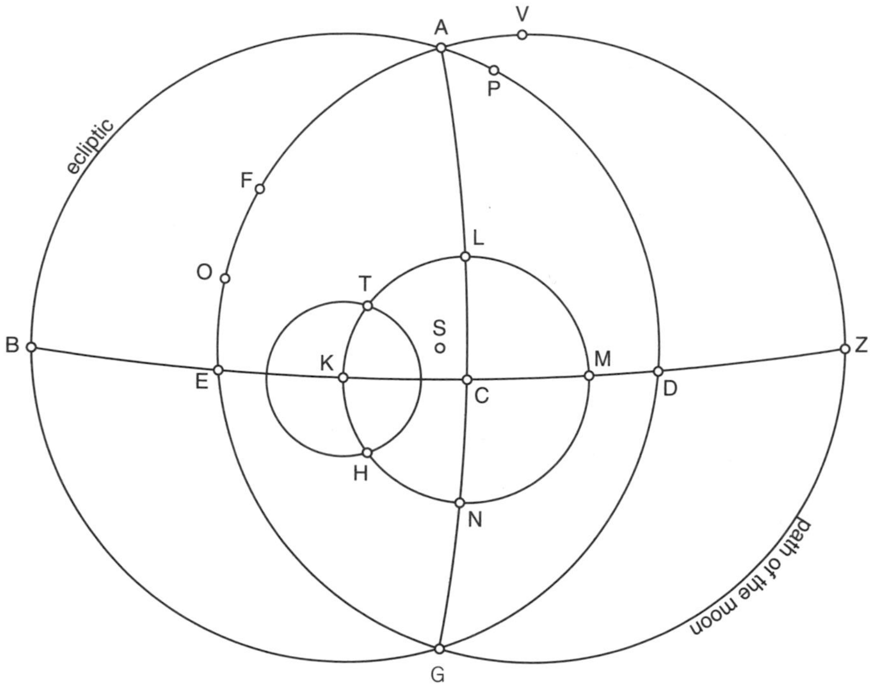

FIGURE 5.12

tor. Even in the starting position, the pole of the Moon at point *T* was not necessarily 90° from the Moon at point *V*. Biṭrūjī's own statement of the dimension of the model had confused small circles and great circles ("its [the Moon's] distance from point A is equal to the distance between points T and K")[172] because while arc *VA* was a great circle arc, arc *TK* was not.[173] Thus, one cannot show that arc *TV* is 90°.

Interestingly, Biṭrūjī's lunar model seemed to entail, without so acknowledging, apparently opposite motions. The motion in longitude was in the direction of the signs, but the motion in anomaly would appear to be sometimes in the opposite direction.[174]

172. Biṭrūjī (trans. and comm. Goldstein), *On the Principles,* vol. 1, p. 148.
173. Cf. ibid., vol. 1, pp. 147–8 and pp. 37–8.
174. Biṭrūjī (trans. and comm. Goldstein), *On the Principles,* p. 38.

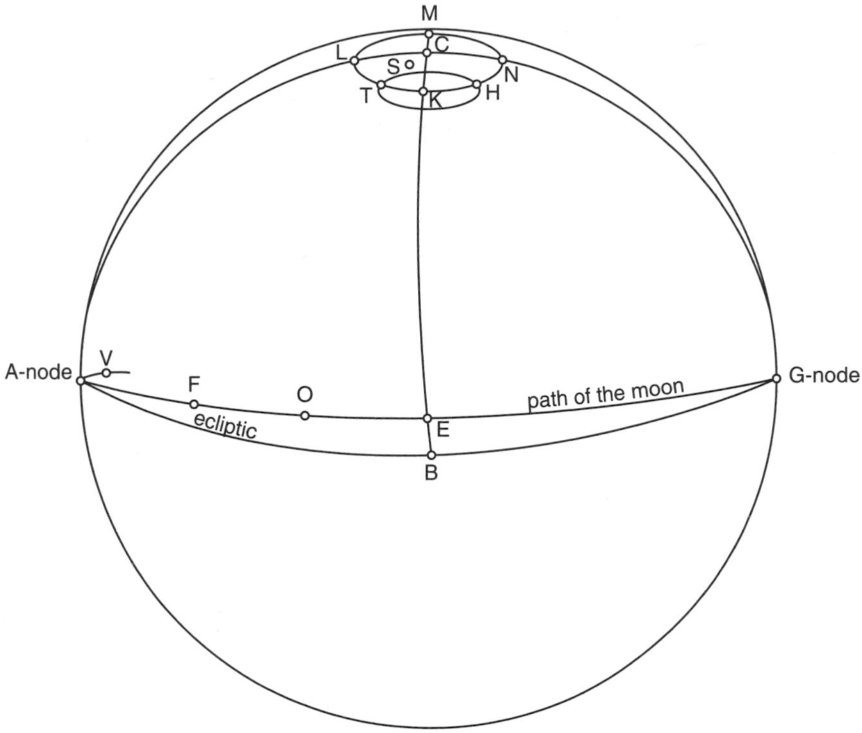

FIGURE 5.13

there are therefore two motions in latitude B.2.I.2

Eclipses occur only when the Moon is at the nodes, where the Moon's circle of latitude intersects the ecliptic. Since the eclipses have been observed at many points on the ecliptic, then the nodes must be moving. So, one of the two motions in latitude is the motion of the nodes, which was where the Moon's path intersected the ecliptic, and the other motion in latitude is what Ibn Naḥmias called the motion of the Moon on the circle of its latitude. Ibn Naḥmias defined that motion as a motion in latitude because the Moon's path took it to the north and to the south of the ecliptic. The nodes are moving backward (0;3° per day) against the background of the zodiac; hence the distinct parameters of the mean daily motion (13;10°) and the mean daily motion with respect to the nodes (13;13°).[175] Even though Ibn Naḥmias (§B.2.I.4) acknowledged that the Moon's mean daily

175. Ptolemy found the motion of the node by subtracting the motion in longitude from the

motion was to the north and south of the ecliptic, he noted that the relatively small 5° inclination of the circle of the Moon's latitude to the ecliptic meant that he would, following Ptolemy, compute the Moon's mean daily motion on the circle of its latitude with respect to the ecliptic.

71v

He found the daily mean motion in longitude to be 13;10,35° approximately B.2.1.4

All of the parameters in this section are rounded from the *Almagest*,[176] and they have fewer significant digits than the values given by Biṭrūjī.[177]

72r

As for according to the second opinion B.2.1.5

In the first opinion, in which the motions that combine to produce the Moon's motions can appear to be in apparently opposite directions, the inclined circle and the ecliptic are imagined on the starred (*mukawkab*) orb. Those orbs would revolve about their own poles with their own motions. According to the second opinion, in which apparently opposite motions were proscribed and in which celestial motions are explained according to lag, these circles are imagined in the orb of the Moon itself. This might have been because, since the motion of the nodes and the motion of the signs could not be the same, the starred orb itself could not move with two apparently opposite directions. As was the case with the solar mean motion, the second opinion (or hypothesis) necessitates a more complex explanation.

73r

it is necessary that the pole of the circle of the latitude B.2.1.9

(See figure 5.14.) This separation of points *R* and *K* is necessary so that the final lag of the pole of the circle of the Moon's latitude is greater than the lag of the pole of the ecliptic by the amount that is the difference of the positions of points *R* and *K*, equal to the motion of the nodes.

After that, the final lag in longitude remains the aforementioned amount B.2.1.10

While Ibn Naḥmias believed that celestial motions could be in apparently opposite directions, he continued to pay attention to the possibility of explaining things through lag (*taqṣīr*). As was the case with the hypothesis for the solar

---

motion in longitude with respect to the node (Neugebauer, *History of Ancient Mathematical Astronomy*, p. 70).

176. Toomer, *Ptolemy's "Almagest,"* p. 179. See also Anatoli, *Almagesṭi*, Paris BNF MS Hébreu 1018, pp. 82–3.

177. Biṭrūjī (trans. and comm. Goldstein), *On the Principles*, vol. 1, p. 145.

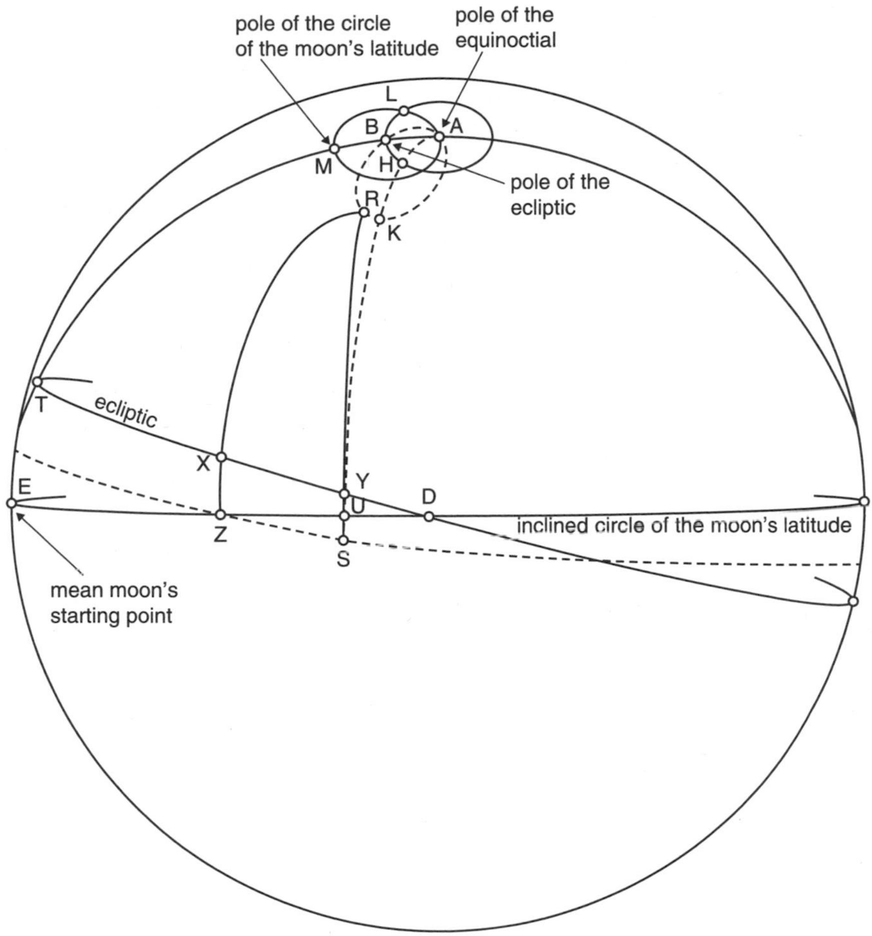

FIGURE 5.14

mean motion, the predictive accuracy of the Moon's mean motion according to the second opinion or hypothesis would, again, not be perfect because the motion in the direction of the signs of the zodiac would vary. In the figure, arc *ZS* from the inclined circle of the Moon's latitude (and with respect to the ecliptic) is supposed to be 13;10°, not 13;13°. The hypothesis accomplished that by saying that the pole of the circle of latitude was not on point *K* but was rather on point *R*, 3' away.

To assess that claim, we solve spherical triangle *HRZ*. *z* (side *HR*) is 5°, *h* (side

RZ) is 90°, and *R* (angle *HRZ*) is 166;50° (i.e., 180°–13;10°). We apply the cosine rule for sides:

$$\cos r = \cos h \cos z + \sin h \sin z \cos R$$
$$\cos r = \sin h \sin z \cos R = \sin 5° \cos 166;50°$$
$$r = 94;53°$$

Now, to find the angle *RHZ* at the pole of the ecliptic, we apply the law of sines in which

$$\sin H / \sin h = \sin R / \sin r$$
$$H = \text{angle } RHZ = 13;13°$$

That calculation shows that the motion from *S* to *Z* requires a motion of 13;10° then *RHZ* + *RHK* = 13;16° about pole *H*. If *RHZ* = 13;13° then *RHZ* + *RHK* = 13;16° which is not a parameter. Generally speaking, then, the model for the Moon's mean motion suffers from the same drawback as the model for the Sun's motion, in that the motion of the pole of the inclined circle of the Moon's latitude in the direction of the motion of the universe would not remain the same. The model for the Moon's mean motion according to the hypothesis of lag is more accurate than the model for the solar mean motion according to the hypothesis of lag because the distance between the poles of the ecliptic and the Moon's circle of latitude is far less than the distance between the poles of the equinoctial and the ecliptic. Any inaccuracy incurred with the solar model would be perpetuated in eclipse predictions.

This hypothesis for the mean lunar motion did not explain how the motion of the pole of the circle of the Moon's latitude remained only 3', that is, the daily motion of the nodes, less than the daily motion of the pole of the ecliptic, even though the lag of the ecliptic in §B.2.I.6 is defined as *twice* its daily motion.

74r

Let this circle be circle KLNO B.2.II.3

(See figure 5.15.) Aside from the parameters, and the fact that, as in the Ptolemaic system, the motion of the lunar epicycle is in the opposite direction of the Moon's mean motion, this hypothesis for the first lunar anomaly is similar to the hypothesis of the circle of the path of the center in the solar model. Ibn Nahmias theorized that the Moon revolves upon on a small circle *KLNO* whose center, point Q, moved on a circle inclined to the circle of latitude.

74v

That is because the degrees of the latitude are more than the degrees of the anomaly . . . The southern extremity that was on point Z *moved by this amount and came to be on point* P B.2.II.5

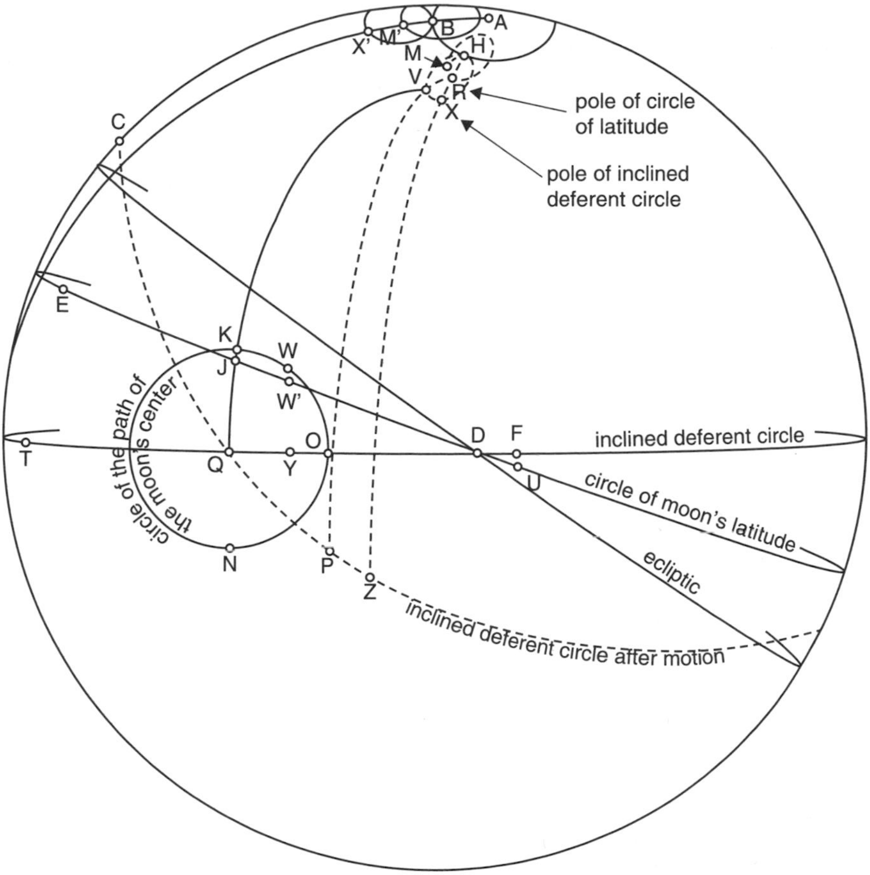

FIGURE 5.15

Ibn Naḥmias's model for the first lunar anomaly attempted to account not just for the 3' motion of the nodes but also for the difference between the motion in anomaly (13;3,54°) and the motion in latitude (13;13,46°). Thus, his model imparted to the pole of the lunar deferent circle and its southern extremity an additional motion on the measure of this difference, which is 10'. The commentary will describe sequentially the components of the Moon's daily motion, but they, in fact, occur simultaneously. Ibn Naḥmias began according to the second opinion (or hypothesis), according to which apparently opposite celestial motions were proscribed. In one day, the pole of the ecliptic goes from point *B* to point *H* with a lag of twice the Moon's daily motion in longitude (13;10°), moving the pole of the circle of the Moon's latitude to point *M*. As well, the pole of the circle of the Moon's latitude lags from point *M* to point *R* an additional 3' to account for the motion of the nodes. It is worth noting both that the daily motion of the nodes was not doubled, although the motion in longitude was, and that the motion of the nodes is in the opposite of its direction in the model for the lunar mean motion.[178] Through the motion (or lag, as the text reads "in reverse") from point *M* to point *R*, the pole of the Moon's inclined deferent circle moves to point *X*, and the pole of the circle of the path of the Moon's center arrives at point *Z*. Then, the pole at *X* moves to point *V* in the direction of the motion of the cosmos, revolving about pole *R* with a motion of 10', the surplus of the motion in latitude over that in anomaly. That moves the pole of the circle of the path of the center from point *Z* to point *P*. Finally, the pole of the circle of the path of the center moves 13;3,54° from point *P* to point *Q* (§B.2.II.7: "with a motion equal to the anomaly, and let it be arc PQ"), the mean motion in anomaly, along with the Moon's motion in anomaly on the circle of the path of the center. At first, I do not take into account the possibility that the alignment of the diameter of the circle of the path of the Moon's center might slant, as Ibn Naḥmias did not believe (cf. §B.2.II.8–9: "So what follows from this thesis is like that which follows from our other thesis") that his model entailed a slant.

Ibn Naḥmias stated that the rotation of point *K* to point *W* about pole *Q* was 13;3,54°; did that rotation place point *W* where it was observed at the intersection between the circle of latitude and the circle of the path of the Moon's center? The following computation of the predictive accuracy of the model for motion in anomaly accepts Ibn Naḥmias's presumption that the motion of the pole of the circle of the path of the Moon's center from points *Z* to *P* to *Q* will place pole *Q* back in circle *TD*. Also, this first computation reflects the figure on 74v of the MS in which the motion of point *K* to point *W* is in the *opposite* direction of the motion of the universe; the text wrote that point *K* "moves about the pole of the circle of the path in the direction of the motion of the universe."

First, we solve triangle *QYW*, with point *Y* being the point on the inclined deferent circle with the same Θ coordinate as point *W*. Ibn Naḥmias did not specify the radius of the circle of the path of the Moon's center, *QW*; I have adopted 5;13°, the parameter for the radius of the epicycle in *On the Principles*.[179] Arc *KQW* is the daily motion in anomaly, 13;3;54°, so *WQY* is 76;6,6°. *WYQ* is a right angle. The law of sines yields side *q* (side *YW*):

$$\sin 5;13°/\sin 90° = \sin q/\sin 76;6,6°$$
$$5;4,50° = q$$

178. Indeed, Ibn Naḥmias may have noticed that wrinkle as he remarked that the motion from *M* to *R* was "in reverse."

179. Biṭrūjī (trans. and comm. Goldstein), *On the Principles*, vol. 1, pp. 39–40. See also, Toomer, *Ptolemy's "Almagest,"* p. 202. Ptolemy gives a parameter of 5;14°. Later, Ibn Naḥmias wrote that the parameter was 5;15°. See *The Light of the World*, §B.2.III.1.

We can also determine w (side QY) with the cosine rule for sides:

$\cos y = \cos q \cos w + \sin q \sin w \cos Y$:
since $Y = 90°$, $\cos y = \cos q \cos w$
$\cos 5;13° = \cos 5;4{,}50° \cos w$
$w = 1;10°$

The next step is to compare side YW with side YW'. We consider triangle DJQ and, first, compute side j (arc DQ). Because Ptolemy defined the Moon's mean motion on its deferent as its motion on the ecliptic, q (side DJ) is the complement of the mean motion, 90–13;10°=76;50°. Angle Q is 90°, and angle D is the radius of the circle of the path of the Moon's center (5;13°). According to the law of sines:

$\sin d/\sin 5;13° = \sin 76;50°$
$d = 5;5°$

Then, we apply the cosine rule for sides:

$\cos q = \cos d \cos j + \sin d \sin j \cos Q$
$\cos 76;50° = \cos j \cos 5;5°$
$j = \text{arc } DQ = 76;47°$

And since arc QY= 1;10°, then arc $DY = 75;37°$

The final step is to compute side YW' in right triangle DYW' (YW' is on great circle arc TWW'Y, an arc that creates a right angle at Y). Angle D (angle W'DY) is 5;13°, the radius of the circle of the path of the Moon's center. We apply an identity from the four-parts formula:[180]

$\cos w' \cos Y = \sin w' \cot d - \sin Y \cot D$
because $Y = 90°$, $\cot 5;13° = \sin 75;37° \cot d$
$d = 5;3;15° = YW'$

Because YW=5;4,50°, the discrepancy after a day would be 0;1,35°.

Using the equation for the slant derived in the comments on §B.1.II.26/X.14, slant = arcsin(sin$r$ sinθ), we find that the slant after one day of mean motion is 1;10,48°. Thus, angle WQY, with the slant, is 74;55,18°. The law of sines yields side q (side YW):

$\sin 5;13°/\sin 90° = \sin q/\sin 74;55{,}18°$
$q = 5;2{,}12°$

Thus, allowing the diameter of the circle of the path of the Moon's center to slant would be a slight improvement as YW with the slant would be slightly closer to YW'.

180. Smart, *Text-Book*, p. 12.

To return to the model for anomaly without the slant, if we specify that the motion from point *K* to point *W* was in the direction of the motion of the cosmos, then arc *DY* would be arc *DQ* + arc *QY* = 77;59°; thus

$\cot 5;13° = \sin 78;23° \cot d$
$d = YW' = 5;6,31°$

In that case, the discrepancy after a day would be 0;1,41°.

To compute the maximum possible discrepancy, we determine where pole *Q* of the circle of the path of the center of the Moon will be when point *K* has revolved 90° about pole *Q*. If point *K* revolves 13;3;54° per day, then it takes 6;53,24 days for point *K* to revolve 90°. In 6;53,24 days, with 13;10° per day of mean motion, pole *Q* has traversed 90;42,36° of mean motion along the circle of latitude as seen from the ecliptic. The first task is to find out where that motion, measured along the circle of latitude, leaves pole *Q* on the inclined deferent circle. Let point *F* be the new position of pole *Q* on the inclined deferent circle, and point *U* be point *F*'s position on the Moon's circle of latitude as measured on the ecliptic. In triangle *DFU*, where angle *D* (angle *FDU*) = 5;13° and *f* (side *DU*) is 0;42,36°, we want to know the measure of *u* (side *DF*). With the law of sines:

$\sin d / \sin 5;13° = \sin 0;42,36°$
$d = 0;3,52°$

Then we apply the rule of cosines for sides:

$\cos f = \cos u \cos d + \sin u \sin d \cos F$
because $F = 90$, $\cos f = \cos u \cos d$
$\cos 0;42,36° = \cos u \cos 0;3,45°$
u = side DF = 0;42,26, meaning pole *Q* is at 90;42,26° of the inclined deferent circle

To determine the displacement of the Moon from the circle of its latitude after point *K* has revolved 90° about pole *Q* and after pole *Q* has revolved 90;42,26° on the inclined deferent circle, we redraw a right spherical triangle *DUF*. Angle *D* is 5;13°, angle *U* is 90°, and side *u* (side *DF*) is 5;13° + 0;42,36° = 5;55,36°. We want to find *d* (side *UF*) and use the law of sines:

$\sin d / \sin 5;13° = \sin 5;55,36° / \sin 90°$
$0;32,16° = d$

This is a noticeable error, on a far greater scale than any deviation incurred in the solar model. This magnitude of this maximum error does not depend on the direction of the motion of the circle of the path of the Moon's center.

75r

Still, we need to add here the concept of the two circles that we mentioned for the Sun that corresponds to this B.2.II.7

Here Ibn Naḥmias has proposed adding the double-circle hypothesis to his model for the first lunar anomaly, probably to keep the Moon in the circle of its observed inclination to the ecliptic of 5°.[181] This is another parallel between the models for the Sun and the Moon. Ibn Naḥmias did not, however, explain, either in the Judeo-Arabic text or in the Hebrew recension, precisely how the double-circle hypothesis would be integrated into the lunar model. The Hebrew recension alluded to the possibility of using the double-circle hypothesis to replace the circle of the path of the Moon's center and thus to account for the entire first lunar anomaly.[182]

We explain this matter according to the first opinion B.2.II.8

(See figure 2.21.) The description of the model that has been analyzed has been according to the second opinion (or hypothesis), in which all of the celestial motions have been explained in terms of lag. Here, Ibn Naḥmias attempts to explain the model according to the first opinion, which means that the motions appear to be in the opposite direction of the daily motion of the universe. I describe the motions sequentially, as does the text, although they should all occur simultaneously. First, the pole of the circle of the path of the Moon moves from point *T* to point *P* through the motion of *X* to *V* by 0;10° per day, the surplus of the motion in latitude over the motion in anomaly. Then, the nodes are said to move in the opposite direction by the same measure (0;10° per day), through the motion of pole *M* to point *R*, although I believe Ibn Naḥmias must have meant 0;3° per day. Then, finally, the Moon has moved on the circle of its path, with the pole of the circle of the path moving on the circle of latitude to *Q*, both by 13;3° per day. When one describes the Moon's motions according to the first opinion (or hypothesis), the diameter of the circle of the path of the center cannot slant.

75v

As for the difference that is between them, it is with regard to the increase or decrease in longitude that is due to the rising times as we showed in the first treatise B.2.II.10

As was the case with the Sun, the difference between the two hypotheses is with regard to rising times as the Moon's longitude was measured, for all practical

181. He also had referred to the double-circle hypothesis from the Sun in *The Light of the World*, §B.2.II.5: "Then, on account of the two deferent orbs for the two aforementioned circles with the Sun, the likes of which need to be our thesis for the Moon as we said previously." The "two deferent orbs" could be an allusion to a Eudoxan couple.

182. *Or ha-ʿolam*, §B2.V.6/X.

purposes, with respect to the ecliptic and not with respect to the circle of the Moon's inclination. For more, see the comments on §A.IV.3. It is interesting that Ibn Naḥmias did not investigate the effect of the slant of the diameter of the circle of the path of the center of the Moon that would occur with the second opinion (or hypothesis), in which apparently opposite motions are proscribed.

the ratio of the radius of the orb of the Moon to the radius of the circle of the path B.2.III.:
In order to have a complete sentence, the translation follows the Hebrew *ʿatah* (now) for the Judeo-Arabic *ān*, which could be a copyist's error for *al-ān* (now). Biṭrūjī's *On the Principles* simply stipulated that the radius of the Moon's polar epicycle was about 5°; there was no corresponding section in which Biṭrūjī attempted to calculate the parameter.[183] Ibn Naḥmias, in attempting to calculate the size of the lunar epicycle, faced a problem similar to the one he faced when attempting to calculate the Sun's eccentricity. The need to preserve the Moon's latitude to the ecliptic entailed a certain size for the epicycle. Thus, the anomaly had to be at its minimum when the Moon's latitude was at its maximum. Ptolemy's lunar model had not been bound by that constraint.

76r

let the circle of the path of the center be BAGE B.2.III.
(See figure 5.16.) Ibn Naḥmias, as he did with the solar model, has attempted to transpose Ptolemy's determination of the diameter of the lunar epicycle to the surface of an orb.[184] Ptolemy took three observations of lunar eclipses, points at which the Moon's longitude matched the Sun's, on a single side of the line of apsides, and then tried to determine the radius of an epicycle that would make all three observations possible. The fact that Ibn Naḥmias referred to the intersection between the circle of the path of the Moon's center and the deferent circle as a line, since the epicycle and the deferent were in the same plane in the *Almagest*, though the intersection would actually be an arc in a homocentric model, is evidence of an effort to adapt the *Almagest*'s calculations to a homocentric system. The measure of the degrees through which the arc (*AGB* or *AB*) on the Ptolemaic epicycle is seen by the observer is 3;24°.[185] The complication is that 3;24° is an angle (*ADM*) subtended in the ecliptic in the *Almagest*, but Ibn Naḥmias needed to compute arc *LN*, the radius of the circle of the path of the center, which is not in the plane of the ecliptic in a

183. Biṭrūjī (trans. and comm. Goldstein), *On the Principles*, vol. 1, p. 144. Goldstein proposed (pp. 37–8) that Biṭrūjī had noted the overlap between the Ptolemaic parameters for the radius of the epicycle, 5;13°, and for the Moon's latitude (5°).

184. Toomer, *Ptolemy's "Almagest,"* pp. 190–202.

185. Neugebauer: *History of Ancient Mathematical Astronomy*, p. 75.

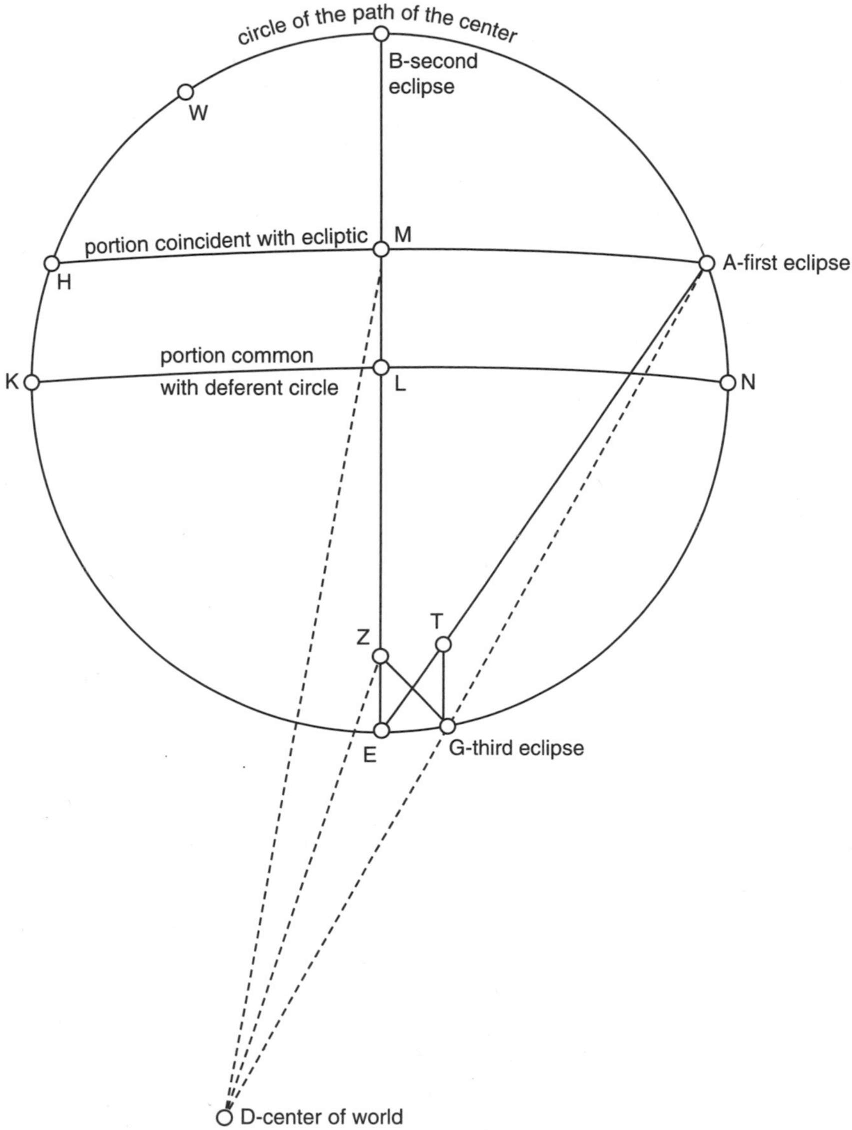
circle of the path of the center
B-second
eclipse
W
portion coincident with ecliptic
M
H
A-first eclipse
portion common
with deferent circle
K
L
N
Z
T
E
G-third eclipse
D-center of world

FIGURE 5.16

homocentric system. Moreover, once Ptolemy's calculations were transferred to a homocentric system, the second and third eclipses were no longer located at the intersection of the circle of the path of the center with either the ecliptic or the Moon's inclined circle. Thus, Ibn Naḥmias's calculations fell apart as he removed the eclipses from the plane of the ecliptic, which is where eclipses are observed to occur.

In addition, when Ibn Naḥmias stated that angles *BEA* and *BEG* were given, he confused plane and spherical geometry. Because *BEA* and *BEG* were angles formed at the circumference of a small circle (on the surface of the orb), their magnitude could not, in fact, be determined in a way commensurate with great circle arcs.

76v

In this way, the ratio that emerged for Ptolemy B.2.III.

Ibn Naḥmias has referred to the ratio of the radius of the deferent to the radius of the epicycle in the Ptolemaic lunar model.[186] Because the time from one solar transit of the meridian to the next was usually not 24 hours exactly, Ibn Naḥmias explained that the time intervals between eclipses had to be adjusted, as Ptolemy explained in *Almagest* III.9, to take these variations into consideration.

After that we return to explain the differences B.2.III

The differences are the measurements of the interval between one eclipse and another. Ptolemy did reiterate the calculation of the dimensions of the epicycle but did so with additional observations.[187] Ibn Naḥmias, on his part, may have wanted to reiterate the calculation since two of the three observed eclipses were pictured as being outside of the circle of latitude.[188] Improving the accuracy of calculations of the lunar mean motions would, theoretically, improve the parameter for the size of the epicycle. But, as was the case with the solar model, there would never be a way for all of the eclipse observations that Ibn Naḥmias used to determine the radius of the circle of the path of the center to fall at the intersection of the circle of the path of the center with the ecliptic.[189]

186. Toomer, *Ptolemy's "Almagest,"* p. 202. Ptolemy gives a parameter of 5;14° but eventually settled on 5;15°. The summary of the *Almagest* in Vatican MS Ebr 392 (8v) has 5;13°. Anatoli (*Almagesṭi*, Paris BNF MS Hébreu 1018, p. 95) has 5;14° and (V.6, p. 58a) 5;15°.

187. Pedersen: *Survey of the Almagest*, p. 177.

188. Instead, Ptolemy (Toomer, *Ptolemy's "Almagest,"* pp. 205–9) corrected calculations of the Moon's mean motions in latitude.

189. Toomer, *Ptolemy's "Almagest,"* pp. 211–6. It is also possible that Ibn Naḥmias was trying to produce a chapter to parallel *Almagest* IV.11, which analyzed Hipparchus's errors in calculating the size of the lunar epicycle.

77r

The fourth chapter, about how the visible variation of the measures of the diameters of the Moon does not necessarily entail the variation of the distances B.2.IV.1

Just before presenting his second lunar model, which accounted for further observed variations in the Moon's motion in anomaly, Ibn Naḥmias critiqued the additional observations that Ptolemy had taken as the foundation for his own second lunar model. Variations in the Earth–Moon distance, and the consequent variations in the Moon's diameter, would result from both the second lunar model that Ptolemy introduced to account for these variations in the lunar anomaly.[190] Ibn Naḥmias conceded observed variations in the Moon's visible diameter but did not allow that those variations were due to variations in the Earth–Moon distance, as variations in the Earth–Moon distance would obviate a cosmos of homocentric orbs. Ibn Naḥmias's commitment to homocentricity was strong enough for him to argue that other astronomers who observed variations in the size of the Moon were mistaken in not attributing those variations to parallax, instrumental and/or observational errors, or atmospheric phenomena that could distort observations. Ibn Naḥmias did not mention that Ptolemy's lunar model predicted unobserved variations in the Moon's size.[191]

In a response appended to the Hebrew recension of *The Light of the World*, Profiat Duran specified Ibn Naḥmias's attribution of observed variations in the Moon's distance from the Earth to vapors[192] as a reason why Ibn Naḥmias's homocentric theories should be dismissed.[193] Duran noted Levi ben Gerson's detection of variations in the radius of the Sun as additional evidence for the existence of epicycles and eccentrics.[194] These observations of Levi ben Gerson

190. Because the first anomaly was independent of the elongation, whereas the second anomaly was related to the elongation, Ptolemy determined that the first anomaly needed to be known before the second and could be known independently of the second. See Pedersen: *Survey of the Almagest*, p. 165. See also Neugebauer: *History of Ancient Mathematical Astronomy*, p. 85.

191. See Lay, "L'Abrégé," p. 53 on the role of observations as a means of verifying theories.

192. See also John North: *Cosmos: An Illustrated History of Astronomy and Cosmology* (Chicago and London: University of Chicago Press, 2008): p. 299. Both Fracastoro and Amico attempted to explain variations in the planets' density by their passage through vapors. Fracastoro and Amico were astronomers in the early sixteenth century affiliated with the University of Padua who may have known of *The Light of the World* (see Morrison, "A Scholarly Intermediary," p. 35, p. 38, pp. 52–3).

193. *Or ha-ʿolam*, Bodleian Canon Misc. 334, fol. 101a–100a. Duran's response is provided, with translation, in this book on pp. 393–98.

194. *Or ha-ʿolam*, Bodleian Canon Misc. 334, fol. 100b. On Levi ben Gerson's observations, see Bernard Goldstein: "Theory and Observation in Medieval Astronomy," in *Isis* LXIII (1972): pp. 39–47, at pp. 41–2. Ibn al-Shāṭir's solar model also took into account observed variations in the diameter of the Sun. See George Saliba: "Theory and Observation in Islamic Astronomy: The Work of Ibn al-Shāṭir of Damascus," in *Journal for the History of Astronomy* XVIII (1987): pp. 35–43.

were key because Ptolemy, whose work was Ibn Naḥmias's standard for predictive accuracy, had not found variations in the Sun's visible diameter. As well, in *Almagest* IX.2, Ptolemy expressed reservations about the observations that led to the computations of retrograde arcs and to the dimensions of the cosmos.[195] Hence, Ibn Naḥmias's own skepticism of certain observations might have been a tendentious interpretation of Ptolemy's reservations about some of his own observations.

Perhaps with these observations of variations in the Sun's observed diameter in mind, Profiat Duran defended the reality of epicycles by citing *Guide* II, 24.[196] And Duran, in his commentary on the *Guide,* rebutted those who rejected epicycles on the grounds that the epicycle's motion did not occur about something at rest: "And this doubt is not difficult to resolve, for the working premise there is that every mobile moves about something at rest, and it is not seen at first that the mobile moves upon a body at rest. But it is proper to explain that it is impossible that it moves without there being a body at rest upon which it moves, even if the body moves upon a thousand [bodies], as long as the lowermost of them all needs to be at rest. And evidence for this is that the man moves in a boat and it is moving, indeed, since the water is at rest."[197]

similarly the measure of each eclipse, its times, and the amount of the lingering B.2.IV.2
In *Almagest* V.17, Ptolemy took into consideration the impact of changing Earth–Moon distances on the computation of lunar parallax.[198] In this passage, Ibn Naḥmias pledged, though he did not actually end up doing so, to modify the computation of parallax so that it would be on the basis of a single Earth–Moon distance. Had Ibn Naḥmias attempted to follow through on his promise, he would have had to contend with something Profiat Duran pointed out in his response to *The Light of the World.* The variations themselves in the size of the Moon correlated with the Earth–Moon distance and neither with parallax nor with weather conditions. In the next section, Ibn Naḥmias addressed and dismissed the correlation between changes in the size of the Moon and changes in the Earth–Moon distance.

195. Toomer, Ptolemy's "Almagest," p. 421. Toomer commented (p. 421, note 8) that this was the only reference to refraction in the *Almagest.*

196. *Or ha-ʿolam,* Bodleian Canon Misc. 334, fol. 100b. Clearly, the debate about Maimonides's position in *Guide* II.24 summarized in the introduction was alive in Duran's (and Ibn Naḥmias's) lifetimes.

197. My translation is based on the Hebrew found in Maimonides: *Seiper Moreh nᵉb̲uk̲im / la-Rab̲ Mosheh bar Maimon, zal ; niṣṣab̲ peirusho . . . peirush Shem Ṭob̲ u-p̲eirush Ep̲odi* (Żolkiew : Gedrucht be Leib Matfes & Berl Lorie, 1860), vol. 2 of 3: 52d/1–2.

198. Toomer, Ptolemy's "Almagest," p. 259ff.

Likewise, we work on the condition that the diameter of the Moon subtends one angle at the sight in each position B.2.IV.2–3

Here, Ibn Naḥmias tried to find other potential explanations for these variations in the Moon's visible diameter besides changes in the Earth–Moon distance. Ibn Naḥmias cited Ptolemy's and Hipparchus's observation of variations in observations of the lunar diameter as evidence that these variations in lunar diameter were not due to variations in the distance of the Moon,[199] for if there were variations in the distance of the Moon, the variations in lunar diameter, according to Ibn Naḥmias, would occur in the same way throughout the history of observations (performed by qualified astronomers such as Ptolemy and, ostensibly, Hipparchus) of the lunar diameter. Consequently, the only plausible reason for the disagreement between earlier astronomers (probably Hipparchus) and Ptolemy would have been irregular meteorological phenomena mentioned in *Almagest* I.3 and the Moon illusion (that the Moon appears larger at the horizon than at the zenith).[200] Indeed, Ptolemy did not have a clear explanation for his disagreement with Hipparchus about whether the equivalence of the solar and lunar diameters was at the Moon's greatest distance or at the Moon's mean distance. By concluding that Hipparchus was correct, Ibn Naḥmias attempted to undercut the Ptolemaic standard for predictive accuracy.

77v

The fifth chapter, about the two anomalies remaining for the Moon B.2.V.1

The first of these remaining two anomalies was the variations in the Moon's size and position that are not predicted by Ptolemy's first lunar model. That first model was successful at predicting the longitudes of eclipses, but the model was less precise at other points, particularly quadratures. Ptolemy's second lunar model was much more successful at predicting and retrodicting all longitudes, but it did so by bringing the Moon much closer to the Earth at certain places through the motion of the center of the Moon's eccentric itself.[201] Even though observations did show variations in the observed lunar diameter, Ptolemy's second lunar model had the effect of predicting that the Moon would get much closer to Earth than it in fact did. In §B.2.V.2–3, Ibn Naḥmias attributed any *observed* variations in the Moon's size to errors in the construction and use of

199. Ptolemy's and Hipparchus's determinations of the Moon's apparent diameter are covered in Toomer, *Ptolemy's "Almagest,"* V.14; see p. 252. The history of Islamic astronomy that intervened between Ptolemy and Ibn Naḥmias saw other determinations of the Moon's apparent diameters that Ibn Naḥmias did not report. See Ragep, *Tadhkira,* p. 460.

200. See Toomer, *Ptolemy's "Almagest,"* p. 39.

201. On the second lunar model, see HAMA, pp. 84–6 and 88–91 (on the inclination [πρόσνευσις] of the epicyclic diameter].

instruments, to atmospheric phenomena, and perhaps to the Moon illusion, as homocentric models precluded any change in the Earth–Moon distance. Still, Ibn Naḥmias would have to modify his lunar model to explain the changes in longitude that Ptolemy had explained only by way of introducing unobserved variations in the lunar diameter.

The second of the remaining two anomalies was most likely the variation in the Moon's motion on its epicycle that Ptolemy explained by allowing the position of the epicycle's apogee to vary according to the prosneusis point. The Hebrew recension (§B.2.V.12/X) contains a slightly expanded discussion of something akin to the prosneusis point, namely that the variations in the longitudinal anomaly could be accounted for by the inclination of the diameter of the circle of the path of the Moon's center. Because the third lunar anomaly was related only to the Moon's velocity on the circle of the path of the center of the Moon, it did not challenge Ibn Naḥmias's commitment to homocentricity.

We find Ibn al-Haytham having explained in the fourth treatise of his book B.2.V.2
This is a reference to Ibn al-Haytham's work on optical problems surrounding astronomical observations, both atmospheric refraction and the Moon illusion.[202] The Moon illusion is the phenomenon that celestial objects, particularly the Moon, appear larger at the horizon than at the zenith. Though Ibn al-Haytham found the psychology of vision to be the main cause of the Moon illusion, atmospheric refraction mattered too. Ibn Rushd discussed problems observing the Moon in his *Talkhīṣ* (Middle commentary) of Aristotle's *Meteorologica* III.3, and Ibn Rushd's comments about there being a halo around the Sun and the Moon drew on the work of Ibn al-Haytham.[203] Thus, Ibn Naḥmias could have learned of Ibn al-Hay-

202. See Ibn al-Haytham (ed. A. I. Sabra): *Kitāb al-Manāẓir: al-maqālāt al-rābiʿa wa-'l-khāmisa* (Kuwait: al-Majlis al-waṭanī li-'l-thaqāfa wa-'l-funūn wa-'l-adab, 2002), vol. 1: pp. 36-87, esp. pp.68-9 and p. 85. See also A. I. Sabra: "Psychology versus Mathematics: Ptolemy and Alhazen on the Moon Illusion," in Edward Grant and J. E. Murdoch (eds.): *Mathematics and Its Application to Science and Natural Philosophy in the Middle Ages* (Cambridge: Cambridge University Press, 1987): pp. 217–47. See also Y. T. Langermann: "Ibn al-Haytham," in Hockey (ed.): *The Biographical Encyclopedia of Astronomers*, pp. 556–7. In the *Optics*, Ibn al-Haytham explained that the Moon illusion was primarily one of "the psychology of perception, though he did allow that thick vapors could sometimes be a secondary factor." On these causes of the Moon illusion, see Sabra, "Psychology versus Mathematics," pp. 240–3. After the *Optics*, Ibn al-Haytham composed a more detailed analysis of the Moon illusion; see A. I. Sabra and Anton Heinen: "On Seeing the Stars: Edition and Translation of Ibn al-Haytham's Risâla fî Ru'yat al-kawâkib," in *Zeitschrift für Geschichte der arabisch-islamischen Wissenschaften* VII (1991/92): pp. 31–72. See also Sabra: "On Seeing the Stars, II. Ibn al-Haytham's 'Answers' to the 'Doubts' Raised by Ibn Maʿdân," in *Zeitschrift für Geschichte der arabisch-islamischen Wissenschaften* X (1995/96): pp. 1–59. On the dating of Ibn al-Haytham's *Risāla fī Ru'yat al-kawākib*, see Sabra, "On Seeing the Stars, II," p. 5.

203. Paul Lettinck: *Aristotle's "Meteorology" and its Reception*, pp. 287–98. Lettinck has drawn on both Ibn Rushd's *Short Commentary* and his *Middle Commentary*.

tham's work via Ibn Rushd's commentaries. Still, this reference to Ibn al-Haytham is significant because references to Ibn al-Haytham's *Optics* before the career of Kamāl al-Dīn Fārisī (d. 1320) were few.[204] A new development is Bilal Ibrahim's finding that Fakhr al-Dīn Rāzī (d. 1210) had access to Ibn al-Haytham's *Optics*.[205]

Ibn Naḥmias has used the issue of the Moon illusion to argue that the task of the lunar model should be just to explain the variations in longitude, outside of eclipses. From his perspective, the Moon illusion rendered observations of variations in the Moon's diameter sufficiently flawed as to make it pointless to have his lunar model address the variations in the Moon's observed size.[206]

Especially when it [the Moon] is at quadratures with the Sun B.2.V.3
The model for the first lunar anomaly did not accurately account for the lunar position outside of syzygies (conjunctions and oppositions, including eclipses) and, according to Ptolemy's observations, the greatest observed deviation from the position predicted by the model for the first lunar anomaly occurred at quadratures (when the Sun and Moon are 90° apart).

78v

Our thesis is, because of this anomaly, that the center of the Moon does not revolve upon this circle itself. B.2.V.4–5
(See figure 5.17.) Ibn Naḥmias has modified his lunar model to account for the variations in longitude entailed by the second Ptolemaic lunar anomaly. The center of the Moon was carried on an additional small circle, *EWD*, whose radius was half of the maximum for the second lunar anomaly. In the figure, at quadratures, the Moon is at point *D*; when the Moon is at points *A* and *G*, that is, at syzygies, the effect of the second anomaly should and will disappear. For the Moon also to be at point *E* at syzygies, then small circle *EWD* must revolve clockwise[207] at twice the rate of small circle *EHST*.[208] Ibn Naḥmias did not explain how small circle

204. A. I. Sabra: "The Commentary That Saved the Text," in *Early Science and Medicine* XII (2007): pp. 117–33. Sabra did not cite this reference by Ibn Rushd. See especially p. 120: "No one in that whole period of more than two hundred years is known to have possessed any substantive knowledge of the book's fundamental theses and arguments."

205. Bilal Ibrahim: "Faḫr ad-Dīn ar-Rāzī, Ibn al-Hayṯam and Aristotelian Science: Essentialism versus Phenomenalism in Post-Classical Islamic Thought," in *Oriens* XLI (2013): pp. 379–431, at pp. 404–11.

206. Previously, Ibn Naḥmias had used lunar parallax and the Moon illusion to explain away variations in the Moon's size that could be explained by the Earth not being at the center of the Moon's path.

207. See *The Light of the World*, §B.2.V.5. "When it was between points *ET* or *EH*, then the increases in the anomaly are over the amount of the first anomaly."

208. *The Light of the World*, §B.2.V.5. "And the [Moon] traverses the circle of its path, I mean circle EWD, twice during the time of a mean month." Even though the Hebrew version also has "Sun," I think that the text must mean "Moon," given the reference to circle EWD.

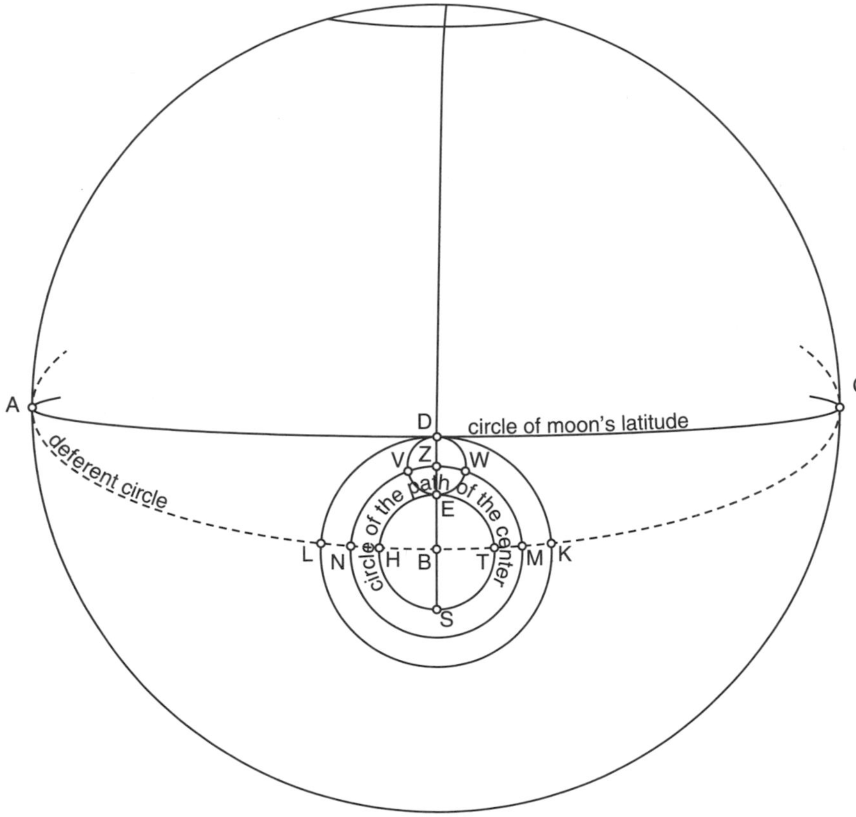

FIGURE 5.17

EWD could revolve twice as quickly as *EHST*. There are also some inconsistencies between the text and the figure in the text. For instance, the end of §B.2.V.5 reads, "As for when the Moon was on point T, namely the point at which there is the greatest increase in the speed from the first anomaly, so at that time point D, in which the Moon is, was on point K." The longitudes would be more accurate if the circle of the moon's latitude passed through *E*, but the text says that the inclination should be increased by the second small circle. But according to the figure in the text, *D* should be at point *E*. These inconsistencies are due to how, in the Ptolemaic system, the two anomalies could combine to yield an equation of 7;40° at quadratures, whereas this was impossible in Ibn Naḥmias's model. Thus, this lunar model would not predict and retrodict the observations very well.

Outside of the quadratures, the Moon would be displaced from the circle of its latitude. I will now calculate the Moon's displacement from its circle of latitude after 45° (θ in this example) of the Moon's motion on small circle *EHST* and 90° of motion on small circle *EWD*. I presume parameters of 5;13° for the radius of small circle *EHST* and 1;13,30° for small circle *EWD*. Taken together, those circles will yield the maximum Ptolemaic equation of 7;40° at $\theta = 90°$, although the equation will not affect the longitude as much as the equation serves to keep the Moon in its circle of latitude. Because the mean motion in anomaly and the mean motion in longitude are not the same, we divide 45° by 13;3,54° and multiply by 13;10° to get the longitude for the center of the circle of the path, point *B*: 45;21°.

(See figure 5.18.) After 45° of motion in anomaly, point *E* has revolved 45° to point *W*, and point *D* has revolved 90° about *Z* to *D'*. The spherical coordinates of *D'* can be calculated with the formulas, found in equations (1) and (2) in the comments on §B.1.II.13: $\Phi(D') = \arccos(\sin r \cos\theta)$ and $\Theta(D') = 45;21° + \arccos[\cos r/\sin(\Phi(\theta))]$. To calculate the Φ coordinate of *D'*, we need to determine arc *BD'* and consider spherical triangle *BZD'*. Angle *BZD'* is 90°, *d'* (side *BZ*) is 6;26,30°, *b* (side *ZD'*) is 1;13,30°, and angle *Z* is 90°. With the cosine rule for sides:

$\cos z = \cos d' \cos b + \sin d' \sin b \cos Z$
$\cos z = \cos 6;26,30° \cos 1;13,30°$
$z = \text{arc } BD' = r = 6;33,25°$
With the law of sines, $\sin z/\sin 90° = \sin 1;13,30°/\sin B$
angle $B = 10;47,26°$ and $\theta = 45° + B = 55;47,26°$
$\Phi(D') = \arccos(\sin r \cos\theta) = 86;19,9°$
$\Theta(D') = 45;21° + \arccos[\cos r/\sin(\Phi(D'))] = 50;46,48°$

Now we need to calculate the Φ coordinate of a point on the circle of the Moon's latitude with $\Theta = 50;46,48°$. We consider spherical triangle *QPG* with angle $P = 90°$, $G = 7;40°$ (the diameter of small circle *EWD* + the radius of small circle *EHT*), and q (side PG) = 39;13,12° (= 90°–50;46,48°). According to the rule of four parts:[209]

$\cos q \cos P = \sin q \cot g - \sin P \cot G$
$\tan g = \sin q \tan G$
$\tan g = \sin 39;13,12° \tan 7;40°$
$g = 4;51,54°$

Thus, $\Phi(Q) = 90° - 4;51,54° = 85;8,6°$. Point *D'* is over a degree from the circle of the Moon's latitude.

If small circle *EWD* must rotate clockwise, then predictive accuracy would be improved by having *EHST* rotate in the direction of the motion of the universe (i.e., counterclockwise at the start). (See the dashed small circle *WD'* in figure 5.18.)

209. Smart, *Text-Book*, p. 12.

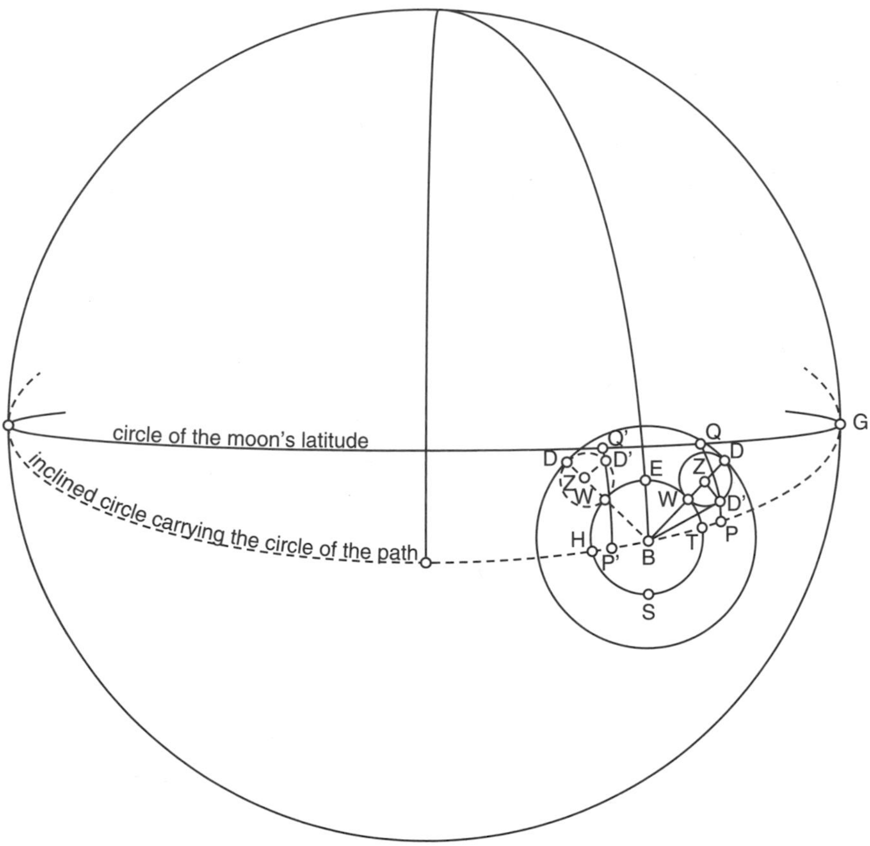

FIGURE 5.18

To determine the Φ and Θ coordinates of D':

Since angle $B = 10;47,26°$, $\theta = 45° - B = 34;12,34°$
As $r = 6;33,25°$, $\Phi(D') = \arccos(\sin r \cos\theta) = 84;34,52°$
$\Theta(D') = 45;21° - \arccos[\cos r / \sin(\Phi(D'))] = 41;39,10°$

Now we need to calculate the Φ coordinate of a point on the circle of the Moon's latitude with $\Theta = 41;39,10°$. We draw spherical triangle *QPG* with angle $P = 90°$, angle $G = 7;40°$ (the diameter of small circle *EWD* + the radius of small circle *EHT*), and q (side *PG*) $= 48;10,50°$ ($= 90° - 41;39,10°$). According to the rule of four parts:

$\cos q \cos P = \sin q \cot g - \sin P \cot G$

$\tan g = \sin q \tan G$
$\tan g = \sin 48;10,50^\circ \tan 7;40^\circ$
$g = 5;43,44^\circ$
Thus, $\Phi(Q) = 90^\circ - 5;43,44^\circ = 84;16,15^\circ$

D' is 0;18,37° from the circle of the Moon's latitude, a discrepancy that would still be observable. Whatever the choice of the direction of motion, in §B.2.V.6 and §B.2.V.11 Ibn Naḥmias proposed additional hypotheses to eliminate this discrepancy.

pole Z moves on the two circles described in the first treatise B.2.V.6
Here, the double-circle hypothesis is used to move the pole of the circle that accounts for the second anomaly in order to eliminate the significant displacement from the circle of the Moon's latitude. Ibn Naḥmias did not specify either the dimensions of the double circles, nor did he calibrate the period of the oscillation of the double circles to compensate for the period of the displacement from the circle of latitude. In §B.2.V.11, he introduced an additional, more detailed modification to eliminate displacements in latitude.

for the repugnancies attached to his [Ptolemy's] hypotheses for these two anomalies B.2.V.7
Ibn Naḥmias held that the repugnancies of Ptolemy's hypotheses for the second and third lunar anomalies, the variations in the Moon's size for the second, inasmuch as it was caused by the motion of the center of the Moon's eccentric, and the prosneusis point for the third were not connected to the impossibility of eccentrics and epicycles.[210] Ibn Naḥmias wrote, "Rather, the impossibility of that which is not feasible befalls it." Such a statement, absent from Biṭrūjī's *On the Principles*, makes the critique of Ptolemy in *The Light of the World* more wide ranging.[211]

210. Lay, "L'Abrégé," pp. 47–8. Ibn Bājja and Ibn Rushd knew of Ibn al-Haytham's *Shukūk*, so there is a context for this reference to such critiques of Ptolemaic astronomy. Note, too, that Ibn Rushd, in his *Qiṣṣur al-Majisṭī* (Paris BNF MS Hébreu 903, 54r–v), pointed out that the eccentric in the Moon did not rotate uniformly about its own center. There were also contemporary systematic critiques of other sciences. See Y. Tzvi Langermann: "Another Andalusian Revolt? Ibn Rushd's Critique of al-Kindī's *Pharmacological Computus*," in Jan P. Hogendijk and Abdelhamid I. Sabra (eds.): *The Enterprise of Science in Islam: New Perspectives* (Cambridge and London: MIT Press, 2003): pp. 351–2. See too George Saliba: "Critiques of Ptolemaic Astronomy in Islamic Spain," in *al-Qantara* I (1999): pp. 3–25.

211. This broader critique could explain why *The Light of the World*, and homocentric astronomy in general, was part of European astronomers' criticisms of Ptolemy in the 15th and 16th centuries. See Michael Shank: "Regiomontanus as a Physical Astronomer: Samplings from *The Defence of Theon against George of Trebizond*," in *Journal for the History of Astronomy* XXXVIII (2007): pp. 325–49, at p. 326. See also p. 327, where Shank discussed "the inaugural lecture of his [Regiomontanus's] course on al-Farghānī at the University of Padua in 1464, in which he mentioned wistfully Averroes's unsuccessful efforts to construct a concentric astronomy."

According to Ibn Naḥmias, the effect of the third anomaly is that the Moon's position in longitude varies from what it should be assuming the motions as they would be otherwise. He wrote that, at certain positions, "The lag in the longitude for the true position of the Moon is more than that necessitated by the remaining anomalies."[212] He noted, correctly, that the third anomaly vanishes at quadratures, when the second anomaly is at its maximum, but he omitted mention of how the third anomaly also vanishes at syzygies.[213]

79v

the anomaly increases B.2.V.9

Ibn Naḥmias has specified that the effect of the third anomaly is greatest at octants. He has also said that his model takes account of the third anomaly by allowing the Moon's motion to increase at certain points. He provided neither a parameter for this variation nor a quantitative analysis.

Because Ibn Naḥmias said that there was no need for an additional orb or an additional motion to account for the third anomaly, the third anomaly would have to follow from the existing motions of circles *HEST* and *EWD* as Ibn Naḥmias wrote, "The anomaly increases because through this movement it speeds up as the Moon approaches point T." Both small circles *EHTS* and *EWD* need to begin by revolving clockwise.

After 45° of mean motion, the spherical coordinates of the Moon are (see comments on §B.2.V.5):

$$\Phi(D') = \arccos(\sin r \cos\theta) = 86;19,9^\circ$$
$$\Theta(D') = 45;21^\circ + \arccos[\cos r/\sin(\Phi(D'))] = 50;46,48^\circ$$

As the comments on §B.2.V.5 showed, the Φ coordinate of a point on the circle of the Moon's latitude with the same Θ coordinate is 85;8,6°. Thus, there is a displacement of over a degree in latitude meaning that Ibn Naḥmias's thesis for the third anomaly did not account for observations.

80r

In this way the Moon is always upon the circle of the latitude B.2.V.11

In this passage, Ibn Naḥmias introduced a modification to compensate for any additional deviations in latitude introduced by the model for the second and third anomalies. The pole of the inclined circle that carries small circle *EWD*, pole *S*, revolves on small circle *SM* so as to remove some of the displacement in latitude outstanding from the model for the second anomaly.

Ibn Naḥmias added that "the movement of pole S upon circle SM is equal

212. *The Light of the World*, §B.2.V.9.

213. Cf. Neugebauer, *History of Ancient Mathematical Astronomy*, pp. 88–9.

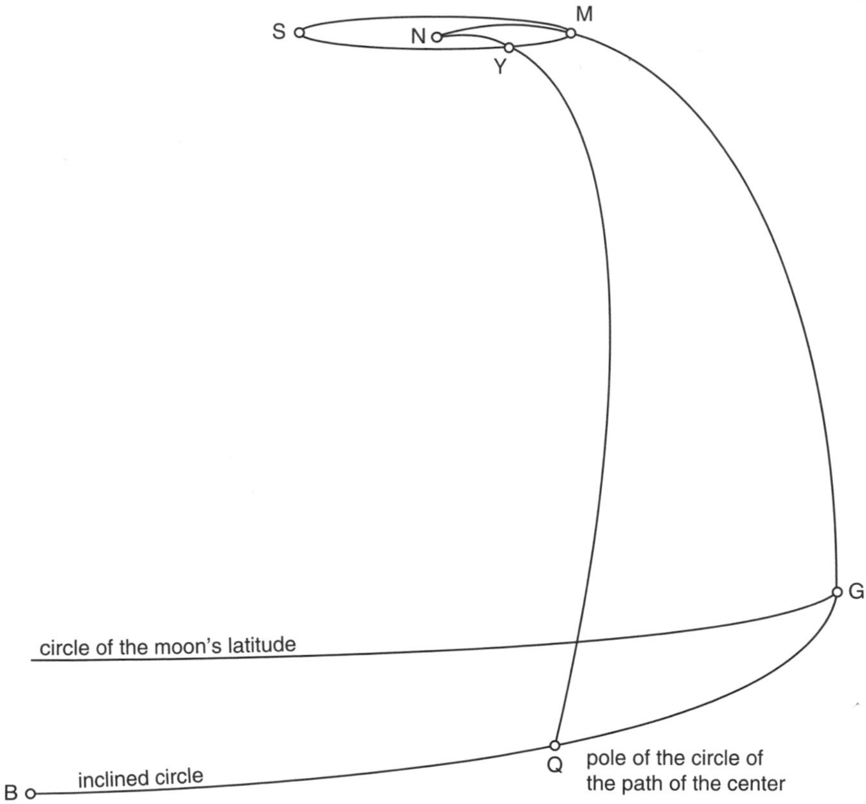

FIGURE 5.19

to the motion of the Moon upon circle EWD." Thus, after 90° of mean motion, pole *B* will be at point *G*, and arc *SB* = arc *MG* = 90°. And because the motion of small circles *EHTS* and *EWD* is the same as before, any effect of small circle *SM* will have disappeared at syzygies. We perform a calculation at an octant so as to assess how well the addition of circle *MS* eliminates the displacement in latitude caused by the third anomaly.

(See figure 5.19.) In the figure, pole *S* has revolved 90° about pole *N* to point *Y*. After 45° of mean motion from point *B*, a solstice, the pole of the circle of the path of the center has moved to point *Q*. We solve spherical triangle *NQG*. Side *n* (side *QG*) is 45°, and we define arc NM as 1°, roughly the maximum displacement from the circle of latitude created by Ibn Naḥmias's model for the second anomaly. Because point *M* is 90° from point *G*, arc *NMG* is 91°, and angle *MGQ* = 90°. We apply the law of cosines for sides:

$$\cos g = \cos n \cos q + \sin n \sin q \cos G$$
$$\cos g = \cos 45^\circ \cos 91^\circ + \sin 45^\circ \sin 91^\circ \cos 90^\circ$$
$$g = 90;42^\circ = NQ$$

If the distance from the pole (Y) to the center of the circle of the path of the center were fixed at 90°, this modification, if calibrated correctly, could have the effect of bringing the Moon closer to the circle of latitude. The outstanding problem, though, is that Ibn Naḥmias's model for the second lunar anomaly will not accord with Ptolemy's observations because there may never be a place where the total equation for the lunar longitude reaches 7;40°.

80v

Since there is no need for an eccentric or an epicyclic orb with them B.3.I.1

(See figure 5.20.) Ibn Naḥmias approved of Ptolemy's model for the precession of the equinoxes, in which the fixed stars were in a single orb that moved from west to east at the rate of 1° per century, inasmuch as there was no eccentric or epicycle in that model.[214] Biṭrūjī's model for precession entailed variations in the rate of precession, but not trepidation. Trepidation is an oscillation in the points of the solstices, entailing variations in the rate of precession *and* changes in the direction of the fixed stars' motions.[215] In Biṭrūjī's model, the equinoxes always precessed in the opposite direction of the motion of the signs of the zodiac. The pole of the ecliptic was inclined to the pole of the equinoctial at an angle equal to the obliquity of the ecliptic. Then, the pole of the ecliptic lagged at an equal velocity on a small circle about the pole of the equinoctial. Pole P_1 is moving on the circle of the path of the pole of the ecliptic. As P_1, the pole of the ecliptic, went to point P_2, and point V_1, the equinox, went to point V_2, a star on the ecliptic moved to point Q, thus closer to pole A. That star's declination has changed. Goldstein's analysis found that the lag of the pole of the ecliptic was uniform with respect to the equinoctial. Thus, the slow westward motion of the equinox was uniform with respect to the equinoctial, but the inclination of the ecliptic to the equinoctial meant that the motion of the equinox with respect to the ecliptic

214. See Toomer, *Ptolemy's "Almagest,"* pp. 327–38. This model is found in *Almagest* VII.2–3. See also the *Almagest* summary in Vatican MS Ebr 392, 28v–29v. See also Anatoli, *Almagesṭi*, Paris BNF MS Hébreu 1018, pp. 171–8.

215. On the origins of trepidation, see F. Jamil Ragep: "Al-Battānī, Cosmology, and the Early History of Trepidation in Islam," in Josep Casulleras and Julio Samsó (eds.): *From Baghdad to Barcelona: Studies in the Islamic Exact Sciences in Honour of Prof. Juan Vernet*, 2 vols. (Barcelona, 1996), vol. 1, pp. 267–98, at pp. 276–7 for Ibn al-Zarqāl. On p. 271, Ragep noted the role of Battānī (d. 923) in understanding Theon's mathematical description of the back-and-forth oscillation of the equinoxes in more physical and cosmological terms. As well, Battānī understood the ecliptic (and, thus, the fixed stars) to oscillate, not just the equinoxes. These views of Battānī (p. 273) reflected the scientific consensus. For more on a trepidation model roughly contemporary with Ibn Naḥmias's, see Mercè Comes, "Ibn al-Hā'im's Trepidation Model," *Suhayl* II (2001): pp. 329–37.

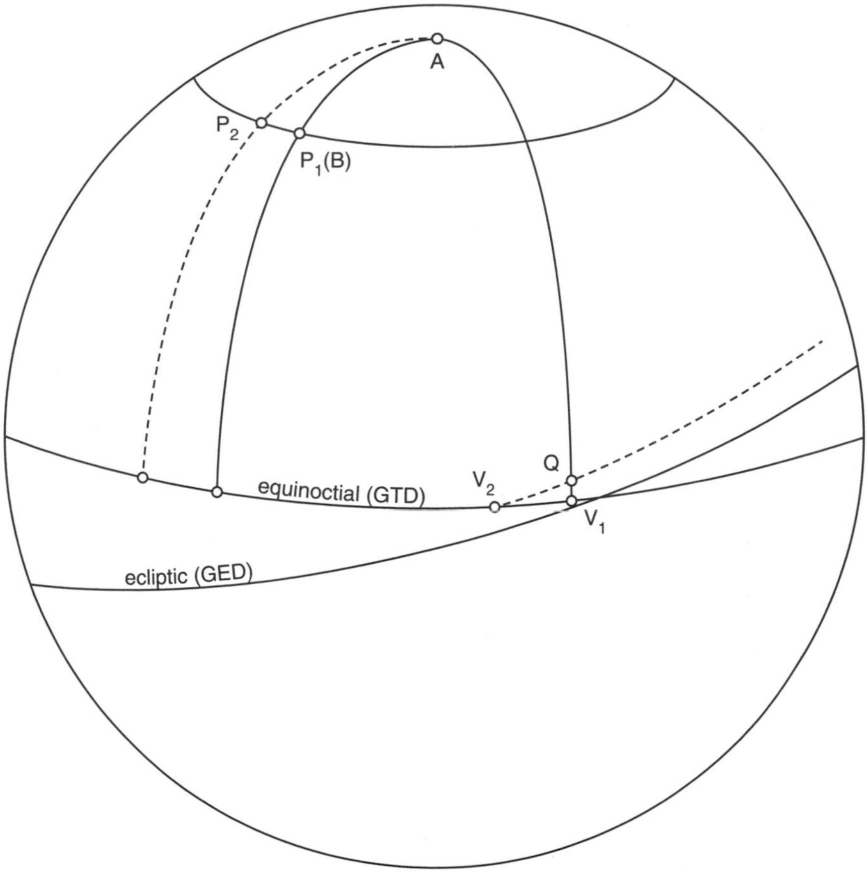

FIGURE 5.20

could not be uniform. For that reason, Goldstein characterized Biṭrūjī's model as one of "variable precession."[216] Ibn Naḥmias criticized Biṭrūjī's model for the motion of the fixed stars for simply correlating variations in the rate of precession with rising times.[217]

J. L. Mancha's recent research has reinterpreted Biṭrūjī's model for precession, finding that it included a "Eudoxan couple," two orbs with inclined poles revolving in opposite directions by the same measure. The key textual grounding for Mancha's interpretation is that Biṭrūjī proposed that the orb of the ecliptic rotate,

216. Biṭrūjī (trans. and comm. Goldstein), *On the Principles*, vol. 1, p. 15. Figure 5.20 is adapted from Biṭrūjī (1971).

217. Biṭrūjī (trans. and comm. Goldstein), *On the Principles*, vol. 1, pp. 16–7.

about its own poles, with the motion of completion.[218] This motion in completion was in the direction of the motion of the universe, that is, in the opposite direction of the lag. According to Mancha's interpretation, there would not be a continuous motion in precession. Rather, the latitude of the equinoxes would vary, along with a smaller motion, in the direction of the signs and opposite it, in longitude. Goldstein's analysis did not acknowledge the role of the motion in completion.[219] Mancha cautioned that there was no evidence that Biṭrūjī understood that his model included a Eudoxan couple or that Biṭrūjī actually intended the motions that would result from the model.[220] This is an important point in that Biṭrūjī wrote that accession and recession would take place only in the direction of the signs.[221] It is interesting, though, to note that Biṭrūjī, in the early chapters of *On the Principles,* did cite Aristotle's statement that the celestial bodies had a gyrational (*lawlabiyya*) motion.[222] Mancha's interpretation of the model meant that the stars would sometimes move in the opposite direction of the signs. Mancha's work has nuanced the argument that an important source of Biṭrūjī's models was models for trepidation and precession such as those of Ibn al-Zarqāl.

The Light of the World did not provide a parameter for precession. Both the *Almagest* summary in the same codex with *The Light of the World* and the Paris MS of the Hebrew *Almagest* translation give 100 years as the period in which the equinoxes precess 1°.[223]

Or it completes with its motion the motion of the uppermost
Ibn Naḥmias cited (but did not name) recent (i.e., post-antique) astronomers who thought that there was no lag in longitude but only a movement (*intiqāl*) in latitude. Biṭrūjī's model for the motion of the fixed stars, according to Mancha's interpretation, would fit that description. As Mancha notes regarding Biṭrūjī's theory of precession, "al-Biṭrūjī considers the changes in declination of the stars to be a well established fact, from which astronomers inferred their motion in longitude, whose rate and direction remain at his time still uncertain."[224] Thus, it

218. Biṭrūjī (trans. and comm. Goldstein), *On the Principles,* vol. 1, pp. 80–1. "But there remains the lag of this sphere about its own poles, for the motion of completion takes place about these poles, which are [fixed] on this [sphere]. Their lag is necessarily in the opposite direction—opposite the universal motion."

219. Biṭrūjī (trans. and comm. Goldstein), *On the Principles,* vol. 1, pp. 15–18.

220. Mancha, "Al-Biṭrūjī's Theory," pp. 157–61.

221. Biṭrūjī (trans. and comm. Goldstein), *On the Principles,* vol. 1, p.80.

222. Biṭrūjī (trans. and comm. Goldstein), *On the Principles of Astronomy,* vol. 1, p. 62. Ibn Naḥmias did not repeat this reference.

223. See the *Almagest* summary in Vatican MS Ebr 392, fol. 28v; and Anatoli, *Almagesṭi*, Paris BNF MS Hébreu 1018, p. 179.

224. Mancha, "Al-Biṭrūjī's Theory," p. 146.

is possible that Ibn Naḥmias was referring to Biṭrūjī, for if Ibn Naḥmias understood Biṭrūjī's model for the fixed stars as Mancha did, then the fixed stars would move to the north and south without an accompanying lag in longitude, meaning that there was no third orb moving the Eudoxan couple at a steady rate.

81r

decreases with the passage of time B.3.I.3

Ibn Naḥmias is saying that the distance of Spica Virginis from the equinox was decreasing when, according to Biṭrūjī's theory of variable precession, it should have been increasing. Ibn Naḥmias took issue with how Biṭrūjī's model entailed the motion of certain stars to the north or south of the equinoxes, a motion that was not corroborated by available observations. Goldstein's analysis of Biṭrūjī's model for precession, too, uncovered a lack of retrodictive accuracy.[225]

The circle of the equinoctial is GTD B.3.I.5

Ibn Naḥmias used this description of Biṭrūjī's model for the fixed stars to demonstrate how Biṭrūjī's model for precession would correspond with observations only at the solstices.[226] Though astronomers had questioned and revised Ptolemy's parameter for precession of 1° per 100 years, there was scant observational evidence for stars moving across the celestial equator, something which Biṭrūjī's model entailed. The figure of Biṭrūjī's model for precession from Goldstein's study shows that stars near the vernal equinox (e.g., points V_1 and V_2) would move north. Ibn Naḥmias's complaint was that, because the lag depends on the daily motion, which is in the opposite direction of precession, the stars would have to move south to appear to go north. Likewise, because precession was the only motion that took place in the opposite direction of the motion of the signs, the equinoxes would have to move in the direction of the signs in order to move in the opposite direction.

82v

each one of those outside of it is in a small circle B.3.II.3

Ibn Naḥmias proposed that each pair of stars be on the same small circle of latitude, with regard to the ecliptic, and that each small circle share the poles of the ecliptic that in turn revolve about the poles of the equinoctial. Thus, these small circles move parallel to the ecliptic. At the end of this paragraph, he remarked how sense perceptions communicate that all of the orbs that might be proposed for the fixed stars have the same motions. This model, though, preserved the possibility that sense perceptions could be mistaken.

225. Biṭrūjī (trans. and comm. Goldstein), *On the Principles,* vol. 1, p. 18.
226. Biṭrūjī (trans. and comm. Goldstein), *On the Principles,* vol. 1, pp. 15–6 and pp. 92–100.

83r

Let the ecliptic be DTZK B.3.II.4

(See figure 5.21 and figure 2.26.) It is worth seeing the lengths to which Ibn Naḥmias went to preserve his skepticism about the observations that implied that all of the fixed stars had the same motion. The points of the winter and summer solstices are carried by circle *QDSX*, whose pole *G* rotates about *A*, the pole of the orb of the equinoctial. Then, he considers the case of a star 5° away from the winter solstice in the direction of the spring equinox. While Biṭrūjī had placed that star in the same orb, Ibn Naḥmias proposed a different model. He placed that star (point *T*) in a separate orb, whose equator was *EHZL* and whose pole was *B*. First, point *T* moves with the orb of the solstices in a circle parallel to the equinoctial through arc *TH*. Then, that orb rotates about its own pole, *B*, and point *T* is carried from point *H* to point *Z*. The result is that point *T* is on point *Z*, which is where Biṭrūjī's model would have placed it. This model preserves the possibility that point *T*'s new position with respect to the equinox and/or the equinoctial is not precisely the same as it had been. Indeed, if we permit, as we did with the models for the solar and lunar mean models, variations in the measure of the second motion, we find that Ibn Naḥmias has allowed for as yet unobserved (but perhaps conceivable) variations in the positions of the fixed stars that could be accounted for by varying the speeds of the motion of each star's individual orb.

83v

With regard to the rising times

Within 45° of the equinoxes, the right ascension of a degree of longitude (on the ecliptic) is less than 1°. Within 45° of the solstices, it is greater than 1°.[227]

As for the motion in accession and recession B.3.II.5

Here, Ibn Naḥmias has made his most important statement on the fixed stars: that what previous astronomers may have perceived to be trepidation was actually the result of some stars moving at a different rate than others. That is why each pair of stars, according to Ibn Naḥmias, would have to be in a separate orb. If two adjacent stars were moving at different rates, the slower might appear to be moving backward with respect to the faster one. That would create the perception of trepidation.

Also it is not impossible that some of these stars have a lag greater than others

Here Ibn Naḥmias justified the need for multiple orbs for the fixed stars by citing the possibility of unobserved variations in their motions. Maimonides, in II.11 of *Guide of the Perplexed*, said multiple orbs for the fixed stars were certainly

227. See Biṭrūjī (trans. and comm. Goldstein), *On the Principles*, vol. 1, p. 24.

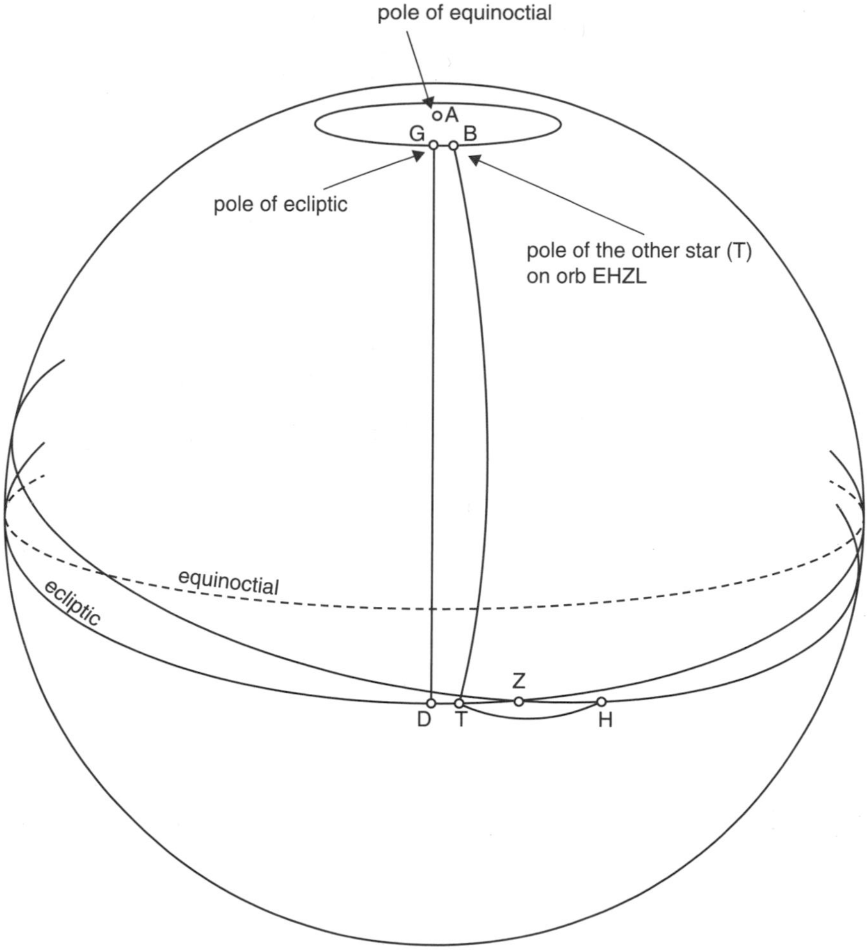

FIGURE 5.21

possible and claimed to be agnostic.[228] There was, in the *Qur'ān* commentaries

228. Maimonides (trans. Pines): *Guide*, vol. 2: p. 274. But, a few lines earlier, Maimonides invoked the principle of economy in the face of multiple mathematically equivalent models: "For if we assume, for instance, that we suppose as a hypothesis an arrangement by means of which the observations regarding the motions of one particular star can be accounted for through the assumption of three spheres, and another arrangement by means of which the same observations are accounted for through the assumption of four spheres, it is preferable for us to rely on the arrangement postulating the lesser number of motions." The *Guide* was not the only precedent for considering multiple orbs for the fixed stars. See Ibn Sīnā (ed. Madkūr): *al-Shifā', al-Samā' wa-'l-*

of Fakhr al-Dīn al-Rāzī (d. 1210) and Niẓām al-Dīn al-Nīsābūrī (d. ca. 1330), a nuanced debate impinging upon the religious implications of such skepticism about the model for the fixed stars.[229] I would categorize Rāzī and Maimonides apart from scholars such as Ṭūsī, Ibn Sīnā, and Nīsābūrī, as the latter three all held that while there was no way to demonstrate conclusively that the fixed stars were in a single orb, there was no evidence that any of them moved with different speeds.[230] Both Rāzī and Maimonides, conversely, were making arguments that depended on skepticism about the motion of the orb(s) of the fixed stars.

The fourth treatise is about the motions of the five planets B.4.0.1

Here, Ibn Naḥmias has outlined a treatise that does not run to completion in the Judeo-Arabic MS. Nor did he discuss this treatise at all in the introduction to *The Light of the World.* The Hebrew recension does not include even the beginning of this treatise, though it does refer to it in the course of an overview of the contents of the text. These divergent approaches to the inclusion of the fourth treatise make it difficult to posit a direct relationship between the two surviving MSS of *The Light of the World.* The location of the treatise on the planets after the treatise on the fixed stars was another way in which *The Light of the World* followed the *Almagest.*

At the beginning of the section on the planets, Ibn Naḥmias made a forceful argument for the correctness of his theories. He said that one of sound *fiṭra* (innate instinct) would not need a proof for the impossibility of the epicycle, and people of unsound *fiṭra* would try to attack his arguments through sophistry. The appeal to *fiṭra* in scientific arguments had a history.[231] Quṭb al-Dīn Shīrāzī

ʿālam (Cairo: General Egyptian Book Organization, 1969), p. 46; and Naṣīr al-Dīn al-Ṭūsī (d. 1274) (Ragep: *Naṣīr al-Dīn al-Ṭūsī's "Memoir on Astronomy,"* pp. 108–9 and p. 389).

229. Carlo Nallino (*ʿIlm al-falak; Tārīkhuh ʿind al-ʿArab fī al-qurūn al-wusṭā* [1911; repr., Beirut: al-Dār al-ʿArabiyya li-'l-kitāb and Awrāq sharqiyya, 1993]: pp. 257–9) called attention to scientists and *mutakallimūn* who acknowledged the possibility of multiple orbs for the fixed stars. See also Morrison: *Islam and Science*, pp. 108–12.

230. Ibn Sīnā ([ed. Madkūr]: *al-Shifāʾ, al-Samāʾ wa-'l-ʿālam*, chap. 6: p. 46) acknowledged the possibility of the fixed stars being in more than one orb. Ṭūsī ([ed. Ragep], *Tadhkira*, p. 389) acknowledged the same thing.

231. Dimitri Gutas: *Avicenna and the Aristotelian Tradition* (Leiden: Brill, 1988): p. 170. See also Sabra, "Science and Philosophy," p. 39, for al-Sharīf al-Jurjānī's defense of astronomy on the basis of *fiṭra*. See also Heidrun Eichner: *The Post-Avicennian Philosophical Tradition and Islamic Orthodoxy. Philosophical and Theological Summae in Context* (Habilitationsschrift: Martin-Luther-Universität Halle-Wittenberg, 2009): p. 184. For a reference in a *tafsīr* to the role of *fiṭra* in grasping scientific concepts, see Robert Morrison: "Natural Theology and the *Qurʾān*," in *Journal of Qurʾānic Studies* XV/1 (2013): pp. 1–22, at p. 12.

attributed the decision not to prefer a seven-orb cosmos to *fiṭra salīma*.[232] At the least, this statement ("true hypothesis") is more evidence that Ibn Naḥmias is taking on the challenge from the *Guide* and resolving the perplexities of theoretical astronomy.

84v

we mention one decisive demonstration for the impossibility of the orb of the epicycle B.4.1.2

Ibn Rushd, in his *Talkhīṣ al-Samā' wa-'l-'ālam*, wrote that we always observe the same face of the Moon, meaning that the Moon is not rolling in any way in its orb, meaning that there can be no epicycle for the Moon.[233] Ibn Naḥmias made the additional claim, as had Gersonides among others, that if the epicycle were not possible for the Moon, it would not be possible for any planet.[234]

85r

the anomaly that is with respect to the Sun disappears B.4.1.4

Ibn Naḥmias has reminded the reader that Ptolemy, when he proposed a model for the planets' motions in longitude (the first anomaly), selected his observations so that they would not be affected by the second anomaly (retrograde motion). In §B.4.0.1, Ibn Naḥmias promised that the third chapter of the fourth treatise would be devoted to the second anomaly, which was retrograde motion. Yavetz's reconstructions of Eudoxus's astronomy pay close attention to how configurations of homocentric orbs could reproduce the retrograde arcs of the planets, and to how the need to account for retrograde motion itself accounted for some of the assumptions that Schiaparelli made in his reconstructions.[235] Ibn Naḥmias's phrasing here showed that he understood that the circles of the path of the center were stand-ins for the epicycle.

In this paragraph, Ibn Naḥmias wrote that the second anomaly was found "when the first returns to what it was." In the middle of a retrograde arc, the planet has the same position as the epicycle center, either at the epicycle's apogee or perigee, meaning that the first anomaly is the only one affecting the planet's position at that time.[236]

232. Shīrāzī, *al-Tuḥfa al-shāhiyya*, Paris BNF MS Arabe 2516, 8v.

233. Ibn Rushd, *Talkhīṣ al-Samā' wa-'l-'ālam*, pp. 237–8. This is a reference to Aristotle's statements in *De Caelo* 290a.

234. For Gersonides's argument against the epicycle, see Goldstein, *The Astronomy of Levi Ben Gerson*, p. 117.

235. Yavetz, "On the Homocentric Spheres," p. 225 and elsewhere.

236. Pedersen, *Survey of the Almagest*, p. 272.

85v

So he found for Saturn in 59 years B.4.I.5

This information is in *Almagest* IX.3.[237] One difference is that whereas Ptolemy listed 3;10° for Mars, Ibn Naḥmias listed 3;6°, and whereas Ptolemy gave 42 revolutions in longitude, Ibn Naḥmias listed 43 (this could be a smudge).[238]

Chapter two, about explaining the true hypothesis proposed for the first anomaly B.4.II.1

This hypothesis is for the anomaly in longitude, though it creates an anomaly in latitude. The planet is on a small circle *GMDL*, whose pole *B* is carried through the ecliptic. The period of the planet's rotation on the small circle equals the period of the pole's rotation through the ecliptic, so the period of the planet's motion in latitude *is* equal to the period of the planet's motion in anomaly, perhaps addressing a weakness of Biṭrūjī's model.[239] This improvement only went so far because there is no available evidence about how Ibn Naḥmias accounted for retrograde motion, the second of the two anomalies. There is no illustration of this model in the Judeo-Arabic MS, and the chapter ends abruptly at the bottom of 85v.

86r

About the particular things that are known through the knowledge of rising times B.4.III.1

This chapter parallels II.9 of the *Almagest.* The rising times are related to the measure of the day or night in that one sums 180° of rising times, beginning from the Sun during the day going backward through the signs or beginning from the degree opposite the Sun for the night, and divides by 15 to get the interval of day or night in equal hours.

According to the outline of this section of the treatise, the particular things related to rising times are a topic of chapter three in the fourth treatise. Because the MS does not contain a heading for chapter three, there must be a page or so missing between 85v and 86r, a conclusion bolstered by how 85v trails off inconclusively. This chapter heading was not mentioned in the list given at the beginning of the fourth section.

On the angles that occur between the ecliptic and the meridian circle B.4.III.3

This section (§B.4.III.3–6), to the end of fol. 86v, parallels *Almagest* II.10.

237. See Toomer, *Ptolemy's "Almagest,"* p. 424.

238. Both Anatoli's translation of *The Almagest* (see Paris BNF MS Hēbreu 1018, p. 197) and the *Almagest* summary in the Vatican Ebr. 392 codex (fol. 30v) listed 42 revolutions in longitude and 3;10°.

239. Biṭrūjī (trans. and comm. Goldstein), *On the Principles,* vol. 1, p. 8.

87r

We know the division of the three remaining quadrants B.4.III.7

This passage follows *Almagest* II.7, including the results of the computations. Ibn Naḥmias has reproduced Ptolemy's demonstration that if one can compute the individual rising times for a single quadrant, one has done so for the other quadrants as well. Although figure 2.30 appears in the Vatican MS, there is no reference to this figure in the text of the MS.

87v

We find [the ratio] in the first 10 B.4.III.11

The parameters are not exactly the same as in the *Almagest,* but these computations correspond to II.7 of the Anatoli translation of the *Almagest.* The set of tens that came first in *The Light of the World* came second in Anatoli's translation.[240]

240. Anatoli, *Almagesṭi,* Paris BNF MS Hébreu 1018, 19a–b.

6

Commentary on the Significant Insertions in the Hebrew Recension of *The Light of the World*

From 124a–123b–a §A.II.5/X

The two arcs ZK KL *from small circle* ZKL

(See figure 4.1.) This demonstration purports to show that the sines of small circle arcs are commensurable with the sines of great circle arcs. In the figure, arcs *AE* and *ED* are on the equator of an orb. The recension's argument may have been that small circle arcs subtend the same angle at the axis of the sphere as the great circle arcs to which they are parallel. But great circle arcs subtend that angle at the center of the sphere while small circle arcs do not. Thus, the alleged relationships between great and small circle arcs hold only when one projects the great and small circles into the plane.

From 123b–a §A.II.7/X

Hence, it has been shown, through the ratio that has been compounded

A compounded ratio, as defined in §A.II.7 of the Judeo-Arabic original, means that with three great circle arcs, the ratio of the first to the third is the product of the ratio of the first to the second and the second to the third. The opposite of compounding (*meḥubbar*/*harkaḇah*) a ratio is separating (*peiruq*).[1] The recension has claimed that the ratio will also hold for small circle arcs if they are parallel

1. See also Euclid (trans., intro., and comm. Thomas Heath): *The Thirteen Books of The Elements* (Mineola, NY: Dover Publications, 1956), vol. 2 of 3: p. 184. There Heath used the terms *componendo* and *separando* for compounding and dividing. Euclid did not provide a direct explanation of these terms in Book Six, Proposition 23 (Ibid., vol. 2, p. 249). He came closest to explaining how one compounds ratios in Book Six, Definition 5 (Ibid., vol. 2, p. 189), which reads, "A ratio is said to be

to great circles, that is, if they were to resemble the latitude parallels. But if the measures of the small circle arcs have to be quantified, the ratios could not be computed with small circle arcs because the rules of spherical trigonometry do not hold for small circle arcs. Thus the recension's assertions about small circle arcs held only if one projects the great and small circles into a plane.

From 121b §A.IV.2/X

And when we propose

(See figure 4.2, which substitutes for figure 2.10.) Here, the recension is attempting to extend the law of six quantities, the Menelaos theorem, to include small circle arcs.[2] Again, the Hebrew word *m^e^ḥubbar*, like *harkab̲ah*, seems to refer to the compounding (i.e., multiplying) and not the composition (addition) of ratios. The recension adds two relationships to the ones found in the Judeo-Arabic original. All of the following pairs of letters denote arcs.

The first relationship added is: SW or $EH/\sin AB = \sin HE/\sin EB\ \sin ZT/\sin TH$ (it is likely that the recension intended "sin SW or sin EH")

The second is: The ratio of SW or $EH/\sin AB$ can also be expressed as $\sin SW/\sin SB = \sin EH/\sin EB\ \sin ZT/\sin WN$
(again, it is likely that the recension intended "sin SW or sin EH")

When the relationships are combined, $\sin EH/\sin EB\ \sin ZT/\sin TH = \sin SW/\sin SB = \sin EH/\sin EB\ \sin ZT/\sin WN$

therefore, $TH = WN$

The reasoning behind the recension's attempt to extend the law of six quantities to small circle arcs does not hold up for a few reasons. First, as SW is a small circle arc, the rules of spherical trigonometry do not hold. Second, both versions of the text state, according to the first figure, that:

$\sin ZA/\sin AB = \sin ZT/\sin TH\ \sin HE/\sin EB$
thus, $\sin ZA/\sin AB = SW$ or $EH/\sin AB$
(or, more likely, $\sin ZA/\sin AB = \sin SW$ or $\sin EH/\sin AB$)
therefore, $\sin ZA = SW$ or EH
(or, more likely, $\sin ZA = \sin SW$ or $\sin EH$)

---

compounded of ratios when the sizes of the ratios multiplied together make some (? ratio, or size)." But Heath believed that this statement was a later interpolation.

2. On the law of six quantities, see Glen Van Brummelen: *Heavenly Mathematics* (Princeton, NJ: Princeton University Press, 2012): pp. 45–51.

Because arc *ZA* was defined as a quadrant, arcs *SW* or *EH* would have to be a quadrant, and inspection shows neither of them to be.

From 120b §B.0.3/X

The fourth treatise on the motions of the five planets

The Hebrew recension, the sole MS of which did not contain a treatise on the planets, nevertheless included a table of contents for a treatise on the planets. Conversely, the sole MS of the Judeo-Arabic original, which included a treatise on the planets that terminated abruptly, did not mention the section on the planets in the table of contents found at the beginning of the text. For the contents of the treatise on the planets' motions, see the commentary on the Judeo-Arabic original on §B.4.0.1. The statement about "the measure of the circle that produces the anomaly related to the Sun" indicated that even the Hebrew recension did not propose solid orbs as movers for all of the motions.

From 117b §B.1.II.20/X

We have explained this matter here, and we transfer it to the lunar model

The recension has referred ("this matter") to the double-circle hypothesis. Like the Judeo-Arabic version, the recension proposed, without providing all of the details, applying the double-circle hypothesis to the Moon's motions in order to remove an outstanding displacement from the circle of the Moon's latitude.

The passage's second sentence, about subdividing an orb, is more evidence for the Hebrew recension's increased attention to proposing physical movers for the orbs. That the double circles might be traced by the motions of the poles of two orbs, with one orb surrounding the other, was only implied but never fully explained in the Judeo-Arabic version. The recension, in contrast, specified that the two circles had to be proposed in two separate orbs, with one surrounding the other. Thus, both poles would travel in small circles with one pole being carried on the circumference of one of the small circles.

Finally, the recension has expressed the belief ("the period of time which is from the greatest motion to the mean motion is less") that the homocentric models can account for the asymmetries of the Ptolemaic model, though they cannot.

The Sun is also found, through this second hypothesis. . . . Thus, we have chosen the first hypothesis §B.1.II.26/X.1–2

These paragraphs note that the circle of the path of the center and the double-circle hypothesis work together to account for variations in motion *and* to keep the Sun in the ecliptic. One way to have the location of greatest speed vary, in order to agree with observations, would be to follow the first opinion, which permitted apparently opposite motions in the heavens, eliminate the circle of the path of the center, and account for the entire anomaly with the double-circle hypothesis. Or the intersection of the ecliptic with the inclined circle carrying the circle of the path of the center could be separated from the equinoxes.

We propose some point or pole or center of a planet on the surface of an orb §B.1.II.26/X.4
This is the beginning of a long expansion of the original version's solar model. The first part (§B.1.II.26/X.5) introduced physical movers to produce a near oscillation on a great circle arc through the double-circle hypothesis. This first proposal presented the double-circle hypothesis as a result of the motion of the poles of orbs (see also §B.1.II.20/X). The lemma introduced three ways in which the near-linear oscillation produced by the double-circle hypothesis that resembles the Ṭūsī couple might result from poles causing motions at the equators of orbs. The first two (§B.1.II.26/X.7–9) are brief. The third (§B.1.II.26/X.10–13) of those ways is akin to a Eudoxan couple (i.e., concentric orbs). Finally, the recension analyzed (§B.1.II.26/X.14–31) the slants that arise from the third way as well as ways to eliminate the departures in latitude arising from the Eudoxan couple. Many of the recension's proposals could be modified to replace the circle of the path of the center entirely.

(See figure 6.1.) To begin with a planar perspective on the physical movers for the double-circle hypothesis that resembles the Ṭūsī couple, in the figure, circle *EZA* revolves 45°, bringing point *E*, the center of circle *GHK* to *E′*. Then circle *GHK*, in its new position, revolves 90° (twice 45°) about center *E′*, meaning that point *K* will always be oscillating on *DAGEK*. The diameter of one of the small circles is half of *DK*. Though the recension alleged that the measure of the diameter can change according to the location of the point of greatest speed, whether a solstice, equinox, or elsewhere, the measure of the diameters of the small circles is constrained by the role that the double-circle hypothesis plays in keeping the Sun in the ecliptic. For the purposes of calculation, one presumes that each orb has a negligible thickness.

(See figure 6.2.) Due to the curved surface of the orb, the actual path of the point will be a narrow hippopede; point *K* will often not lie precisely on arc *GE*.[3] The three approaches to producing a near oscillation on a great circle arc through the motions of solid orbs and not just their poles, beginning in §B.1.II.26/X.7, attempt to produce the oscillation by having the poles of orbs drive the motion of great circles at or near the orbs' equators. The wording of the Hebrew recension suggests that the approaches that produce the near-linear oscillation at the orbs' equators, most importantly the third way, which was the Eudoxan couple, were perceived by the recension to be equivalent to the aforementioned double-circle hypothesis.[4] Mathematically, though, the proposals beginning in §B.1.II.26/X.7 are not as precise as the double-circle hypothesis that resembles the Ṭūsī couple.

3. For the finding that the actual path of the point is a pinched hippopede, see George Saliba and E. S. Kennedy: "The Spherical Case of the Ṭūsī Couple," in *Arabic Sciences and Philosophy* I (1991): pp. 285–91. See also the comments on §B.1.II.20.

4. Ragep has argued for similarities between the Ṭūsī couple and Ibn al-Haytham's attempt to

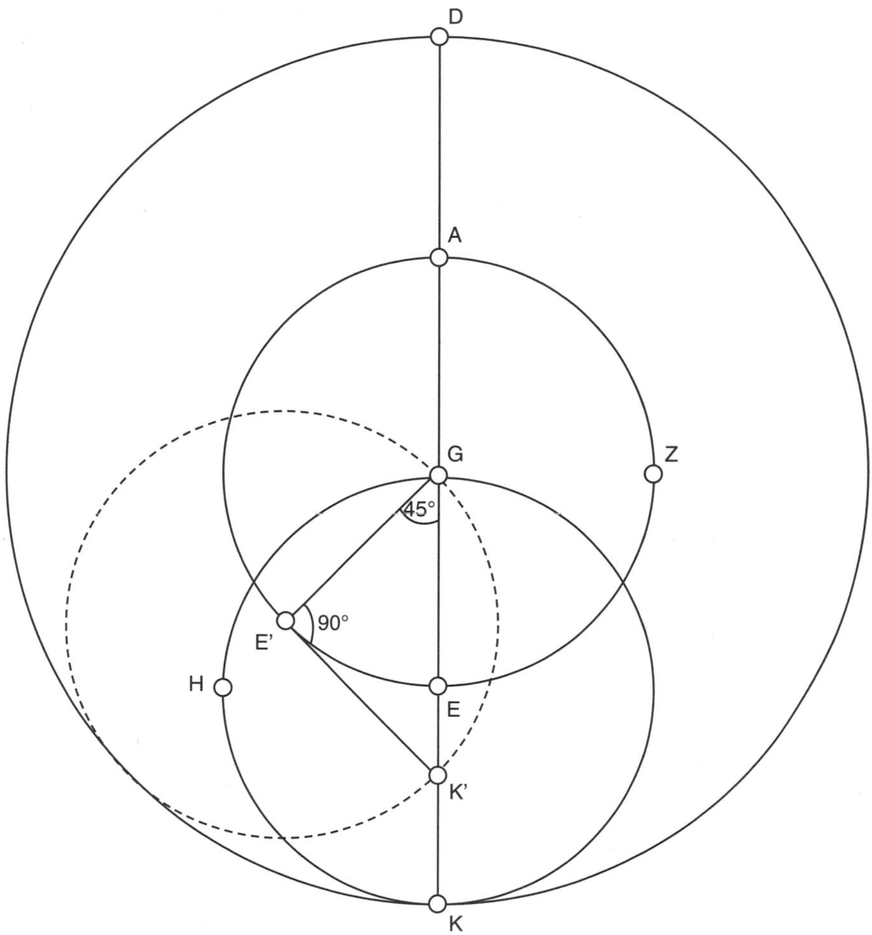

FIGURE 6.1

And if we propose that the mobile is on arc KD §B.1.II.26/
(See figures 4.3 and 6.1.) A slant is the change of a small circle's diameter's (itself X.5
a great circle arc) alignment with respect to the equator of the orb that carries it (see commentary on §B.1.II.10 in chapter 5). The Judeo-Arabic version showed that the slant could be a result of the model for the solar anomaly, with the circle of the path of the center, or the slant could result from these hypotheses for creating an oscillation. In this paragraph, the recension investigates the slant resulting

correct the faults of the Ptolemaic theory for planetary latitudes with concentric epicycles. See F. J. Ragep (ed., trans., and comm.): *Naṣīr al-Dīn al-Ṭūsī's "Memoir on Astronomy" (al-Tadhkira fī 'ilm al-hay'a)* (New York and Berlin: Springer-Verlag, 1993): pp. 450–6.

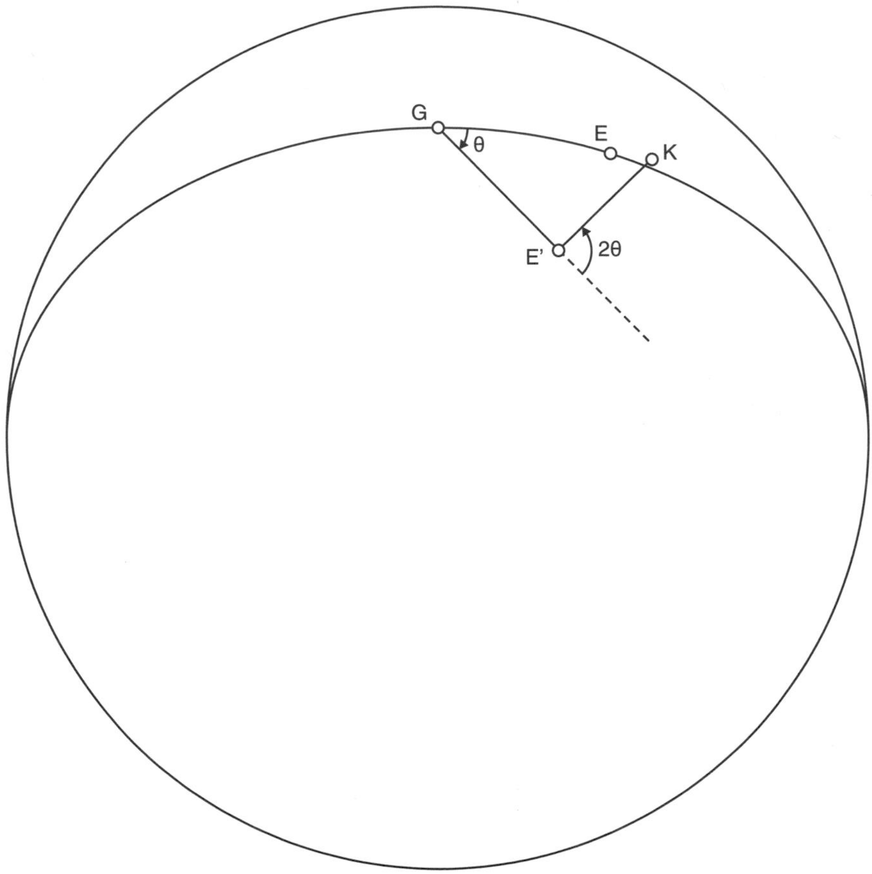

FIGURE 6.2 Adapted from Saliba and Kennedy (1991).

from the hypothesis of the double circles. In the figure, arc *GEK* is the diameter of a small circle; if point *E* moved 90° to point *Z*, point *K* would be superimposed on point *G* after small circle *GHK* rotates 180° in the opposite direction. The diameter that was superimposed on arc *GEK* would be perpendicular to arc *DGK* and therefore slanted by 90°.

From 116a

the above-mentioned diameter, rather the arc upon which it

The recension has not explained how the entire diameter, or the arc to which it is attached, could remain superimposed on arc *DGK*. The recension must have intended that the diameter ascends and descends on the arc in the sense that *K*, an endpoint of the diameter, is always on arc *DK*.

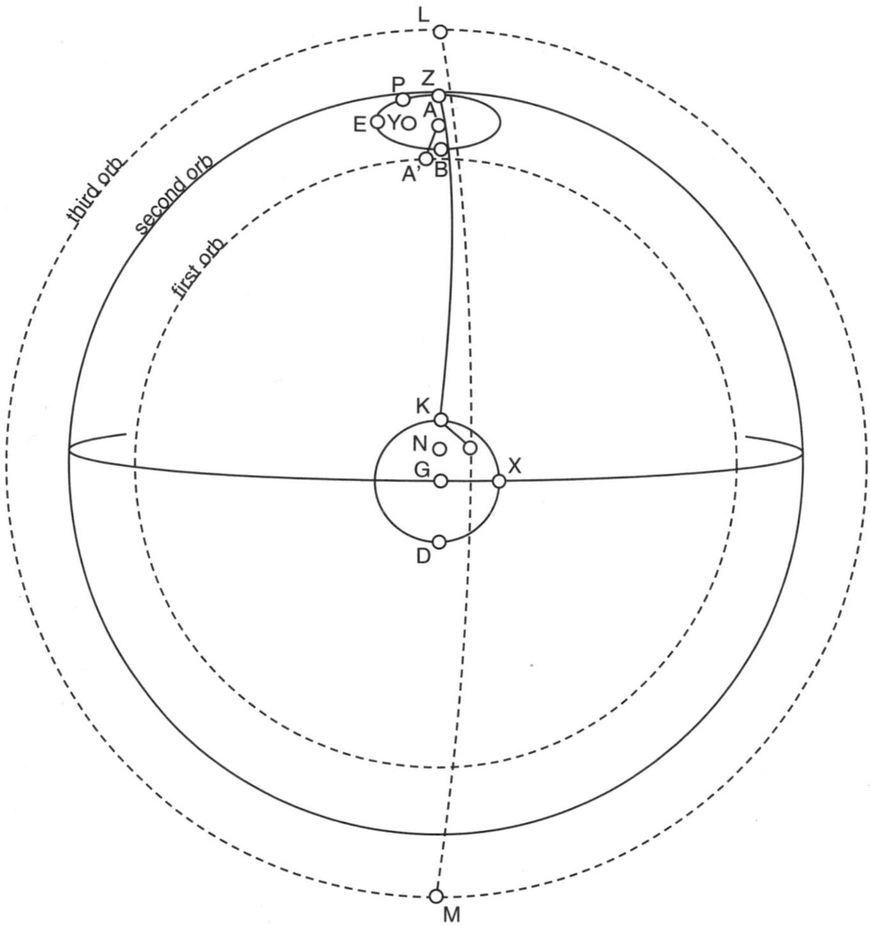

FIGURE 6.3

The first of the three ways §B.1.II.26/X.7

These three ways are the recension's answer to the question of how the near oscillation of a point on a great circle arc at the orbs' equator can be the result of the motion of the orbs' poles. The hypothesis of the slant emerges as a result of the third way, which appears in §B.1.II.26/X.10, with the slant coming up in §B.1.II.26/X.14 For the purposes of calculation, one presumes that each orb has a negligible thickness.

(See figure 6.3.) In the first of the three ways, the recension proposed that there is a third orb in which point *K* and the point opposite (not shown) are fixed in such a way that they move only on a portion of arc *LM* that faces arc *ZABKGD*.

Point *K* and the point opposite it are in the second orb. When the motion of the first, lowest, orb about pole *A* causes pole *Z* on the second orb to move to point *E* and then to point *B*, pole *K*, because it is fixed in the third orb, will oscillate only on arc *KGD*. The direction of the oscillation of point *K* on arc *KGD* will depend on whether pole *Z* is in small circle arc *ZEB* or not. The recension did not explain how the poles of one orb were attached to the other orbs.

The oscillation of point *K* on arc *KGD* will not be uniform. On the arc from point *K* to point *G*, the point's motion through the second half will be faster than its motion through the first, and on the arc from point *G* to point *D*, its motion through the first half will be faster than its motion through the second. To demonstrate this, draw great circle arc *GP*, where angle *ZAP* is 45°. Arc *GP* intersects arc *EA* at point *Y*, and because arcs *GE*, *GY*, and *GA* are 90°, then angle *PYA* is 90°, and angle *PAY* is 45°. In these calculations, we assume that the radius of small circles *BEZ* and *KXD* is 5°. According to the law of sines:

$\sin 5° / \sin 90° = \sin PY / \sin 45°$
$\sin 5° \sin 45° = \sin PY$
$3;32° = \text{arc } PY$
Since arc $GY = 90°$, $GP = 93;32°$

Then, we bisect arc *KG* at point *N*, making arc *GN* 2;30°. If point *K* had moved all the way to point *N* through the motion of pole *Z* to point *P*, then arc *PN* would be 90°. But then the sum of two sides of a triangle, arcs *PN* and *GN*, would be less than the third side, arc *GP*.

The second way §B.1.II.2

(See figure 6.4.) In the figure, poles *T* and *H* in the second orb are the poles of great circle *DGK*. Poles *T* and *H* are fixed in a lower orb, the first orb. Then, point *K* and the point opposite it are affixed to a third orb with pole *Z*. Pole *Z* is fixed in a fourth orb with pole *A*. The distance between pole *Z* and point *K* will remain a quadrant because point *K* (in the second orb) is affixed to the orb whose pole is *Z*. When pole *Z*, on the third orb, revolves about point *A* on the fourth orb, toward point *E*, point *K* would tend to move to point *X*. But, instead, circle *DGK* revolves about poles *T* and *H*, moving point *K* toward point *G*. X.8

As was true with the first approach, the oscillation back and forth is produced along great circle arcs. In addition, both the first and second ways entail incomplete revolutions for certain orbs. The difference between the first and second ways is that in the second way, point *K* is fixed in the third orb; points *D* and *G* are in the second orb. In the first way, one would have to propose a groove along arc *LM* (in figure 6.3) so that point *K* would stay on track. In this second way, an orb moves back and forth without needing such a groove. This question of how to keep the oscillating point on track was not confined to *The*

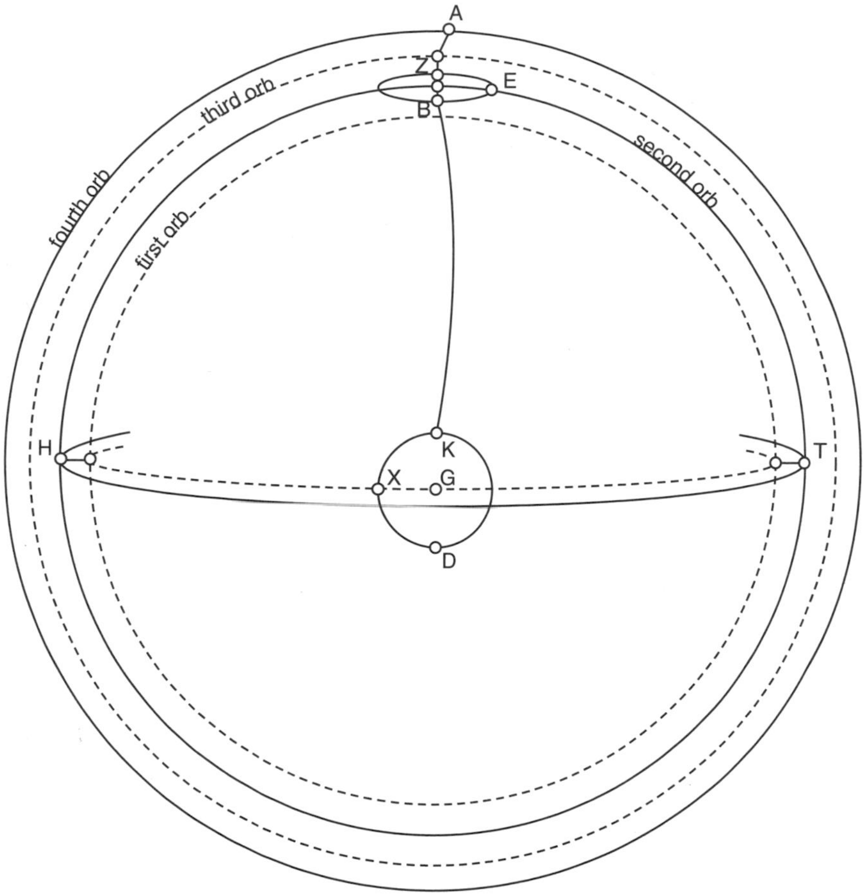

FIGURE 6.4

Light of the World. Noel Swerdlow, in his analysis of Regiomontanus's homocentric astronomy, observed that in Regiomontanus's reciprocation mechanism, an oscillating point would stray from the great circle arc because nothing was keeping it on track.[5] I have argued elsewhere that astronomers at Padua would

5. See Noel Swerdlow: "Regiomontanus's Concentric-Sphere Models for the Sun and Moon," in *Journal for History of Astronomy* XXX (1999): pp. 1–23, at p. 17. "To expect *S* to know to remain in the ecliptic while *L* moves about a circle is to expect a great deal." I thank Noel Swerdlow for apprising me of this article. Swerdlow has noted, as well, that Regiomontanus's reciprocation mechanism could have been a modification of Biṭrūjī's theories. On that, see ibid., p. 4 and p. 17.

have been particularly interested in the Hebrew recension of *The Light of the World* because it contained solutions to these problems with Regiomontanus's reciprocation mechanism.[6]

As was the case with the first approach, and for the same reasons, the oscillation of point *K* on arc *KGD* will not be uniform. On the arc from point *K* to point *G*, the second half will be faster than the first, and on the arc from point *G* to point *D*, the oscillation of point *K* in the first half will be faster than the oscillation of point *K* in the second half.

The third way, it being the most complete §B.1.II.2

(See figure 4.5.) Here, the recension proposed an oscillation of a point from point *K* to point *G* always through complete revolutions, that is, without the requirement, which was necessary in the first two approaches, that some of the orbs oscillate back and forth without completing a revolution. Nor did this third proposal rely on fixing the oscillating point in a track or groove. The core of this approach requires only two orbs, one with pole *Z* and points *D*, *G*, and *K*, and another with pole *A*. For the purposes of calculation, one presumes that each orb has a negligible thickness. The first orb is fixed in the second at a point (*A*′) corresponding to pole *A* on the second orb. First, the second orb revolves about pole *A*, causing pole *Z* to revolve a certain number of degrees, 90 in the figure in the MS, about pole *A* to arrive at point *E*. As a result, the position of point *K* becomes point *M*, with a latitude from *T* equal to arc *KG*. Then, the first orb whose pole is *Z* rotates about pole *Z* to bring point *K* from point *M* over to point *G*. The preceding sequential description serves just to help the reader envision how the proposal functions. All of the motions would, in fact, occur simultaneously. As the author of the Hebrew recension acknowledged ("there occurs for point *K* a slight deviation"), this solution will not function precisely. This is because arcs *GA* and *GT* are 90°, and arc *GE*, if *E* is truly the pole of great circle arc *QMG*, must also be 90°. Thus, for arc *MG* to be 90°, point *M* must be on arc *EAT*. Therefore, after 90°, 180°, and 270° of pole *Z*'s revolution on small circle arc *BEZ*, the model functions. X.10

(See figure 6.5.) After pole *Z* has revolved 45° on small circle arc *BEZ*, however, there will be a deviation as the recension intimated. The recension was correct to note, though, that the deviation will not be large. The calculation presumes that the radius of small circle *BEZ* is 2;30° and that the thickness of the orbs is negligible. We solve triangle *A'NY'*, with point *N* being the location of point *K* once *Z* has revolved 45° to *P*, but before *K* has revolved, in the opposite direction,

6. See "A Scholarly Intermediary between the Ottoman Empire and Renaissance Europe," *Isis* CV (2014): pp. 32–57, at pp. 45–6.

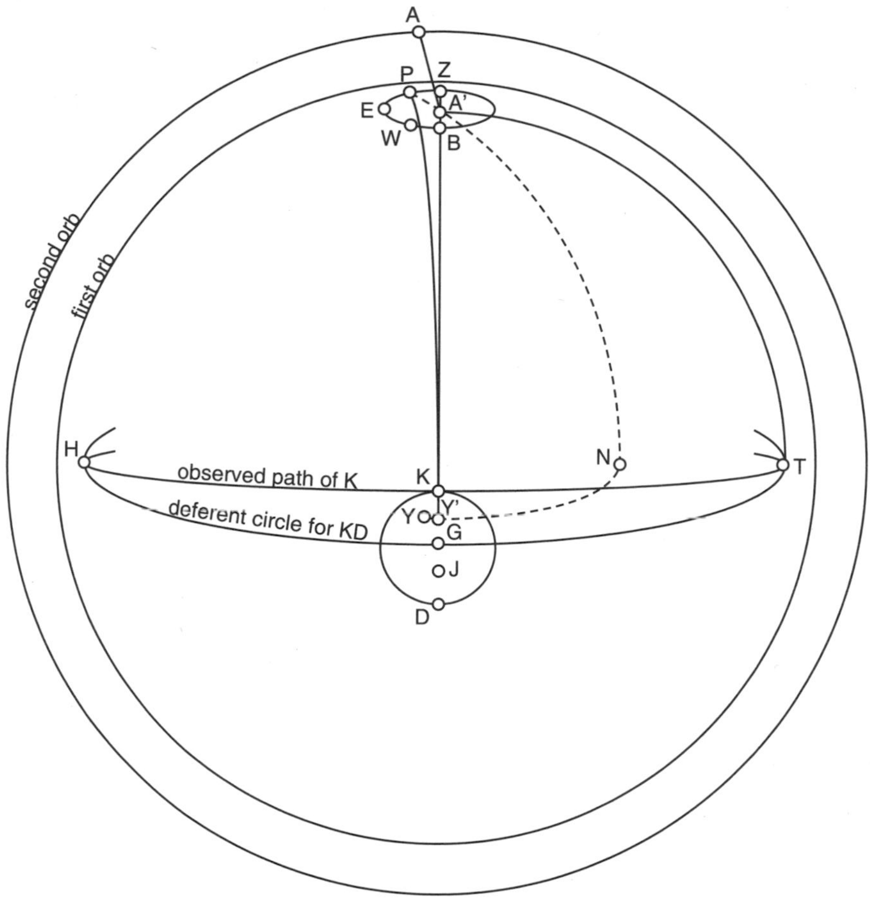

FIGURE 6.5

about pole *Z* at *P*. Angle *A'* is 45°, *y'* (side *A'N*) is 87;30°, and angle *N* is 90°. Then, we apply the law of cosines:

$\cos Y' = -\cos A' \cos N + \sin A' \sin N \cos y'$
Because $N = 90°$, $\cos Y' = \sin A' \sin N \cos y'$
$\cos Y' = \sin 45° \sin 90° \cos 87;30°$
$Y' = 88;14°$ and angle $A'Y'P$ is 1;46°

Then, following the sine law:

$\sin 87;30° / \sin 88;14° = \sin a' / \sin 45°$
a' (side NY') = 44;58°

Thus, for point K to oscillate precisely on arc KGD, one orb's revolution could not be precisely equal to the other. Were the two orbs to revolve equally, point Y would be displaced from arc DGK by 0;2°. There is no indication that the author of the recension calculated this deviation as a function of the measure of the angular revolution of pole Z on small circle.

Analysis of this hypothesis will show that it is equivalent to part of Schiaparelli's reconstruction of Eudoxus's models; the oscillating point (K) will trace a hippopede on the surface of an orb. Hence, this hypothesis constitutes what Ragep has termed a Eudoxan couple, a pair of concentric orbs, with the pole of one inclined to the pole of the other, revolving in opposite directions at the same rate.[7] Graph 6.1 illustrates this hippopede from the Eudoxan couple, which should be compared with the much narrower hippopede traced by the double-circle hypothesis.

The classic analyses of the hippopede in ancient astronomy take Schiaparelli's reconstructions of Eudoxus's models as their starting point.[8] The following discussion, however, begins by analyzing the recension's model with spherical trigonometry and works backward to Schiaparelli's reconstruction of Eudoxus's models. Though there is no evidence that the author of the recension carried out such a systematic analysis, such an analysis is worthwhile because it illustrates the vast degree to which Eudoxan principles of modeling were present in *The Light of the World* and that a medieval scholar equipped only with spherical trigonometry could have plotted a hippopede. If the author of the recension did assess the models' predictive accuracy (cf. §B.1.II.26/X.12): "there occurs for point K a slight deviation to the right and to the left from line DK"), the author could easily have done so by calculating positions with spherical trigonometry.

(See figure 6.6.) In the figure, we first solve spherical triangle ABC. Side c is r. According to the law of cosines:

$$\cos C = -\cos A \cos B + \sin A \sin B \cos c$$
$$\cos C = -\cos\theta \cos(180° - \theta) + \sin\theta \sin(180° - \theta) \cos r$$
$$\cos C = -\cos\theta{*}(\cos 180° \cos\theta + \sin 180° \sin\theta) + \sin\theta\,(\sin 180° \cos\theta - \cos 180° \sin\theta){*}\cos r$$
$$\cos C = \sin^2\theta{*}\cos r + \cos^2\theta$$
$$C = \arccos(\sin^2\theta{*}\cos r + \cos^2\theta) \qquad (1)$$

7. Ragep, *Tadhkira*, pp. 451–2.

8. Schiaparelli demonstrated his parametric equations, in modern notation, in Giovanni Schiaparelli: *Scritti sulla storia della astronomia antica* (Bologna: N. Zanichelli, 1925–7): vol. 2 of 3: pp. 3–112 (the essay appeared originally in 1877), at pp. 48–53. See also, Thomas Heath: *Aristarchus of Samos: The Ancient Copernicus* (Oxford: Clarendon Press, 1913; Mineola, NY: Dover Publications, 1981, reprinted 2004): pp. 204–5; and Otto Neugebauer: "On the 'Hippopede' of Eudoxus," in *Scripta Mathematica* XIX (1953): pp. 225–29; reprinted in *Astronomy and History* (New York, Heidelberg, and Tokyo: Springer-Verlag, 1983): pp. 305–9.

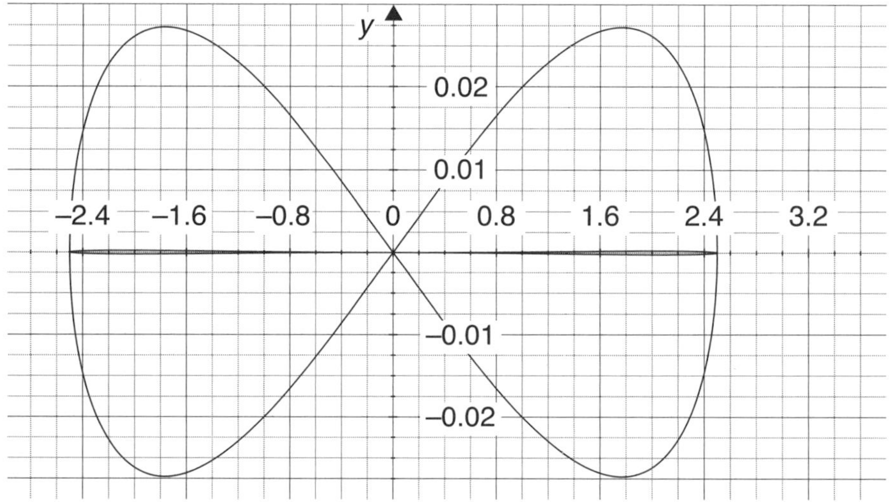

DETAIL OF HIPPOPEDE

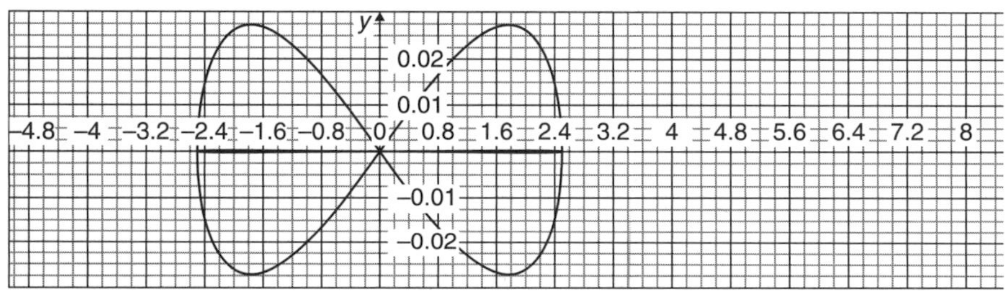

GRAPH 6.1

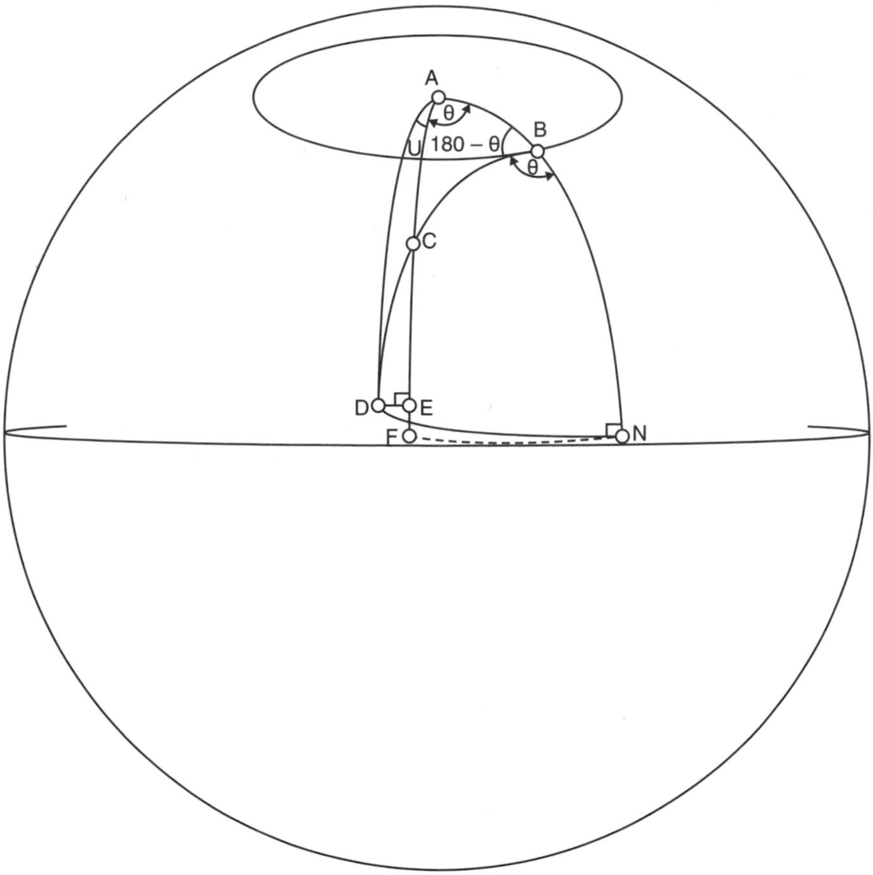

FIGURE 6.6

We can use the law of cosines to solve for *a*:

$$\cos A = -\cos B \cos C + \sin B \sin C \cos a$$
$$\cos\theta = -\cos(180° - \theta) \cos C + \sin(180° - \theta) \sin C \cos a$$
$$\cos\theta = -(\cos 180° \cos\theta + \sin 180° \sin\theta) \cos C + (\sin 180° \cos\theta - \cos 180° \sin\theta) \sin C \cos a$$
$$\cos a = \cos\theta\,(1 - \cos C)/\sin\theta \sin C$$
$$a = \arccos((\cos\theta - \cos\theta \cos C)/\sin\theta \sin C) \qquad (2)$$

We take point *E* to be the origin on the surface of the orb; we can calculate the displacement of point *D* from an oscillation on a great circle arc along the y-axis. In spherical triangle *CDE*, angle *C* is congruent to angle *C* in spherical triangle

ABC. Side *CD* is $90° - a$. Because angle *E* is 90°, the displacement is equal to side *DE*. According to the law of sines:

$$\sin(90° - a)/\sin 90° = \sin DE/\sin C$$
$$\cos a = \sin DE/\sin C$$
substituting equation (2) for a: $DE = \arcsin(\sin C\,(\cos\theta - \cos\theta\,\cos C)/\sin\theta\,\sin C)$
$$DE = \arcsin(\cos\theta\,(1 - \cos C)/\sin\theta)$$
$$DE = \arcsin(\cos\theta\,(1 - \sin^2\theta\,\cos r - \cos^2\theta)/\sin\theta)$$
$$DE = \arcsin[\cos\theta\,(1 - \sin^2\theta\,\cos r - (1 - \sin^2\theta))/\sin\theta]$$
$$DE = \arcsin[\cos\theta\,\sin^2\theta\,(1 - \cos r)/\sin\theta] = \arcsin[\sin 2\theta\,\sin^2 r/2]$$

Arc *DE* is one of the spherical coordinates that describes the position of point *D*. To determine the angle of arc *EF*, the other spherical coordinate of point *D*, one calculates arc *AE* with the following steps. Angle *U* is angle *DAE*, and arc $BD = 90°$.

A. We consider triangle *ABD* and apply the rule of cosines for sides:[9]

$$\cos b = \cos a\,\cos d + \sin a\,\sin d\,\cos B$$
$$\cos b = \sin 90°\,\sin r\,\cos(180° - \theta)$$
$$\cos b = -\sin r\,\cos\theta = \cos AD \qquad (3)$$

B. In that same triangle, because[10] $\sin b\,\cos A = \cos a\,\sin d - \sin a\,\cos d\,\cos B$

$$\text{then } \sin b\,\cos(U + \theta) = -\sin 90°\,\cos r\,\cos(180° - \theta)$$
$$\sin AD\,\cos(U + \theta) = \cos r\,\cos\theta$$

C. In that same triangle, we apply the law of sines:

$$\sin AD/\sin\theta = 1/\sin(U + \theta)$$
$$\sin(U + \theta)\,\sin AD = \sin\theta$$

D. Multiply the equation in §B by cosθ:

$$\cos\theta\,\sin AD\,(\cos U\,\cos\theta - \sin U\,\sin\theta) = \cos r\,\cos^2\theta$$

E. Multiply the equation in §C by sinθ:

$$\sin\theta\,\sin AD\,(\sin U\,\cos\theta + \cos U\,\sin\theta) = \sin^2\theta$$

F. Add the equations from §D and §E:

$$\sin AD\,\cos U = \cos r\,\cos^2\theta + \sin^2\theta \qquad (4)$$

9. See Heath, *Aristarchus*, pp. 204–5. Steps *A* through *G* resemble the demonstration that Heath provided.

10. W. M. Smart: *Text-Book on Spherical Astronomy*, 6th ed. (Cambridge: Cambridge University Press, 1960): p. 10.

G. We consider triangle *ADE* and apply an analogue[11] of the law of cosines for sides:

$\sin a \cos E = \cos e \sin d - \sin e \cos d \cos A$
because angle $E = 90°$, $\sin e \cos d \cos A = \cos e \sin d$
$\sin e \cos A = \tan d \cos e$
$\sin AD \cos U = \tan d \cos e$
Substituting equation (4) for $\sin AD \cos U$ and equation (3) for $\cos e$ (which is $\cos AD$), $\tan d = \sin AD \cos U / \cos e = (\cos r \cos^2\theta + \sin^2\theta)/{-\sin r \cos\theta}$
because *EF* is the complement of *d*, then $\tan EF = \cot d = -\sin r \cos\theta/(\cos r \cos^2\theta + \sin^2\theta)$
$EF = \arctan(-\sin r \cos\theta/(\cos r \cos^2\theta + \sin^2\theta))$

The equations for arcs *DE* and *EF* defined the angles that those arcs made at the center of an orb with respect to an origin at *E*. Schiaparelli's reconstruction, though, projected the hippopede into the plane.[12] We redefine the spherical coordinates of point *D* with respect to origin *F*. Angle *EFD* (angle *F*) is the azimuthal (Θ) coordinate, and arc *FD* is the distance from the pole (Φ). Then we apply the law of sines to triangle *AFD*:

$\sin DF/\sin U = \sin AD/\sin F$
$\sin DF \sin F = \sin AD \sin U$ (5)

Then we apply the rule of cosines for sides:

$\cos AD = \cos DF \cos 90° + \sin DF \sin 90° \cos F$
$\cos AD = \sin DF \cos F$ (6)

Assuming a unit sphere, to convert spherical coordinates to cartesian coordinates, we apply the formulas $x = \cos\Theta \sin\Phi$ and $y = \sin\Theta \sin\Phi$. Thus:

$x = \cos\Theta \sin\Phi = \cos F \sin DF = \cos AD$ [From equation (6)]
From equation (3), $\cos AD = -\sin r \cos\theta$ (7)
$y = \sin\Theta \sin\Phi = \sin DF \sin F = \sin AD \sin U$
From equation (5), $\sin DF \sin F = \sin AD \sin U$ (8)

Now, we consider spherical triangle *ABD* and apply the following identity:[13]

$\sin a \cos B = \cos b \sin d - \sin b \cos d \cos A$
$\cos(180° - \theta) = \cos AD \sin r - \sin AD \cos r \cos(U + \theta)$

11. Smart, *Text-Book*, p. 10.
12. Heath, *Aristarchus*, p. 205.
13. Smart, *Text-Book*, p. 10.

we apply equation (7): $\cos(180° - \theta) = (-\sin r \cos\theta) \sin r - \sin AD \cos r \cos(U + \theta)$

$$\sin AD \cos(U + \theta) = \cos\theta \cos r \qquad (9)$$

We apply the law of sines:

$$\sin AD/\sin(180° - \theta) = 1/\sin(U + \theta)$$
$$\sin AD \sin(U + \theta) = \sin\theta \qquad (10)$$

We multiply equation (9) by $-\sin\theta$ and equation (10) by $\cos\theta$, and add them, to yield:

$$\cos\theta \sin AD \sin(U + \theta) - \sin\theta \sin AD \cos(U + \theta) = \sin\theta \cos\theta - \sin\theta \cos\theta \cos r$$
$$\sin AD [\cos\theta \sin(U + \theta) - \sin\theta \cos(U + \theta)] = \sin\theta \cos\theta (1 - \cos r)$$
$$\sin AD \sin U = \sin^2(r/2) \sin(2\theta) \qquad (11)$$

From equation (11), $\sin AD \sin U$ is the y-coordinate; from equation (7), we have the x-coordinate. Schiaparelli's equations for the projection of the hippopede into a plane, for a sphere of R=1, were $x = \sin r \cos\theta$ and $y = -\sin^2 r/2 \sin 2\theta$.[14] The double-circle hypothesis yields a narrower hippopede (see, again, graph 6.1). At $\theta = 45°$, there is a maximum displacement of 0;1,37° with the Eudoxan couple, versus a maximum displacement of 0;0,6° with the double-circle hypothesis at $\theta = 55°$.

From 115b

(See figure 4.4.) *Know that between the two points* DG *and between the two points* GK §B.1.II.26/X.12

This comment indicates that the operation of this hypothesis was known, at least to an extent. As was the case with the hypothesis of the slant, inclined poles meant imperfect alignments.

a great circle arc equal to arc DGK §B.1.II.26/X.13

For the same reason there is a slight deviation of point *K* from arc *DGK*, point *T* will not trace arc *STM*.

For it is clear that whenever the point is near the edges of the arc

(See figure 6.5.) The recension does not explain how it arrived at this conclusion about the variations in the point's speed. We can determine the location of point *Y'* on arc *KG* in order to determine how far *K* has moved toward *G* after the first 45° of mean motion. In triangle *YA'N*, *a'* (side *NY'*) is 44;58° (see comments on §B.1.II.26/X.10), angle *A'* is 45°, and angle *N* is 90°. Following the law of sines:

14. See Heath: *Aristarchus*, pp. 190–224, esp. 204–5. The technique for projecting the hippopede into the plane follows pp. 204–5.

$\sin 44;58° / \sin 45° = \sin n / \sin 90°$
n = side $A'Y' = 88;3°$

Since arc *A'K* is 87;30° and arc *A'G* is 90°, point *Y* is less than halfway from point *K* to point *G*, meaning that its fastest motion is between 45° and 135° of θ. Modern techniques confirm the recension's statement about the rate of change of the point's speed. We take the equation for the x-coordinate of a hippopede projected into the plane:

$x = \sin r \cos\theta$

Since the inclination of one pole to the other is fixed as *r*, the variable in the equation is cosθ and its derivative is –sinθ. Since the recension's comment pertained to absolute rates of change, angles 90° and 270° yield an absolute value of 1 for the slope of f(*x*)= cosθ. When θ is 90° or 270°, point *K* will be at point *G*, the midpoint of arc *DK*, the point of its fastest motion. See also the comments on §B.1.II.26.

then its diameter slants through these two motions §B.1.II.2
(See figure 4.4.) Here, the recension explained how the slant imparted to point *K* would affect the alignment of the diameter of small circle *KND* with respect to the circle carrying the circle of the path of the center. The recension held that arc *SN* on the small circle *KND*, equal to arc *KP*, is the measure of revolution that would be transferred to the circle of the path of the center through the slant. When it comes to calculating positions, the slant can be redefined from small circle arc *NS* to angle *NGS*. X.14

We determine the measure of the slant after pole *Z* rotates 90° to point *E*. The maximum of the angle (*NGS*) demarcated by arc *NS* will be equal to the radius of small circle *ZEB* (or small circle *KND*). To demonstrate that, consider triangle *MGT*. Arc *TG* is 90°, and arc *MG* must be 90° as well because the motion of point *K* from *M* to *G* is 90°. Since arc *EAMT* is a colure, then arc *MT* = arc *EA* = 2;30°. Then, if angle *MGT* is 2;30°, angle *NGS* is 2;30°, the radius of small circles *ZEB* and *KND*. The slant is also equal to the distance of the pole, rotating on small circle *BEZ*, from great circle arc *ZABKGD*.

Because the angle at the center of the path of the circle, again small circle *NGS*, is also the complement of angle *AGM* (in figure 4.4) or angle *A*'Y'N (in figure 6.5), we can describe the slant as follows (see, again, figure 6.5):

Angle *A'* is 45°, *y'* (side *A'N*) is 87;30°, and angle N is 90°. Then, we apply the law of cosines:

$\cos Y' = -\cos A \cos N + \sin A \sin N \cos y'$
$\cos Y' = \sin A \sin N \cos y'$
y' is always 87;30° (i.e., 90° – r), so $TY'N = 90° - Y' = 90° - \arccos(\cos(90° - r)\sin\theta) = \arcsin(\sin r \sin\theta)$

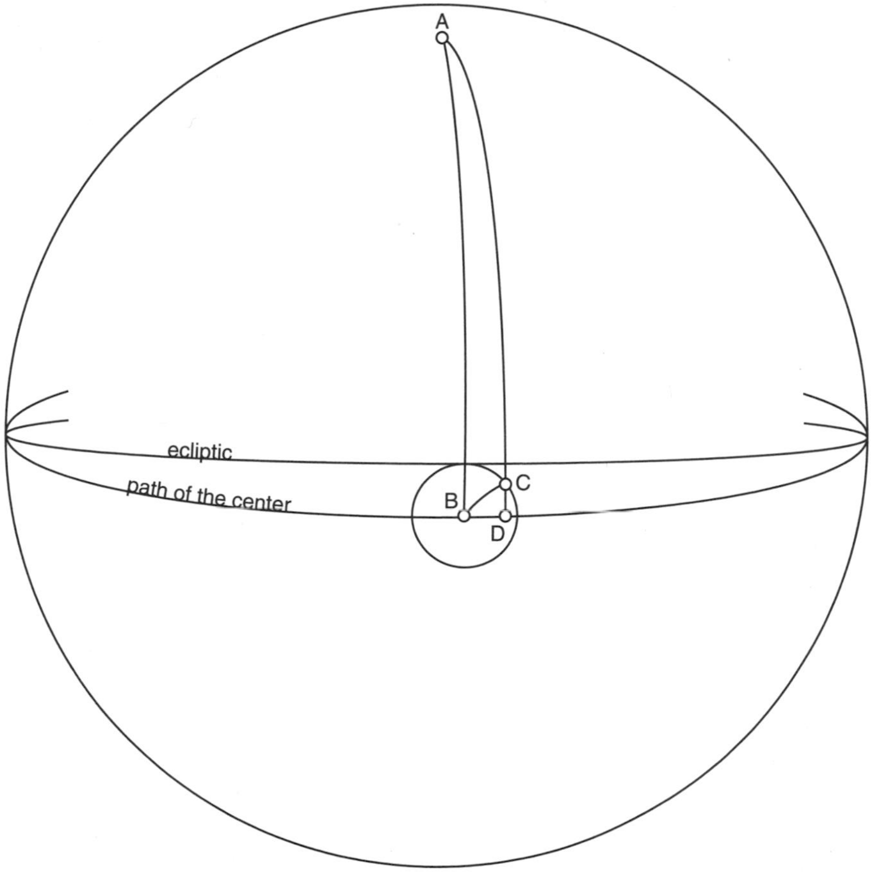

FIGURE 6.7

(See figure 6.7.) One can analyze the effect of the slant on the position of the Sun on the circle of the path of the center. The Φ coordinate of point *C* on the circumference of the circle of the path of the center is the distance *b* (side *AC*) where angle *CBA* is the sum of θ and the slant [which is arcsin(sin*r* sinθ)]. Using the cosine law for sides:

$\cos b = \cos c \cos a + \sin c \sin a \cos B$

$\cos b = \cos 90° \cos r + \sin 90° \sin r \cos(\arcsin(\sin r \sin\theta) + \theta)$

$\cos b = \sin r \cos(\arcsin(\sin r \sin\theta) + \theta)$

$b = \arccos[\sin r \cos(\arcsin(\sin r \sin\theta) + \theta)]$ = Φ coordinate of *C* corrected for the slant

Because the slant is arcsin(sin*r* sinθ) , then the effect of the slant on the Φ coordinate of *C* will be greatest when θ = 90° and 270°.

Φ(C) for θ = 90°, with the maximum slant of 2;30°, is 90;6,36°. Without the hypothesis of the slant, Φ(C) for θ = 90° is 90°. The hypothesis of the slant would seem to account for the maximum displacement (see comments on §B.1.II.13) incurred in Ibn Naḥmias's solar model with the circle of the path of the center of the Sun, but the recension has presented the slant as a by-product of the Eudoxan couple, not of the circle of the path of the center.

Also, we want the latitude that occurs for pole K §B.1.II.26/X.16

(See figure 4.4) Now, the recension has presented a series of modifications to the third approach, the Eudoxan couple. Here, the recension added two more orbs in order to remove the oscillation on great circle arc *KGD* while doubling the slant. The pole of the third orb faces pole *A* of the second orb, and the pole of the fourth orb faces pole *Z* of the first orb. While all four orbs move simultaneously, imagining the motions sequentially is easier. My comments presume that each orb is of negligible thickness so that a point on one orb could be said to be on a great circle arc with a point in another orb. After point *K* has moved to point *G*, that is, after θ = 90°, with pole *Z*, the pole of point *K*, moving to point *E*, the second, upper set of orbs moves the pole of the circle of the path of the center at point *G* to point *V* by the pole on the third orb facing pole *A* moving through the quadrant *AN* about a pole on the fourth orb facing pole *Z*. Then, the pole that is at point *E'* on the fourth orb revolves about *A'* to arrive at point *B'*. *E'*, *A'*, and *B'* are not indicated in the MS. Point *B'*, the new location of the pole of the circle of the path of the center, is on a great circle arc with point *Z* and is also 90° from point *K*. The new location of the pole of point *K*, *B'*, is 5° away from arc *ZABKGD*, doubling the slant. Although the comment on §B.1.II.26/X.14 showed that the hypothesis of the circle of the path of the center was precise, this investigation of doubling the slant means that the recension's author did not see the slant as a corollary of the hypothesis of the circle of the path of the center.

(See figure 6.8.) At points other than 90°, 180°, and 270°, even if the proposal functions as alleged, there is still a slight deviation between predicted and real positions. We will assess whether the pole at point *J* is 90° away from point *K* after 45° of mean revolution (θ). In figure 6.8 , the poles of the first and fourth orbs are aligned, as are the poles of the second and third orbs. In the first motion (1), pole *Z* of point *K* revolves 45° about pole *A* on the second orb with the second orb's motion and moves to point *E*, corresponding with the motion of pole *Z'* to point *E'* on the fourth orb. In the second motion (not shown), the first orb revolves 45° about pole *E*; through these first two motions, the pole of the circle of the path of the center that was at point *K* moves toward point *G*. In the third motion (3), pole *A'* from the third orb revolves 45° about its pole *Z'* on the fourth orb to arrive at point *U*. That motion also moves pole *Z* of point *K*, on the first orb, from point *E*

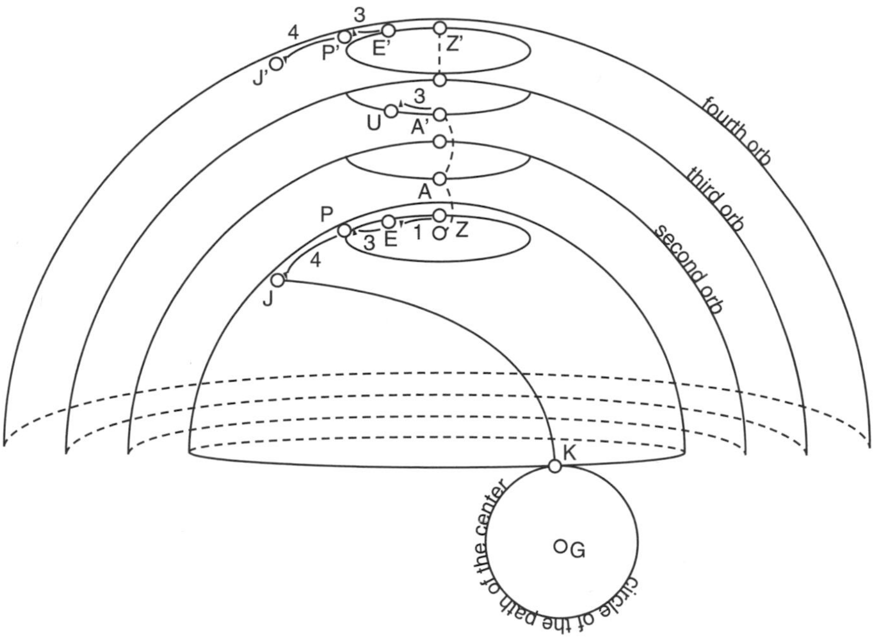

FIGURE 6.8

to point *P*, corresponding to point *P′* on the fourth orb. In the fourth motion, the third orb revolves 45° about pole *A′*, now at point *U*. That fourth motion moves pole *Z* of point *K*, which had been at point *P*, to point *J*, corresponding to point *J′* on the fourth orb. With pole *K* now back at point *K* on the circle of the path of the center, the goal is to determine the length of great circle arc *JK* in spherical triangle *JUK*. First, we can determine side *KU* by drawing triangle *KUZ′*. Angle *Z′* is 45°, *u* (side *KZ′*) is 90°, and *k* (side *UZ′*) is 2;30°. We apply the cosine rule for sides:

$\cos z' = \cos u \cos k + \sin u \sin k \cos Z'$
$\cos z' = \sin u \sin k \cos Z'$
$\cos z' = \sin 2;30^\circ \cos 45^\circ$
$z' = 88;14^\circ = \text{side } KU$

Then, we use the cosine rule for sides to solve for *U* (angle *KUZ′*):

$\cos u = \cos k \cos a + \sin k \sin a \cos U$
$0 = \cos 2;30^\circ \cos 88;14^\circ + \sin 2;30^\circ \sin 88;14^\circ \cos U$
$U = 134;57^\circ$
As angle *Z′UJ′* is 90°, as it is the sum of the two 45° revolutions of *Z′*, angle $KUJ' = 360^\circ - KUZ' - 90^\circ = 135;3^\circ$

We examine spherical triangle *KUJ'*. Angle *U* is 135;3°, *k* (side *UJ*) is 5°, and *j* (side *KU*) is 88;14°. We apply the cosine rule for sides:

$$\cos u = \cos j \cos k + \sin j \sin k \cos U$$
$$\cos u = \cos 88;14° \cos 2;30° + \sin 88;14° \sin 2;30° \cos 135;3\ °$$
$$u = 90;0,12°$$

This modification imparts a minimal deviation. But presuming the same inclination of the poles, a Eudoxan couple imparts a near oscillation on *KGD* that is much greater than a doubled slant.

From 115a

If we want the opposite of this §B.1.II.26

(See figure 6.9.) There is no figure in the MS of the Hebrew recension that illustrates this modification of the third approach. In order to eliminate the slant and to double the oscillation in latitude, the recension proposed a slightly different configuration of two upper orbs. The figure shows only circles of the paths of the poles of the two upper orbs. The small circles have radii of arc *B'A'* = arc *A'Z'*. Upper pole *A'* in the third orb revolves on a small circle about upper pole *B'* in the fourth orb, in the opposite direction of the revolution of pole *Z*. As pole *A'* revolves, its new position (point *A''* will be on the side of arc *GB'A'Z'* opposite point *E'*. Thus, point *E'* on the upper orb will have a new position, *E''* A revolution of the third orb about the pole at *A''* by a similar measure in the opposite direction should supposedly bring the pole that had been at point *E''* back to the great circle arc *Z'A'B'G* and eliminate the slant. X.20

To analyze this proposal, we presume $\theta = 45°$. As pole *A'* from the third orb rotates around pole *B'* from the fourth orb, away from point *E'*, we assess whether *A'* will reach a great circle arc, passing through pole *B'*, which is 85° away from point *G*. Point *E''* will not, however, be on that great circle. We consider triangle *A''B'G* after 90° of mean motion. Angle *B'* is 45°, *a''* (side *B'G*) is 85°, and *g* (side *B'A''*) is 5°. We use the rule of cosines for sides:

$$\cos b' = \cos g \cos a'' + \sin g \sin a'' \cos B'$$
$$\cos b' = \cos 5° \cos 85° + \sin 5° \sin 85° \cos 90°$$
$$b' = 85;1,7°$$

This modification will produce a near oscillation on a great circle arc at 90°, 180°, and 270°. There will be a small deviation, and thus a small slant, at other points, as *E''* is 5° from *A''*, meaning that *E''* will not be exactly 90° from *G*.

From 114b

I say that it is possible for us to propose §B.1.II.26

(See figure 4.5.) Here the recension proposes another way to cause a point to slant, in this case point *T*, without imparting to that point any deviation in latitude. The X.22

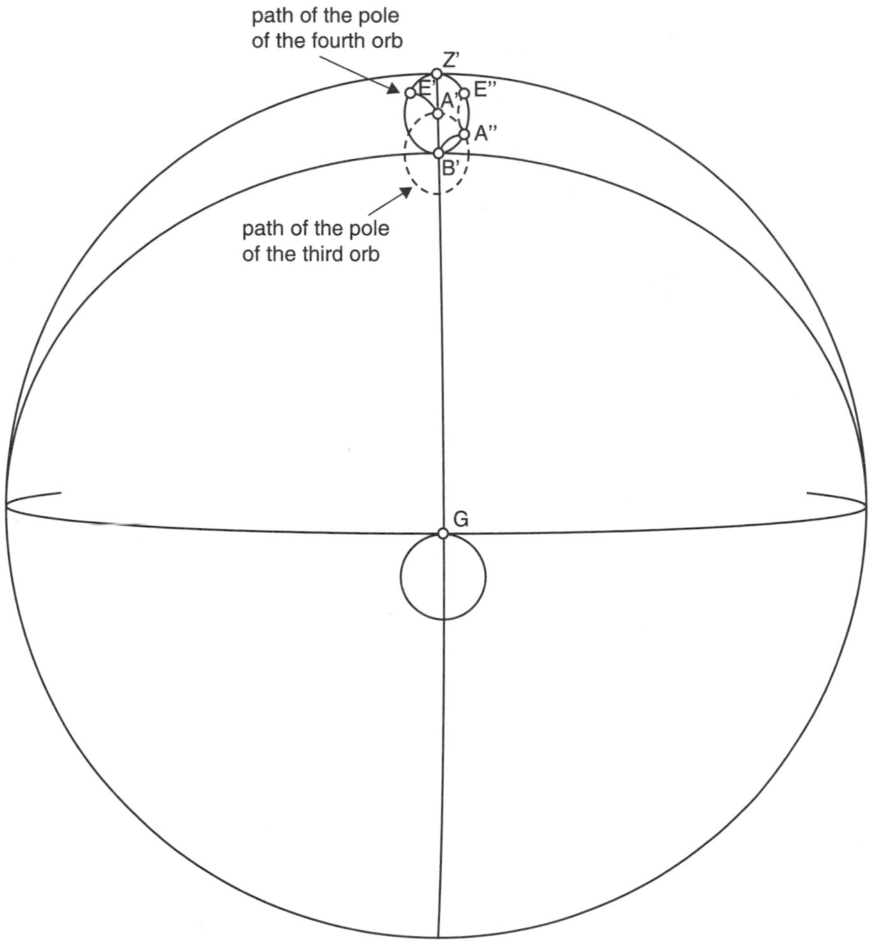

FIGURE 6.9

oscillations on great circle arc *AHB* could be caused by the double-circle device, or a Eudoxan couple, the third approach described earlier for imparting a linear oscillation through the motions of orbs.

In this first version of the modification (§B.1.II.26/X.23: *The first is our thesis that there are two orbs moving pole* E. . .), there is a pair of orbs moving pole *E* in latitude to the south, and another set of orbs moving pole *E* in latitude toward the north. That is, one additional pair of orbs causes a pole to oscillate along arc *BHA* to cause point *T* to slant to point *W* and another, additional, pair of orbs oscillates in the opposite direction on *BHA* to cause that same pole to slant from point *W* to point *Z*. These orbs are probably located in the vicinity of pole *E*. The slant is greatest after 90°, 180°, and 270° of mean motion. According to the analysis of the model for the solar anomaly without the double-circle hypothesis, the greatest deviations from the ecliptic were also at those points.

But because *BHA* is a great circle arc perpendicular to great circle arc *ETH*, and because arc *EH* is 90°, then the distance from any location on arc *BHA* to pole *E* is 90°. Thus, a slant can be imparted to diameter *LEN* without proposing orbs that move pole *E* to point *K* or pole *E* to point *C*. A set of orbs creating a near oscillation on great circle arc *HB* would also impart a slant on the measure of angle *TEZ* to the circle of the path of the center.

The second way is our thesis §B.1.II.26/X.24

(See figure 6.10.) In this second way, one of the two sets of orbs is supposed to move pole *H* of pole *E* of circle *LMNT*, the circle of the path of the center, in one direction on arc *AHB*, and the other set of orbs moves it in the other direction on *A'H'B'*. Together, both sets of orbs are supposed to impart a slant, in the same direction, to small circle *LMNT*, without pole *E* moving on arc *DEG*, as the motion imparted to pole *E* by pole *H* moving on arc *AHB* is canceled out by the pole's motion on arc *A'H'B*. Again, these orbs, for the purposes of calculation, have no thickness.

In the figure, arcs *AHB* and *A'H'B'* intersect at point *H/H'*, for if these arcs were superimposed on each other, one set of orbs would undo both pole *E*'s motion on arc *DEG and* the slant. Thus, lower arc *AHB* is oriented with point *B* closer to arc *DEG*. That way, the movement of pole *H* to point *B* is intended to move pole *E* to point *Q*. Upper arc *A'H'B'* is oriented with point *A'* closer to arc *DEG*. That way, the movement of pole *H'*, which faces pole *H*, to point *B'* is intended to move pole *E* to point *X*. One problem with this model is that one must presume that pole *E* will stay on arc *DEG* in an imaginary track, a problem that has arisen in the comments on §B.1.II.26/X.8. Nothing would seem to prevent pole *E* from moving on arc *TEM*.

That drawback aside, and though there is no quantitative analysis in *Or ha-'olam*, the model can function to cancel out the motions on arc *DEG* if we determine the angle of the inclination of arcs *AHB* and *A'H'B'*, measured by angle *BHY*, so that arc *BP* is 90°.

Arcs *HE* and *HP* are 90°, so angle *EHP* equals arc *PE*. If *EL*, the radius of the circle of the path of the center, is 2;30°, and points *L*, *Q*, *P*, and *E* are equidistant, then arc *EP* is 0;49,48°. Since angle *PHB* is 90°, then angle *EHB* is 89;10,12°. Since angle *EHD* is 90°, then angle *BHY* is 0;49,48°, which is the inclination of lower arc *AHB* to the colure (arc *GHYD*). The angle between arcs *AHB* and *A'H'B'* would be 1;39,36°.

Now we need to determine, given that inclination of arc *AHB* to arc *A'H'B'*, if pole *H* is at point *A*, whether arc *AP* would be 90°. Again, arc *HP* is 90°, and angle

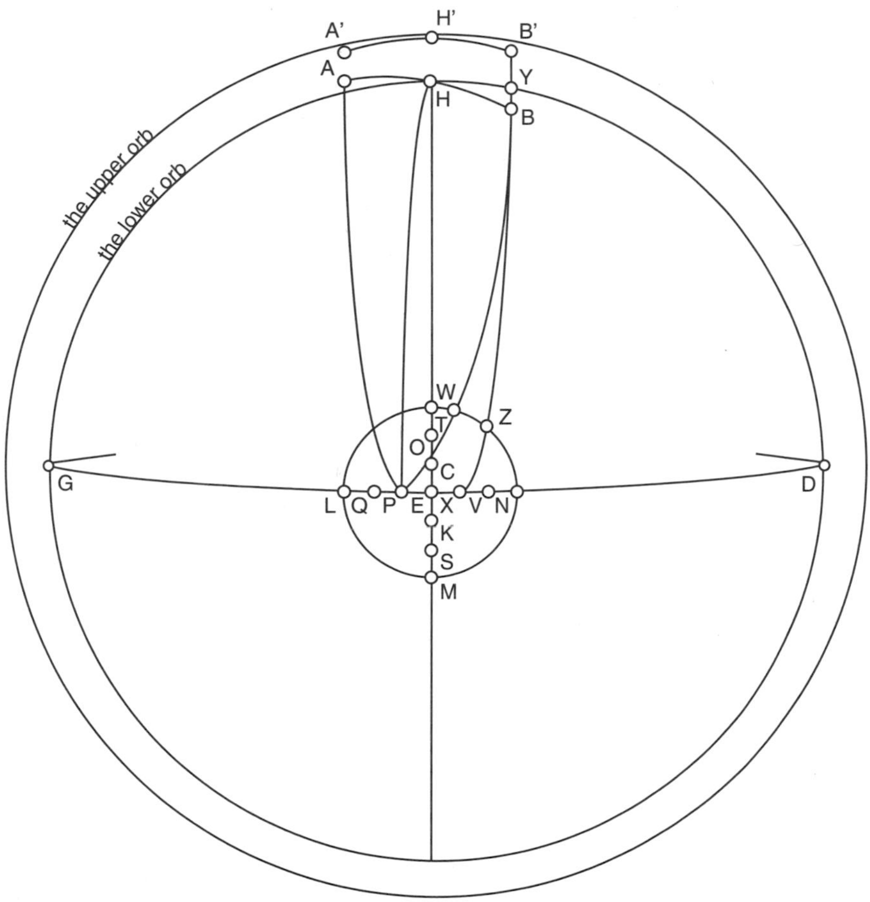

FIGURE 6.10

AHP is 90° because angle *BHP* is 90°; thus, arc *AB* is 2;30°. We consider triangle *AHP* and apply the rule of cosines for sides:

> $\cos h = \cos p \cos a + \sin p \sin a \cos H$
> Because $H = 90°$ and a (side *HP*) = 90°, $\cos h = 0$
> Therefore, h (side *AP*) = 90°

This version of the model would seem to function in the first quadrant. The recension did not explain sufficiently what would happen in the second quadrant. The text reads "pole H returns, through the first two orbs, to *B'*," but the text did not explain how pole *H* would begin the quadrant at point *H* since pole *H* had concluded the first quadrant at point *B*. Since the text reads that "In the third

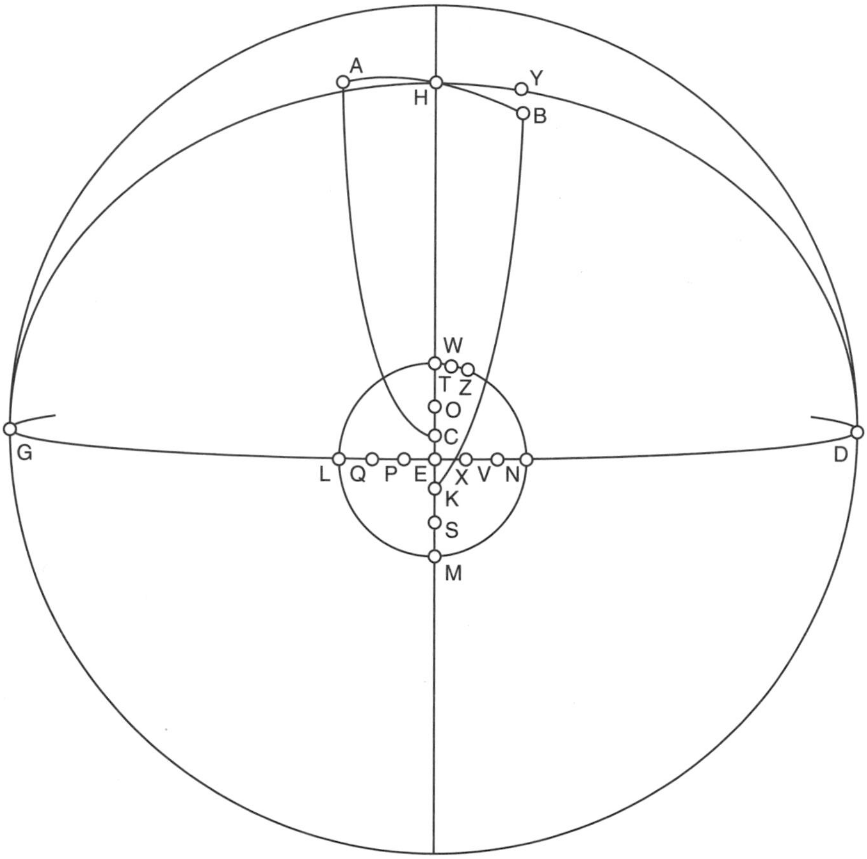

FIGURE 6.11

quadrant, pole *H* moves through the first two orbs from *H* to *A*," it is plausible that, in the second quadrant, the text should have read "pole *H* returns, through the first two orbs, to *H*." In that case, the text would not have explained how the pole's motion in the opposite direction, that is from point *B* to *H*, rather than from *H* to point *B*, nevertheless made the equator of the orb move from point *P* to point *Q* and not from point *P* back to *E*.

From 114a

The third way is our thesis §B.1.II.26

(See figure 6.11.) This is the recension's third proposal to produce a slant. The X.27

mention of the "first figure" of the lemma was likely a reference to the third way of producing a near oscillation on a great circle arc through the motions of orbs. The great circle arc is arc *TEM*, and the slant's measure is twice arc *TZ*.

There are two ways that this motion might occur. One would be to have great circle arc *HAB* facing arc *TEM* on an upper orb. Then, one could use the double-circle hypothesis presented earlier to have point *E* oscillate up and down arc *TEM* without introducing an additional slant. The advantage of this interpretation would be that the realignment of arc *HAB* forces pole *E*, of the circle of the path of the center, to oscillate on the arc facing arc *TEM*. The second possibility, though, would be for arc *HAB* to be almost perpendicular to *TEM*, as shown in figure 6.11. A slight inclination of arc *HAB* to the colure would cause pole *E* to move north and south of arc *DEG*. In such a case, a second additional orb with arc *H'A'B'* would not be necessary. Once again, the recension has not explained how pole *E* would remain on track on arc *TEM* as opposed to on arc *LEN*.

The small deviation created for the Sun from the ecliptic in latitude . . . does not reach [113b] 3" of latitude §B.1.II.26/X.30

At $\theta = 54°$, there is the greatest discrepancy between the Φ coordinate predicted by the circle of the path of the center, along with the hypothesis of the slant, and the Φ coordinate of the corresponding position on the ecliptic. See the comments on §B.1.II.26/X.14 for how to determine the Φ coordinate of a point on the circle of the path of the center that has been affected by the slant, and see also the comments on §B.1.II.13 for how to determine the Φ coordinate of the corresponding point on the ecliptic. The maximum displacement is 0:0,3,14°. Whether the source of the slant is the Eudoxan couple or the model for the solar anomaly with the circle of the path of the center, the recension may have recognized the precision of the hypothesis of the slant. Even more interesting is the possible correspondence between the results of the commentary's mathematical analysis and the recension's assertion, which may indicate that the author of the recension did somehow assess the models' predictive accuracy. Still, the recension was wrong to say that the discrepancy is greatest when the solar anomaly is greatest, that is, at $\theta = 90°$ and $\theta = 270°$. Also, if the author of the recension understood fully how precise the hypothesis of the circle of the path of the center combined with the hypothesis of the slant were, it is unclear why, then, the recension would try to explain how the slant could be doubled. Thus, the recension's interest in doubling the slant suggests that the recension preferred to present the hypothesis of the slant as an independent way of generating variable motions, distinct from the circle of the path of the center.

From 113b

My intent is to say that the smallness of the slant §B.1.II.26/X.31

In the second hypothesis, the hypothesis in which all of the motions are understood to be in a single direction, the recension has now said that there were actually two slants, arising from each of the two motions that the model for the Sun's mean motion comprised. The two slants could be said to be opposite (*hepek*) in that they were angled in opposite directions.

Since all this has already been shown §B.1.II.26

This passage proposes that the solar model from the Judeo-Arabic version, with the circle of the path of the center and the double-circle hypothesis, can be replaced by a model in which the mean motion, of the pole of the circle of the path, is about the poles of the ecliptic, and the double-circle hypothesis, Eudoxan couple, or the hypothesis of the slant is introduced to account for anomaly without the circle of the path of the center. This resembles, or more precisely foreshadows, the later homocentric models of Amico.[15] The text of the recension states that the measure of the two circles is equal to the greatest anomaly, which would not be the case were the circle of the path of the center also contributing to the Sun's motion in anomaly. X.32

It is interesting that the Judeo-Arabic version of *The Light of the World* used the asymmetry of how the time period from the greatest motion to the mean motion is less than the time period from the mean motion to the smallest as a reason not to jettison the circle of the path of the center of the Sun (see comments on §B.1.II.26). Now the recension alleges that a solar model without the circle of the path of the center can account for this asymmetry. Because the commentaries have shown that the period of oscillation of either the double-circle hypothesis or Eudoxan couple is symmetrical, I am not sure how the promised asymmetry would result.

When the Sun moves through quadrant TN *and becomes distant in latitude from the ecliptic* §B.1.II.26 X.33

(See figure 6.12.) This version of the hypothesis of the slant is another attempt to cause the diameter of the circle of the path of the center to slant without imposing a deviation in latitude. As pole *B* moves to point *K*, point *T* moves to point *N*. Simultaneously, the orb of the deferent rotates 90° about pole *B* to put pole *E*, of the circle of the path of the center, at point *G*. When pole *E* is on point *G*, point *T* will not be in the ecliptic (arc *DTG*).

(See figure 6.13.) Point *N* is point *T*'s new position after revolving 90° on small circle *LTM*. We draw spherical triangle *GNX*, with point *X* being where a great circle arc passing through point *N* meets great circle arc *GTD* at a 90° angle. Angle *G* is 2;30°, as arcs *GT* and *GE* are 90° and arc *ET* is 2;30°, and *n* (side *XG*) is 2;30°, as it is the radius of small circle *LTM*. One problem with this proposal

15. Noel Swerdlow: "Aristotelian Planetary Theory in the Renaissance: Giovanni Battista Amico's Homocentric Spheres," in *Journal for the History of Astronomy* III (1972): p. 41.

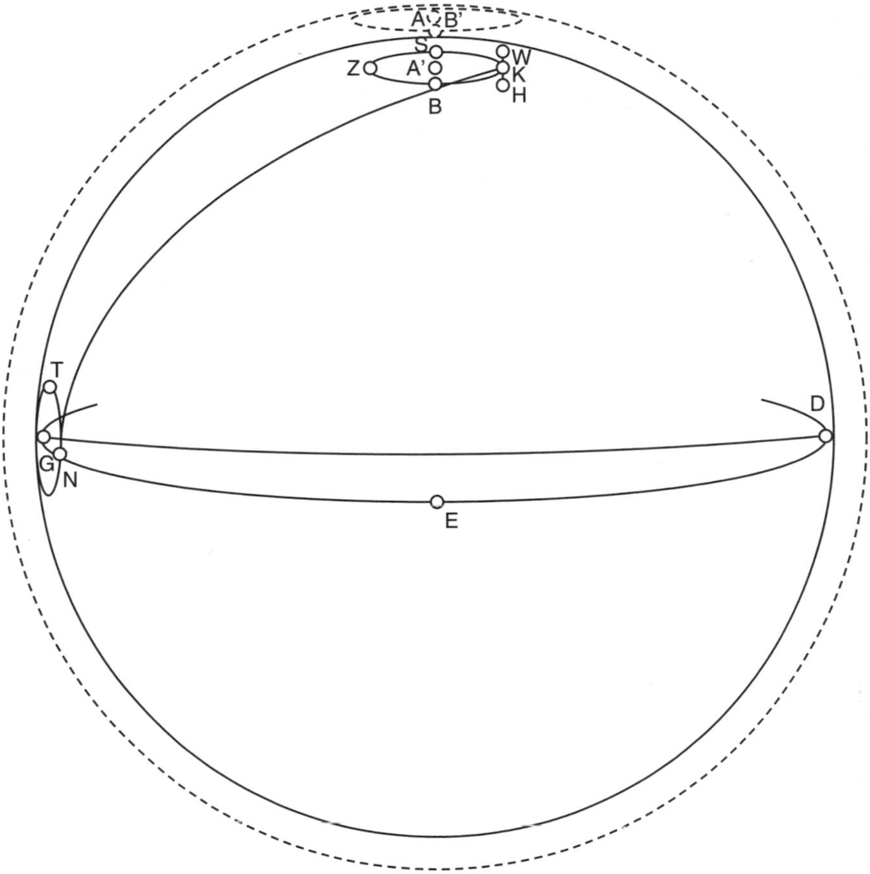

FIGURE 6.12

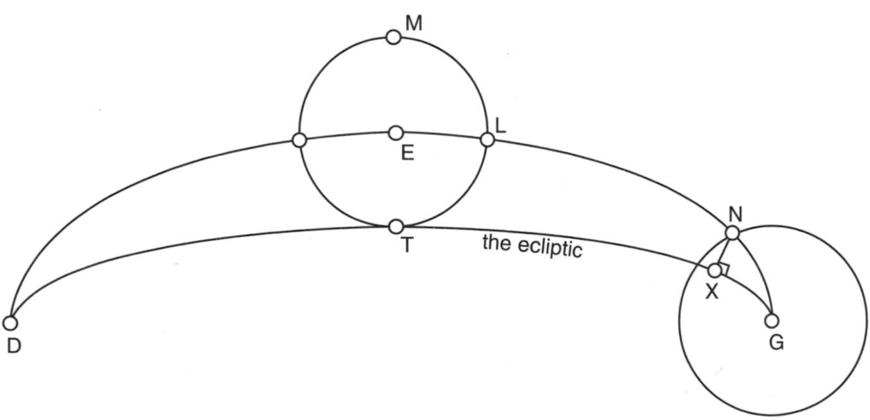

FIGURE 6.13

for creating a slant is that arc *KG* is not 90° because *AG* is 90°; if *KG* were 90°, it would pull the pole of the circle of the path to the north of the ecliptic. Second, when pole *B* is at point *S* and pole *E* of the circle of the path of the center is 180° away from its starting position, the distance from the pole of the deferent to the pole of the circle of the path of the Sun will not be 90°. These problems could be ameliorated by emending the figures in the MS and commentary by switching points *B* and *S*. Later in the presentation of the model, the recension comments, "it is my intention to say myself that its mean motion is the motion of point *K* about pole *A* upon circle *BKSZ*." If the motion of the pole of the deferent began at point *K*, switching points *B* and *S* would not improve predictive accuracy.

our thesis is that the pole of the deferent revolves upon great circle arc WKH §B.1.II.26
(See figure 6.12.) This oscillation of the pole of the circle of the path of the center X.34
on arc *WKH* is intended solely to eliminate the Sun's deviation from the ecliptic due to the slant that results from the motion of pole *B* (of the deferent *GED*) in the direction of point *K*. Arc *WKH* is a segment of the arc running from the pole of the deferent to the pole of the circle of the path of the center, with half of arc *WKH* being equal to the great circle arc whose chord is the largest slant. This improvement would be akin to placing double circles at the pole of the orb rather than at the equator. So, when pole *B* is at point *K*, and the slant is at its maximum, then it would move from point *K* in the direction of point *W*. When pole *B* oscillates on great circle arc *WKH*, presuming that the midpoint of arc *WKH* is fixed on small circle *BKSZ*, the constant angular distance from the pole of the deferent to the pole of the circle of the path of the center means that the pole of the circle of the path of the center will no longer be on great circle arc *GED*, except after 90°, 180°, and 270° of mean motion.

An oscillation on arc *WKH*, if arc *WKH*'s midpoint is on the circumference of small circle *BKSZ*, will not eliminate the slant because the slant appears whenever the pole of the deferent deviates from great circle arc *METBAOS* (cf. figure 4.6). The ensuing proposal of an oscillation on arc *BAS* (see §B.1.II.26/X.39: "rather it revolves on arc *BAS*") rather than on *WKH*, could serve to keep the Sun in the ecliptic without having to take into account the additional deviation resulting from the slant.

From 112b

Indeed, through the hypothesis of the two circles §B.1.II.26
(See figure 6.14.) This modification of the solar model furnishes another way to X.39
account for the Sun's mean motion without two motions of varying measures. The recension has proposed that the Sun's mean motion, the motion of the center of the circle of the path of the center, is due to the oscillation of the pole of the deferent orb on great circle arc *BAS*. The need to maintain a distance of 90°

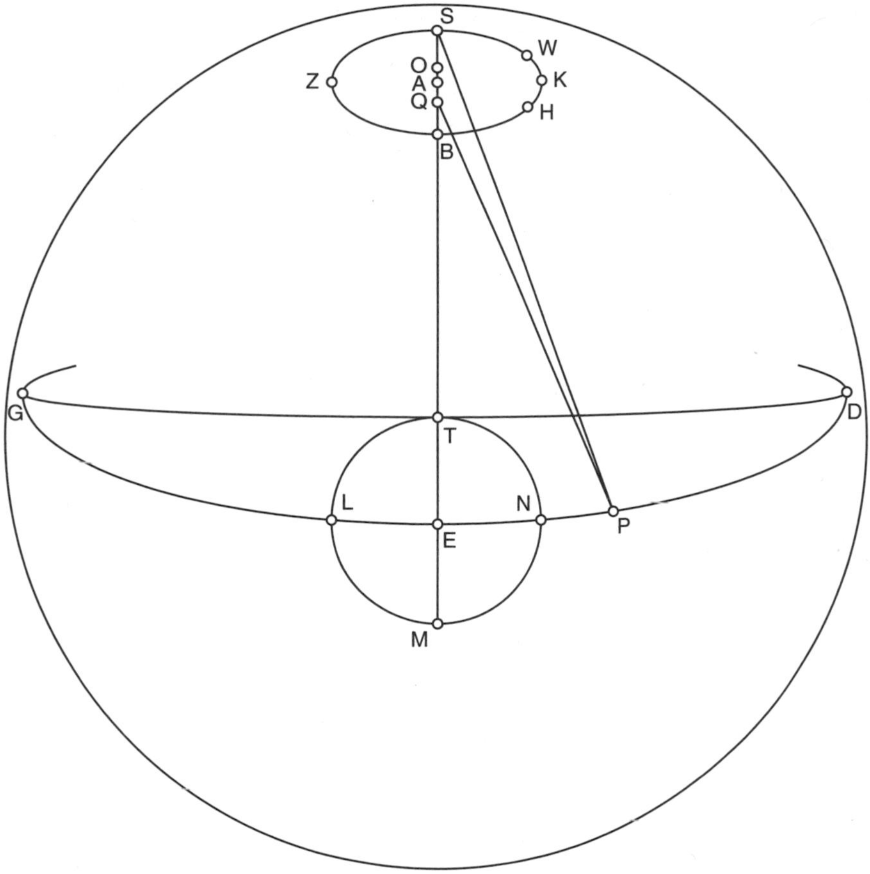

FIGURE 6.14

between the pole of the deferent and the center of the circle of the path of the center (*E*) would drive the center of the circle of the path of the center around great circle *EPDG*. This modification of the solar model does, however, depend on the center of the circle of the path of the center being in a groove or track lest *E* oscillate vertically.

In order to compensate for the deviation in latitude arising from the motion of the Sun on the circle of the path of the center, the recension also proposed that the pole of the ecliptic move on great circle arc *OA*, again probably through a double-circle hypothesis. The recension has specified that the pole of the ecliptic is in the orb of the Moon, probably so that the variations in the ecliptic's latitude are not transmitted upward to the rest of the solar model. But the recension did not investigate the effect of these variations in the ecliptic's latitude on the lunar model.

The goal of the mathematical analysis will be only to assess whether the distance from the pole's new position to the pole of the circle of the path of the center of the Sun will be 90° after 45° of mean motion. Because the analysis deals only with mean motion, the analysis does not take into account the additional oscillation of the pole of the ecliptic on *OA*. Again, points *B* and *S* are inverted in figure 6.14 in comparison to the way they are pictured in the MS.

We presume 45° of rotation for θ, and that pole *E* of the circle of the path of the center has advanced 45° toward point *D* or point *G*; once the pole of the circle of the path of the center oscillates on *BAS*, the direction of motion on arc *DEG* does not matter. Point *P* is the new location of pole *E*. Point *Q* is the new position of pole *B*, after 45° of rotation with double circles of r = 2;30°, on arc *BAS*. Arc *EP* is 45°, and angle *QEP* is 90°; point *Q* will be 1;46 ° (cf. comments on §B.1.II.26/X.10) away from point *B* in the direction of point *S*. Arc *QE*, then, will be 91;46°. We consider spherical triangle *QEP*; angle *E* is 90°, *p* (side *QE*) is 91;46°, and *q* (side *EP*) is 45°. We apply the rule of cosines for sides:

$$\cos e = \cos q \cos p + \sin q \sin p \cos E$$
$$\cos e = \cos 45^\circ \cos 91;46^\circ$$
$$e = \text{arc } QP = 91;15^\circ$$

Thus, an oscillation on arc *BAS*, the measure of which was constrained by the size of the circle of the path of the center, will not keep the mean Sun, that is, the pole of the circle of the path of the center, on its observed path. The measure of this deviation, 1;15°, also far exceeds the deviation from the ecliptic caused by the Sun's revolution on the circle of the path of the center. Whatever this proposal's merit, it does show, again, that the recension entertained different ways of accounting for the Sun's mean motion, as well as for its anomaly.

our thesis is that the measure of the greatest slant is equal to the degrees of the greatest anomaly §B.1.II.2 X.41
The recension has proposed accounting for the entire anomaly with the slant, just as it suggested accounting for the entire anomaly with the double-circle hypothesis.

From 110a

It has already become clear from what has preceded that the number of orbs of the Sun, according to the hypothesis of the slant, is five orbs §B.1.III. 10/X
This passage makes two interesting points. First, the deferent orb moves about its own poles with the mean motion, rather than its own pole moving about the pole of the equinoctial as the Judeo-Arabic version proposed. Second, the recension

has placed the circle of the path of the center in the same orb as the deferent. Thus, there is no separate mover for the circle of the path of the center, indicating that there were motions for which the recension did not propose a mover. We find, too, with the lunar model (§B.2.V.12/Xb) that the circles responsible for the second and third lunar anomalies, as well as the circle of the path of the center of the Moon are also in the lowermost orb.

From 103b

And if a small slant occurred §B.2.V.12/X

The recension is saying that the hypothesis of the slant could be used to eliminate some of these discrepancies. It is particularly notable that the recension understood that the oscillations from the slant might be made to correspond to the oscillations of the prosneusis point and that Ibn Naḥmias's observation that the greatest effect of the third lunar anomaly is at trines is virtually correct.[16]

From 103b–a

It is known from what has preceded that the orbs of the Moon §B.2.V.12/XB

This insertion is interesting because it outlined the sum total of the orbs necessary, from the point of view of the recension, to account for the motions of the Moon. The five principal orbs are the following: (a) the orb carrying the center of the Moon, which also has the circle responsible for the second and third anomalies; (b) the orb that moves the center of the Moon and which contains the circle of the path of the center; (c) the orb that carries the poles of the circle of the path and that moves them in longitude; (d) the orb that moves the extremes of the deferent by the surplus of the motion in latitude over the anomaly; (e) the orb that moves the extremes of the deferent circle and the circle of latitude by the 3' per day. The recension listed only three compounds of orbs: (a) a compound of three orbs for the slant, (b) a compound of two orbs to eliminate the latitude created by the orb proposed for the second and third anomalies, and (c) a compound that would be necessary if a slant was needed for the second and third anomalies. Though the recension said that this compound would be the fourth, I cannot identify the third compound.

From 101a

As for the stars that are outside the ecliptic §B.3.II.4/X

This addition in the Hebrew recension is significant for two reasons. First, the Hebrew recension conceded that some fixed stars actually have been observed to preserve their positions with respect to one another, as the recension went on

16. Olaf Pedersen (ed. Alexander Jones): *A Survey of the Almagest* (New York, Heidelberg, Berlin: Springer-Verlag, 2011): p. 192. The actual elongations are 57° and 123°.

to say that the stars outside of the ecliptic preserve their positions with respect to one another to a certain point. Thus, the Hebrew recension, perhaps authored by Ibn Naḥmias himself, was less skeptical than the Judeo-Arabic original, of all past observations. Second, the author of the recension stated that an underlying element of all of the models in *The Light of the World* was a hypothesis (or hypotheses) that produced an oscillation along a great circle arc in that the hypotheses from the solar and lunar motions could be utilized in explaining the motions of the fixed stars. From the point of view of the author of the recension, other models lacked any such theoretical innovation.[17] The presence of hypotheses in *The Light of the World* to account for an oscillation along a great circle arc was, in all likelihood, an important reason for the text's passage to Renaissance Europe.[18]

17. J. L. Mancha's scholarship has argued that Biṭrūjī's model for the fixed stars did, in fact, include a Eudoxan couple but acknowledged that Biṭrūjī might not have interpreted his own model that way. See J. L. Mancha: "Al-Biṭrūjī's Theory of the Motions of the Fixed Stars," in *Archive for History of Exact Sciences* LVIII (2004): pp. 157–61.

18. Morrison, "A Scholarly Intermediary," pp. 45–6.

7

Hebrew Text of Profiat Duran's Response to *The Light of the World*

101a

הגיעוני השמים החדשים אשר נטו ידי החכם המעלה, ר׳ יוסף ן׳ נחמיש, אשר ראוי שישבח על זריזותו וחריצותו והשתדלו, מה שנכספו נפשות החשובים לידיעתו, עם שלא עזרהו במציאות להשלים כונתו. וכבר הרגיש בזה והתגבר על המוחשות ורצה להמשיך המציאות אחר הידיעות. ועשה פרק אחד לבאר כי חלוף שעור קטר הירח אצל ההבטה לא יחייב בהכרח חלוף מרחק הירח מהארץ. ואמר שם כי הוא יניח שמרכז הירח במרחק א׳ תמיד ממרכז הארץ. וזה יסתרהו

100b

החוש אצל המביטים כלם אשר היו מימי קדם כי הורגש אצלם במה שאין ספק בו. כי בהיות הירח בהתחלת תנועת החלוף אשר כפי גלגל המזלות, ובהתחלת תנועת החלוף ג״כ אשר כפי השמש, וזה בקצת הדבוק וההקבלות, כמות שיעור חלוף ההבטה אז בשיעור א׳, והיותר גדול לפי בטלמיוס הוא נ״ד דקים. ובהיות הירח בקפ׳ מעלה מאלה השתי תנועות יחד, וזה בקצת הרבועים יהיה שיעור חלוף הבטתה אז בשיעור א׳, והיותר גדול לפי בטלמיוס מעלה ומ״ד דקים. לא חלק א׳ מהמביטים בחלוף זה החלוף עם שחלקו בכמותו האחרונים מהם. ויצא אֲמות זה ממבטי בטלמיוס בשנת י״ו משנות אדו ננוס הנזכר במג׳סטי. וכבר הסכים בחלוף זה החלוף אברהים אלזרקאל אשר היה מהמיוחדים במביטים כמו שאמר עליו ן׳ רשד במביטים שלו.

ובהיות העניין כן אם יתן ר׳ יוסף סבת זה עם הנחת מרחק הירח מהארץ א׳ כמו שהניח גם אני אודהו וראש וקציו ללמודיים אקראהו. ועם פלפולו אפלא עמו איך שגה בזה? ואשר עוד אפלא עליו אמרו כי קטר הירח יעשה מיתר לזוית אחת בעצמה בכל מקום. ונתן סבת החלוף הנראה בזה שנוי באויר או לחותו חייב אותו הירח בעתים ידועים. וכמה הוא רחוק מן ההקש השכלי שיהיה זה החלוף מתדרג והולך על סדר אחד מהתחלת שתי התנועות הנזכרות עד קפ׳ מעלה מהן וזה בכל אקלים ובכל זמן, ויהיה שנוי האויר מתדמה תמיד. וכבר כתב נור החכמים השלם ר׳ לוי ז״ל את זה מופת חותך על מציאות יציאת המרכז. וכן ג״כ הורגש מחלוף קטר השמש בהתחלת תנועת החלוף לקטרו בקפ׳ מעלה ממנה.

ואשר חשב לחזק סברתו כי אברכס ובטלמיוס לא הסכימו בשיעור קטר הירח, כי בטלמיוס נתן קטרה שוה לקטרה בהיותה בהתחלת תנועת בחלוף אשר כפי גלגל המזלות, ואברכס בהיותה במעבר האמצעי, אין לו עקר. כי הקירוב שביניהם אפשר שייחוס לכלי המבט או לצחות האויר במבט אברכס לפי דרכו וגם לפי האמת כי במבטיהם במקום מיוחד היו ובזולת השנות רב. ואשר אמר שכבר [התבארה][1] מניעת יציאת המרכז מי יתן ידיעתי במופת על זה.

ואל שני המקריים האלה רצוני התחלפות שיעורי הקטרים בכוכבים ובהתחלפות ההבטה אשר הם במופת על מציאות יציאת המרכז רמז הרמ֗ פרק כד֗ חלק שני מספר המורה כשאמר היש שם דבר יתכן עמו שתהיה התנועה הסבובית שוה שלמה ויראה בה מה שיראה. ירצה התחלפות שיעורי הקטרים והתחלפות ההבטים המושגים בראיות. ורמז בו גכ֗ לשני המקריים האחרים אשר הם המהירות והאיחור פעם והנדודות פעם והיושר פעם עם שלמותה. וירצה בראיה בשני אלה הראיה השכלית, והטיב באמרו התנועה הסבובית, כי עניין חכמי התכונה [בתנועות][2] ההם הוא בבחינת הסבוב. ולזה יגבילו שיעורי התנועות ההם בזויות אשר אצל המרכז. ובזאת הבחינה תהיו על דרך משל תנועת הירח יותר מהירה מהשמש עם שעל דרך האמת תנועת השמש יותר מהירה מתנועת הירח. ולפי שכבר יתחלפו כמות הנלקה ואמרו בהיות הירח בהתחלת תנועת החלוף אשר כפי גלגל המזלות לכמותו ואמרו בהיותו בקפ֗ מעלה ממנה, והיה זה במופת על התחלף מרחק הירח מהארץ בעת זולת עת.

לזה אמר הרב זל֗ בפרק הנזכר ויעיר על אמיתת זה אמיתת הלקיות הנמנות בשרשים ההם. והיה אלו רצה החכם ר֗ יוסף להבין זה מכונת הרב לא היה תמה במאמרו כמו שכתב בפתיחת חבורו. וכבר שתק מהדבור בספק המתחייב לכונתו מחלוף שיעורי חלוף ההבטה לפי שלא יכול ליחס זה אל עובי האויר וצחותו, כמו שענה בחלוף שיעורי הקטרים, כי המקומות האמתיים הנראים יוגבלו במרכז הכוכב

100a

לבד.

והנה אחשוב כי לא נעלם דבר מכל זה מלפניו, אבל הספיק לו השתדלות בהמציא תכונה תהיה בה התנועה הסבובית שוה שלמה, וימשכו ממנה שני המקריים האחרונים הנזכרים. וכמו שהשתדלו בזה זולתו, ואולי הוא השלים תכונה יותר. ולא אוכל לומר בזה מאמר פוסק כי לא עיינתי במאמרו מדין העיון הראוי לנודדת מחשבתי בענינים אחרים. והנה במה שראיתי בו הביא הקדמה אחת שנפלאתי עליה הפלא ופלא, והיא ההקדמה הא֗ מהקדמות הלמודיות אשר הביא בפרק השני. אמר כשרצינו לחלק מספר ידוע על היחס הנקרא בעל אמצעי ושתי קצוות וכו֗, ולא אדע אם ההקדמה הזאת לו או לקחה מזולתו. ואיך שיהיה היא הקדמה בטלה, כי מציאות מספר נחלק על היחס ההוא נמנע כמו שהתבאר זה מו֗ מיג֗ לאקלידס למשכילים. והמופת שעשה בהקדמה זו יצטדק בקוים כמו שעשה אקלידוס במאמר השני. וכן עשה. הוא העתיק משלו בתמונה מהמספר ונטה עליה קו.

1. הכתוב: התבארת.
2. הכתוב: התנועות.

8

Translation of the Hebrew Text of Profiat Duran's Response to *The Light of the World*

101a

THE RESPONSE OF EFOD TO THE BOOK OF RABBI JOSEPH IBN NAḤMIAS

The new heavens spread out by the hands of the excellent scholar Rabbi Joseph Ibn Naḥmias, who should be praised for his industry and his diligence and his effort, brought me to what the souls of the eminent sought to know, although it did not help him, in reality, to complete his aim. He felt this, and it overpowered sense perceptions so that he desired that reality proceed from what was known.[1] He produced one chapter to show that the observed variation in the lunar diameter does not necessarily entail a variation in the distance of the Moon from the Earth. He said there that his thesis was that the center of the Moon is always at a single distance from the center of the Earth. The senses contradict this,

100b

according to all observers from the days of yore, without any doubt about what they sensed. For when the Moon is at the beginning of the motion in anomaly with respect to the orb of the ecliptic, and also at the beginning of the motion in anomaly with respect to the Sun, this being at conjunction and oppositions, the observed anomaly is thus a single measure, with the largest according to Ptolemy

1. That is, Ibn Naḥmias prioritized metaphysics over that which is known through sense perception.

being 54'.[2] When the Moon is at 180° of these two motions put together, this being at the time of quadratures, the measure of the observed anomaly at that time thus a single measure, with the largest according to Ptolemy being a degree and 44 minutes. Not a single observer disputed the variation in this anomaly, even though recent astronomers argued about its measure. Its verification emerged from Ptolemy's observations in year 16 of [illegible],[3] who was mentioned in the *Almagest*. Ibrāhīm al-Zarqāl, one of the outstanding observers, as Ibn Rushd said, has confirmed the variation of this anomaly through his observations.

With the matter being thus, if R. Joseph could give the reason for this [the variations in the lunar anomaly], with the thesis of a single distance of the Moon from the Earth, as was his thesis, then I, too, would thank him and read it, from start to finish, to the mathematical [astronomers]. Yet I am astonished by his casuistry; how did he err with this? I am even more astonished by his saying that the diameter of the Moon is made the chord of a single angle itself in each place. He gave the reason for this visible variation [in the Moon's diameter] the changes in the air or in its humidity that the Moon necessitates at known times. How far he is from reasoning intellectually that this anomaly proceeds by degrees according to a single order from the beginning of the two aforementioned motions until 180° of them, in each clime and at each time, with the changes in the air always being uniform! The perfect light of the scholars R. Levi, may his memory be as a blessing, already wrote this as a decisive proof in favor of the existence of the eccentric.[4] It [eccentricity] was also sensed from the variation of the diameter of the Sun at the beginning of the motion in anomaly to its diameter at 180° of that.[5]

And the one who opines, to strengthen his explanation, that Hipparchus and Ptolemy did not agree about the measure of the diameter of the Moon, for Ptolemy gave its diameter as equal to its [the Sun's] diameter at the beginning

2. Literally: the quantity of the measure of the observed anomaly is a single measure at that time. Duran is arguing that the Moon's anomalistic motion is not constant, implying that the source of these variations would be changes in the distance of the Moon from the Earth.

3. Ptolemy mentioned the observational basis for the discovery of the second lunar anomaly in *Almagest* V.3 (see Gerald Toomer, *Ptolemy's "Almagest"* [London, 1984]: pp. 223–5). Of the two observations Ptolemy mentioned, one comes from Epiphi 16, of the third Kallippic cycle (August 7, -127). The other observation was during the reign of Antoninus (February 9, 139). The text may have combined the "16" from the earlier observation with a version of Antoninus. The maximum equation for the second anomaly is 2;40°, about a degree from the parameter mentioned in the text. Likewise, the maximum equation for the first anomaly is 5;1°. These discrepancies between Ptolemy's actual parameters and the text may explain the inaccuracies in the report of this observation. My best transcription of the illegible term in the MS is "Ado nnos."

4. For Gersonides's argument that the Moon could not be moving on an epicycle see Goldstein, *The Astronomy of Levi Ben Gerson*, p. 117.

5. Gersonides measured variations in the solar diameter. See Goldstein, *The Astronomy of Levi ben Gerson*, pp. 141–2.

of the motion in anomaly which is according to the orb of the ecliptic, and Hipparchus when it [the Moon] was at its mean passage,[6] has no real point. Their [the measures'] proximity could be attributed to the observational instruments or to the purity of the air in the observation of Hipparchus, both according to his method and the truth, for they hold, with their observations, a special place, let alone the manifold repetition [of the observations]. And the one who said that the impossibility of the eccentric has been shown—who will give me knowledge of its demonstration?

Rabbi Moses hinted at these two cases, that is to say the variations of the diameters of the planets and the observed variations that are in the demonstration for the existence of the eccentric in Chapter 24 of the second part of the book the *Guide* when he said: is there something with which it is possible that the motion in revolution be equal and perfect and that there is seen in it that which is seen?[7] He meant the variations of the measures of the diameters and the perceived variation of observations. He also hinted in it [the *Guide*] at the two other cases [that prove the eccentric] which are the speed at one time and the slowness at another time and retrograde motion at one time and direct motion at one time, all with the motion being perfect. He desired intellectual evidence for these two [cases], and he was correct when he said motion in revolution, for the concern of astronomers in those motions was with the criterion of revolution. Thus, they define the measures of motions by the angles that are at the center. And by this criterion, for example, the motion of the Moon is faster than the Sun even though, truly, the motion of the Sun is faster than the motion of the Moon. Because the amount eclipsed might vary, from when they said that the Moon was at the beginning of the motion in anomaly according to the orb of the ecliptic, to its amount when they said it was at 180° of it [that motion], this is a demonstration for the change of the distance of the Moon from the Earth over time.

Thus the Rabbi, may his memory be as a blessing, said in the aforementioned chapter: what attests to the truth of this is always the truth of the eclipses calculated with those hypotheses.[8] And if the scholar R. Joseph had wanted to

6. Thus, if Hipparchus and Ptolemy found the measures of the observed diameter to be equal, but at different places in the orbit, then the distance of the Moon from the Earth might vary, implying the existence of an eccentric. See Goldstein, *The Astronomy of Levi Ben Gerson*, p. 157. For Ptolemy's reference to his predecessors, see Toomer: *Ptolemy's "Almagest,"* V.14, p. 252).

7. Cf. Moses Maimonides (Shlomo Pines trans.): *The Guide of the Perplexed* (Chicago and London: University of Chicago Press, 1963): "Is it in any way possible that motion should be on the one hand circular, uniform, and perfect, and that on the other hand the things that are observable should be observed in consequence of it?"

8. Cf. Maimonides (trans. Pines), *Guide*, p. 326: "The truth of this is attested by the correctness of the calculations—always made on the basis of these principles—concerning the eclipses and the exact determination of their times as well as of the moment when it begins to be dark

understand this of the intention of the Rabbi [Maimonides], he would not have been astounded by what he [Maimonides] said as he [Ibn Naḥmias] wrote in the introduction to his composition. He had already been rendered mute by the doubt that followed from his intention regarding the variation of the measures of the observed anomaly, since he could not attribute this to the thickness of the air and its purity, like his response to the variations in the measures of the diameters, because the true visible positions are determined by the center of the planet

100a

alone.

Indeed I think that nothing of this was hidden from him, but that it was enough for him to try to create a model in which the rotational motion is perfect and equal, and from which the last two aforementioned cases follow. And just as others besides him tried to do this, perhaps he completed more of a model. I am unable to judge because I did not take a proper interest in that which he said due to my thoughts wandering with other concerns. Behold in that which I saw of it, he introduced one preliminary about which I was most astonished, it being the first of the mathematical preliminaries that he introduced in the second chapter. He said: when we want to divide a known number into the ratio known as a middle term and two extremes, etc.; I do not know if this preliminary is his or if he took it from someone else. And how [would it be taken from elsewhere] as it is an invalid axiom? For the existence of a number divided according to that ratio is forbidden, as this was explained in the sixth proposition of the thirteenth book by Euclid for the enlightened. The proof that he made for this preliminary is justified for lines as Euclid did in the second treatise. So that is what he did: he transposed his example from numbers and spread out a line upon them in the figure.[9]

and of the length of time of the darkness." See p. 51a in the 1959 Hebrew translation of the *Guide* (Moses Maimonides [Samuel Ibn Tibbon trans., Yehudah Eben-Shemu'el annotations]: *Seper Moreh nebukim le -Rabbeinu Mosheh ibn Maymon* [Jerusalem: Mosad ha-Rab Quq, 1959]; Judeo-Arabic (cf. Maimonides [ed. S. Munk]: *Dalālat al-ḥā'irīn* [Jerusalem, 1929]: p. 228): וישהד עלי צחה דלך צחה אלכסופאת אלמחסובה בתלך אלאצול דאימא. Duran has argued that Ibn Naḥmias did not value mathematical astronomy to the same extent that Maimonides did.

9. See the comments on §A.II.1.

GLOSSARY OF JUDEO-ARABIC, HEBREW, AND ENGLISH TECHNICAL TERMS

The transliterations of the Judeo-Arabic words follow standard Arabic morphology. The entries are alphabetized according to the Arabic roots. Because the Judeo-Arabic and Hebrew versions of The Light of the World *do not overlap perfectly, not all terms appear in both languages. I have indicated the first appearance of the term in the Judeo-Arabic text.*

’-s-d

al-asad = Leo §B.4.III.8

’-ṣ-l

’aṣl (pl. uṣūl) = ‘iqqar, y^{e}sod, shoresh = hypothesis §0.1

’-l-f

ta’allaf, i’talaf = hitḥabbeir = to be compounded §A.II.11

mu’allaf = m^{e}ḥubbar = compounded §A.II.7

’-h-l

ahl al-‘ilm al-riyāḍī = ba‘alei ha-ḥok̲mah ha-limmudit = practitioners of mathematics §0.1

ahl al-‘ilm al-ṭabī‘ī = ba‘alei ha-ḥok̲mah ha-ṭib̲‘it = practitioners of physics §0.1

b-kh-r

bukhār = heb̲el= vapors §A.I.1

b-r-h-n

barhana = heib̲i’ mop̲eit = to demonstrate, to prove §A.II.4

burhān (pl. barāhīn) = moḇeit = proof, demonstration §0.2

b-ṣ-r

baṣar = r^{e}'ut = sight §A.I.1
    (ʻind al-) baṣar = (eiṣel ha-) habbaṭah = visible §0.2

b-y-n

bayyan = bei'eir = to show, to explain §0.2

tabayyan = hitba'eir, nitba'eir = to be shown, to be explained, to become clear §0.2

bayān, tabayyun = bi'ur = explanation §B.0,2

t-m-m

tamām = tashlum = completion §A.II.2

t-w-'-m

al-taw'amān = ha-t^{e}'omim = Gemini §B.1.III.5

th-k-n

thiḵan = ʻoḇi= thickness §A.III.4

th-w-r

al-thawr = Taurus §B.4.III.6

j-d-y

jady = g^{e}di = Capricorn §0.5

j-r-m

jirm (pl. ajrām) = geshem, gerem = body §0.7

j-s-m

mujassam = m^{e}gushsham = solid §A.II.8

j-z-'

juz' (pl. ajzā') = ḥeileq = degree, part §0.8

tajzi'a = ḥilluq = division §A.I.4

juz'ī = ḥelqi = particular, individual §0.18

j-w-z

majāz = maʻaḇar = passage §A.I.1
    al-majāz al-awsaṭ = ha-maʻaḇar ha-emṣaʻi = mean passage §A.I.1

j-y-b

jayb (pl. juyūb) = beqaʻ = sine §A.II.4

ḥ-d-b

muḥaddab = gaḇnuni = convex §0.15

ḥ-r-f

inḥaraf = hitʻaqqeim, nitʻaqqeim = to slant, to be inclined §B.2.I.8

inḥirāf = ʿiqqum = slant, inclination §B.0.2

ḥarf (pl. ḥiraf) = zeir = edge §A.III.5

ḥ-r-k

taḥarrak = hitnoʿeiʿa = to move §0.3

muḥarrik = meiniʿa = mover §0.4

ḥaraka = t^{e}nuʿah = movement §0.1

ḥ-l-q

dhāt al-ḥalaq = baʿal ha-ṭabbaʿot = armillary sphere (literally, ringed instrument) §A.III.2

ḥ-m-l

ḥamal = ṭaleh = Aries §0.5

ḥ-w-t

al-ḥūt = Pisces §B.4.III.6

ḥ-w-r

miḥwar = b^{e}ri'aḥ = axis, pin §A.III.4

ḥ-w-ṭ

muḥīṭ = maqqiḇ = circumference §A.I.4

kh-r-j

khurūj (al-markaz) = y^{e}ṣi'at (ha-merkaz) = eccentricity §B.4.I.4

istikhrāj = hoṣa'ah = derivation, extraction §A.IV.5

khārij (al-markaz) = yoṣei' (ha-merkaz) = eccentric §0.1

kh-ṭ-ṭ

khaṭṭ = tei'eir = to trace §A.III.5

khaṭṭ (pl. khuṭūṭ) = qaw = line §A.I.1

kh-l-f

ikhtilāf = ḥilluḇ = anomaly, variation §0.1
ikhtilāf al-manẓar = shinnuy ha-mar'eh = parallax §B.2.IV.1

mukhtalif = m^{e}ḥullaḇ = different §0.6

khilāf = hep̱eḵ = opposite §0.1

khalf = hep̱eḵ = reverse §B.1.II.11

kh-l-w

khalā' = reiqut = void §0.3

kh-y-l

takhayyal = ṣuyyar = to be imagined §0.5

d-l-l

istadall = heiḇi' r^{e}'ayah = to infer §B.2.I.1

dalīl (pl. dalā'il) = r^{e}'ayah = evidence §0.1

d-r-k

adrak = hissig = to perceive §B.1.I.3

idrāk = hassagah = perception §0.15

d-l-w

al-dalw = Aquarius §B.4.III.8

d-w-r

dār = soḇeiḇ = to revolve, to rotate §A.I.1

dā'ira (pl. dawā'ir) = 'iggulah = circle §0.5

dā'irat (niṭāq) al-burūj = 'iggulat (ḥagorat) ha-mazalot = the ecliptic §0.5

dā'irat al-mamarr = 'iggulat ha-ma'aḇar = the circle of the path §B.1.I.2

al-dā'ira al-mārra bi-nuqṭatay al-inqilābayn = ha-'iggulah ha-'oḇeret bi-n^{e}qudot ha-hippuḵim = the solstitial colure §A.IV.1

dā'irat nuṣf al-nahār = 'iggulat ḥeiṣi ha-yom = the meridian circle §A.I.3

madār = sibbuḇ = revolution §A.III.2

tadwīr (pl. tadāwīr) = haqqap̱ah = epicycle §0.1

r-b-'

rub' (pl. arbā') = reḇa', roḇa'= quadrant, quarter §0.5

r-ṣ-d

raṣad = 'iyyein = to observe §A.III.3

raṣad (pl. arṣād)= 'iyyun = observation §0.10

r-f-'

irtifā' = goḇah = altitude §A.I.1

r-k-b

rakkab = hirkiḇ = to attach §A.III.2

murakkab = murkaḇ = compounded §A.I.2

r-m-y

al-rāmiy = Sagittarius §B.4.III.8

z-h-r

al-zuhara = nogah = Venus §0.5

z-ḥ-l

zuḥal = shabb^{e}tay = Saturn §B.4.0.1

z-w-l

zawāl = hasarah = deviation, departure §0.10

z-w-y

zāwiya = zawit = angle §0.6

s-r-ṭ-n

al-saraṭān = sarṭan = Cancer §0.5

s-ṭ-r

dhāt al-masāṭir = ba‘al ha-sharbiṭim = parallactic rods §A.III.4

s-m-k

al-simāk al-a‘zal = ha-simaḵ al-a‘zal = Spica Virginis §B.3.I.3

s-m-t

samt al-ra’s = noḵaḥ ha-ro’sh = zenith §A.I.1

s-m-w

al-samā’ = ha-shamayim = the heavens §0.1
  wasaṭ al-samā’ = emṣa‘ ha-shamayim = mid-heaven §A.I.1
  fī wasaṭ al-samā’ = b^{e}-emṣa‘ ha-shamayim = culminating §A.I.1

s-n-b-l

al-sunbula = Virgo §B.4.III.6

s-w-‘

sā‘a = sha‘ah = hour §A.IV.4
  sā‘a istiwā’iyya = sha‘ah shawah = equinoctial hour §A.IV.4
  sā‘a zamāniyya = sha‘ah z^{e}mannit = seasonal hour §A.IV.4

s-w-y

istiwā’ = hashwa’ah = equinox §A.I.1
  khaṭṭ al-istiwā’ = qaw ha-shaweh = celestial equator

musāw^{in}, mutasāw^{in} = shaweh = equal §A.II.8

mustaw^{in} = mishtawweh = uniform §0.8

s-y-r

sayr = mahalaḵ = speed, direct motion §0.17

masīr = mahalaḵ = path, direct motion, speed §0.17

sh-b-h

tashabbuh = hitdammut = resemblance §0.13

mutashābih = mitdammeh = similar, uniform §0.6

ashadd shubh^{an} = yoteir domeh = has the strongest resemblance §A.I.1

shabīh = domeh = similar §B.1.I.2

sh-r-y

al-mushtarī = ṣedeq = Jupiter §B.4.0.1

sh-ṭ-b

dhāt al-shuṭba al-sayyāra = ba'al ha-yad ha-mitno'ei'at = dioptra §A.III.5

sh-n-'

(ghāyat al-) shinā'a = (taḵlit ha-) zarut = extremely hideous §0.6

ṣ-l-ḥ

aṣlaḥa = tiqqein = to correct §B.1.III.3

iṣlāḥ = tiqqun = correction §B.1.III.5

(takallam bi-) iṣṭilāḥ = dibbeir . . . hiskim = to speak in conventional, technical language §0.1

ṣ-w-r

ṣūra (pl. ṣuwar) = ṣurah = form, figure §A.I.1

ḍdd

ḍādd = (li-h^{e}yot) hep̱eḵ = to oppose §0.1

mutaḍādd = hap̱ki = opposing §0.1

ṭ-b-'

ṭabī'ī = ṭiḇ'i = physical, natural §0.1

ṭ-l-'

ṭala' = zaraḥ = to rise §A.I.1

ṭulū' = z^{e}riḥah = rising §A.I.1

maṭla' (pl. maṭāli') = miṣ'ad = rising time, point of ascension (= m^{e}qom z^{e}riḥah) §0.18

ṭ-w-l

ṭūl = oreḵ = longitude §0.7

ẓ-h-r

ẓahur = nir'eh, nitba'eir = to appear, result §0.1

'-d-l

(nuqṭat al-) i'tidāl = (n^{e}qudat ha-) shiwwuy = equinox §0.6

mu'addil al-nahār = m^{e}shawweh ha-yom = equinoctial §0.5

'-r-ḍ

'arḍ = roḥaḇ = latitude §0.7

'arḍ = 'oḇi = width §0.4

'-ṭ-r-d

'uṭārid = Mercury §B.4.0.1

'-ṭ-f

in'iṭāf = 'aqallaton = curve §B.2.V.2

‘-q-r-b

al-‘aqrab = Scorpio §B.4.III.6

f-r-j

munfarij = nirwaḥ = obtuse §B.1.III.2

f-l-k

falak (pl. aflāk) = galgal = orb §0.1
falak al-burūj = galgal ha-mazalot = the ecliptic orb §0.5
al-falak al-mukawkab = galgal ko<u>k</u><u>b</u>ei she<u>b</u>et = starred orb §0.14

q-b-l

muqābala, istiqbāl = haqbalah = = opposition §A.I.1

al-iqbāl wa-’l-idbār = ha-iḥur w^{e}-ha-q^{e}dimah = trepidation §B.3.I.2

q-d-m

muqaddama = haqdamah = preliminary, lemma §A.I.1

muqaddim = qodeim = premise, antecedent §B.1.II.2

q-ṣ-r

qaṣṣar = qiṣṣeir = to lag §0.13

taqṣīr = qiṣṣur = lag §0.12

q-ṭ-b

quṭb (pl. aqṭāb) = qoṭe<u>b</u> = pole §A.III.2

q-ṭ-r

quṭr (pl. aqṭār) = qoṭer, ala<u>k</u>son = diameter §A.II.2

q-ṭ-‘

qaṭa‘ = ḥata<u>k</u> = to cut, to traverse §A.I.1

inqaṭa‘ = nitḥattei<u>k</u> = to intersect §A.I.1

qiṭā‘, quṭ‘a = ḥati<u>k</u>ah = segment §A.IV.3

taqāṭu‘ = ḥittu<u>k</u> = intersection §A.I.3

q-‘-r

muqa‘‘ar = ‘aqmumi = concave §0.15

q-l-b

munqalab, inqilāb = hippu<u>k</u> = solstice §A.I.1

q-w-s

qaws (pl. aqwās) = qeshet = arc §A.II.2
al-qaws = Sagittarius §B.4.III.8

q-w-m

mustaqīm = yashar = straight §A.I.1

qā’im (‘alā) = niṣṣa<u>b</u> (‘al) = perpendicular (to) §0.6

q-w-y

quwwa = ko'aḥ = power §0.12

q-y-s

qiyās (pl. aqyisa) = heqqeish = syllogism, analogy, measurement §A.I.1

k-r-w

kura (pl. kuran) = kaddur, kiddur, kiddor = sphere §0.5

k-s-f

inkasaf = nilqah = to be eclipsed §0.7

kusūf = laqut = eclipse §0.7

k-w-k-b

kawkab (pl. kawākib) = kok̲ab̲ = planet, star §0.4

k-y-f

kayfiyya = eik̲ut = manner §0.9

m-r-kh

mirrīkh = Mars §B.4.0.1

m-s-s

mumāss = m^{e}mushshash = tangent §0.15

mumāssa = mishshush = tangency §0.15

m-l-'

mil' = milluy = plenum §0.3

m-y-l

mayl (pl. muyūl) = n^{e}ṭiyyah = slant, declination, inclination, latitude §A.I.1

mā'il = noṭeh = inclined, slanted §0.18

(al-kura) al-mā'ila = (ha-kaddur) ha-noṭeh = *sphaera obliqua* §A.IV.3

n-s-b

nisba = 'erek̲ = ratio §0.7

n-ṣ-b

(al-kura) al-muntaṣiba = (ha-kaddur) ha-yashar = *sphaera recta* §0.18

n-ṭ-q

minṭaqa (pl. manāṭiq), niṭāq = ḥagorah = equator §B.3.I.4

n-q-ṭ

nuqṭa (pl. nuqaṭ) = n^{e}qudah = point §0.5

n-q-l

anqal= he'etiq = to move (transitive) §0.5

intaqal = ne'etaq = to move, to be moved §0.13

nuqla = haʿtaqah = movement §B.0.3

n-h-y

intahā ilā = higgiʿa = to reach §0.14

nihāya = takl̲it = extremity §A.I.4

h-y-ʾ

hayʾa = t^{e}k̲unah = model §0.9

w-t-r

wattar, awtar = yitteir = to subtend, to cut off (with an arc) §B.1.III.2

watr (pl. awtār) = yeter = chord §A.II.2

w-z-y

muwāz^{in} = n^{e}k̲oḥi = parallel §A.III.2

w-ḍ-ʿ

waḍaʿ = sam = to propose, to position, to be a thesis §0.11

waḍʿ (pl. awḍāʿ) = hannaḥah, haṣṣaʿah = thesis, position §0.1

mawḍūʿ = munnaḥ = set forth, proposed §0.5

w-h-m

tawahham = ḥashab̲, ṣiyyeir, dimmah = to imagine §0.11

tawahhum = dimyon = imagination §B.2.IV.2

y-w-m

al-ayyām bi-layālihā = ha-yamim b^{e}-leiloteihem = the nychthemerons §B.2.III.3

BIBLIOGRAPHY

MANUSCRIPT SOURCES

Abū al-Khayr, Yiṣḥaq ben Samuel: *Peirush al-Fargani*, Bodleian MS Neubauer 2015.

al-Shīrāzī, Quṭb al-Dīn: *al-Tuḥfa al-shāhiyya*, Paris BNF MS Arabe 2516.

al-Shīrāzī, Quṭb al-Dīn: *Fa'alta fa-lā talum*, Istanbul Fatih MS 3175/2.

al-Shīrāzī, Quṭb al-Dīn: *Nihāyat al-idrāk fī dirāyat al-aflāk*, Istanbul Pertev Paşa MS 381.

Anatoli, Jacob: *Almagesṭi*. Paris BNF MS Hébreu 1018.

Anonymous: *Compendium of the Almagest*, Vatican MS Ebr 392.

Ibn Aflaḥ, Jābir (trans. Jacob Ben Makir): *Qiṣṣur al-Majisṭi*, Paris BNF MS Hébreu 1025.

Ibn Tibbon, Moses: *Seiper ha-yesodot*, Vienna Nationalbibliothek Cod. Hebr. 194.

PUBLISHED SOURCES

al-Bīrūnī (ed. and trans. Marie-Thérèse Debarnot): *Kitāb Maqālīd 'ilm al-hay'a* (Damascus: Institut Français de Damas, 1985).

al-Biṭrūjī (trans. and comm. Bernard R. Goldstein): *Al-Biṭrūjī: On the Principles of Astronomy* (New Haven and London: Yale University Press, 1971), 2 vols.

al-Ghazālī (trans. Michael Marmura): *The Incoherence of the Philosophers* (Provo, UT, and London: Brigham Young University Press, 1997).

Avi-Yonah, Reuven: "Ptolemy vs. al-Biṭrūjī: A Study in Scientific Decision-Making in the Middle Ages," *Archives internationales d'histoire des sciences* XXXV (1985): pp. 124–47.

Bartolocci, Giulio: *Bibliotheca magna rabbinica de scriptoribus* (Rome: Sacrae Congregationis de Propaganda Fide, 1675–94), vol. 4.

Bausani, Alessandro: "Die Bewegungen der Erde im *Kitāb Ikhwān aṣ-ṣafā'*: Ein Vor-Philolaisch-Pythagoräisches System?" *Zeitschrift für Geschichte der arabisch-islamischen Wissenschaften* I (1984): pp. 88–99.

Bellver Martínez, José: "El Lugar del *Iṣlāḥ al-Maŷisṭī* de Ŷābir b. Aflaḥ en La Llamada 'Rebelión Andalusí Contra la Astronomía Ptolemaica,'" *al-Qanṭara* XXX (2009): pp. 83–136.

Berman, Lawrence: "Samuel ben Judah of Marseilles," in Alexander Altmann (ed.): *Jewish Medieval and Renaissance Studies* (Cambridge: Harvard University Press, 1967): pp. 289–320.

Berman, Lawrence: "Brethren of Sincerity, Epistles of," in Michael Berenbaum and Fred Skolnik (eds.): *Encyclopaedia Judaica*, 2nd ed. (Detroit: Macmillan Reference, 2007): vol. 4: pp. 170–1.

Blau, Joshua: *A Grammar of Mediaeval Judaeo-Arabic* [in Hebrew] (Jerusalem, 1961).

Calvo, Emilia: "Abū ʿAbd Allāh Muḥammad ibn Muʿādh al-Jayyānī," in Thomas Hockey et al. (eds.): *The Biographical Encyclopedia of Astronomers* (New York: Springer, 2007): pp. 652–3.

———: "Astronomical Theories Related to the Sun in Ibn al-Hāʾim's *al-Zīj al-Kāmil fī al-taʿālīm*," *Zeitschrift für Geschichte der arabisch-islamischen Wissenschaften* XII (1998): pp. 51–111.

———: "Jābir ibn Aflaḥ," in Thomas Hockey et al. (eds.). *The Biographical Encyclopedia of Astronomers* (New York: Springer, 2007): pp. 581–2.

Carmody, Francis: *"De motibus celorum"; Critical Edition of the Latin Translation of Michael Scot* (Berkeley: University of California Press, 1952).

Casulleras, Josep: "The Contents of Qāsim ibn Muṭarrif al-Qaṭṭān's *Kitāb al-Hayʾa*," in Maribel Fierro and Julio Samsó (eds.): *The Formation of al-Andalus, Part 2: Language, Religion, Culture and the Sciences* (Aldershot, UK: Ashgate, 1998): pp. 339–58.

Chabás, José and Bernard R. Goldstein: *The Alfonsine Tables of Toledo* (Dordrecht, Boston, and London: Kluwer Academic Publishers, 2003).

Comes, Mercè: "Ibn al-Hāʾim's Trepidation Model," *Suhayl* II (2001): 329–37.

———: "The Accession and Recession Theory in al-Andalus and the North of Africa," in Josep Casulleras and Julio Samsó (eds.): *From Baghdad to Barcelona: Studies in the Islamic Exact Sciences in Honour of Prof. Juan Vernet* (Barcelona, 1996), vol. 1: pp. 349–64.

———: "The Possible Scientific Exchange between the Courts of Hulaghu of Maragha and Alphonse 10th of Castille," in Nasrollah Pourjavady and Ziva Vesel (eds.): *Science, techniques et instruments dans le monde Iranien* (Brepols: Peeters, 2004): pp. 29–50.

Crescas, Hasdai: (intro. E. Shweid): *Seiper Or Ha-shem* (Jerusalem: Makor, 1970; facsimile edition of the Ferrara, 1555 edition).

———: (trans., notes, and preface Éric Smilevitch): *Lumière de l'éternel* (Or Hachem) (Paris: Éditions Hermann, 2010).

Di Bono, Mario: "Copernicus, Amico, Fracastoro, and Ṭūsī's Device: Observations on the Use and Transmission of a Model," *Journal for the History of Astronomy* XXVI (1995): pp. 133–54.

Dobrzycki, Jerzy: "The Medieval Theory of Precession," *Studia Copernicana* XLIII (2010): pp. 15–60.

Dreyer, J. L. E.: *A History of Astronomy from Thales to Kepler* (Cambridge: Cambridge University Press, 1906; repr., Mineola, NY: Dover Publications, 1953).

Eichner, Heidrun: *The Post-Avicennian Philosophical Tradition and Islamic Orthodoxy. Philosophical and Theological Summae in Context* (Habilitationsschrift: Martin-Luther-Universität Halle-Wittenberg, 2009).

Endress, Gerhard and Dimitri Gutas (eds.): *A Greek and Arabic Lexicon: Materials for a Dictionary of the Mediaeval Translations from Greek into Arabic* (Leiden: E. J. Brill, 1992–): vol. 1, fascicle 2.

Endress, Gerhard: "Mathematics and Philosophy in Medieval Islam," in Jan P. Hogendijk and Abdelhamid I. Sabra (eds.): *The Enterprise of Science in Islam: New Perspectives* (Cambridge: MIT Press, 2003): pp. 121–76.

Euclid (intro., trans., and comm. Thomas L. Heath): *The Thirteen Books of the Elements* (Cambridge: Cambridge University Press, 1908; repr., Mineola, NY: Dover Publications, 1956), 3 vols.

Forcada, Miquel: "Astronomy, Astrology, and the Sciences of the Ancients in Early Al-Andalus," *Zeitschrift für Geschichte des arabisch-islamischen Wissenschaften* XVI (2005): pp. 1–74.

———: "Ibn Bājja and the Classification of the Sciences," *Arabic Sciences and Philosophy* XVI (2006): pp. 287–307.

———: "Ibn Ṭufayl," in Thomas Hockey et al. (eds.): *The Biographical Encyclopedia of Astronomers, Springer Reference* (New York: Springer, 2007, p. 572).

———: "La ciencia en Averroes," in *Averroes y los averroísmos: actas del III Congreso de Filosofia Medieval* (Zaragoza: Sociedad de Filosofía Medieval, 1999): pp. 49–102.

Frank, Richard: "The Science of *Kalām*," *Arabic Sciences and Philosophy* II (1992): pp. 7–37.

Freudenthal, Gad: "'Instrumentalism' and 'Realism' as Categories in the History of Astronomy: Duhem vs. Popper, Maimonides vs. Gersonides," *Centaurus* XLV (2003): pp. 227–48.

———: "Maimonides on the Knowability of the Heavens and Their Mover (*Guide* 2:24)," *Aleph* VIII (2008): pp. 151–7.

———: "The Medieval Astrologization of Aristotle's *Biology*: Averroes on the Role of the Celestial Bodies in the Generation of Animate Beings," *Arabic Sciences and Philosophy* XII (2002): pp. 111–37.

——— (ed.): *Science in Medieval Jewish Cultures* (Cambridge and New York: Cambridge University Press, 2012).

———: "Towards a Distinction between the two Rabbis Joseph ibn Joseph ibn Naḥmias" [in Hebrew], *Qiryat Sep̱er* (1988–9): pp. 917–9.

Gauthier, Leon (ed. and trans.): *Ḥayy ibn Yaqẓān* (Beirut: Imprimerie Catholique, 1936).

Glasner, Ruth: *Averroes' "Physics": A Turning Point in Medieval Natural Philosophy* (Oxford: Oxford University Press, 2009).

———: "The Evolution of the Genre of Philosophical-Scientific Commentary: *Hebrew Supercommentaries on Aristotles's 'Physics,'*" in Freudenthal (ed.): *Science in Medieval Jewish Cultures* (Cambridge and New York: Cambridge University Press, 2012): pp. 182–206.

———: *A Fourteenth-Century Scientific-Philosophic Controversy: Yedaiah Ha-Penini's Treatise on Opposite Motions and Book of Confutation* [in Hebrew] (Jerusalem: World Union of Jewish Studies, 1998).

———: "Levi Ben Gershom and the Study of Ibn Rushd in the Fourteenth Century," *Jewish Quarterly Review*, new series, LXXXVIII (1995): pp. 51–90.

———: "The Peculiar History of Aristotelianism among Spanish Jews," in Resianne Fontaine, Ruth Glasner, Reimund Leicht, and Giuseppe Veltri (eds.): *Studies in the History of Culture and Science: A Tribute to Gad Freudenthal* (Leiden and Boston: Brill, 2011): pp. 361–81.

Goldstein, Bernard R.: "The Arabic Version of Ptolemy's *Planetary Hypotheses*," *Transactions of the American Philosophical Society*, new series, LVII (1967), no. 4: pp. 3–55.

———: "Astronomy among Jews in the Middle Ages," in Freudenthal (ed.), *Science in Medieval Jewish Cultures* (Cambridge and New York: Cambridge University Press, 2012): pp. 136–46.

———: "Astronomy and the Jewish Community in Early Islam," *Aleph* I (2001): pp. 17–57.

———: "Astronomy in the Medieval Spanish Jewish Community," in L. Nauta and A. Vanderjagt (eds.): *Between Demonstration and Imagination: Essays in the History of Science and Philosophy Presented to John D. North*: (Leiden: Brill: 1999): pp. 225–41.

———: *The Astronomy of Levi ben Gerson (1288–1344)* (New York, Berlin, Heidelberg, Tokyo: Springer-Verlag, 1985).

———: "Levi ben Gerson's Analysis of Precession," *Journal for the History of Astronomy* VI (1975): pp. 31–41.

———: "On the Theory of Trepidation according to Thābit ibn Qurra and al-Zarqāllu and Its Implications for Homocentric Planetary Theory," *Centaurus* X (1964): 232–47.

———: "Scientific Traditions in Late Medieval Jewish Communities," in G. Dahan (ed.): *Les Juifs au regard de l'histoire: Mélanges en l'honneur de M. Bernhard Blumenkranz* (Paris: Picard, 1985): pp. 235–47.

———: "The Medieval Hebrew Tradition in Astronomy," *Journal of the American Oriental Society* LVIII (1965): pp. 145–8.

———: "The Role of Science in the Jewish Community in Fourteenth-Century France," *Annals of the New York Academy of Sciences* CCCXIV (1978): pp. 39–49.

Goldstein, Bernard R. and Noel Swerdlow: "Planetary Distances and Sizes in an Anonymous Arabic Treatise Preserved in Bodleian Ms. Marsh 621," *Centaurus* XV (1971): pp. 135–50.

Gutas, Dimitri: *Avicenna and the Aristotelian Tradition* (Leiden: Brill, 1988).

Gutwirth, Eleazar: "Hispano-Jewish Attitudes to the Moors in the Fifteenth Century," *Sᵉp̱arad* XLIX (1989): pp. 237–62.

———: "History, Language, and the Sciences in Medieval Spain," in Gad Freudenthal (ed.): *Science in Medieval Jewish Cultures* (New York and Cambridge: Cambridge University Press, 2012): pp. 511–28.

Hamm, Elizabeth A.: "Ptolemy's Planetary Theory: An English Translation of Book One, Part A of the *Planetary Hypotheses* with Introduction and Commentary" (Ph.D. dissertation, University of Toronto, 2011).

Heath, Thomas: *Aristarchus of Samos: The Ancient Copernicus* (Oxford: Clarendon Press, 1913; repr., Mineola, NY: Dover Publications, 2004).

Heribert M. Nobis (ed.): *Nicolaus Copernicus Gesamtausgabe* (Hildesheim: H.A. Gerstenberg, 1974–).

Herz-Fischler, Roger: *A Mathematical History of the Golden Number* (Mineola, NY: Dover Publications, 1998).

Hugonnard-Roche, Henri: "Remarques sur l'évolution doctrinale d'Averroès dans les commentaires au *De Caelo*; Le problème du mouvement de la terre," *Mélanges de la Casa de Velázquez* XIII (1977): pp. 103–17.

Ibn Bājja: "Fī al-Faḥṣ 'an al-quwwa al-nuzū'iyya," in 'Abd al-Raḥmān Badawī (ed.): *Rasā'il Falsafiyya li-'l-Kindī wa-'l-Fārābī wa-Ibn Bājja wa-Ibn 'Adī* (Beirut: Dār al-Andalus, 1983): pp. 147–56.

Ibn Falaquera, Shem Tov ben Joseph (ed. and comm. Yair Shiffman): *Moreh ha-Moreh* (Jerusalem: World Union of Jewish Studies, 2001): pp. 284–5.

Ibn al-Haytham (ed. A. I. Sabra): *Kitāb al-Manāẓir: al-maqālāt al-rābi'a wa-'l-khāmisa* (Kuwait: al-Majlis al-waṭanī li-'l-thaqāfa wa-'l-funūn wa-'l-adab, 2002), 2 vols.

Ibn Rushd (ed. Arthur Hyman): *Averroes' "De Substantia Orbis"* (Cambridge, MA and Jerusalem: The Medieval Academy of America and the Israel Academy of Sciences and Humanities, 1986).

———(Charles Genequand intro. and trans.): *Ibn Rushd's "Metaphysics": A Translation with Introduction of Ibn Rushd's Commentary on Aristotle's "Metaphysics," Book Lam* (Leiden: E. J. Brill, 1984).

———: "Kitāb al-Samā' wa-'l-'ālam," in *Rasā'il Ibn Rushd* (Hyderabad: Dā'irat al-ma'ārif, 1947).

———(ed. Maurice Bouyges): *Tafsīr mā ba'd al-ṭabī'a* (Beirut, 1948).

———(Jamāl al-Dīn al-'Alawī ed. and intro.): *Talkhīṣ al-Samā' wa-'l-'ālam* (Fes: Kulliyyat al-ādāb, 1984).

———(ed. Maḥmūd Qāsim, comm. Charles Butterworth and Aḥmad 'Abd al-Majīd Harīdī): *Talkhīṣ Kitāb al-Maqūlāt* (Cairo: General Egyptian Book Organization, 1980).

———(ed. Maḥmūd Qāsim; edition completed by Charles Butterworth and Aḥmad 'Abd al-Majīd Harīdī): *Talkhīṣ Kitāb al-Qiyās* (Cairo: General Egyptian Book Organization, 1983).

Ibn Sīnā (ed. Ibrāhīm Madkūr, Muḥammad Madwar, and Imām Ibrāhīm Aḥmad): *Al-Shifā': 'Ilm al-Hay'a* (Cairo: General Egyptian Book Organization, 1980).

——— (ed. Ibrāhīm Madkūr): *al-Shifā', al-Samā' wa-'l-'ālam* (Cairo: General Egyptian Book Organization, 1969).

——— (trans. and notes Asad Q. Ahmed): *Avicenna's "Deliverance": Logic* (Karachi and New York: Oxford University Press, 2011).

Ibn Tibbon, Samuel (ed. and intro. Resianne Fontaine): *Otot ha-Shamayim: Samuel Ibn Tibbon's Hebrew Version of Aristotle's "Meteorology"* (Leiden, New York, and Köln: Brill, 1995).

Ibn Ṭufayl (ed. Fārūq Sa'd): *Ḥayy ibn Yaqẓān*, 5th ed. (Beirut: Dār al-āfāq al-jadīda).

——— (trans., intro., and notes Lenn Evan Goodman): *Ibn Tufayl's "Hayy Ibn Yaqzān"* (Chicago and London: University of Chicago Press, 2009).

Ibrahim, Bilal: "Faḫr ad-Dīn ar-Rāzī, Ibn al-Hayṯam and Aristotelian Science: Essentialism versus Phenomenalism in Post-Classical Islamic Thought," *Oriens* XLI (2013): pp. 379–431.

Ikhwān al-Ṣafā': *Rasā'il* (Beirut: Dār Ṣādir, n.d.), 4 vols.

Ilan, Nahem: "Know What You Shall Say Back to Epiqoros" [in Hebrew], in David Doron and Joshua Blau (eds.): *Heritage and Innovation in Medieval Judeo-Arabic Culture* [in Hebrew] (Ramat Gan: Bar Ilan Press, 2000): pp. 9–26.

———: "'R^{e}dip̱at ha-emet' w^{e}-' Dereḵ la-rabbim': 'iyyunim b^{e}-mishnat R. Yisra'el Yisr^{e}'eli mi-Ṭoledo" (Hebrew University, Ph.D. dissertation, 1999).

Kaspi, Joseph (ed. Werbluner): *Commentaria hebraica in R. Mosis Maimonidis* (Frankfurt, 1848: repr., in *Sh^{e}losha qadmonei m^{e}p̱ar^{e}shei "ha-Moreh"* [Jerusalem, 1968]).

Kennedy, E. S.: "A Survey of Islamic Astronomical Tables," *Transactions of the American Philosophical Society*, new series, XLVI (1956), part two: pp. 123–77.

———: "Alpetragius's Astronomy," *Journal for the History of Astronomy* IV (1973): pp. 134–36.

———: "Review of Carmody, *al-Biṭrūjī's De motibus celorum*," *Speculum* XXIX (1954): pp. 246–51.

Kennedy, E. S., and Victor Roberts: "The Planetary Theory of Ibn al-Shāṭir," *Isis* L (1959): pp. 227–35.

King, David A. and Julio Samsó, with Bernard R. Goldstein: "Astronomical Handbooks and Tables from the Islamic World (750–1900): an Interim Report," *Suhayl* II (2001): pp. 9–105.

Kozodoy, Maud: "A Study of the Life and Works of Profiat Duran" (Ph.D. dissertation, The Jewish Theological Seminary, 2006).

Kraemer, Joel: "Maimonides and the Spanish Aristotelian School," in Mark D. Meyerson and Edward D. English (eds.): *Christians, Muslims and Jews in Medieval and Early Modern Spain* (Notre Dame, IN: University of Notre Dame Press, 1999): pp. 40–68.

Langermann, Y. Tzvi: "A Compendium of Renaissance Science: *Ta'alumot Hokhma* by Moshe Galeano," *Aleph: Historical Studies in Science and Judaism* VII (2007): pp. 283–318.

———: "Another Andalusian Revolt? Ibn Rushd's Critique of al-Kindī's *Pharmacological Computus*," in Jan P. Hogendijk and Abdelhamid I. Sabra: *The Enterprise of Science in Islam: New Perspectives* (Cambridge and London: MIT Press, 2003).

———: "Arabic Writings in Hebrew Manuscripts: A Preliminary Re-Listing," *Arabic Sciences and Philosophy*, 6 (1996): 137–60.

———: "Ibn al-Haytham" in Thomas Hockey et al. (eds.): *The Biographical Encyclopedia of Astronomers* (New York: Springer, 2007): pp. 556–57.

———: "Ibn Kammūna and the New Wisdom of the Thirteenth Century," *Arabic Sciences and Philosophy* XV (2005): pp. 277–327.

———: "Medicine, Mechanics and Magic from Moses ben Judah Galeano's *Ta'alumot Ḥokhmah*," *Aleph: Historical Studies in Science and Judaism* IX (2009): 353–77.

———: "Medieval Hebrew Texts on the Quadrature of the Lune," *Historia Mathematica* XXIII (1996): pp. 31–53.

———: "My Truest Perplexities," *Aleph* VIII (2008): pp. 301–17.

———: "On the Beginnings of Hebrew Scientific Literature and on Studying History through 'Maqbiloṯ' (Parallels)," *Aleph* II (2002): pp. 169–89.

———: "Sa'adya and the Sciences," in Y. Tzvi Langermann (ed.): *The Jews and the Sciences in the Middle Ages* (Aldershot, Great Britain and Brookfield, VT: Ashgate, 1999): pp. 1–21 (second article).

———: "Science in the Jewish Communities of the Iberian Peninsula: An Interim Report," in Langermann: *The Jews and the Sciences in the Middle Ages* (Aldershot, UK and Brookfield, VT: Ashgate, 1999): pp. 1–54.

———: "Some Astrological Themes in the Thought of Abraham Ibn Ezra," in Isadore Twersky and Jay M. Harris (eds.): *Rabbi Abraham Ibn Ezra: Studies in the Writings of a Twelfth-Century Jewish Polymath* (Cambridge, MA, and London: Harvard University Press, 1993): pp. 28–85.

———: "The 'True Perplexity': The *Guide of the Perplexed* Part II, Chapter 24," in Joel Kraemer (ed.): *Perspectives on Maimonides: Philosophical and Historical Studies* (The Littman Library, Oxford University Press, 1991): pp. 159–74.

Lay, Juliane: "*L'Abrégé de l'Almageste*: Un Inédit d'Averroès en Version Hébraïque," *Arabic Sciences and Philosophy* VI (1996): pp. 23–61.

Lettinck, Paul: *Aristotle's "Meteorology" and its Reception in the Arab World* (Leiden, Boston, Köln: Brill, 1999).

Lettinck, Paul: *Aristotle's "Physics" and its Reception in the Arabic World: With an Edition of the Unpublished Parts of Ibn Bājja's Commentary on the "Physics"* (Leiden, New York: Brill, 1994).

Levinger, Jacob S., Irene Garbell, and Colette Sirat: "Duran, Profiat," in Michael Berenbaum and Fred Skolnik (eds.): *Encyclopaedia Judaica*, 2nd ed. (Detroit: Macmillan Reference, 2007), vol. 6: pp. 56–58.

Lévy, Tony: "Gersonide, Commentateur d'Euclide," in Gad Freudenthal (ed.): *Studies on Gersonides* (Leiden, New York, Köln: E. J. Brill, 1992): pp. 83–147.

Lorch, Richard: "The Astronomy of Jābir ibn Aflaḥ," *Centaurus* XIX (1975): pp. 85–107.

———: "Review of 'al-Biṭrūjī: *On the Principles of Astronomy*' by Bernard Goldstein," *Archives internationales d'histoire des sciences* XXIV (1974): pp. 173–5.

Maimonides, Moses (ed. and trans. Moses Hyamson): *The Book of Knowledge* (Jerusalem, 1965).

——— (ed. Hüseyin Atay): *Dalālat al-ḥā'irīn* (Ankara: Ankara University Press, 1972; repr., Cairo: Maktabat al-thaqāfa al-dīniyya, n.d.).

——— (ed. S. Munk): *Dalālat al-ḥā'irīn* (Jerusalem: Y. Yunovitz, 1929).

——— (trans. Shlomo Pines, intro. Leo Strauss): *The Guide of the Perplexed* (Chicago and London: University of Chicago Press, 1963), 2 vols.

———: *Seiper Moreh n*ᵉ*bukim/la-Rab Mosheh bar Maimon, zal; niṣṣab peirusho . . . peirush Shem Ṭob u-peirush Epodi* (Żolkiew : Gedrucht be Leib Matfes & Berl Lorie, 1860), 3 vols.

———: (Samuel Ibn Tibbon trans., Yehudah Eben-Shemu'el annotations): *Seiper "Moreh nebukim" le -Rabbeinu Mosheh ibn Maymon* (Jerusalem: Mosad ha-Rab Quq, 1959).

Mancha, José Luis: "Al-Biṭrūjī's Theory of the Motions of the Fixed Stars," *Archive for History of Exact Sciences* LVIII (2004): pp. 143–82.

———: "Demonstrative Astronomy: Notes on Levi ben Geršom's Answer to *Guide* II.24," in Fontaine et al (eds.): *Studies in the History of Culture and Science: A Tribute to Gad Freudenthal* (Leiden and Boston: Brill, 2011): pp. 323–46.

———: "Ibn al-Haytham's Homocentric Epicycles in Latin Astronomical Texts of the XIVth and XVth Centuries," *Centaurus* XXXIII (1990): pp. 70–89.

———: "Right Ascensions and Hippopedes: Homocentric Models in Levi ben Gerson's *Astronomy*; I. First Anomaly," *Centaurus* XLV (2003): pp. 264–83.

———: "The Theory of Access and Recess in Levi ben Gerson's Astronomy and its Sources," *Aleph* XII (2012): pp. 37–64.

Meyerson, Mark: *A Jewish Renaissance in Fifteenth-Century Spain* (Princeton: Princeton University Press, 2004).

Mimura, Taro: "Quṭb al-Dīn Shīrāzī's Medical Work, *al-Tuḥfa al-Sa'diyya* (Commentary on volume 1 of Ibn Sīnā's *al-Qānūn fī al-Ṭibb*) and its Sources," *Tārīkh-e 'Elm* X (2013): no. 2, pp. 1–13.

Morelon, Régis: "La Version Arabe Du Livre Des Hypothèses De Ptolémée," *Mélanges Institut Dominicain d'Études Orientales du Cairo* XXI (1993): pp. 7–85.

Morrison, Robert: "An Astronomical Treatise by Mūsā Jālīnūs Alias Moses Galeano," *Aleph: Historical Studies in Science and Judaism* X/2 (2011): pp. 315–53.

———: "Andalusian Responses to Ptolemy in Hebrew," in Jonathan Decter and Michael Rand (eds.): *Precious Treasures from Hebrew and Arabic* (Piscataway, NJ: Gorgias Press, 2007): pp. 69–86.

———: *Islam and Science: The Intellectual Career of Niẓām al-Dīn al-Nīsābūrī* (Oxon: Routledge, 2007).

———: "Natural Theology and the Qur'ān," *Journal of Qur'ānic Studies* XV (2013): pp. 1–22.

———: "Quṭb al-Dīn al-Shīrāzī's Hypotheses for Celestial Motions," *Journal for the History of Arabic Science* XIII (2005): pp. 21–140.

———: "A Scholarly Intermediary between the Ottoman Empire and Renaissance Europe," *Isis* CV (2014): 32–57.

———: "Science (Medieval)," in Norman A. Stillman (ed.): *Encyclopedia of Jews in the Islamic World* (Leiden: E. J. Brill, 2010).

———: "The Solar Model in Joseph Ibn Joseph Ibn Nahmias' *The Light of the World*," *Arabic Sciences and Philosophy* XV (2005): pp. 57–108.

Nallino, Carlo: *'Ilm al-falak; Tārīkhuh 'ind al-'Arab fī al-qurūn al-wusṭā* (1911; repr., Beirut: al-Dār al-'Arabiyya li-'l-kitāb and Awrāq sharqiyya, 1993).

Narboni, Moses (ed. J. Goldenthal): *Der Commentar des Rabbi Moses Narbonensis zu dem Werke "More Nebuchim" des Maimonides* (Vienna, 1862; repr., in *Sh^{e}losha qadmonei m^{e}p̱ar^{e}shei "ha-Moreh"* [(Jerusalem, 1968]).

Neugebauer, Otto: *The Exact Sciences in Antiquity* (Providence, RI: Brown University Press, 1957; repr., Dover Publications, 1969).

———: *History of Ancient Mathematical Astronomy* (Berlin, New York: Springer-Verlag, 1975), 3 vols.

———: "On the 'Hippopede' of Eudoxus," *Scripta Mathematica* XIX (1953): pp. 225–29; repr. in *Astronomy and History* (New York, Heidelberg, and Tokyo: Springer-Verlag, 1983): pp. 305–9.

Neugebauer, Otto, and Noel Swerdlow: *Mathematical Astronomy in Copernicus' "De Revolutionibus"* (New York and Berlin: Springer-Verlag, 1984), 2 vols.

Offenberg, A. K.: "The First Printed Book Produced at Constantinople," *Studia Rosenthaliana* III (1969): pp. 96–112.

Olszowy-Schlanger, Judith: "The Science of Language among Medieval Jews," in Freudenthal (ed.), *Science in Medieval Jewish Cultures* (Cambridge and New York: Cambridge University Press, 2012): pp. 359–424.

Pedersen, Olaf (ed. Alexander Jones): *A Survey of the Almagest* (New York, Heidelberg, Berlin: Springer-Verlag, 2011).

Peruzzi, Enrico: *La nave di Ermete: la cosmologia di Girolamo Fracastoro* (Florence: Olschki, 1995).

Pines, Shlomo: "Études sur Awḥad al-Zamān Abu'l-Barakāt al-Baghdādī," in *Studies in Abū 'l-Barakāt al-Baghdādī: Physics and Metaphysics* (Jerusalem: Magnes Press; Leiden: E. J. Brill, 1979): pp. 1–95.

———: "Ibn al-Haytham's Critique of Ptolemy," in *Actes du dixième congrès international des sciences* I, no. 10 (Ithaca, 1962; Paris, 1964): pp. 547–50.

———: "Notes on Abu'l Barakāt's '*Celestial Physics*,'" in *Studies in Abū 'l-Barakāt al-Baghdādī: Physics and Metaphysics* (Jerusalem: Magnes Press; Leiden: Brill, 1979).

Proclus (attributed): "*Kitāb al-īḍāḥ li-Arisṭūṭālīs fī al-khayr al-maḥḍ*," in ʿAbd al-Raḥmān Badawī (ed.): *al-Aflāṭūniyya al-muḥdatha ʿind al-ʾArab* (Kuwait: Wikālat al-maṭbūʿāt, 1977): pp. 1–33.

Ptolemy: (trans., comm. Gerald Toomer): *Ptolemy's Almagest* (London: Duckworth, 1984).

Puig, Roser: "The Theory of the Moon in the *Al-Zīj al-kāmil fī-l-taʿālīm* of Ibn al-Hāʾim (*circa* 1205)," *Suhayl* 1 (2000): pp. 71–99.

Ragep, F. Jamil: "Al-Battānī, Cosmology, and the Early History of Trepidation in Islam," in Josep Casulleras and Julio Samsó (eds.): *From Baghdad to Barcelona: Studies in the Islamic Exact Sciences in Honour of Prof. Juan Vernet* (Barcelona, 1996), vol. 1: pp. 267–98.

———: "ʿAlī Qushjī and Regiomontanus: Eccentric Transformations and Copernican Revolutions," *Journal for the History of Astronomy* XXXVI (2005): 359–71.

———: "Copernicus and His Islamic Predecessors: Some Historical Remarks," *History of Science* XLV (2007): pp. 65–81.

———: "Freeing Astronomy from Philosophy: An Aspect of Islamic Influence on Science," *Osiris*, 2nd series, XVI (2001): pp. 49–71.

———: "Ibn al-Haytham and Eudoxus: The Revival of Homocentric Modeling in Islam," in Charles Burnett, Jan P. Hogendijk, Kim Plofker, and Michio Yano (eds.): *Studies in Honour of David Pingree* (Leiden: E. J. Brill, 2004): pp. 786–809.

——— (ed., trans., and comm.): *Naṣīr al-Dīn al-Ṭūsī's "Memoir on Astronomy" (al-Tadhkira fī ʿilm al-hayʾa)* (New York and Berlin: Springer-Verlag, 1993).

———: "The Origins of the Ṭūsī Couple Revisited," *Proceedings of the "Scientific and Philosophical Heritage of Naṣīr al-Dīn al-Ṭūsī" Conference*, February 23–24, 2011 (Tehran, Iran: Miras-e Maktoob), in press.

———: "Ṭūsī and Copernicus: The Earth's Motion in Context," *Science in Context* XIV (2001): 145–63.

Richler, Benjamin (ed.), Paleographical and codicological descriptions by Malachi Beit-Arié (with Nurit Pasternak): *Hebrew Manuscripts in the Vatican Library: Catalogue* (Vatican City: Biblioteca Apostolica Vaticano, 2008).

Roberts, Victor: "The Solar and Lunar Theory of Ibn al-Shāṭir: A Pre-Copernican Copernican Model," *Isis* XL (1957): pp. 428–32.

Robinson, James T.: "The First References in Hebrew to al-Biṭrūjī's *On the Principles of Astronomy*," *Aleph* III (2003): pp. 145–63.
Rosen, Edward: "Copernicus and Al-Bitruji," *Centaurus* VII (1960): pp. 152–6.
Sabra, A. I.: "The Andalusian Revolt against Ptolemaic Astronomy: Averroes and al-Biṭrūjī," in Everett Mendelsohn (ed.): *Transformation and Tradition in the Sciences* (Cambridge and New York: Cambridge University Press, 1984): pp. 133–53.
———: "The Commentary That Saved the Text," *Early Science and Medicine* XII (2007): pp. 117–33.
———: "Configuring the Universe: Aporetic, Problem Solving, and Kinematic Modeling as Themes of Arabic Astronomy," *Perspectives in Science* VI (1998): pp. 288–330.
———: "Ibn al-Haytham's Treatise: Solutions of Difficulties Concerning the Movement of *Iltifāf*," *Journal for the History of Arabic Science* III (1979): pp. 388–422.
———: "On Seeing the Stars, II. Ibn al-Haytham's 'Answers' to the 'Doubts' Raised by Ibn Ma'dân," *Zeitschrift für Geschichte der arabisch-islamischen Wissenschaften* X (1995/96): pp. 1–59.
———: "Psychology versus Mathematics: Ptolemy and Alhazen on the Moon Illusion," in Edward Grant, and J. E. Murdoch (eds.): *Mathematics and Its Application to Science and Natural Philosophy in the Middle Ages* (Cambridge: Cambridge University Press, 1987): pp. 217–47.
———: "Science and Philosophy in Medieval Islamic Theology," *Zeitschrift für Geschichte der arabisch-islamischen Wissenschaften* XIII (1994): pp. 1–42.
Sabra, A. I. and Anton Heinen: "On Seeing the Stars: Edition and translation of Ibn al-Haytham's Risâla fî Ru'yat al-kawâkib," *Zeitschrift für Geschichte der arabisch-islamischen Wissenschaften* VII (1991/92): pp. 31–72.
Saliba, George: "Critiques of Ptolemaic Astronomy in Islamic Spain," *al-Qantara* I (1999): pp. 3–25.
———: "Early Arabic Critique of Ptolemaic Cosmology: A Ninth-Century Text on the Motion of the Celestial Spheres," *Journal for the History of Astronomy* XXV (1994): pp. 115–41.
———: "The First Non-Ptolemaic Astronomy at the Marāgha School," *Isis* LXX (1979): pp. 571–6.
———: *A History of Arabic Astronomy* (New York: NYU Press, 1994).
———: *Islamic Science and the Making of the European Renaissance* (Cambridge and London: MIT Press, 2007).
———: "The Role of the *Almagest* Commentaries in Medieval Arabic Astronomy: A Preliminary Survey of Ṭūsī's Redaction of Ptolemy's *Almagest*," *Archives Internationales d'Histoire des Sciences* XXXVII (1987): pp. 3–20.
———: "Theory and Observation in Islamic Astronomy: The Work of Ibn al-Shāṭir of Damascus," *Journal for the History of Astronomy* XVIII (1987): pp. 35–43.
——— (ed. and intro.): *The Astronomical Work of Mu'ayyad al-Dīn al-'Urḍī (Kitāb al-Hay'a): A Thirteenth-Century Reform of Ptolemaic Astronomy* (Beirut, 1990).
Saliba, George and E. S. Kennedy: "The Spherical Case of the Ṭūsī Couple," *Arabic Sciences and Philosophy* I (1991): pp. 285–91.

Samsó, Julio: "Abraham Zacut and José Vizinho's *Almanach Perpetuum* in Arabic (16th–19th C.)," *Centaurus* XLVI (2004): pp. 82–97.

———: "Biṭrūjī: Nūr al-Dīn Abū Isḥāq [Abū Ja'far] Ibrāhīm ibn Yūsuf al-Biṭrūjī," in Thomas Hockey et al. (eds.): *The Biographical Encyclopedia of Astronomers* (New York: Springer, 2007): pp. 133–34.

———: "The Early Development of Astrology in al-Andalus," *Journal for the History of Arabic Science* III (1979): pp. 228–43.

———: "A Homocentric Solar Model by Abū Ja'far al-Khāzin," *Journal for the History of Arabic Science* I (1977): pp. 268–75.

———: *Las Ciencias de los Antiguos en al-Andalus* (Almeria: Fundación Ibn Ṭufayl, 2011).

———: "On al-Bitrujī and the Hay'a Tradition in al-Andalus," in Julio Samsó: *Islamic Astronomy and Medieval Spain* (Aldershot, Hampshire, UK: Variorum, 1994): article XII.

———: "Trepidation in al-Andalus in the 11th Century," in Samsó: *Islamic Astronomy and Medieval Spain* (Aldershot, Hampshire, UK: Variorum, 1994): VIII.

Schiaparelli, Giovanni: *Scritti sulla storia della astronomia antica* (Bologna: N. Zanichelli, 1925–7), 3 vols.

Sezgin, Fuat: *Geschichte des arabischen Schrifttums* (Leiden: E. J. Brill, 1978), vol. 6.

Shank, Michael: "Regiomontanus as a Physical Astronomer: Samplings from *The Defence of Theon against George of Trebizond*," *Journal for the History of Astronomy* XXXVIII (2007): pp. 325–49.

———: "The 'Notes on al-Biṭrūjī' Attributed to Regiomontanus: Second Thoughts," *Journal for the History of Astronomy* XXIII (1992): pp. 15–30.

Smart, W. M.: *Text-Book on Spherical Astronomy*, 6th ed. (Cambridge: Cambridge University Press, 1960).

Steinschneider, Moritz: *Die hebräischen Übersetzungen des Mittelalters und die Juden als Dolmetscher* (Berlin, 1893; repr., Graz, 1956).

———: *Mathematik bei den Juden* (Hildesheim: G. Olms, 1964).

Stroumsa, Sarah: *Maimonides in His World: Portrait of a Mediterranean Thinker* (Princeton: Princeton University Press, 2009).

Swerdlow, Noel: "Aristotelian Planetary Theory in the Renaissance: Giovanni Battista Amico's Homocentric Spheres," in *Journal for the History of Astronomy* III (1972): pp. 36–48.

———: "Regiomontanus's Concentric-Sphere Models for the Sun and Moon," *Journal for the History of Astronomy* XXX (1999): pp. 1–23.

Ta-Shma, Israel Moses: "Israeli, Israel," in Michael Berenbaum and Fred Skolnik (eds.): *Encyclopaedia Judaica*, 2nd ed. (Detroit: Macmillan Reference 2007)

———: "Naḥmias, Joseph ben Joseph," in Michael Berenbaum and Fred Skolnik (eds.): *Encyclopaedia Judaica*, 2nd ed. (Detroit: Macmillan Reference, 2007).

Toomer, Gerald J.: "The Solar Theory of az-Zarqāl: A History of Errors," *Centaurus* XIV (1969): pp. 306–36.

Van Brummelen, Glen: *Heavenly Mathematics: The Forgotten Art of Spherical Trigonometry* (Princeton, NJ and Oxford, UK: Princeton University Press, 2013).

Van der Waerden, B. L.: "The Heliocentric System in Greek, Persian, and Hindu Astronomy," in David A. King and George Saliba (eds): *From Deferent to Equant* (New York: Annals of the New York Academy of Sciences, 1987): 525–45.

Villuendas, M. V.: *La trigonometria europea en el siglo XI : Estudio de la obra de Ibn Mu'ad: "el Kitab mayhulat"* (Barcelona: Memorias de la Real Academia de Buenas Letras de Barcelona, 1979).

Wilkinson, Robert: *Orientalism, Aramaic, and Kabbalah in the Catholic Reformation* (Leiden, Boston: Brill, 2007).

Winter, Tim (ed.): *The Cambridge Companion to Classical Islamic Theology* (Cambridge: Cambridge University Press, 2008).

Yafūt, Sālim: "Ibn Bājja wa-'ilm al-falak al-Baṭlamyūsī," in Sālim Yafūt (ed.): *Dirāsāt fī tārīkh al-'ulūm wa-'l-ibistīmūlūjiyyā* (Rabat, 1996): pp. 65–73.

Yavetz, Ido: "A New Role for the Hippopede of Eudoxus," *Archive for History of Exact Sciences* LVI (2001): pp. 69–93.

———: "On Simplicius' Testimony Regarding Eudoxan Lunar Theory," *Science in Context* XVI (2003): pp. 319–29.

———: "On the Homocentric Spheres of Eudoxus," *Archive for History of Exact Sciences* LII (1998): pp. 221–78.

Zimmermann, Fritz W.: *Al-Fārābī's Commentary and Short Treatise on Aristotle's "De Interpretatione"* (New York and Oxford: Oxford University Press for The British Academy, 1981).

Zonta, Mauro: "La Tradizione Ebraica dell'*Almagesto* di Tolomeo," *Henoch* XV (1993): pp. 325–50.

———: "Medieval Hebrew Translations of Philosophical and Scientific Texts: *A Chronological Table*," in Freudenthal (ed.), *Science in Medieval Jewish Cultures* (Cambridge and New York: Cambridge University Press, 2012): pp. 17–73.

INDEX

Note from the author: I have left off the 'al' when it is at the beginning of Arabic names and the 'ha' when it is at the beginning of Hebrew names. Technical terms are also indexed through the glossary of technical terms. References beginning with § refer to the relevant sections of the texts, translation, and commentary as explained on p39.

Abū al-Khayr, Isaac ben Samuel, 276
Almohads, 7
Alfonso X
  Alfonsine Tables, 37 n144
  And Hülegü, 23 n93, 301 n143
Amico, Giovanni Battista, 2 n6, 28, 42, 301, 386
  And vapors, 337 n192
Anatoli, Jacob
  Hebrew *Almagesṭi*, 13 n48, 357
    On chords, 283 n89, 285 n102
    Parameters, 319 n164, 336 n186
Andalusia, 7, 12, 13, 22, 265 n6, 267, 276, 301 n143
Andalusian Revolt (see under Sabra, A. I.)
Aquarius, 36, §B.4.III.8–10
Aragon, 5
Aries, 36, §0.5–6, §A.IV.4–5, §B.1.II.29, §B.4.III.7–8
Aristarchos of Samos, 282
Aristotle
  And Eudoxus, 10 n33, 19–21
  And Ibn Rushd, 10
  Astronomy, 3, 10, 171, 271
  Celestial bodies' gyrational motion, 350
  Opposite motions in the heavens, 273–5
  Planets' rolling motions, 265
  Writings
    De Caelo, 102, 175, 268, 355 n233
    Metaphysics, 11, 271, 265 n3, 271, 280
    Meteorology, 272 n34, 274, 340
    Physics, 278 n63
Aristyllus, §B.3.I.4
Armillary sphere, §A.III.1–2, §B.2.V.2
Asher ben Yeḥi'el, 5
Averroës (See Ibn Rushd)
Avner of Burgos, 23 n93

Baghdādī, Abū al-Barakāt, 278
Bar Ḥiyya, Abraham
  Ṣurat ha-Areṣ, 13–4
Barayta di-Sh^{e}*mu'el*, 17
Bellver Martinez, José, 8 n21
Bible
  Ecclesiastes, 17, §B.4.I.1
  Esther, 5
  Ezekiel, 17
  Isaiah, 49, §0.16
  Jeremiah, 5
  Proverbs, 5
  Psalms, 49

Bīrūnī
al-Qānūn al-Mas'ūdī, 282
Earth's diurnal rotation, 282–3
trigonometry, 286 n108
Biṭrūjī
And *De Revolutionibus*, 4
And Galeano's *Puzzles of Wisdom*, 43
And Gersonides' *The Wars of the Lord*, 14
And Ibn Ṭufayl, 23 n95
And Ibrāhīm b. Sīnān, 22
And *Y*e*sod 'Olam*, 14
Eudoxan couples, 21–2, 37 n142
Celestial motions in a single direction, 24, §0.2, §0.9
Desire (*tashawwuq*) (See under desire)
Homocentric approach, 3–4, 19, §0.9
Inspiration for homocentric approach, 19–22
Lag (*taqṣīr*), 24, §0.12
In lunar model, 33
In model for the fixed stars, 37, §B.3.I.1–3, §B.3.I.6
On the Principles of Astronomy, 2
And *The Light of the World*, 6–7, 29, 48
Circle of the path, 26
Hebrew responses to, 17
Lunar model, 12 n42, 24 n98, 33, 323–4
Radius of the lunar epicycle, 330 n179, 334
Model for the fixed stars, 24 n98, §B.0.3, §B3.I.1, §B.3.I.1, §B.3.I.6–7, 351–2
Eudoxan couple in, 21, 37
Trepidation, §B.3.I.2
Obliquity of the ecliptic, 288
Organization of, 280, 283, 286
Planetary model, 356
Predictive accuracy vs. homocentricity, 44, 48, §0.10, 265–7, 300, 309
Reception in the Latin West, 43 n159
Solar model, 30–1, §B.1.II.28–9, 293–4, 367n5
Maximum solar anomaly, 318
Maximum solar equation, 319
Trepidation models as Biṭrūjī's source, 295, 314
Bonjorn, Jacob ben David, 14

Calippus, 19
Calvo, Emilia, 7 n20
Cancer, §0.5, §A.IV.2, §B.4.III.8
Capricorn, §0.5–6, §A.IV.2, §B.2.I.8, §B.4.III.8, §B.4.III.10–1
Chords, §A.II.1–4, §B.1.II.26/X.14, §B.1.II.26/X.31, §B.1.II.26/X.34, §B.1.II.26/X.39, §B.1.III.6–7
Climes, §A.IV.1, §A.IV.4–5, §B.4.III.1–2, 396
Colure, §A.I.3, §A.IV.2, §B.2.I.5, §B.1.II.26/X.14, 382, 385
Comes, Mercè, 37 n140, 301 n143, 348 n215
Concupiscence (*shahwa*), 279 n66
Copernicus, Nicholas
And Moses Galeano/Mūsā Jālīnūs, 41 n154, 42 n158
Connection with Islamic societies, 44 n164, 282 n84
De Revolutionibus, 4, 43
Crescas, Ḥasday
Opposite motions in the heavens, 274

Declinations, partial, §A.IV.1
Desire (*tashawwuq*), 24 n98–9
Dioptra, §A.III.5, 288
Dioscorides (see under Ibn Shapruṭ, Ḥasday)
Duran, Profiat, 5 n12
Commentary on the *Guide for the Perplexed*, 16, 41, 338
Response to *The Light of the World*, 1, 16, 40–1, 283, 337–8, 393–8

Earth-Moon distance (see under Moon)
Earth's diurnal rotation (see under Bīrūnī, Ikhwān al-Ṣafā' and Ragep)
Ecliptic (see under Sun, Moon, Fixed stars, and Biṭrūjī)
Elisha The Greek, 23 n93
Endress, Gerhard, 18 n75, 265 n6, 269
Equinoctial hours (see under hours)
Equinoxes, 30, §0.6, §A.I.3, §A.V.1–2
In Ibn Naḥmias' solar model
Greatest solar anomaly, §B.1.III.1–2, §B.1.III.8, 319–20
Solar anomaly, §B.1.II.3–6, §B.1.II.12, §B.1.II.28–9, 301, 312, 362
Solar mean motion, §B.1.I.2–4
In Ibn Naḥmias' lunar model, §B.2.III.3, §B.4.I.5, §B.4.III.4–6
Precession of the, 20n 84, 36–7, §B.3.I.1–4, §B.3.I.6, §B.3.II.4
Trepidation, 7 n20, 20–2, 36–8, §B.3.I.2, §B.3.II.5, §B.3.II.4/X,
Euclid
And chords, §A.II.1–4
Compounding ratios, §A.II.7, 359 n1
Ratio of mean and extreme parts, §A.II.1

The Elements, §A.II.2, 283 n90, 398. (*See also* under Gersonides and Ibn Tibbon, Moses)
Eudoxus, 10 n33, 19–23 (see also under Schiaparelli)
  Eudoxan couple (see under hypotheses)
  Reconstructions of Eudoxus' models, 48, 271, 314–6, 355, 370

Farghānī
  Elements of Astronomy, 276, 346 n211
Fārisī, Kamāl al-Dīn, 341
Fiṭra, §B.4.I.1, 354–5
Fixed stars
  Observations of the, 38, §B.2.V.2, §B.3.I.4
  Motions, 37, §0.13–5, §A.I.1, §B.3.II.4/X
    Precession
      Biṭrūjī's model, 21–2, 37, §B.3.I.1, §B.3.I.1–7
      Ibn Naḥmias' model, 22, 38, §B.3.II.1–4
    Trepidation, 36, §B.3.I.2, §B.3.II.5
Forcada, Miquel, 276, 277 n53–4
Fracastoro, Girolamo, 2 n6, 337 n192
France, 15 n55, 274 (see also under Provence)
Freudenthal, Gad, 5, 41 n152, 267 n14, 268, 279 n64, 290

Galeano, Moses (a.k.a. Mūsā Jālīnūs)
  And Christian scholars, 43
  And homocentric astronomy, 42
  And Ibn al-Shāṭir, 43, 48
  And *The Light of the World*, 42, 48
  In the Veneto, 42–4, 48
  Ta'alumot Ḥokmah (Puzzles of Wisdom), 43
Gemini, §B.1.III.5, §B.4.III.8
Gerson, Levi ben (see Gersonides)
Gersonides
  And Profiat Duran, 1 n1
  demonstration in astronomy, 14
  Eudoxan couples, 21
  Instruments, 14
  Knowledge of Pseudo-Ṭūsī Commentary on Euclid's *Elements*, 301 n143
  Observations, 41, 271, 337
  Opposite motions in the heavens, 274
  Rejecting homocentric models, 14, 23
  Rejection of the epicycle, 14, 268, 355, 396 n4
  Trepidation, 37
  Wars of the Lord, 14
Ghazālī
  Incoherence of the Philosophers, 10
Glasner, Ruth, 2 n4, 274 n42, 274 n44
God, 8 n24, 17 n69 (see also *Guide of the Perplexed* [under Maimonides] and Bible)
  Enabling knowledge of sub-lunar world, §0.8
  Knowable from the study of the heavens, 267 n14
Goldstein, Bernard
  Analysis of Biṭrūjī's models, 277
    Lunar model, 323, 334 n183
    Model for the fixed stars, 349–51
    Solar model, 316
  Gersonides' astronomy, 267 n15
  Historiography of science in Jewish cultures, 6 n15, 13, 274 n42
  On the source of Biṭrūjī's models, 20–1
  Polar epicycle, 26, 293 n128, 294
Guide of the Perplexed (see under Maimonides)
Gutas, Dimitri, 269

Hebrew recension of *The Light of the World*, 5, 46 (See also under Sun, Moon, Fixed Stars, Hypothesis, and Duran, Profiat: *Response*)
  Date of composition, 1 n1
  Eudoxan couple (see also under hypotheses), 22–3, 29, 32, 48, §B.1.II.26/Z.10–13, 316
  In the Veneto, 42–3, 368
  Need for physical movers, 26, 29, 295, 303, 306, 316, §B.1.II.20/X, §B.1.II.26/X.4–5, §B.2.V.12/Xb
  References to planetary models, 289–90, 354, §B.0.3/X
  Slant (see also under hypotheses), 28–9, 32, 295
Heraclides, 282
Hipparchus
  length of year, §A.V.1
  lunar diameter, §B.2.IV.3, 266, 339, 396–7
  lunar epicycle, 336 n189
  precession, §B.3.I.4
Hippopede (see under Schiaparelli)
Hours
  Equinoctial, §A.IV.4–5, §B.4.III.1–2, §B.4.III.8, §B.4.III.12
  Seasonal, §A.IV.4–5, §B.4.III.1
Hypothesis (see also under Sun and Moon)
  circle of the path of the center, 26–34
    Lunar first anomaly, §B.2.I.3–6, §B.2.II.2–9
      Maximum lunar anomaly, §B.2.III.1–2
    Lunar second anomaly, §B.2.V.4, §B.2.V.11; in the recension: §B.2.V.12/Xb

Hypothesis, circle of the path of the center *(continued)*
Planetary model, §B.4.II.1
Solar anomaly, §B.1.II.3–28, §B.1.II.26/X.1–3, §B.1.II.26/X.22–8, §B.1.II.26/X.32–3, §B.1.II.26/X.36
Maximum solar anomaly, §B.1.III.1–4, §B.1.III.7–9, §B.1.III.2/X, §B.1.III.10/X
Double circles, 26–29, 32, 34, 42, 48
Giovanni Battista Amico, 42
Lunar first anomaly, §B.2.II.7, §B.1.II.20/X
Lunar second anomaly, §B.2.V.6
Solar anomaly, §B.1.II.15–26, §B.1.II.26/X.1–5, 370, 381–2, 386, 388–90
Compared to Eudoxan Couple, 375
Eccentrics (and eccentricity), 3–4, 7, 9 n29, 11–16, 41, 44 n165, §0.1–3, §0.7–11, §0.15, §B.0.2, 396–7
Lunar diameter, §B.2.IV.1–3
Lunar second anomaly, §B.2.V.1–2, 345
Maximum solar anomaly, §B.1.III.1
Model for the fixed stars, §B.3.I.1
Planetary model, §B.4.I.4
Solar anomaly, §B.1.II.27, 316–9
Epicycles, 4, 7, 9 n29, 11–12, 14–6, 26, 28 n113, 31–3, 35–6, 39, 41, 44 n165, §0.1–3, §0.6, §0.9, §0.15
Lunar model, §B.2.I.4
Lunar first anomaly, 328–30
Maximum lunar anomaly, §B.2.III.1
Lunar diameter, §B.2.IV.1–3
Lunar second anomaly, §B.2.V.1–2,
Lunar third anomaly, §B.0.2, §B.2.V.7–8, §B.2.V.12/X
Model for fixed stars, §B.3.I.1
Planetary model, §B.4.0.1, §B.4.I.2–4
Solar anomaly, §B.1.II.13, §B.1.II.27, 294, 316
Maximum solar anomaly, 319
Eudoxan Couple, 19–23, 29, 32, 37 n142, 42, 314, 333 n181
As concentric epicycles, 21–2, 362 n4
Model for fixed stars, 350–1
Solar anomaly in the recension, §B.1.II.26/X.10–13, 316, 362, 378–81, 385–6
Opposite motions, 17 n69, 24–5, 31, §0.1–6, §0.9–11, §0.17, 288
In Biṭrūjī's lunar model, 324
Lunar anomaly, §B.2.II.10
Lunar mean motion, §B.2.I.5
Model for the fixed stars, §B.3.I.1
Solar anomaly, §B.1.II.11, §B.1.II.15, 296, 309, 312, §B.1.II.26/X.38
Solar mean motion, §B.1.I.4–5
Slant, 28–9, 31–2 (See also under prosneusis)
maximum solar anomaly, §B.1.III.2/X, §B.1.III.8/X, §B.1.III.10/X
solar anomaly in the recension, §B.1.II.20/X, §B.1.II.26/X.5–7, §B.1.II.26/X.14–41
solar mean motion, §B.1.II.11
lunar third anomaly, §B.2.V.6/X, §B.2.V.12/X-Xb
Translating the Arabic *aṣl,* 23–4, 266, 268–71

Iberia/Iberian Peninsula, 1–2, 4–6, 13–4, 22–3, 42, 274 n42
Ibn Aflaḥ, Jābir
Andalusian Revolt, 8
Evidence agaist Earth's motion, 281
Menelaos theorem/configuration, 285–6
Iṣlāḥ al-Majisṭī, 286
Ibn al-Haytham (See also under Ibn Bājja)
al-Shukūk ʿalā Baṭlamyūs (Doubts about Ptolemy), 11 n37, 12, 345 n210
Concentric epicycles (as a Eudoxan Couple), 21–3, 269 n20, 362 n4
Critique of equant, 12
Kitāb al-Manāẓir (Optics), §B.2.V.2
Maqāla fī ḥarakat al-iltifāf (Treatise on the Motion of *iltifāf*), 21–2
Moon illusion, §B.2.V.2
Risāla fī Ru'yat al-kawākib (On Seeing the Stars), 340 n202
Ibn al-Muthannā, 13
Ibn al-Shāṭir
And Copernicus, 43–4, 48
In *Puzzles of Wisdom,* 43, 48
Rejecting eccentrics, 44 n165
Solar model, 337 n194
Ibn al-Zarqāl
Connection to homocentric astronomy, 20–1, 314, 316, 350
Model for trepidation, 20–1, 36, §B.3.II.4/X, 314, 348 n215, 350
Observations, 396

Solar model, 7 n20
Ibn Bājja
And *Al-Shukūk 'alā Baṭlamyūs,* 12, 345 n210
Andalusian Revolt, 8, 265, 267
Critique of Ibn al-Haytham's mathematical astronomy, 18 n75
Opposite motions in the heavens, 273
Physics, 279 n63
Ibn Ezra, Abraham
Scriptural commentary, 18
Translation of Ibn al-Muthannā's commentary on Khwārizmī's tables, 13
Ibn Falaquera, Shem
Lost ancient astronomy, 15
Ibn Gabirol, Solomon
Keter, 18
Ibn Mu'ādh (see under Jayyānī)
Ibn Mūsā, Muḥammad
Motion of two concentric orbs, 280
Ibn Naḥmias family, 5–6
Ibn Naḥmias, Joseph (see also under Sun, Moon, Planets, Fixed Stars, Hebrew Recension, and Hypotheses)
Authorship of *The Light of the World,* 5, 46, 101, §B.4.I.1, 263
Homocentric astronomy, 19–23, 44
Knowledge of *The Compendium of "The Almagest",* 11, 12 n39
Life span, 1 n1, 5
Other writings, 6
Response to the *Guide,* 16, §0.3, §0.8
Skepticism of observations, 10 n35, 38–9, 268, §B.2.IV.1–3, §B.3.II.1
Ibn Naḥmias, Joseph (the elder), 5–6
Ibn Rushd (Averroës)
Andalusian Revolt, 7, 265, 267, 345 n210
Argument against the planets' rolling motion, §B.4.I.2, 265
Homocentric astronomy, 10, 18, 316
Ibn Rushd's knowledge
Of Ibn al-Haytham's *Shukūk* , 11–12, 345 n210
Of moon illusion and Ibn al-Haytham's *Optics,* 340–1
Knowledge of Ibn Rushd in the *Mashriq* (Islamic East), 23 n93
Orbs' desire, 279
Opposite motions in the heavens, 274
Permissibility of the epicycle and eccentric, 11, 265 n5, 276
Rejection of the epicycle and eccentric, 11, §B.4.I.2, 271–2, 276
Writings
Compendium of 'The Almagest,' 11, 276, 345 n210
De Substantia Orbis, 273 n40
Great Commentary on 'The Metaphysics,' 11, 271–2
Kitāb al-Athār al-'ulwiyya, 272 n34
Kitāb al-Kawn wa-'l-fasād, 272 n34
Kitāb al-Samā' wa-'l-'ālam, 272 n34, 273 n41, 279
Talkhīṣ al-Samā' wa-'l-'ālam, 273 n41, 279, 355
Talkhīṣ Kitāb al-Maqūlāt (Commentary on *The Categories*), 273
Talkhīṣ Kitāb al-qiyās (Commentary on the *Prior Analytics*), 282
Ibn Shapruṭ, Ḥasday, 13
Ibn Sīnā (Avicenna)
Al-Qānūn fī al-ṭibb, 23 n. 93
Length of year, §A.V.2
Motion of fixed stars, 38 n145, 354
Ibn Sīnān, Ibrāhīm
Eudoxan couples, 21–2
Kitāb fī ḥarakāt al-shams, 22
Trepidation, 22 n. 90
Ibn Tibbon, Moses
Hebrew translation of Biṭrūjī's *On the Principles,* 17
Hebrew translation of Euclid's *Elements,* 284 n96
Ibn Tibbon, Samuel
And Biṭrūjī's *On the Principles,* 17
Hebrew translation of *The Guide of the Perplexed,* 15 n55
Hebrew version of Aristotle's *Meteorology,* 272 n34
Ibn Ṭufayl, Abū Bakr
Andalusian Revolt, 8, 267
Ḥayy ibn Yaqẓān, 277
Homocentric astronomy, 23 n95, §0.9, 277
Possible acceptance of eccentrics and/or epicycles, 277
Ibrahim, Bilal, 341
Ījī, 'Aḍud al-Dīn, 290
Ikhwān al-Ṣafā'
Lag, 278 n62
Motion of the Earth, 282
Ilan, Nahem, 1 n2, 5 n11

Israeli, Isaac, 14
Israeli, Israel, 1 n2, 5 n10–11, 14 n50–1
 In *Mashriq* (Islamic East), 301 n143
Italy, 2–4, 43–4
 Padua, 28, 42–4, 48, 337 n192, 346 n211
 Veneto, 41–4, 48

Jālīnūs, Mūsā (see Galeano, Moses)
Jayyānī, Ibn Mu'ādh
 Kitāb Majhūlāt qisī al-kura, §A.II.7
Judeo-Arabic
 Linguistic characteristics, 45
Jupiter, 19, §B.4.0.1, §B.4.I.4–5, §B.4.II.1, §B.0.3/X

Kaspi, Joseph, 15
Kennedy, E. S.
 On the source of Biṭrūjī's models, 20
 Spherical Ṭūsī couple, 27 n110, 308, 362 n3
Khāzin, Abū Ja'far, 7 n16
Khwārizmī (see under Ibn al-Muthannā and Ibn Ezra)
King David, §0.8, §B.4.I.1
King Solomon, §B.4.I.1
Lag (*taqṣīr*) (see also under Biṭrūjī and Ikhwān al-Ṣafā'), 16, 24–5, 28, §0.12–4, §0.17, §A.I.3, §A.V.1, §B.0.1
 Lunar model, 33–4
  Maximum lunar anomaly, §B.2.III.2, §B.2.III.4
  Mean motion, §B.2.I.1, §B.2.I.6–10
  Lunar first anomaly, §B.2.II.4–5
  Lunar second anomaly, §B.2.V.5, §B.2.V.9
 Model for the fixed stars, §B.3.II.2, §B.3.II.4–5; §B.3.II.4/X
 Solar model, 30–1
  Mean motion, §B.1.II.1–10
  Solar anomaly, §B.1.II.7, §B.1.II.10, §B.1.II.21–30
  Maximum solar anomaly, §B.1.III.1–6, §B.1.III.9–10

Langermann, Tzvi, 2 n5, 9 n28. 18 n71, 20 n84, 23 n93, 45, 266, 267 n14 & n16, 275–6
Lay, Juliane, 11 n37, 12 n39, 276
Leo, §B.4.III.8
Lunar model (see under Moon)

Maimonides
 Andalusian Revolt, 8
 Attaining true knowledge in astronomy, 14, 268, 276, 316
 Economy in modelling, 354 n228
 Fixed stars, 38 n145
 Guide of the Perplexed, 8–10, 15–6, §0.8, 266–7, 280, 314, 338 n196, 353–4
  Commentaries on the *Guide*, 15–6
 Homocentric astronomy, 9–10, 16–7, §0.3, 266–8, 314
 Incommensurability of mathematical and physical approaches, 9–10, 266–7, 397–8
 Opposite motions in the heavens, 25 n104
 Permissibility of the epicycle and the eccentric, 11
Mancha, José Luis, 21, 37 n142–3, 350–1, 392 n17
Mars, 19, §B.4.0.1, §B.4.I.4–5, §B.4.II.1, 275, §B.0.3/X
Mathematics, practitioners of, §0.1
Menelaos theorem/configuration, 285, 286 n106, 360
Mercury, 4 n9, §B.4.0.1, §B.4.I.2, §B.4.I.4–5, §B.0.3/X
Meridional armillary, §A.III.1, 287, 317
Mileus, §B.3.I.4
Mishnah, 17
Moon illusion, 339–41
Moon (see also under hypotheses)
 Conjunctions, §B.2.V.1, §B.2.V.3–5, §B.2.V.8–10; §B.2.V.6/X, 395
 Diameter, 10 n35, 33, 35, §A.III.5, §B.0.2, §B.2.IV.1–3, 266, 341
 Earth-Moon distance, 35, §B.2.IV.1–3, 271, 339–40
 Eclipses, 9 n29, 18 n71, 32, §0.7, §A.I.1
  And lunar first anomaly, §B.2.I.2–4, §B.2.II.1, §B.2.II.6, §B.2.II.9–10
  And lunar second and third anomalies, §B.2.V.1, §B.2.V.10
  And maximum lunar anomaly, §B.2.III.1–4
  Observations, §B.2.IV.2–3
 Inclined circle carrying the circle of the path of the center, 26, 28
 Lunar first anomaly, model for, 34–5, §B.2.II.1–10
 Lunar mean motion, model for, 32–4, §B.2.I.1–10
 Lunar second anomaly, model for, 35, §B.2.V.1–6

Lunar third anomaly, 35–6, §B.2.V.7–10
Maximum lunar anomaly, §B.2.III.1–5
Motion in latitude, 32–4, §B.2.I.1–10, §B.2.II.1–9, §B.2.V.11–12, §B.2.V.12/X
Nodes, 33–4, §B.2.I.1–4, §B.2.I.7, §B.2.I.10, §B.2.II.8–10, 326, 328–30
Octants, 35 n135, 346–7
Oppositions, §A.I.1, §B.1.I.5, §B.2.III.1, §B.2.V.3–5, §B.2.V.8–9
Orbs, §B.2.V.12/Xb
Parallax, 35 n133, 41, §B.2.I.4, §B.2.IV.1–2, §B.2.V.1, §B.4.III.3
Prosneusis (inclination), 28 n113, §B.2.V.1, §B.2.V.10
Quadratures, 33–4, §B.2.V.1–5, §B.2.V.8–11, §B.2.V.6/X, §B.2.V.12/X, 396
Syzygies, 34, 341–2, 346–7
Mutakallimūn, 8, 290, 354 n229

Narboni, Mosheh, 15
Nīsābūrī, Niẓām al-Dīn, 354
Nychthemerons, §B.2.III.3

Obliquity (inclination) of the ecliptic, 30, 37, §A.III.1, §B.1.I.1, §B.1.I.2, 288, 292, 316, 348
Ottoman Empire, 5 n12, 42

Padua (see under Italy)
Parallactic rods, §A.III.4, 288
Parecliptic, §B.2.I.5, §B.2.I.7, §B.2.III.2, §B.2.V.2/Xb, 280
Penini, Yedaiah, 273, 274 n42 & n44 (see also under Hypotheses; Opposite motions in the heavens)
Physical evidence, §0.1, 281
Physics, practitioners of, §0.1
Pisces, 36, §B.4.III.6–7, §B.4.III.10–11
Planets (see separate listings for Jupiter, Mars, Mercury, Moon, Saturn, Sun, Venus)
Polar epicycle (see under Biṭrūjī, Circle of the path)
Precession (see under Equinoxes)
Proclus, 279
Provence, 4, 13–4, 274 n42, 301 n143 (See also under France)
Prosneusis, 12, 28 n113, 33, 35§B.0.2, §B.2.V.1, 340, 345, 391
Ptolemy
Epicycle and eccentric, 3–4, 9, §0.7, 272
Equant, 7 n16, 11–12
Lack of opposition in celestial motions, §0.11–12, 268, 274
And Ibn Naḥmias' solar model, §B.1.II.11, §B.1.II.15, §B.1.II.27, §B.1.II.29
And Ibn Naḥmias' lunar model, §B.2.I.7
Motion of the earth, §A.I.2
Obliquity of the ecliptic, §A.IV.1
Observations, 30, 38, §B.2.V.2, 266, 277, §B.1.II.26/X.30, §B.2.V.12/X
Planetary Hypotheses, 3 n8, 269–70
Arabic translations of, 269
Hebrew translations of, 13n48, 269 n25
Predictive accuracy, 9, 12, 26, 35, §0.7, 265–8
Tables, 30, 44, §A.IV.5, §B.1.II.27
The Almagest
Arabic recensions and translations of, 13 n48, 27 n110
Chords, §A.II.2–4
Hebrew translations of, 13 n48
Instruments, 287–8
Lunar model, 33–5, B.2.V.1–3
Earth-moon distance, §B.2.IV.1
Lunar diameter, 35, §B.2.IV.1–3, §B.2.V.1–2, 266, 396–7
Parameters, 32, §B.2.I.1, §B.2.I.4, §B.2.I.8, §B.2.III.1–4, 395–7
Prosneusis, 12, 33, 35, §B.2.V.1, §B.2.V.7, §B.2.V.10
Model for the fixed stars, 36, §B.3.I.1–6, §B.3.II.4/X
Parameters, 37 n144
Planetary models, §B.4.I.2–5
Planetary sizes and distances, 10 n35, 266 n11, 275
Planets' motions in latitude, 22
Solar model, §B.0.2, §B.1.II.3, §B.1.II.20/X
Parameters, 30, §B.1.I.3, §B.1.III.1–3, §B.1.III.5

Qabīṣī, 9
Qalonymos ben Qalonymos, 13 n48, 269 n25

Ragep, Jamil
Biṭrūjī and Eudoxus, 22 n90
Earth's diurnal rotation, 282
Eudoxan couples, 19, 370
Homocentric astronomy
And the Ṭūsī Couple, 362 n4

Ibrāhīm b. Sīnān, 22
Origins of Ṭūsī couple, 301 n144
Translating *aṣl*, 269 n20
Trepidation, 36 n137, 348 n215
Rāzī, Fakhr al-Dīn, 38 n145, 341, 354
Regiomontanus, 35 n134, 42, 44, 345 n211, 367–8
Repugnancies, 12, 36, §B.1.II.30, §B.2.V.7
Ringed Instrument (see armillary sphere)
And Ibn Naḥmias' lunar model, §B.2.II.10, §B.2.III.3
And Ibn Naḥmias' model for the fixed stars, §B.3.II.4–5
And Ibn Naḥmias' solar model, §B.1.I.4
And Rising times, §0.18, §A.IV.1, §A.IV.3–5, §B.4.III.1–4, §B.4.III.7–12
Robinson, James, 17

Saadia Gaon, 18
Sabra, A. I.
Andalusian Revolt, 7–12, 18
On the reception of Ibn al-Haytham's *Optics*, 341
Translation of *aṣl*, 268
Sagittarius B.4.III.8
Saliba, George
Ibn al-Shāṭir and eccentrics, 44 n165
On an 11th century Andalusian critic of Ptolemy 12
Spherical Ṭūsī Couple, 308, 362 n3
Ṭūsī's *Recension of "The Almagest,"* 27 n110
Saturn, 19, §B.4.0.1, §B.4.I.4–5, §B.4.II.1, §B.0.3/X, 280
Schiaparelli, Giovanni (see also under Eudoxus and Hypotheses: Eudoxan couple)
Reconstruction of Eudoxus' models, 19–20, 355, 370, 374–5
Scorpio, §B.4.III.6, §B.4.III.8
Seasonal hours (see under hours)
Shank, Michael, 2 n6, 42 n159, 345 n211
Shahwa (See Concupiscence)
Shem Ṭob, Shem Ṭob ibn, 16
Shīrāzī, Quṭb al-Dīn
al-Ṭuḥfa al-sa'diyya, 23 n93
Fa'alta fa-lā talum, 279
Fixed stars on Saturn's parecliptic and seven-orb cosmos, 279–80, 354–5
Nihāyat al-idrāk fī dirāyat al-aflāk, 269 n24
Translating *aṣl*, 269
Trepidation, 295
Simplicius
Report of Eudoxus' astronomy, 19, 21
Solar model (see under sun)
Solstices, 30, §A.I.1, §A.I.3, §A.III.1, §A.IV.1, §A.V.1
Lunar model, §B.2.I.6, §B.2.I.8–9, §B.2.II.3–4,
Model for the fixed stars, §B.3.I.1–6, §B.3.II.2–4
Planetary model, §B.4.I.5, §B.4.III.4–5
Solar model, §B.1.I.2–4, §B.1.II.4, §B.1.II.28–9. §B.1.III.1–5, §B.1.III.8, §B.1.II.26/X.1–2, 362
Sphaera Obliqua, §0.18, §A.IV.1, §A.IV.3, §B.4.III.9, §B.4.III.12
Sphaera Recta, §0.18, §A.I.1, §A.IV.1, §A.IV.3, §A.IV.5, §B.4.III.1–2, §B.4.III.9, §B.4.III.11–12
Spica Virginis, §B.3.I.3, §B.3.I.6
Starred orb, §0.14, §B.2.II.2, §B.2.II.5, §B.2.II.7
Stars, fixed (See under Biṭrūjī, Fixed stars, Goldstein, Ibn Naḥmias, Ibn Sīnā, Ptolemy, and Shīrāzī)
Steinschneider, Moritz, 6 n15, 41 n153, 45–6, 269 n25
Stroumsa, Sarah, 7 n18,
Sun (see also under hypotheses)
Anomaly, model for, §B.1.II.1–30
Circle of the path of the center, §B.1.II.3–9, §B.1.II.13, §B.1.II.17, §B.1.II.21, §B.1.III.1–2, §B.1.III.9
Double circles, 27, 32, 42, 48, §B.1.II.15–25, 270, 313–4, §B.1.II.26/X.1–5, 361, 388–90
Eudoxan Couple, 29, 32, 42, §B.1.II.26/X.10–21
Error analysis, §B.1.II.13–4
Inclined circle (carrying the circle of the path of the center), 26, 28, §B.1.II.2, §B.1.II.4, §B.1.II.9, §B.1.II.21
Slant, 28–9, 31–2, §B.1.II.11, §B.1.II.20/X, §B.1.II.26/X.5, §B.1.II.26/X.7, §B.1.II.26/X.14–26, §B.1.II.26/X.29–37, §B.1.II.26/X.39, §B.1.II.26/X.41
Asymmetries in motion, §B.1.II.26, 300, 309, 313–4, 361
Diameter, 41, §B.2.IV.1, §B.2.IV.3, 396
Individual anomalies, §B.1.III.9–10
Maximum anomaly, §B.1.III.1–8
Slant and, §B.1.III.2/X, §B.1.III.8/X, §B.1.III.10/X

Mean motion, model for, §B.1.I.1–5
Motion in latitude, §B.1.I.3–4, §B.1.II.26/X.15, §B.1.II.26/X.23, §B.1.II.26/X.35, §B.1.II.26/X.37–41
Ṣurat ha-areṣ (see under Bar Ḥiyya)
Swerdlow, Noel, 42 n159, 367

Tables (see under Bonjorn, Ibn al-Muthannā, and Ptolemy)
Taqṣīr (see under lag)
Tashawwuq (see desire)
Taurus, §B.4.III.6, §B.4.III.8
Time-degrees, 178 n46–7, §A.IV.4, §B.4.III.1, §B.4.III.7–8, §B.4.III.12
Timocharus, §B.3.I.4
Toledo, 5
Trepidation (see under equinoxes)
Ṭūsī, Naṣīr al-Dīn
Homocentric astronomy, 22 n90
Fixed stars, 38 n145, 354
Knowledge of Ṭūsī in the Islamic West, 23 n93, 301 n143
Ṭūsī couple, 22, 29, 316, 362
in *Taḥrīr al-Majisṭī*, 27, 48, 301, 306–8
in *al-Tadhkira fī 'ilm al-hay'a*, 27, 48, 301, 306–8

Vapors, §A.I.1, 337, 340 n202,
Veneto (see under Italy)
Venus, §0.4–5, §B.4.0.1, §B.4.I.2, §B.4.I.4–5, §B.0.3/X
Virgo, §B.4.III.6–7

Yavetz, Ido, 19 n79, 20

Zacut, Abraham, 6
Zarqāl, Ibrāhīm (see Ibn al-Zarqāl)
Zarqālluh (see under Ibn al-Zarqāl)
Zodiac (see under ecliptic)